Mathematics Mechanization and Applications

Mathematics Mechanization and Applications

Edited by

Xiao-Shan Gao
Institute of Systems Science, Academia Sinica
Beijing, China

and

Dongming Wang
Laboratoire LEIBNIZ Institut IMAG
Grenoble, France

ACADEMIC PRESS
A Harcourt Science and Technology Company
San Diego San Francisco New York Boston
London Sydney Tokyo

This book is printed on acid-free paper.

Academic Press
A Harcourt Science and Technology Company
Harcourt Place, 32 Jamestown Road, London NW1 7BY, UK
http://www.academicpress.com

Academic Press
A Harcourt Science and Technology Company
525 B Street, Suite 1900, San Diego, California 92101-4495, USA
http://www.academicpress.com

ISBN 0-12-734760-7

A catalogue for this book is available from the British Library

Printed and bound in Great Britain by MPG Books Ltd, Cornwall, UK

00 01 02 03 04 05 MP 9 8 7 6 5 4 3 2 1

CONTENTS

PREFACE

Mathematics mechanization consists of mechanizing mathematical activities, such as calculation, reasoning and discovery, so that these activities may be carried out mechanically and step by step on machines and electronic devices in order to reduce the mental labour and paper-and-pencil work of mathematicians and mathematics users. Modern research on mathematics mechanization has been devoted to the development of theory and methods, design and implementation of software tools, and their applications to diverse problems in science, engineering and industry. Such research in China has been actively conducted since the later 1970s under the leadership of Professor Wen-tsün Wu.

The term *mathematics mechanization* was derived by Wu from the classical article "Toward Mechanical Mathematics" by H. Wang published in 1960. Wu has upheld mathematics mechanization and extended its connotation gradually since the early 1980s. The scope of mathematics mechanization in Wu's interpretation is much beyond applied logic deduction addressed in Wang's work. Wu insists that mathematics is to become mechanized and automated step by step in various specialized subareas, and algorithmic studies of mathematics using computing technologies will be the major trends of mathematical development in the information age.

Wu pointed out that the principal concern of mathematics in ancient China was mechanization, in contrast to the axiomatization of mathematics (mainly Euclidean geometry) originating in ancient Greece. Although the two distinct systems of thought and methods have both influenced the development of mathematics, axiomatic methods now occupy a dominant position in modern mathematics, especially pure mathematics. Ancient Chinese mathematicians were mainly interested in solving concrete problems arising in reality or from necessity, which led naturally to the study of arithmetic operations and algebraic equations. Actually, equation solving was the central theme in ancient Chinese mathematics, as was theorem proving in ancient Greek mathematics. Two thousand years ago, Chinese mathematicians had already discovered methods for finding the square and cubic roots and for solving systems of linear equations (a method similar to the one named after Gauss). They were able to solve simultaneous systems of non-linear equations in four unknowns in the 13th century. In fact, elimination methods for systems of linear and non-linear equations are some of the masterpieces of ancient Chinese mathematics representing the thought of mechanization. Even the invention of calculus benefited perhaps also a lot from the Chinese mechanical thought brought to Europe by the Arabs.

After a thorough study of the history of Chinese mathematics and inspired by the thought of mechanization he had observed, Wu stopped resolutely his research in topology and started exploring mathematics mechanization and its modern realization on computers in the late 1970s. An unexpected result of his investigations, Wu's method of mechanical theorem proving in elementary geometry was introduced at that time. The algebraic tool underlying this method has been further developed and is now known as the Wu–Ritt's characteristic set method. On various occasions, Wu has stressed the deep influence of ancient Chinese mathematics on his discovery of the method.

For the last two decades, extensive research has been done by Wu and his students and associates along the line of mechanization with emphasis on theorem proving and equation solving, two typical and important activities in mathematics. Several other groups in China have followed up the research of Wu. Continuing the application-oriented tradition of ancient Chinese mechanical mathematics, a large portion of the research in the Chinese groups is connected to the applications of theory and methods to real world problems, ranging from the analysis of robotics and mechanisms to non-linear programming and chemical equilibrium computation.

Implementation on modern computers is an indispensable step to verify the feasibility of methods and ideas of mechanization. Wu and his early students working in this area experienced a difficult period with implementational practice. In the process of writing the first few papers, Wu could not get access to a computer — he used pencil and paper to calculate polynomials of hundreds of terms. Even in the later 1980s, the computing equipment for Wu's group was still quite rudimentary. The early programs of Wu and his students for polynomial operation and elimination were written on very primitive computers mostly using Fortran. A turning point occurred in 1990 when the Chinese Ministry of Science and Technology and Academia Sinica funded Wu's research group by establishing the Mathematics Mechanization Research Center. Since then, Wu's group has received large grants for several research projects including "Machine Proof and Its Applications" (1992–1997), "Mathematics Mechanization and Its Applications" (1997–2001) and "Mathematics Mechanization and Platform for Automated Reasoning" (1998–2002) — the last is one of the 15 national key projects in basic research.

This book provides a uniform presentation of major developments, carried out mostly in Chinese research groups, on algorithms and software tools for mechanizing algebraic equation solving and geometric theorem proving together with their applications to problems in science and engineering. It also covers most of the results accomplished in the above-mentioned projects, of which most of the contributors are participants. The reader will find the remarkable impact of Wu's work and ideas on the research by Chinese groups described in this book. The book is divided into four parts with 20 chapters written by experienced researchers. Each chapter is devoted to surveying and expounding the main results in one selected subject.

The present book contains surveys for most of the theoretical and practical work done by Wu and members of his extended group in the past decade. However, for some unforeseen reasons several significant pieces of work are not reviewed. They include Wu's work on constructive treatment of Chern classes for varieties with singularities and on the construction of well-behaved bases (Wu 1991a), the work of Yang et al. (1992, 1996) on the resultant approach for algebraic equation solving and geometric theorem proving and of Yang (1998) for automated inequality proving, the work of Xu et al. (1996a,b) on the application of Wu's method in computer vision, and the work of Gao and Chou (1998a,b) on automated diagram construction.

Part of the work reviewed in this book has remained either unpublished, or published only in Chinese journals or even only in the Chinese language which are hardly accessible to Western readers, so it may still remain unknown in the West. One of our motivations for preparing this book is to provide Western researchers, practitioners, educators and students in relevant areas with an easily accessible source of references for the researches of mathematics mechanization made in Chinese groups. Meanwhile, we hope to draw their attention to some of the old and new findings by Chinese mathematicians. Our primary goal will have been achieved if the reader finds this book helpful for him or her to look up bibliographic information about Chinese researchers and to understand better what these researchers have been doing under the banner of mathematics mechanization. We shall be satisfied if the present book will also contribute to enriching the scientific literature in the interaction between computers and mathematics.

The scope of mathematics mechanization has considerable overlap with that of two well-established areas of research: symbolic computation and automated reasoning. Similar and important work in these areas has been done by many other researchers across the world. For example, we may mention the well-known methods of Gröbner bases and cylindrical algebraic decomposition, various other elimination methods based on resultants and triangular sets, and an enormous amount of work related to these methods. This book is clearly distinguished by presenting deliberately the work and results from Wu's extended Chinese group. A reader from another group finding his own work only briefly mentioned or even ignored should not feel surprised!

When completing this book, we were celebrating the 80th birthday of Professor Wu and his achievements of two decades in mathematics mechanization. On this occasion and on behalf of the other contributors, we would like to dedicate this book to him and to thank him for bringing us to this great academic adventure. We wish him full of happy time in the coming years.

Thanks to the fine suggestion of an anonymous referee, we take the pleasure to append the following short biography of Professor Wu.

Professor Wen-tsün Wu is 80

Born in Shanghai in May 1919, Wu studied mathematics at Shanghai Jiao-Tong University from 1936 to 1940. He was a high school teacher in Shanghai from 1940 to 1945 when academic activities in China were interrupted by the Sino-Japanese War. In 1945, he entered into the newly founded Institute of Mathematics, Academia Sinica and began mathematical research under S.-S. Chern. In 1947, he went to France and studied topology under C. Ehresmann at the University of Strasbourg. After his state thesis was completed in 1949, he moved to Paris working at the CNRS with H. Cartan. Wu's work and contributions, including the famous Wu formula and Wu class, were considered major advances and have become classical results of topology.

After his return to China in 1951, Wu was first a professor at Peking University and then rejoined the Institute of Mathematics, Academia Sinica, continuing to work in algebraic and differential topology until the Cultural Revolution. He made important contributions to the theory of imbedding in Euclidean spaces, Pontrjagin class theory and rational homotopy theory.

Wu's research and publications in topology were made impossible during the Cultural Revolution (1966–1976). In that period, Wu occupied himself in studying the history of Chinese mathematics. He was able to reconstruct the proofs for many classical results of Chinese mathematics in an ancient style, that is quite different from the usual Euclidean approach, and provide evidences showing that Chinese mathematics has its own tradition, that is based on computation as contrast to the axiomatic and proof model of Western mathematics.

He also pointed out the immense contribution of ancient Chinese mathematics and its mechanical character to the development of modern mathematics.

Following the mechanical thought of ancient Chinese mathematics, Wu began to work on mechanical theorem proving in geometry in 1976 and published his first paper the year after. Since then he has systematically developed a general method for solving systems of algebraic equations and investigated its applications to automated geometric reasoning and many other subjects from inverse robot kinematics to non-linear optimization. This method is now widely known as Wu's method.

Wu was director of the Division of Mathematical and Physical Sciences, Academia Sinica from 1992 to 1994 and president of the Chinese Mathematical Society from 1985 to 1987. He is one of the founders of the Institute of Systems Science, Academia Sinica and has been its honorary director since 1983. He founded the Mathematics Mechanization Research Center in 1990 and was its director until this spring when he took up honorary director.

Wu received the First Prize of the National Science Awards of China in 1956 and became the youngest member of Academia Sinica in 1957. He is recipient of the Mathematics Award from the Third World Academy of Sciences (1990) and has been a member of the Academy since 1991. He is also recipient of several other distinguished awards including the Jia-Geng Cheng Award for mathematical and physical sciences from the Jia-Geng Cheng Foundation (1993), the Distinguished Scientist Award from Hong Kong Qiu-Shi Foundation (1994), and the Herbrand Award for distinguished contributions to automated reasoning from the International Conference on Automated Deduction (1997).

Wu has been chief scientist for several large national research projects since 1992. He has delivered addresses at many congresses, served as editors for numerous journals, books and book series, and held positions and honorary titles in various other academic organizations at both national and international level.

Beijing and Paris
July 1999

Xiao-Shan Gao
Dongming Wang

LIST OF CONTRIBUTORS

Chou, Shang-Ching
Department of Computer Science
Wichita State University
Wichita, KS 67260
USA
chou@cs.twsu.edu

Fan, Engui
Institute of Mathematics
Fudan University
Shanghai 200433
China

Feng, Guochen
Department of Mathematics
Jilin University
Changchun 130023
China
fgc@mail.jlu.edu.cn

Gao, Xiao-Shan
Institute of Systems Science
Academia Sinica
Beijing 100080
China
xgao@mmrc.iss.ac.cn

Hou, Xiaorong
Chengdu Institute of Computer Applications
Academia Sinica
Chengdu 610041
China
xhou@hotmail.com

Li, Hongbo
Institut für Informatik und Praktische Mathematik
Christian-Albrechts-Universität zu Kiel
Preusserstraße 1-9
D-24105 Kiel
Germany
hli@informatik.uni-kiel.de
hli@mmrc.iss.ac.cn

Li, Zhibin
Department of Computer Science
Lanzhou University
Lanzhou 730000
China
lizb@lzu.edu.cn

Li, Ziming
Institute of Algorithms and Scientific Computing
German National Research Center for Information Technology
D-53754 St. Augustin
Germany
Ziming.Li@gmd.de
zmli@mmrc.iss.ac.cn

Liao, Qizheng
Department of Mechanical Engineering
Beijing University of Posts and Telecommunications
Beijing 100876
China
qzliao@bupt.edu.cn

Liu, Zhuo-Jun
Institute of Systems Science
Academia Sinica
Beijing 100080
China
zliu@mmrc.iss.ac.cn

Lin, Dongdai
Institute of Systems Science
Academia Sinica
Beijing 100080
China
ddlin@mmrc.iss.ac.cn

Lu, Zhengyi
Department of Mathematics
Sichuan University
Chengdu 610064
China
nic2601@pridns.scuu.edu.cn

Ma, Shilong
Department of Computer Science
Beijing University of Aeronautics
and Astronautics
Beijing 100083
China
scning@Math03.math.ac.cn

Ren, Hongju
Department of Mathematics
Jilin University
Changchun 130023
China

Wang, Ding-Kang
Institute of Systems Science
Academia Sinica
Beijing 100080
China
dwang@mmrc.iss.ac.cn

Wang, Dongming
Laboratoire LEIBNIZ
Institut IMAG
46, avenue Félix Viallet
38031 Grenoble Cedex
France
Dongming.Wang@imag.fr
Dongming.Wang@lip6.fr

Wang, Shi-kun
Institute of Applied Mathematics
Academia Sinica
Beijing 100080
China
xyswsk@pku.edu.cn

Wu, Jinzhao
Max-Planck-Institut für Informatik
Im Stadtwald
D-66123 Saarbrücken
Germany
wu@mpi-sb.mpg.de

Wu, Ke
Institute of Theoretical Physics
Academia Sinica
Beijing 100080
China
wuke@itp.ac.cn

Wu, Wen-tsün
Institute of Systems Science
Academia Sinica
Beijing 100080
China
wtwu@mmrc.iss.ac.cn

Zhang, Hongqing
Department of Applied Mathematics
Dalian University of Technology
Dalian 116024
China
zhanghq@dlut.edu.cn

Zhang, Jing-Zhong
Chengdu Institute of Computer Applications
Academia Sinica
Chengdu 610041
China
zjz@guangztc.edu.cn

Zhang, Shugong
Department of Mathematics
Jilin University
Changchun 130023
China
sgzh@mail.jlu.edu.cn

Zhi, Lihong
Department of Computer Science
Ehime University
Matsuyama 790-8577
Japan
lzhi@hpc.cs.ehime-u.ac.jp

Zhou, Yunshi
Department of Mathematics
Jilin University
Changchun 130023
China
zys@public.cc.jl.cn

Part I

Polynomial System Solving

Chapter 1

THE CHARACTERISTIC SET METHOD AND ITS APPLICATIONS

Wen-tsün Wu

This first chapter starts with a brief historical review on polynomial equation solving in ancient China and a brief introduction to the characteristic set method, known as *Ritt–Wu's*, *Wu–Ritt's* or *Wu's method*, that has been developed by the author since the late 1970s. Some variants of this general method are presented that incorporate several specialized techniques to enhance the computational performance. A few selected applications of the method are discussed. This serves to illustrate its applicability and practical efficiency for polynomial equation solving. Part of the material presented here will be used for many other applications described in later chapters.

1.1 Introduction

1.1.1 Polynomial equation solving in ancient China

There are two main activities of mathematics: theorem proving and equation solving. While theorem proving originates from ancient Greek mainly in so-called Euclidean geometry, equation solving was the main concern of mathematics in ancient China.

Chinese ancient mathematics has the following notable features.

- Instead of calculations of pencil-and-paper type, the ancient Chinese made all computations using manipulating rods on counting boards or anything of the sort that served the purpose.

- The Chinese already possessed, in very remote times, the most perfect *place-valued* decimal system. This allowed them to represent the integers by properly arranged rods placed in due positions on the board. In particular, the integer 0 in, or as, a decimal integer was represented just by leaving some empty place in the appropriate position.

MATHEMATICS MECHANIZATION AND APPLICATIONS
ISBN 0-12-734760-7

- Results were usually expressed in the form of separate problems, each of which was divided into several items, as follows.

  (1) Statement of the problem with numerical data.

  (2) Numerical answer to the problem.

  (3) "Shu," or the method of arriving at the result. It is most often just what we call today the "algorithm," sometimes also a formula or a theorem.

  (4) Sometimes "Zhu," or demonstrations, which explain the reasoning underlying the method in (3). Actually, it is a "proof" when "Shu" is a formula or a theorem.

  (5) "Cao," meaning drafts, containing details of the calculations for arriving at the final result, appeared first in some classics of the Tang Dynasty (618–907 AD). It was then widely adopted in writings of the Song and Yuan Dynasties (938–1368 AD). This was possible because printing techniques were well-developed in those days.

  Note that the numerical values in (1) play no role at all in the method, which is so general that any other numerical value of the same type could be substituted equally well.

- The main concern of our ancestors was the solving of problems arising in reality or from necessity which naturally led to the solving of equations. Thus instead of being devoted to theorem proving as in the ancient Greek mathematics, equation solving was the central theme of Chinese ancient mathematics.

- For geometry, instead of developing a deductive system for theorem proving, measurement and geometry problem solving were emphasized based on some different procedure of proof.

- In a word, the ancient Chinese mathematics was highly computational and *mechanical* in character. It seems that it was developed in its own way, independently of Western mathematics which was developed in a way dating back to the ancient Greeks. It may be considered as a *mechanical* system quite different from the *axiomatic* system of Greek origin which governs modern pure mathematics in the present day.

Chinese ancient mathematicians gave general methods of solving systems of linear equations, similar to the *Gaussian elimination method*, more than

2000 years ago. This method was summarized in the classic work "Nine Chapters of Arithmetic," completed in the third or second century BC. which is a comprehensive discourse on traditional Chinese mathematics. Its influence on subsequent developments in Chinese mathematics is no less profound than that of Euclid on Western mathematics. The treatise was not due to a single author or a few authors as usual in the present day. It seems rather to be a collection of achievements due to several generations of intelligent scholars, learned people, and officials in charge of astronomical, financial, reconstructive and plannifying affairs, etc. Most of the results and methods in the book were already in current use long before the Qin Dynasty (221–207 BC).

In the Song-Yuan Dynasties (938–1368 AC), there also appeared methods for finding numerical solutions of polynomial equations of high degree in a single unknown. It would be natural to unify these two kinds of methods together to consider systems of polynomial equations of high degree not restricted to a single unknown. This was indeed the case, as seen from the following example.

Example 1.1:

Consider a set of three equations in three unknowns taken from "Jade Mirror of Four Elements" by Shi-jie Zhu (1303). These equations arose from a geometrical problem and in modern notation may be written in the following form:

$$\begin{aligned} xyz - xy^2 - z - x - y &= 0, \\ xz - x^2 - z - y + x &= 0, \\ z^2 - x^2 - y^2 &= 0. \end{aligned}$$

The problem is to determine z. Zhu's method runs now briefly as follows. From geometry x cannot be 0 so x will be divided out freely during the procedure below. In first eliminating y from the given equations and then reducing successively degrees of x in the equations thus obtained, one will get finally two equations in x and z alone which are both linear in x. Eliminating x from these two last equations yields a single equation in z alone:

$$h(z) = z^4 - 6z^3 + 4z^2 + 6z - 5 = 0.$$

Zhu now chooses the positive solution $z = 5$ of the last equation as the solution required. The details of computation may be found in the excellent treatise by Hoe (1977).

Zhu stops at the solution $z = 5$ for the above problem. However, one may go a step further in taking any one of the two equations in x, z and linear in x

and also any one, say the second, from the original set of equations to form a system of three equations below:

$$\begin{aligned} z^4 - 6z^3 + 4z^2 + 6z - 5 &= 0, \\ x(7 + 3z - z^2) - (6 + 7z + 3z^2 - z^3) &= 0, \\ y - (xz - x^2 - z + x) &= 0. \end{aligned}$$

We may then solve the above equations successively in single unknowns z, x, and y in succession to get solutions for the triple (z, x, y).

Since the Ming Dynasty (1368–1644 AC), the brilliant development of Chinese ancient mathematics in algorithmic fashion has declined and has not exerted any further influence on the mathematical world.

For a detailed discussion of the ancient Chinese mathematics and a discussion of polynomial equation solving in ancient times beyond China, see Chaps. 1 and 2 of Wu (1999).

1.1.2 Methods of polynomial equation solving

In the 17th century, Descartes proposed a *program* of problem solving which may be schematized by the following diagram:

$$\begin{aligned} \text{Solving of any problem} &\longrightarrow \text{Solving of a problem in mathematics} \\ &\longrightarrow \text{Solving of a problem in algebra} \\ &\longrightarrow \text{Solving of a system of algebraic equations} \\ &\longrightarrow \text{Solving of a single algebraic equation.} \end{aligned}$$

Although the Descartes scheme appears absurd, the importance of Descartes' thought is indisputable. In fact, in recent years, that polynomial equation solving may play an important role in all kinds of problems has been gradually well-recognized by the scientific world. Diverse methods have appeared to deal with the problem of solving an arbitrary system of polynomial equations $PS = 0$ with $PS \subset \boldsymbol{K}[\boldsymbol{x}]$, $\boldsymbol{x} = (x_1, \ldots, x_n)$ and $\boldsymbol{K}$ a field of characteristic 0, most frequently the field $\boldsymbol{Q}$ of rational numbers.

There are mainly two kinds of methods for *polynomial equation solving*: numerical ones and symbolic ones. For numerical methods, we may cite the generalized Newton–Raphson method and its diverse variants, the homotopy continuation method, etc. These methods start directly from the given numerical data in the equations and apply some limiting process in order to converge toward some solution of the given polynomial equations. In general these methods will give only one solution among the set of all possible solutions and are thus *local* in character in the main. The homotopy continuation method, based on some topological considerations, under certain circumstances may give the set of all possible solutions and is thus sometimes *global* in character. This is particularly true in the case of *eigenvalue* determination, and T. Y. Li, one of the pioneers who developed the method, has shown how this may be true in some general cases. In any case all these methods give only solutions in approximate values and suffer from error as well as stability control. In contrast to numerical methods, symbolic methods treat at the outset all the approximate numerical data as indeterminates and only symbolic computations are to be used during the procedure. The solutions will be given in some acceptable algebraic form and all the possible solutions may be expressed in this way so that the methods are *global* in character. Among such symbolic methods we may cite in particular the Buchberger method based on Gröbner basis construction of an ideal (see, e.g., Buchberger 1985). The method has been proved to be quite successful in the zero-dimensional case (over $\boldsymbol{Q}$). To the knowledge of the author, the more general case is still under serious study. See also Lazard (1991) and Kobayashi (1988) for examples of other methods.

1.1.3 The characteristic set method

Along the lines of thought furnished by our ancestors we have introduced in Wu (1984a, 1986a) concepts and techniques from modern mathematics, mainly borrowed from the two classic texts of J. F. Ritt (1932, 1950), in order to reformulate and to make precise and rigorous the procedure originated from Shijie Zhu's work in the year 1303. The outcome is a general *characteristic set method* described in the form of the *well-ordering principle* in Sect. 1.2.2 and *zero decomposition theorems* in Sect. 1.2.3. With the principle and theorems, the problem of *polynomial equation solving* is seen to be completely settled, at least in the theoretical sense.

Briefly, the method of characteristic sets is an elimination procedure which may be used to transform the zero set of a polynomial set in general form to that of a polynomial set in triangular form. It has been extended to cover algebraic ordinary and partial differential equations (see Wu 1989b and Chap. 6). Since this method was introduced, other similar elimination methods have been proposed to improve the efficiency of computation for polynomial equation solving.

The method of characteristic sets has diverse applications, of which one of the most successful and influential is for automated theorem proving and discovering in geometry (see Chaps. 5 and 6). Theoretically, it may serve as a basis to solve many problems from and related to computational algebraic geometry (see Chap. 5 of Wu 1999). This method has also been applied to many problems from real-world applications including

- analysis of robotics and mechanisms (Chap. 18),
- surface fitting in CAGD (Chap. 19),
- implicitization and parametrization for algebraic curves and surfaces (Chap. 14),
- constructing limit cycles and establishing center-focus conditions for differential dynamical systems (Chap. 15),
- finding analytical solutions for ordinary and partial differential equation systems (Chaps. 16 and 17),
- first-order theorem proving (Chap. 11),
- computer vision problems (Kapur and Mundy 1988; Gao and Cheng 1998),
- problems from theoretic physics (Chap. 4) and celestial mechanics (Sect. 1.4.2),
- non-linear programming and optimization problems (Sect. 1.4.3), and
- chemical equilibrium computation (Wu 1990b).

The rest of this chapter is organized as follows. The general method of characteristic sets is introduced in the following section. In Sect. 1.3, we shall review several methods of polynomial equation solving. Three applications of the characteristic set method are discussed in Sect. 1.4.

1.2 The method of characteristic sets

1.2.1 Ascending sets

Let $\boldsymbol{K}$ be a field of characteristic zero, $\boldsymbol{x} = (x_1, x_2, \dots, x_n)$ a set of variables with $\boldsymbol{x}_k = (x_1, \dots, x_k)$, and $\boldsymbol{K}[\boldsymbol{x}]$ the set of polynomials in the variables $\boldsymbol{x}$ with coefficients in $\boldsymbol{K}$. Polynomials mentioned below will be understood to be in $\boldsymbol{K}[\boldsymbol{x}]$ unless otherwise stated. The following natural ordering for the variables $\boldsymbol{x}$ will be taken unless otherwise stated: $x_1 \prec x_2 \prec \cdots \prec x_n$.

For any non-zero polynomial $P \in \boldsymbol{K}[\boldsymbol{x}]$, denote by

- $\deg(P, x_i)$ the *degree* of P in x_i,
- $\mathrm{cls}(P)$ the *class* of P, that is the greatest subscript c for which x_c occurs actually in P (and otherwise 0 if P is a non-zero constant in $\boldsymbol{K}$),
- $\mathrm{lv}(P)$ the *leading variable* of P, that is x_c if $\mathrm{cls}(P) = c > 0$ (and otherwise undefined), and let
- $\deg(P) := \deg(P, \mathrm{lv}(P))$, and
- *index-set* of $P := [t\ c\ d]$ with $t =$ the number of terms in P, $c = \mathrm{cls}(P)$ and $d = \deg(P)$.

The symbol ":=" is understood as "defined to be."

For a non-constant polynomial P of class $c > 0$, P can be written in the following *normal form*

$$P = Ix_c^d + \text{lower degree terms in } x_c,$$

in which $c = \mathrm{cls}(P) > 0, d = \deg(P) > 0$, and the coefficients of x_c-powers all belong to $\boldsymbol{K}[\boldsymbol{x}_{c-1}]$. The polynomial I is called the *initial* of P; in notation: $I = \mathrm{ini}(P)$. The formal derivative $\partial P/\partial x_c$ is called the *separant* of P; in notation: $\mathrm{sep}(P) = \partial P/\partial x_c$.

Let P and Q be two non-zero polynomials. We say that P is of *lower ordering* than Q or Q is of *higher ordering* than P if either $\mathrm{cls}(P) < \mathrm{cls}(Q)$ or $\mathrm{cls}(P) = \mathrm{cls}(Q)$ and $\deg(P) < \deg(Q)$. In notation: $P \prec Q$ or $Q \succ P$. It is clear that the binary relation $\prec$ introduces a partial ordering among all non-zero polynomials in $\boldsymbol{K}[\boldsymbol{x}]$.

A polynomial Q is *reduced* with respect to P if $\mathrm{cls}(P) = c > 0$ and $\deg(Q, x_c) < \deg(P)$. A polynomial set AS is an *ascending set* if its polynomials are all non-zero and can be arranged in a sequence

$$\mathit{AS}: \quad A_1, A_2, \ldots, A_r \tag{1.1}$$

such that $\mathrm{cls}(A_1) < \mathrm{cls}(A_2) < \cdots < \mathrm{cls}(A_r)$ and $\mathrm{ini}(A_j)$ is reduced with respect to A_i for each pair $i < j$. A polynomial set AS is a *triangulated set* or a *weak ascending set* if its polynomials are all non-zero and can be arranged in a sequence as (1.1) such that $\mathrm{cls}(A_1) < \mathrm{cls}(A_2) < \cdots < \mathrm{cls}(A_r)$. A *trivial* ascending set or triangulated set is a polynomial set consisting of only one non-zero constant polynomial.

Let

$$\mathit{AS}': \quad A'_1, A'_2, \ldots, A'_{r'},$$
$$\mathit{AS}'': \quad A''_1, A''_2, \ldots, A''_{r''},$$

be two ascending sets (whose polynomials are arranged) in ascending order. AS' is said to be of *lower ordering* than AS'' or AS'' is of *higher ordering* than AS' if either (o_1) or (o_2) below holds true:

(o_1) There is some k such that $A'_1 \sim A''_1, \ldots, A'_{k-1} \sim A''_{k-1}$, while $A'_k \prec A''_k$.

(o_2) $r' > r''$ and $A'_1 \sim A''_1, \ldots, A'_{r''} \sim A''_{r''}$.

In notation: $AS' \prec AS''$ or $AS'' \succ AS'$.

It is clear that the binary relation $\prec$ introduces a partial ordering among all ascending sets. A polynomial Q is *reduced* with respect to an ascending set AS if Q is reduced with respect to all the polynomials in AS.

Lemma 1.1:

The ordering of an ascending set will be lowered by adjoining to it a polynomial reduced with respect to it. More precisely, by adjoining a polynomial reduced with respect to an ascending set, there may be extracted an ascending set of lower ordering than the original one.

Lemma 1.2:

A sequence of ascending sets steadily lower in ordering is finite.

A *basic set* of a polynomial set PS is any ascending set of lowest ordering contained in PS. It is evident that any two basic sets of a polynomial set are of the same ordering.

Let F and G be non-zero polynomials with $c = \mathrm{cls}(F)$ and $I = \mathrm{ini}(F)$. By simple division we have

$$I^s G = QF + R,$$

for some non-negative integer s and polynomials Q, R with R reduced with respect to F. The exponent s will be chosen to be as small as possible to make such a formula hold. The polynomial R is then uniquely determined, and is called the *(pseudo-) remainder* (in Ritt's sense) of G with respect to F. In notation: $R = \mathrm{prem}(G, F)$. Note that the remainder defined here is different from the pseudo-remainder in current use.

For an ascending set (or triangulated set) AS as given by (1.1) in which the A_i are arranged in ascending order and for any non-zero polynomial G, there is a uniquely determined polynomial R reduced with respect to AS such that

$$I_G G = \sum_i Q_i A_i + R$$

for some polynomials Q_i and some power-product I_G of initials of the polynomials A_i in AS. The uniquely determined polynomial R is called the *remainder* of G with respect to AS. In notation: $R = \text{prem}(G, AS)$. The above formula is called the *remainder formula* of G with respect to AS.

1.2.2 Characteristic sets and well-ordering principle

Let PS be a polynomial set. We first set $PS = PS_0$ and choose an arbitrary basic set BS_0 of PS. If there are still polynomials in PS_0 but not in BS_0, let us form the remainders of these polynomials with respect to BS_0. Suppose that there are non-zero remainders forming a non-empty set RS_0. Then we form a new polynomial set $PS_1 = PS_0 \cup BS_0 \cup RS_0$. Choose now an arbitrary basic set BS_1 of PS_1. By Lemma 1.1 $BS_0 \cup RS_0$ is of lower ordering than BS_0. Continuing in this way, we will get successively PS_i, BS_i, RS_i for $i = 1, 2, \ldots$ with

$$BS_0 \succ BS_1 \succ BS_2 \succ \cdots.$$

By Lemma 1.2 the sequence can only be finite so that up to a certain stage m we should have $RS_m = \varnothing$. The final basic set BS_m is then an ascending set specifically related to the given polynomial set PS; let BS_m be denoted by CS. The above procedure can be exhibited in the form of the following scheme:

$$\begin{array}{rclllllll} PS & = & PS_0 & PS_1 & \cdots & PS_i & \cdots & PS_m & \\ & & BS_0 & BS_1 & \cdots & BS_i & \cdots & BS_m = CS & \\ & & RS_0 & RS_1 & \cdots & RS_i & \cdots & RS_m = \varnothing. & \end{array} \qquad \text{(S)}$$

For any polynomial set PS, we define

$$\text{prem}(PS, CS) := \{\text{prem}(P, CS) |\ P \in PS\}$$

and denote, by $\text{Zero}(PS)$, the set of all common zeros of the polynomials in PS. A formal definition of $\text{Zero}(PS)$ is given in (1.2). Let D be any polynomial; we have

$$\text{Zero}(PS/D) := \text{Zero}(PS) \setminus \text{Zero}(D).$$

Theorem 1.3:

The procedure in the scheme (S) will end in a finite number of steps at some stage m with corresponding remainder set $RS_m = \varnothing$. The corresponding basic set $BS_m = CS$ verifies (c_1) $\text{prem}(PS, CS) = \{0\}$ *and* (c_2) $\text{Zero}(PS) \subset \text{Zero}(CS)$.

Any ascending set *CS* gotten from a polynomial set *PS* by means of some procedures in accordance with the scheme (S) is called a *characteristic set* of *PS*. More generally, any ascending set *CS* verifying the properties (c_1) and (c_2) in Theorem 1.3 is a *characteristic set* of the polynomial set *PS*.

Remark 1.1:

The notion and terminology of characteristic sets have their origin in the works of J. F. Ritt (1932, 1950). However, as pointed out to the present author by D. Kapur, Ritt defined his characteristic set only for ideals while the characteristic set here is defined for polynomial sets, or for bases of ideals.

The following theorem occupies a central position in our whole theory.

Theorem 1.4: (Well-ordering principle)

Let CS be a characteristic set of a polynomial set PS. Then:

$$\mathrm{Zero}(CS/IP) \subset \mathrm{Zero}(PS) \subset \mathrm{Zero}(CS),$$
$$\mathrm{Zero}(PS/IP) = \mathrm{Zero}(CS/IP),$$
$$\mathrm{Zero}(PS) = \mathrm{Zero}(CS/IP) \cup \textstyle\bigcup_i \mathrm{Zero}(PS \cup \{I_i\})$$

where I_i is the initial of the polynomial C_i in CS, and IP is the product of the initials of the polynomials in CS (or initial-product of *CS for short).*

Several variants of the well-ordering principle with the aim of enhancing the efficiency may be found in Wu (1989c, 1999; see also Wang 1992b). For example, we may replace $PS_1 = PS_0 \cup BS_0 \cup RS_0$ in scheme (S) by $PS_1 = BS_0 \cup RS_0$.

Example 1.2:

Let us recall the polynomials of Zhu as in Example 1.1. Using the notations therein and with ordering $z \prec x \prec y$ ($z = x_1, x = x_2, y = x_3$) we have $PS = PS_0 = \{P_1, P_2, P_3\}$ and $BS_0 = \{P_2\}$, for which the polynomials P_i are reproduced below:

$$P_1 = xyz - xy^2 - z - x - y,$$
$$P_2 = xz - x^2 - z - y + x,$$
$$P_3 = z^2 - x^2 - y^2.$$

It is readily found that

$$R_{01} = \mathrm{prem}(P_1, BS_0) = x(-x^4 + 2x^3z + 2x^3 - x^2z^2 - 5x^2z - x^2 + 3xz^2 + 3xz + x - 2z^2 - z - 2),$$
$$R_{02} = \mathrm{prem}(P_3, BS_0) = x(-x^3 + 2x^2z + 2x^2 - xz^2 - 4xz - 2x + 2z^2 + 2z).$$

Hence $RS_0 = \{R_{01}, R_{02}\} \neq \varnothing$. Let us proceed as in scheme (S), so that

$$PS_1 = BS_0 \cup RS_0 = \{P_0, R_{01}, R_{02}\}.$$

The procedure will end at stage $m = 5$, and we have the final $RS_5 = \varnothing$ and $PS_5 = BS_5 = CS = \{C_1, C_2\}$ with C_1, C_2 given below:

$$C_1 = x(z-1)^2(z+1)(z-5), \qquad C_2 = -y - x^2 + xz + x - z.$$

It is readily verified that all the remainders of P_i with respect to CS are 0, so that $CS = \{C_1, C_2\}$ is a characteristic set of PS. By the well-ordering principle, we have

$$\text{Zero}(PS) = \text{Zero}(CS/I_1) \cup \text{Zero}(PS \cup \{I_1\}),$$

where $I_1 = (z-1)^2(z+1)(z-5)$.

1.2.3 Zero decomposition theorems

Let $\boldsymbol{K}$ be a field of characteristic 0, $\omega_1, \ldots, \omega_i, \ldots$ an infinite sequence of independent indeterminates over $\boldsymbol{K}$, Ω the field of rational functions of ω_i over $\boldsymbol{K}$, and Ω^* the algebraic closure of Ω. We call Ω^* the *universal field* over $\boldsymbol{K}$. An *extension field* of $\boldsymbol{K}$ is any finite extension field of $\boldsymbol{K}$ contained in Ω^*.

For an extension field $\boldsymbol{K}^*$ over $\boldsymbol{K}$, let $\text{TrDeg}(\boldsymbol{K}^*/\boldsymbol{K})$ denote the transcendence degree of $\boldsymbol{K}^*$ over $\boldsymbol{K}$. An *extended point* in $\boldsymbol{K}^n$ is a point in $\boldsymbol{K}^{*n}$ with $\boldsymbol{K}^*$ some extension field of $\boldsymbol{K}$. The *defining field* of an extended point $\Xi = (\xi_1, \ldots, \xi_n)$ in $\boldsymbol{K}^n$ is $\boldsymbol{K}(\Xi) = \boldsymbol{K}(\xi_1, \ldots, \xi_n)$. The *dimension* over $\boldsymbol{K}$ of an extended point Ξ in $\boldsymbol{K}^n$ is $\text{Dim}_{\boldsymbol{K}}\Xi = \text{TrDeg}(\boldsymbol{K}(\Xi)/\boldsymbol{K})$.

For an extended point $\xi = (\xi_1, \ldots, \xi_n)$, let us assume that ξ_{p_j} $(j = 1, \ldots, r)$ are all the elements such that ξ_{p_j} are algebraic over $\boldsymbol{K}(\xi_1, \ldots, \xi_{p_j-1})$. Then we may find a set of polynomials (Wu 1999)

$$P_1(x_1, \ldots, x_{p_1}), \ldots, P_r(x_1, \ldots, x_{p_r})$$

such that $P_i(\xi_1, \ldots, \xi_{p_i}) = 0$ for $i = 1, \ldots, r$ and P_i are the polynomials with lowest possible degree in x_{p_i} to satisfy the above equations. We call $P_1, \ldots, P_r$ the *defining equations* of ξ.

An extended point Ξ' in $\boldsymbol{K}^n$ is called a *specialization* over $\boldsymbol{K}$ of an extended point Ξ in $\boldsymbol{K}^n$ if for any polynomial $P \in \boldsymbol{K}[\boldsymbol{x}]$ with $P(\Xi) = 0$, we have also $P(\Xi') = 0$. The set of all specializations over $\boldsymbol{K}$ of an extended point Ξ in $\boldsymbol{K}^n$ is denoted by $\text{Spec}_{\boldsymbol{K}}(\Xi)$.

For any polynomial set $PS \subset \boldsymbol{K}[\boldsymbol{x}]$, let

$$\text{Zero}(PS) := \{\boldsymbol{x} \text{ is an extended point in } \boldsymbol{K}^n |\ P(\boldsymbol{x}) = 0, \forall P \in PS\}. \qquad (1.2)$$

For any polynomial D, the zero set $\text{Zero}(PS/D) = \text{Zero}(PS) \setminus \text{Zero}(D)$ is called a *quasi-algebraic variety* over K. The zero decomposition theorems are to determine the structure of (quasi-) algebraic varieties using ascending sets.

Theorem 1.5: (Zero decomposition theorem I)

There is an algorithm which permits one to determine, for any given polynomial set PS, in a finite number of steps a finite set of ascending sets CS_k such that

$$\text{Zero}(PS) = \bigcup_k \text{Zero}(CS_k/IP_k), \tag{1.3}$$

where each IP_k is the initial-product of CS_k. Moreover, we have $\text{prem}(PS, CS_k) = \{0\}$ *for all k.*

Proof:

By the well-ordering principle (Theorem 1.4), we have

$$\text{Zero}(PS) = \text{Zero}(CS/IP) \cup \bigcup_i \text{Zero}(PS \cup \{I_i\}),$$

where CS is a characteristic set of PS, IP is the initial-product of CS, and I_i are the initials of the polynomials in CS. One may continue the process by applying the well-ordering principle to each $PS \cup \{I_i\}$. According to Lemma 1.1, the new ascending sets obtained in this way will be of lower ordering than that of CS. By Lemma 1.2, the procedure will end in a finite number of steps.

Theorem 1.6: (Zero decomposition theorem II)

There is an algorithm which permits one to determine for any given polynomial set PS in a finite number of steps a finite set of ascending sets CS_k such that

$$\text{Zero}(PS) = \bigcup_k \text{Zero}(CS_k/ISP_k),$$

and $\text{prem}(PS, CS_k) = \{0\}$ *for all k, where each ISP_k is the product of the initials and separants of the polynomials in CS_k.*

The zero decomposition theorems can be modified in many ways to enhance the computational efficiency. The main techniques used include subeliminant sequence of polynomial pairs (Wu 1999), subresultant sequences (Wu 1999; Wang 1998a,d), factorization (Wu 1999; Chou and Gao 1990b, 1992a; Wang 1992d, 1994), top-down elimination procedures (Kapur and Wan 1990), Seidenberg's elimination procedures (Wang 1993a,b, 2000a), and new types of ascending chains (Wu 1984a,b; Chou and Gao 1990b; Wang 1993b).

1.2.4 Variety decomposition theorems

An *affine algebraic variety* or simply a *variety* $\mathcal{V}$ in $\boldsymbol{K}^n$ or $A^n_{\boldsymbol{K}}(\boldsymbol{x})$ (or over $\boldsymbol{K}$ in $A^n_{\boldsymbol{K}}(\boldsymbol{x})$) is the zero set of some polynomial set PS in $\boldsymbol{K}[\boldsymbol{x}]$, i.e., $\mathcal{V} = \text{Zero}(PS)$. An algebraic variety $\mathcal{V}_1$ is a *subvariety* of an algebraic variety $\mathcal{V}_2$ if $\mathcal{V}_1 \subset \mathcal{V}_2$. $\mathcal{V}_1$ is a *proper subvariety* of $\mathcal{V}_2$ if $\mathcal{V}_1 \subset \mathcal{V}_2$ and $\mathcal{V}_1 \neq \mathcal{V}_2$. An algebraic variety $\mathcal{V}$ in $\boldsymbol{K}^n$ is *irreducible* over $\boldsymbol{K}$ if $\mathcal{V}$ is not the union of two proper subvarieties in $\boldsymbol{K}^n$. We have the following criteria of irreducibility for algebraic varieties.

Criterion Theorem A:

An algebraic variety $\text{Zero}(PS)$ *in* $\boldsymbol{K}^n$ *is irreducible over* $\boldsymbol{K}$ *if and only if from* $P_1P_2 = 0|\text{Zero}(PS)$ *it follows that either* $P_1 = 0|\text{Zero}(PS)$ *or* $P_2 = 0|\text{Zero}(PS)$, *where* $P = 0|\text{Zero}(PS)$ *means that* P *vanishes on* $\text{Zero}(PS)$.

Criterion Theorem B:

An algebraic variety $\mathcal{V}$ *in* $\boldsymbol{K}^n$ *is irreducible over* $\boldsymbol{K}$ *if and only if there exists on* $\mathcal{V}$ *an extended point* Ξ *in* $\boldsymbol{K}^n$ *such that* $\mathcal{V} = \text{Spec}_{\boldsymbol{K}}(\Xi)$.

It is easy to prove the converse: $\text{Spec}_{\boldsymbol{K}}(\Xi)$ is irreducible for all extended point Ξ.

Any extended point Ξ in $\boldsymbol{K}^n$ satisfying $\text{Zero}(PS) = \text{Spec}_{\boldsymbol{K}}(\Xi)$ is called a *generic point* of the irreducible variety $\text{Zero}(PS)$. The *dimension* of an irreducible algebraic variety in $\boldsymbol{K}^n$ is defined as the dimension over $\boldsymbol{K}$ of any generic point of that variety.

The following result is well-known.

Theorem 1.7:

Every algebraic variety $\mathcal{V}$ *can be represented as an irredundant union of irreducible algebraic varieties.*

The irreducible algebraic varieties mentioned in Theorem 1.7 are called the *irreducible components* of $\mathcal{V}$. The main task of this section is to give a constructive proof for the above theorem.

Consider a non-trivial ascending set in ascending order

$$AS: \qquad A_1, A_2, \ldots, A_r, \tag{1.4}$$

in which

$$A_i = I_{i,0}x_{c_i}^{d_i} + \cdots + I_{i,j}x_{c_i}^{d_i-j} + \cdots + I_{i,d_i}, \quad i = 1, \ldots, r,$$

are the normal representations of A_i. The ascending set $AS_k: A_1, A_2, \ldots, A_k$ is called the kth *partial ascending set* of AS.

The ascending set *AS* in (1.4) is *irreducible* in stage k $(1 \leq k \leq r)$ if the following are true:

(1) For $k = 1$, A_1 is irreducible in $\boldsymbol{K}[\boldsymbol{x}_{c_1}]$.

(2) For $k > 1$, *AS* is irreducible in stage $k - 1$. Moreover, let $\boldsymbol{K}_{c_k - 1}$ be the extension field of $\boldsymbol{K}$ with x_j, for $1 < j < c_k$ and $j \neq$ any c_i, adjoined to $\boldsymbol{K}$ as independent transcendental elements, and $x_{c_i}, i < k$, adjoined to $\boldsymbol{K}$ successively by the irreducible equations $A_i = 0$, for $i = 1, \ldots, k - 1$. Then A_k, as a polynomial in x_{c_k}, is primitive in $\boldsymbol{K}[\boldsymbol{x}_{c_k}]$ and is irreducible in $\boldsymbol{K}(\boldsymbol{x}_{c_k - 1})[x_{c_k}]$,

The ascending set *AS* in (1.4) is *irreducible* if it is irreducible in stage r. In the above definition, we actually determine a solution for the equations $A_i = 0, i = 1, \ldots, r$, which is called a *generic point* of *AS*.

Example 1.3:

Consider the ascending set $AS:\ A_1, A_2$, with $A_1 = x_2^2 + x_1,\ A_2 = x_3^2 + x_1$. Then *AS* is clearly irreducible in stage 1 but not so in stage 2 since

$$A_2 \equiv (x_3 + x_2)(x_3 - x_2) + A_1$$

so that $A_2 \equiv (x_3 + x_2)(x_3 - x_2) \bmod_{\boldsymbol{K}} A_1$.

Theorem 1.8:

Let AS be an irreducible ascending set with generic point Ξ. Then:

(1) $AS = 0$ *is a set of defining equations of* Ξ.

(2) *For any non-zero polynomial* P, $\operatorname{prem}(P, AS) = 0$ *iff* $P(\Xi) = 0$.

(3) *Let* $\operatorname{PD}(AS) := \{P \in \boldsymbol{K}[\boldsymbol{x}] | \operatorname{prem}(P, AS) = 0\}$. *Then* $\operatorname{PD}(AS)$ *is a prime ideal.*

(4) $\operatorname{Zero}(\operatorname{PD}(AS)) = \operatorname{Spec}_{\boldsymbol{K}}(\Xi)$ *is an irreducible variety.*

We call $\operatorname{PD}(AS)$ in (3) the *prime ideal associated* with *AS*, and $\operatorname{Zero}(\operatorname{PD}(AS))$ in (4) the *irreducible variety associated* to *AS*; the latter is denoted by $\operatorname{Var}[AS]$. For any polynomial set *PS*, $\operatorname{Ideal}(PS)$ denotes the ideal generated by the polynomials in *PS*.

Theorem 1.9:

Based on Chow's theory of Chow forms, there is an algorithm which computes, from any irreducible ascending set IRR, in a finite number of steps a finite basis CB of $\mathrm{Var}[\mathit{IRR}]$ *or* $\mathrm{PD}(\mathit{IRR})$ *such that*

$$\mathrm{Var}[\mathit{IRR}] = \mathrm{Zero}(\mathit{CB}), \qquad \mathrm{PD}(\mathit{IRR}) = \mathrm{Ideal}(\mathit{CB}).$$

Other methods for computing bases of $\mathrm{PD}(\mathit{IRR})$ can be found in Chou (1988) and Wang (1989e). Now we may state the theorem of variety decomposition.

Theorem 1.10:

There is an algorithm which computes, for any given polynomial set PS, a finite number of irreducible ascending sets IRR_k *in a finite number of steps, such that*

$$\mathrm{Zero}(\mathit{PS}) = \bigcup_k \mathrm{Zero}(\mathit{IRR}_k/\mathit{IP}_k),$$

in which IP_k *is the initial-product of* IRR_k *for each* k. *The above zero relation is also equivalent to the following one:*

$$\mathrm{Zero}(\mathit{PS}) = \bigcup_k \mathrm{Var}[\mathit{IRR}_k].$$

Moreover, redundant varieties in the second formula can be removed by additional computations so that the union will become non-contractible and is unique up to an order.

Note that the proof of this theorem needs an algorithm of factorization of polynomials over algebraically extended fields of $\boldsymbol{Q}(u_1, \ldots, u_t)$ for indeterminates u_i (Wu 1984a,b). For an introduction to such factorization algorithms, see Chap. 13.

The second half of Theorem 1.10 can be generalized as follows.

Theorem 1.11: (Unmixed decomposition theorem; Gao and Chou 1993a)

The zero decomposition (1.3) *in Theorem 1.5 is equivalent to the following one:*

$$\mathrm{Zero}(\mathit{PS}) = \bigcup_k \mathrm{Zero}(\mathrm{QD}(\mathit{CS}_k)),$$

where

$$\mathrm{QD}(\mathit{CS}_k) := \left\{ P \in \boldsymbol{K}[\boldsymbol{x}] \;\middle|\; \begin{array}{r} \exists \text{ initial power product } J \text{ of } \mathit{CS}_k \\ \text{such that } JP \in \mathrm{Ideal}(\mathit{CS}_k) \end{array} \right\}.$$

Moreover, every $\mathrm{Zero}(\mathrm{QD}(\mathit{CS}_k))$, *if not empty, is an unmixed algebraic variety (i.e., its irredundant irreducible components all have the same dimension).*

Note that the above version of the decomposition theorem has the following implications.

(1) Those components Zero(QD(CS_k)), for which $|CS_k| > |PS|$, are redundant and can be removed automatically by using the affine dimension theorem (Hartshorne 1977). In the process of theorem proving, this may simplify the computation a lot.

(2) Since the bases of QD(CS_k) can be computed for any ascending set, the second formula gives an unmixed decomposition of the variety without using algorithms of factorization.

QD(CS_k) is also called the *saturation* of Ideal(CS_k) with respect to IP_k, a concept originated from commutative algebra (see the equivaltent notation sat in Definition 2.13). Many examples of irreducible decomposition can be found in Chou and Gao (1992a), Wang (1992a) and Wu (1999).

The reader may refer to Chou (1988), Cox et al. (1992), Mishra (1993), Wang (1989–97, 2000a) and Wu (1994a, 1999) for other detailed expositions of the characteristic set method and Chap. 20 for a review of its implementations.

1.3 Polynomial equation solving

1.3.1 Principle based on characteristic sets

Let us consider a polynomial set $AS \subset \boldsymbol{K}[\boldsymbol{x}]$ of some particular form. Suppose that the indeterminates $x_1, \ldots, x_n$ can be renamed as $u_1, \ldots, u_d, y_1, \ldots, y_r$ $(r + d = n)$ such that the polynomials in AS can be arranged in normal forms below:

$$A_1(\boldsymbol{u}, y_1), A_2(\boldsymbol{u}, y_1, y_2), \ldots, A_r(\boldsymbol{u}, y_1, \ldots, y_r),$$

where $\boldsymbol{u}$ stands for $u_1, \ldots, u_d$. Then the polynomial set AS is a *triangulated set*. It is clear to see that an ascending set is always a triangulated set.

Principle PES. A set of polynomial equations of the form $AS = 0$ in which AS is a triangulated set may be considered as solved for $y_1, \ldots, y_r$ as definite algebraic functions of $u_1, \ldots, u_d$, so far as the initials are all non-zero. If definite values of the u's are known, then definite values of the y's may also be determined for the solutions, whenever the initials are not turned into 0. In a word, Zero(AS/IP), with IP the initial-product of AS, and more generally Zero(AS/IPG) for any polynomial $G \in \boldsymbol{K}[\boldsymbol{x}]$ may be considered as a well-determined set.

The *characteristic set method* provides an effective device for converting an arbitrary polynomial set into well-determined sets, and thus a complete settlement for the problem of *polynomial equation solving* at least theoretically. This method has been used to solve many problems which will be reported in later chapters of the book. But for difficult problems, the size (in particular the number of terms) of the polynomials occurring in the procedures often increases rapidly so that the polynomials may soon become too large and out of control. An alternative partial remedy to the size-explosion problem is a *method of condensed coefficients.* For this purpose we shall first introduce some systematic notations for the polynomials and coefficients involved as follows.

Let the given polynomial set $PS \subset \boldsymbol{K}[\boldsymbol{x}]$ consist of polynomials $P_1, \ldots, P_k$. Starting from PS let us form successively, according to the algorithm indicated in the zero decomposition theorem I by remainder-formation, polynomials P_{k+1}, P_{k+2}, etc. In a polynomial P_j of the sequence we shall denote by c_t^j the coefficient of a monomial

$$\boldsymbol{x}^t = x_n^{d_n} \cdots x_1^{d_1},$$

with $t = (d_n, \ldots, d_1)$. We may also group some terms in the polynomial P_j with the same part $\boldsymbol{x}^u, u = (d_n, \ldots, d_k)$, or $(d_n, \ldots, d_k, 0, \ldots, 0)$, with $k-1$ 0's, where $k > 1$, into a single term. The corresponding coefficient of $\boldsymbol{x}^u$ will then be denoted as c_u^j, so that

$$c_u^j = \sum_v c_{uv}^j \boldsymbol{x}'^v,$$

in which $\boldsymbol{x}' = (x_{k-1}, \ldots, x_1)$ and the summation is to be extended over integer tuples $v = (d_{k-1}, \ldots, d_1)$. The coefficients like c_u^j or c_{uv}^j will henceforth be called *condensed coefficients* of the respective polynomials.

Example 1.4:

Let us consider Example 1.2 again and assume that the coefficients c^j's for $j = 1, 2, 3$ are independent constant indeterminates. By simple manipulation the polynomial set $PS = PS_0 = \{P_1, P_2, P_3\}$ is seen to be equivalent to the set $PS_1 = \{P_4, P_5, P_6\}$ with P_4, etc. given in condensed form below:

$$\begin{aligned} P_4 &= x_3^2 + c_{10}^4 x_3 + c_{02}^4 x_2^2 + c_{01}^4 x_2 + c_{00}^4, \\ P_5 &= x_3 x_2 + c_{10}^5 x_3 + c_{02}^5 x_2^2 + c_{01}^5 x_2 + c_{00}^5, \\ P_6 &= c_{10}^6 x_3 + c_{02}^6 x_2^2 + c_{01}^6 x_2 + c_{00}^6. \end{aligned}$$

In the above polynomials, we have

$$\begin{aligned} c_{10}^i &= c_{101}^i x_1 + c_{100}^i, \\ c_{02}^i &= c_{020}^i, \ c_{01}^i = c_{011}^i x_1 + c_{010}^i, \\ c_{00}^i &= c_{002}^i x_1^2 + c_{001}^i x_1 + c_{000}^i \end{aligned}$$

for $i = 4, 5, 6$. Remark that all the c^i's ($i = 4, 5, 6$) are definite rational functions of the c^j's for $j = 1, 2, 3$. For the sake of simplicity we shall consider the c^i_{ghk}'s for $i = 4, 5, 6$ as independent indeterminates in what follows.

Form now the remainders of P_4 and P_5 with respect to P_6, all considered as polynomials in x_3. We get then two further polynomials

$$\begin{aligned}
P_7 &= c^7_{030}x_2^3 + c^7_{02}x_2^2 + c^7_{01}x_2 + c^7_{00},\\
P_8 &= c^8_{040}x_2^4 + c^8_{03}x_2^3 + c^8_{02}x_2^2 + c^8_{01}x_2 + c^8_{00},
\end{aligned}$$

where

$$\begin{aligned}
&c^7_{030} = -\ c^6_{02}, \qquad c^7_{02} = c^6_{10}c^5_{02} - c^6_{02}c^5_{10} - c^6_{01},\\
&c^7_{01} = c^6_{10}c^5_{01} - c^6_{01}c^5_{10} - c^6_{00},\\
&c^7_{00} = c^6_{10}c^5_{00} - c^6_{00}c^5_{10},\\
&c^8_{040} = (c^6_{02})^2, \qquad c^8_{03} = 2c^6_{02}c^6_{01},\\
&c^8_{02} = (c^6_{10})^2c^4_{02} - c^6_{10}c^6_{02}c^4_{10} + 2c^6_{02}c^6_{00} + (c^6_{01})^2,\\
&c^8_{01} = (c^6_{10})^2c^4_{01} - c^6_{10}c^6_{01}c^4_{10} + 2c^6_{01}c^6_{00},\\
&c^8_{00} = (c^6_{10})^2c^4_{00} - c^6_{10}c^6_{00}c^4_{10} + (c^6_{00})^2.
\end{aligned}$$

The above equations, which express the condensed coefficients of the polynomials in terms of those of the preceding polynomials, will henceforth be called the *auxiliary equations*. Applying subresultant sequence computation to P_7, P_8, we get a polynomial P_9 of class 2 and degree 1, and a resultant P_{10} deprived of x_2. Expressing all the c's in P_9 and P_{10} in terms of the c^i's for $i = 4, 5, 6$ by means of the above auxiliary equations about c^7_* and c^8_*, it turns out that both P_9 and P_{10} have factors of the form c^6_{02} and $c^6_{10} = c^6_{101}x_1 + c^6_{100}$. Now remove these factors and denote the quotient polynomials thus obtained still by P_9 and P_{10}. By verification we see that the remainders of P_4, P_5, P_6 with respect to the ascending set $CS = \{P_{10}, P_9, P_6\}$ are all 0. It follows that CS is a characteristic set of $PS_1 = \{P_4, P_5, P_6\}$ and hence also of $PS = \{P_1, P_2, P_3\}$ so far as the removed factors c^6_{02} and c^6_{10} are considered as non-zero. Remark that if the coefficients of polynomials in CS are expressed in terms of those of the original coefficients of P_1, P_2, P_3 and clear of fractions, then the number of terms would be of order of billions.

Our investigations on the characteristic set method for solving systems of polynomial equations with diverse applications have been described in a series of papers (Wu 1987b, 1988, 1989a,c,d,e, 1990a–d, 1991b, 1992a–e, 1993a–d, 1994b,d, 1995a,b,c). Systematic experiments with the method have been reported in Aubry and Moreno Maza (1999) and Wang (1991d, 1992c, 1996a).

1.3.2 A hybrid method

As we have remarked in the previous section, there are two kinds of general methods for polynomial equation solving, symbolic ones and numerical ones; the basic field $\boldsymbol{K}$ being understood to be $\boldsymbol{R}$ (the field of real numbers) or $\boldsymbol{C}$ (the field of complex numbers) unless otherwise stated. Both kinds of methods have their own merits and demerits. Unlike numerical ones, symbolic methods have the advantage of being *exact* in character. Consider, however, the simple problem of solving a system of three quadratic equations in three unknowns x, y, z. Whether the coefficients of these equations are parametric or approximate numerical, for symbolic methods one has to replace the numerical values by indeterminates. Using our method for this problem, it is found that the final equations, in particular the one in x alone, will be either difficult to be determined or too large to have any interest or to be of any usefulness. Symbolic methods like ours therefore have the disadvantage of encountering very likely too big polynomials in the process of computation. However, if one replaces the intermediate coefficients of the given equations by their original numerical values, then the final polynomial, being of degree 8 in x, may have 9 terms only in the worst case. This shows that symbolic methods may be quite impracticable, especially for problems arising from practice. On the other hand, the usual numerical methods can hardly give the totality of all possible solutions. It seems thus that one solution is to use *hybrid methods*, in keeping the advantages of both kinds of methods, symbolic as well as numerical, while avoiding their disadvantages.

Remark that if we treat equations with numerical coefficients by a symbolic method, then there will be accumulated errors of coefficients in computations which will render the final results unreliable. Remark also that factorization is almost inevitable in symbolic methods, while factorization of polynomials with approximate numerical coefficients lacks exact meaning. Consequently, in order to give a satisfactory hybrid method for polynomial equation solving, we have to settle first the following two problems.

Problem H1:

The error estimation and error control during the hybrid computations.

Problem H2:

The factorization of polynomials with approximate numerical coefficients.

These two problems will be treated separately below.

Estimation of roots of a numerical algebraic equation. It is well-known that the roots of a polynomial equation in a single unknown are continuous in the coefficients and are highly sensitive to even slight variation of the coefficients. To be precise, let us cite below a theorem due to A. M. Ostrowski.

Theorem 1.12: (Ostrowski 1960)

Consider two polynomials

$$f(z) = a_0 z^n + \cdots + a_n, \qquad g(z) = b_0 z^n + \cdots + b_n.$$

Assume that $a_0 a_n \neq 0$ *and let the roots of* $f(z) = 0$ *be* $z_1, \ldots, z_n$. *Suppose that for a certain positive* τ *with* $4n\tau^{\frac{1}{n}} \leq 1$, *we have*

$$|b_i - a_i| \leq |a_i|\tau$$

for $i = 1, \ldots, n$. *Then the* n *roots* $y_1, \ldots, y_n$ *of* $g(z) = 0$ *can be ordered in such a way that*

$$|\frac{y_i}{z_i} - 1| < 8n\tau^{\frac{1}{n}}$$

for each $i = 1, \ldots, n$.

This theorem shows that for an equation, say of degree 4, in order to get roots accurate up to two decimal places, we require that the coefficients of the equation should be accurate up to perhaps 8 or more decimal places. Examples show that it is impossible to improve such an estimation of errors. It thus seems that the evaluation of errors becomes practically useless. In spite of this there is no reason to be discouraged.

In the sequel we shall restrict ourselves to the case of $\boldsymbol{K} = \boldsymbol{R}$, the case of $\boldsymbol{K} = \boldsymbol{C}$ being entirely similar. For our purposes, we shall treat the problem in the following way. For an approximate numerical value, say a, let us write it in the form $a_0 + t_0$, where a_0 is a *known exact rational* number close to a, while t_0 is a sufficiently small number, *unknown* but *exact*. For example, we may write π in the form $\pi = \frac{31416}{10000} + t$ in which $\frac{31416}{10000}$ and t are both *exact* numbers, t being an unknown number with magnitude $|t| < 10^{-4}$.

We call the form $a := a_0 + t_0$ an *exact representation* of the approximate number a, where a_0 is a known exact rational number, and t_0 an unknown exact number with $|t_0| < 10^{-N}$ and N a positive integer. The number a_0 is called the *essential part* and t_0 the *tiny part* of a, and N is called the *precision degree* of the representation.

Let us consider a system of polynomial equations $PS = 0$ with $PS \subset \boldsymbol{R}[x_1, \ldots, x_n]$. An *exact representation to precision degree* N of a polynomial set PS is the polynomial set gotten from PS by replacing each approximate numerical coefficient c_i in any polynomial of *PS* by its exact representation $c_{i0}+t_i$ to the same prescribed precision degree N, where c_{i0} are the essential parts and t_i are the tiny parts with all $|t_i| < 10^{-N}$. For the exact representation of *PS* in the above definition, t_i and N are called the *tinies* and the *precision degree* of the representation, respectively, and any linear combination of products of k tinies with definite rational coefficients is called a *tiny* of *order* k.

Let us now solve *PS* = 0 with approximate numerical coefficients in the following way. First fix a precision degree N and a positive integer s. Transform the given polynomial set into its exact representation to be denoted still by *PS*. Now solve *PS* = 0 by some symbolic method, say by the characteristic set method. To determine such a characteristic set *CS* of *PS* let us proceed in the usual way by the characteristic set algorithm. During the procedure we shall, however, set any tiny of order greater than s to be 0. Let the first polynomial of the characteristic set *CS* thus gotten be

$$C_1 = (a_0 + T_0)x_1^n + \cdots + (a_n + T_n), \tag{1.5}$$

in which T_k are tinies of order less than or equal to s and a_k are exact known rational numbers; similarly for the other polynomials C_i in *CS*. The polynomial set $\{C_i\}$ obtained above is called an *approximate characteristic set* to *order s* of the given polynomial set *PS*. Using other methods in the same way as in Sects. 1.2.2 and 1.2.3, we may get *approximate* decompositions for Zero(*PS*), etc., which we will not go into here.

Suppose that in (1.5) $a_0 a_n \neq 0$. Let C_{10} be the polynomial given by

$$C_{10} = a_0 x_1^n + \cdots + a_n. \tag{1.6}$$

We treat the other polynomials C_i in the same way to get polynomials C_{i0}. Let us solve the equation $C_{10} = 0$ of (1.6) instead of the equation $C_1 = 0$ of (1.5), and then the other polynomial equations $C_{i0} = 0$ in succession. The theorem of Ostrowski, where it may be applied, will give a rough estimate of the errors in the solutions. Thus the above provides us with a hybrid method for polynomial equation solving with error estimates roughly governed by Ostrowski's theorem.

Remark 1.2:

We keep only terms which have degrees $\leq s$ in the computation. Thus, if the t_i's are m in number, then the number of terms in C_{10} will be at most $(n+1)(m+1)$ for $s = 1$ and $(n+1)(1 + m\frac{m+3}{2})$ for $s = 2$. It is similar for other polynomials occurring in the procedure. In this way the polynomials will be controlled to have a reasonable number of terms and the computations may be carried out up to the end without much difficulties.

Remark 1.3:

Suppose that none of the essential parts of any initial become 0 during the procedure, then the above hybrid method will raise no theoretical difficulties and the Ostrowski theorem will furnish means of error estimate and control, at least theoretically. It is, however, not clear under what situations some initial in the procedure may have its essential part 0. Nevertheless it will be shown below that for practical problems this may rarely happen.

We come now to show how to settle the difficulties arising from exact representation of numerical data coefficients in the characteristic set method as pointed out in the above remark.

For this purpose let us consider a polynomial set $PS \subset \boldsymbol{R}[x_1, \ldots, x_n]$ whose zero set is to be determined. We try to do this by determining first a characteristic set of PS in accordance with the scheme (S) in Sect. 1.2.2. First treat the approximate numerical data occurring in the polynomials of PS as independent indeterminates. Proceeding as in scheme (S), let us arrange the polynomials of the given set PS as well as the successively introduced non-zero remainders in a definite order according to their order of appearance, say

$$P_1, P_2, \ldots, P_M.$$

Let us now fix a precision degree N and replace each approximate numerical data coefficient c_i by its exact representation $c_{i0} + t_i$ to the precision degree $N' > N$ with c_{i0} the essential part and t_i the corresponding tiny. Proceed in the same way as before in treating the tinies t_i as independent indeterminates and arrange the polynomials of PS as well as the successive new non-zero remainders in the definite order according to their order of appearance; we will get a sequence of polynomials below which are clearly in 1-1 correspondence with the preceding one:

$$P'_1, P'_2, \ldots, P'_M.$$

In fact, it is clear that each P'_k is gotten from the corresponding P_k by replacing each indeterminate c_i in P_k with its exact representation $c_{i0} + t_i$.

An exact representation $c_i = c_{i0} + t_i$ is *pathological* at stage k if for some monomial in x_i's in P'_k, its coefficient as a polynomial in the tinies has its part not involving any tiny equal to 0. In this case such a monomial is called a *pathological monomial*. An exact representation $c_i = c_{i0} + t_i$ is *healthy* if it is pathological at no stage in the sequence P'_i. When the exact representation P'_i is healthy, we may proceed as before and we will get a sequence in the same form as P_i' and the approximate characteristic set is of the same form as that gotten from P_i' even if all the tinies are set to be 0 during the procedure. Our problem is thus reduced to seeking for some *healthy* representation for the approximate numerical data coefficients c_i of the polynomials in the given polynomial set *PS* with a given precision degree N.

Theorem 1.13: (Wu 1993c, 1999)

Let $c_i = c_{i0} + t_i$ be any exact representation of the approximate numerical data coefficients of the polynomials in the given set PS $\subset \boldsymbol{R}[x_1, \ldots, x_n]$ with a precision order N. Then for any sufficiently small $\epsilon > 0$ there are healthy representations $c_{i0}' + t_i'$ with essential parts $c_{i0}' = c_{i0} + d_i$ and tinies $t_i' = t_i - d_i$ of precision degree N such that $|d_i| < \epsilon$. Moreover, the points $(d_1, \ldots, d_r)$ may be arbitrarily chosen from a subset open and dense in some sufficiently small neighborhood of the origin in the space $\boldsymbol{R}^r(t_1, \ldots, t_r)$.

Clearly, it is almost certain that in randomly choosing such a point $(d_1, \ldots, d_r)$ with d_i sufficiently small, the modified exact representation $c_i = c_{i0}' + t_i'$ with $c_{i0}' = c_{i0} + d_i, t_i' = t_i - d_i$ will be healthy. It follows that the above Theorem 1.13 actually gives a solution to Problem H1 in practice.

Approximate factorization of multivariate polynomials. For symbolic methods, factorization of multivariate polynomials is actually unavoidable. However, if the coefficients of a polynomial are in approximate numerical values, then it is difficult to decide whether the given polynomial is factorizable or not. The notion of approximate factorization occurred in some early writings of H. Zassenhaus and more recently it appears also in some papers by Sasaki et al. (1991). We shall introduce the notion of *approximate factorization* in the following way.

Let $f \in \boldsymbol{Q}[x, \boldsymbol{y}, t_1, \ldots, t_r]$ be a polynomial in exact representation, where $\boldsymbol{y} = (y_1, \ldots, y_l)$ is a set of parameters and t_i are tinies also treated as parameters. First consider the case in which $\boldsymbol{y}$ is absent while the basic field is a general field $\boldsymbol{K}$ of characteristic 0.

An *approximate factorization* of $f \in \boldsymbol{K}[x, t_1, \ldots, t_r]$ up to *order* s as a polynomial in x is a product of the form

$$f = gh \mod T^{s+1}, \tag{1.7}$$

in which g, h are polynomials in $\boldsymbol{K}[x, t_1, \ldots, t_r]$ of positive degree in x, and $\mod T^{s+1}$ means that (1.7) is an equality up to some tiny of order greater than s.

In order to approximately factorize f let us first remark that, by a preliminary almost linear transformation in x and t_i, we may without loss of generality lay down the following supposition:

(S) The leading coefficient in x of $f \in \boldsymbol{K}[x, t_1, \ldots, t_r]$ is free of t_i.

Under the supposition (S) the polynomial f can be written in the form

$$f = f_0 + f_1 + \cdots + f_m, \tag{1.8}$$

in which f_j is homogeneous of degree j in $t_1, \ldots, t_r$ for each j.

Let us introduce a new parameter u, replace each t_i by ut_i and consider the polynomial thus obtained from a polynomial $P \in \boldsymbol{K}[x, t_1, \ldots, t_r]$ as a polynomial $P^* \in \boldsymbol{K}[x, t_1, \ldots, t_r, u]$. Then (1.8) can be rewritten in the form

$$f^* = f_0 + uf_1 + \cdots + u^m f_m. \tag{1.8*}$$

We recover f from f^* by setting u equal to 1.

Lemma 1.14:

Factorizing f as in (1.7) is equivalent to factorizing f^* in the form below:

$$f^* = g^* h^* \mod u^{s+1}, \tag{1.9}$$

in which g^*, h^* are polynomials in $\boldsymbol{K}[x, t_1, \ldots, t_r, u]$ of degrees in x both greater than 1, and $\operatorname{mod} u^{s+1}$ means that (1.9) is an equality up to terms each having a factor in u of degree greater than s, or $f^* - g^*h^*$ is a multiple of u^{s+1}.

As usual, we base our method of the above approximate factorization in the form of (1.9) on Hensel construction as exhibited in van der Waerdern (1930).

Theorem 1.15: (Wu 1993c)

Suppose that in (1.8) f_0 satisfies (S) and g_0, h_0 in $\boldsymbol{K}[x]$ are known with

$$f_0 = g_0 h_0, \tag{1.10}$$

and that the degrees in x of f_0, g_0, h_0 are respectively n, p, q all greater than 0 with $n = p + q$. Suppose further that g_0, h_0 have a common factor D in x. Then for any given s there is an algorithm which permits one to decide whether f as a polynomial in $\boldsymbol{K}[x, t_1, \ldots, t_r]$ is approximately factorizable up to order s, and in the affirmative to determine such an approximate factorization as in (1.7). Moreover, if g_0 and h_0 are relatively prime, then such an approximate factorization is always possible and may be directly given by the algorithm.

Consider now the problem of approximate factorization of polynomials in exact representation $f \in \boldsymbol{Q}[x, \boldsymbol{y}, t_1, \ldots, t_r]$ involving parameters $\boldsymbol{y} = (y_1, \ldots, y_l)$ with tinies $(t_1, \ldots, t_r)$ as before. By an almost linear transformation if necessary we may suppose without loss of generality that the leading coefficient of f as a polynomial of x is free of $\boldsymbol{y}$ and $t_1, \ldots, t_r$. Let us introduce the extension field $\boldsymbol{K} = \boldsymbol{Q}(\boldsymbol{y})$ and consider f as a polynomial in $\boldsymbol{K}[x, t_1, \ldots, t_r]$. Then f satisfies the supposition (S) and Theorem 1.15 may be applied. If we turn $\boldsymbol{K}$ back to $\boldsymbol{Q}(\boldsymbol{y})$ and write the possible approximately factorized polynomials in the form of rational functions in $x, \boldsymbol{y}, t_1, \ldots, t_r$, then there will appear in the denominators some polynomials in $\boldsymbol{y}$ alone. By clearing of fractions we then get the following theorem as a corollary of Theorem 1.15.

Theorem 1.16:

Suppose that for $f \in \boldsymbol{Q}[x, \boldsymbol{y}, t_1, \ldots, t_r]$ in (1.8) each f_k is homogeneous of degree k in $t_1, \ldots, t_r$, while $f_0 \in \boldsymbol{Q}[x]$ is deprived of $\boldsymbol{y}$ and $t_1, \ldots, t_r$. Suppose that f_0 satisfies (1.10) in which g_0 and h_0 in $\boldsymbol{Q}[x]$ are known polynomials with degrees in x both positive. Then there is an algorithm which permits one to decide whether there exist polynomials $I, I_1, I_2 \in \boldsymbol{Q}[\boldsymbol{y}]$ and polynomials $g, h \in \boldsymbol{Q}[x, \boldsymbol{y}, t_1, \ldots, t_r]$ such that for any given s

$$If = gh \mod T^{s+1}, \tag{1.7$'$}$$

in which the parts in g and h not involving any $t_1, \ldots, t_r$ are $I_1 g_0, I_2 h_0$ and $I_1 I_2 = I$.

In application of the above theorem to actual characteristic set computations of a polynomial set $PS \subset \boldsymbol{R}[x_1, \ldots, x_n]$, we may encounter the problem of approximate factorization of some polynomial f during the procedure. The variable x in f will then be some x_m while the $\boldsymbol{y}$ will be some of the variables $x_1, \ldots, x_{m-1}$ preceding x_m in the natural variable ordering. It follows that the polynomial I in $(1.7)'$ may be absorbed in the initial of f, which thus causes no trouble for further computations. It follows that Theorems 1.15 and 1.16 actually furnish a solution to Problem H2.

1.3.3 A projection theorem on quasi-varieties

Classical elimination theories are used to give necessary and sufficient conditions for the existence of solutions for a set of homogeneous polynomial equations. It was proved that for a set of homogeneous polynomial equations whose coefficients are indeterminates, this kind of condition can be represented as the zero set of an ideal of polynomials in the coefficients of the equations (Hodge and Pedoe 1952). In the more frequently used affine case, this is not true. For instance, the condition for the equation $ax - 1 = 0$ to have solutions is $a \neq 0$. It is clear that to express such existential conditions, one needs to use quasi-varieties. For this purpose, we introduce the concept of projection.

Let $\boldsymbol{E}$ be an extension field of $\boldsymbol{K}$, and *PS* and *DS* be polynomial sets in $\boldsymbol{Q}[x_1, \ldots, x_n, t_1, \ldots, t_m]$. The *projection* of $\mathcal{Z} = \text{Zero}(PS/DS)$ with $t_1, \ldots, t_m$, $\text{Proj}_{t_1,\ldots,t_m} Z$, is defined to be

$$\{(a_1, \ldots, a_n) \in \boldsymbol{E}^n |\ \exists (b_1, \ldots, b_m) \in \boldsymbol{E}^m \text{ such that } (a_1, \ldots, a_n, b_1, \ldots, b_m) \in \mathcal{Z}\}.$$

In what follows, we briefly explain the method of computing projection proposed in Wu (1990a).

Theorem 1.17:

The projection of a quasi-variety is also a quasi-variety which can be determined explicitly in an algorithmic manner.

First, we consider the case of $m = 1$, i.e., only one projection variable exists. Below we shall arrange the variables involved in the order $x_1 \prec \cdots \prec x_n \prec t = t_1$.

A *simple quasi-variety* of *type* d in the affine space $\boldsymbol{K}^{n+1}$ is a quasi-variety of the form $\mathcal{Z} = \text{Zero}(XS \cup \{F\}/GH)$ in which $XS \subset \boldsymbol{K}[\boldsymbol{x}], F, G \in \boldsymbol{K}[\boldsymbol{x}, t], H \in \boldsymbol{K}[\boldsymbol{x}]$ and $\deg(F, t) = d$ with $d > 0$ while F is lacking for $d = 0$.

By the zero decomposition theorems, we can prove that the projection of any quasi-variety can be reduced to that of simple quasi-varieties.

Lemma 1.18:

In $\boldsymbol{K}^{n+1}$ *any quasi-variety* $\text{Zero}(PS/GH)$, *in which* $PS \subset \boldsymbol{K}[\boldsymbol{x}, t], G \in \boldsymbol{K}[\boldsymbol{x}, t]$ *while* $H \in \boldsymbol{K}[\boldsymbol{x}]$, *is the union of simple quasi-varieties each of form* $\text{Zero}(XS \cup \{F\}/Q)$ *for which* Q *is divisible by the initial of* F *in variable* t.

Let us consider now three particular cases of simple quasi-varieties for which either F or G is lacking.

Lemma 1.19:

Let $A = \sum_{i=0}^{n} A_i t^i \in K[\boldsymbol{x}, t]$ *where* A_i *are polynomials in the* x_i; *then*

$$\text{Proj}_t \text{Zero}(A) = \bigcup_{i=1}^{n} \text{Zero}(PS_i/A_i) \bigcup \text{Zero}(\varnothing/A_0),$$

where $PS_i = \{A_n, \ldots, A_{i+1}, A\}$.

Lemma 1.20:

Let $B = \sum_{i=0}^{n} B_i t^i \in \boldsymbol{K}[\boldsymbol{x}, t]$ *where* B_i *are polynomials in the* x_i*; then*

$$\mathrm{Proj}_t \mathrm{Zero}(\varnothing/B) = \bigcup_{i=0}^{n} \mathrm{Zero}(\varnothing/B_i).$$

Lemma 1.21:

Let P *and* Q *be polynomials in* $\boldsymbol{K}[\boldsymbol{x}, t]$ *such that* $d = \deg(P, t) > 0$ *and* $\deg(Q, t) > 0$*; then* $\mathrm{Proj}_t \mathrm{Zero}(P/QI) = \mathrm{Proj}_t \mathrm{Zero}(\varnothing/RI)$ *where* I *is the initial of* P *and* $R = \mathrm{prem}(Q^d, P)$.

Now the proof of the projection theorem becomes evident since we can reduce the projection of general quasi-varieties to simple quasi-varieties and can find the projection of simple quasi-varieties by Lemmas 1.19–1.21.

Wang (1993a,b) proposed to incorporate the projection process into the elimination procedure; see the following chapter for relevant discussions. For the application of the projection theorem to the implicitization of rational parametric equations, see Chap. 14.

1.4 Applications

This section reports three applications of the polynomial equation solving methods introduced in the preceding section. Other applications are presented in the following chapters.

1.4.1 Solving of problems in enumerative geometry

As illustrations for our general method of polynomial equation solving let us consider problems of enumerative geometry. Such problems can be formulated in the following general form.

Problem EG:

Find the number of geometrical entities of certain type verifying some prescribed geometrical conditions.

In general, by introducing suitable coordinates, the geometrical entities in question can be expressed by points in some affine spaces $\boldsymbol{K}^n(\boldsymbol{x})$, where $\boldsymbol{x} = (x_1, \ldots, x_n)$, $\boldsymbol{K}$ is a field of characteristic 0, and the coordinates $\boldsymbol{x}$ verify some polynomial equations. The prescribed geometrical conditions in question will also in general be expressed in the form of polynomial equations in the coordinates $\boldsymbol{x}$, together with eventually some other parameters. In any case, we shall restrict ourselves to Problem EG of the above characterization. The solving of such problems thus leads to the solving of some polynomial equations which may then be treated by our general methods of polynomial equation solving.

The classical enumerative geometry forms a fascinating chapter of modern algebraic geometry. The solving of the corresponding problems is based on the *counting principle* of the theory of algebraic correspondence (Hodge and Pedoe 1952). Owing to the method used the problems treated have some restrictions below. First, the geometrical entities are restricted to be in *projective* spaces over algebraically closed fields of characteristic 0, mainly the field of complex numbers. The geometrical conditions are also restricted to be of a projective character. Secondly, the number of solutions to be determined holds only in the *generic* cases and the number has to be counted based on the notion of *multiplicity*. In non-generic cases there is no general way of treatment.

Using the equation solving method described in this section, we can give not only the number of solutions but also the determining equations of these geometric objects. The following are three examples solved by this method (Wu 1999).

Apollonius Problem:

Determine the number of circles touching three given circles in the ordinary plane.

Malfatti Problem:

Given a triangle in the ordinary plane, determine the number of triples of circles touching each other and each touching two sides of the triangle.

Lines on cubic surface in projective space. A cubic surface S_3 in a projective space $P^3_{\boldsymbol{K}}(Z)$ with $Z = (z_0 : z_1 : z_2 : z_3)$ is given by the equation

$$\begin{aligned} S = {} & a_1 z_0^3 + a_2 z_1^3 + a_3 z_2^3 + a_4 z_3^3 + a_5 z_0^2 z_1 + a_6 z_0^2 z_2 + a_7 z_0^2 z_3 \\ & + a_8 z_0 z_1^2 + a_9 z_1^2 z_2 + a_{10} z_1^2 z_3 + a_{11} z_0 z_2^2 + a_{12} z_1 z_2^2 + a_{13} z_2^2 z_3 \\ & + a_{14} z_0 z_3^2 + a_{15} z_1 z_3^2 + a_{16} z_2 z_3^2 \\ & + a_{17} z_0 z_1 z_2 + a_{18} z_0 z_1 z_3 + a_{19} z_0 z_2 z_3 + a_{20} z_1 z_2 z_3 = 0. \end{aligned}$$

Lines-on-Cubic-Surface Problem:

Determine the number of lines lying on the cubic surface S_3, supposing that there are only a finite number of lines on it.

It can be shown by means of the *counting principle* in the theory of algebraic correspondence that the number of lines lying on a cubic surface, if not infinite, is always 27, with multiplicities taken into account, see e.g. van der Waerdern

(1945). Applying our method to this problem, we may determine the *exact* number (i.e. without taking multiplicities into account) of such lines on an arbitrary cubic surface and also the lines themselves.

For this purpose, let us remark that the cubic surface S_3 should not be a cone and should have no plane as a component since otherwise there will be an infinity of lines lying on S_3. This being understood, we shall take a coordinate tetrahedron $U_0U_1U_2U_3$ as follows. First we may take a point U_3 on S_3 such that the tangent plane T_3 at U_3 will intersect S_3 in a non-degenerate cubic curve C_3. Next we may take a point U_2 on C_3 such that the tangent plane T_2 of S_3 at U_2 is different from T_3 which intersects C_3 in a further point U_1 besides U_2. The tangent plane T_1 of S_3 at U_1 can contain neither U_2 nor U_3 since otherwise the line U_1U_2 (or U_1U_3) will meet the cubic surface S_3 in 4 points, two at U_1 and two at U_2 (or U_3), which is absurd. The tangent plane T_1 of S_3 at U_1 will then intersect T_2 in a line which will intersect S_3 at 3 points, two being at U_1, and the third one will be taken to be U_0.

From the construction, T_3 is the plane $U_1U_2U_3$, T_2 is the plane $U_0U_1U_2$ and T_1 contains the line U_0U_1. It follows respectively that $a_{15} = a_{16} = 0$, $a_{11} = a_{12} = 0$, and $a_8 = 0$. Consider now an arbitrary line L with Plücker coordinates $(p_{01} : p_{02} : p_{03} : p_{12} : p_{13} : p_{23})$. The line L, if not in any one of the coordinate planes $z_i = 0, i = 0, 1, 2, 3$, will intersect these planes at the points

$$P_0 = (0 : p_{01} : p_{02} : p_{03}), \qquad P_1 = (-p_{01} : 0 : p_{12} : p_{13}),$$
$$P_2 = (-p_{02} : -p_{12} : 0 : p_{23}), \qquad P_3 = (-p_{03} : -p_{13} : -p_{23} : 0).$$

We shall suppose that L lies wholly on S_3. By construction, the line U_2U_3 is not on S_3 so that $p_{01} \neq 0$. Then L is on S_3 if and only if for any u the point $Q_u = P_0 + u.P_1$ is on S_3. It follows that for L lying wholly on S_3 it is necessary and sufficient that $P_{ui} = 0, i = 0, 1, 2, 3$ with the polynomials P_{ui} given below:

$$\begin{aligned}
P_{u0} = {} & p_{03}p_{02}^2a_{13} + p_{03}p_{02}p_{01}a_{20} + p_{03}p_{01}^2a_{10} + p_{02}p_{01}^2a_9,\\
P_{u1} = {} & p_{13}p_{02}^2a_{13} + p_{13}p_{02}p_{01}a_{20} + p_{13}p_{01}^2a_{10} + 2p_{12}p_{03}p_{02}a_{13}\\
& + p_{12}p_{03}p_{01}a_{20} + p_{12}p_{01}^2a_9 - p_{03}^2p_{01}a_{14} - p_{03}p_{02}p_{01}a_{19}\\
& - p_{03}p_{01}^2a_{18} - p_{02}p_{01}^2a_{17},\\
P_{u2} = {} & 2p_{13}p_{12}p_{02}a_{13} + p_{13}p_{12}p_{01}a_{20} - 2p_{13}p_{03}p_{01}a_{14} - p_{13}p_{02}p_{01}a_{19}\\
& - p_{13}p_{01}^2a_{18} + p_{12}^2p_{03}a_{13} - p_{12}p_{03}p_{01}a_{19} - p_{12}p_{01}^2a_{17}\\
& + p_{03}p_{01}^2a_7 + p_{02}p_{01}^2a_6 + p_{01}^3a_5,\\
P_{u3} = {} & -p_{13}^2p_{01}a_{14} + p_{13}p_{12}^2a_{13} - p_{13}p_{12}p_{01}a_{19} + p_{13}p_{01}^2a_7 + p_{12}p_{01}^2a_6.
\end{aligned}$$

Besides we have the relation $P_L = 0$ for the Plücker coordinates where

$$P_L = p_{23}p_{01} - p_{13}p_{02} + p_{12}p_{03}.$$

We have then a polynomial set $PS = \{P_{u0}, P_{u1}, P_{u2}, P_{u3}, P_L\}$, and the lines on S_3 will be given by the set Zero(PS).

Let us arrange the variables $p_{ij}, i < j$ in the order $p_{01} \prec p_{02} \prec p_{03} \prec p_{12} \prec p_{13} \prec p_{23}$. Then a characteristic set CS may be found to consist of 5 polynomials $C_i,\ i = 1, \ldots, 5$, with C_1 irreducible; their index-sets are given below:

$$[15895\ p_{02}\ 27],\ [4\ p_{03}\ 1],\ [1245\ p_{12}\ 1],\ [10\ p_{13}\ 1],\ [3\ p_{23}\ 1].$$

The initials are $I_i, i = 1, \ldots, 5$, with $I_1 = a_5^2 a_6 a_9 a_{13} a_{15}$. It follows that in the generic case for which $I_i \neq 0,\ i = 1, \ldots, 5$, and the discriminant of C_1 with respect to p_{02} is also not equal to 0, the number of lines on S_3 is exactly 27 and the lines themselves are given by Zero(PS). The non-generic case may also be studied in the same way if we like by setting some of the I_i and/or the discriminant of C_1 to be 0.

1.4.2 Central configurations in planet and vortex motions

Let $J_n = \{1, \ldots, n\}$, $J_n^2 = \{(i,j) |\ i,j \in J_n, i \neq j\}$, and $m_1, \ldots, m_n$ be the *masses* of n particles moving under mutual Newtonian gravitational attractions. Let $r_1, \ldots, r_n$ be the positions of these masses at a certain moment, with $r_i \neq r_j$ for $i \neq j$, and

$$\mathcal{C}: \quad \begin{bmatrix} m_1 & m_2 & \cdots & m_n \\ r_1 & r_2 & \cdots & r_n \end{bmatrix} \tag{1.11}$$

a *configuration* formed of masses m_i at positions $r_i, i \in J_n$. $\mathcal{C}$ is a *rigid configuration* (respectively a *central configuration*) with respect to the masses $m_1, \ldots, m_n$ if there are initial velocities not all zero of the masses m_i such that under the Newtonian gravitational attractions the configurations formed by the masses during the motion will remain *congruent* (respectively *similar*) to the initial one.

As a direct consequence of Newtonian mechanics for a configuration $\mathcal{C}$ of the form (1.11), the center of the masses $m_1, \ldots, m_n$ may be considered to be fixed during the motion.

An *inertial coordinate system* associated to a rigid or a central configuration $\mathcal{C}$ as in (1.11) is a Cartesian coordinate system for which the origin is at the fixed center of the masses $m_1, \ldots, m_n$. Again as a direct consequence of Newtonian mechanics, we have the following proposition.

Theorem 1.22:

For a rigid or a central configuration $\mathcal{C}$ of the form (1.11), we have

$$\sum_i m_i r_i = 0 \tag{1.12}$$

in any associated inertial coordinate system.

Two rigid configurations with respect to the same masses $m_1, \ldots, m_n$ belong to the same *configuration class* if they have the same center of mass and there exists an orientation-preserving similarity transformation which keeps the center of mass fixed and transforms one configuration to the other.

Clearly a rigid configuration is also a central configuration with the same set of masses. The converse is known to be true, for which we omit the proof, however. In what follows only inertial coordinate systems are considered.

Let $q_3(n) = q_3(n; m_1, \ldots, m_n)$ be the number of classes of rigid configurations with given masses $m_1, \ldots, m_n$; $q_2(n) = q_2(n; m_1, \ldots, m_n)$ the number of classes of rigid configurations with given masses $m_1, \ldots, m_n$ for which the masses m_i are situated in the same plane; and $q_1(n) = q_1(n; m_1, \ldots, m_n)$ the number of classes of rigid configurations with given masses $m_1, \ldots, m_n$ for which the masses m_i are situated in the same line. The following results are known. For given masses $m_1, \ldots, m_n$ we have:

$q_1(2) = q_2(2) = q_3(2) = 1;$

$q_1(3) = 3$ (Euler 1767);

$q_2(3) = q_3(3) = 5$ (Lagrange 1772);

$q_1(n) = \dfrac{n!}{2}$ (Moulton 1910);

$q_3(n) > q_2(n)$ for each $n > 4$ with some particular sets of $m_1, \ldots, m_n$ (Waldvogel 1972);

$$q_1(n; m_1, \ldots, m_n) \leq q_2(n; m_1, \ldots, m_n) \leq q_3(n; m_1, \ldots, m_n).$$

Wintner Conjecture:

$q_3(n; m_1, \ldots, m_n)$ and hence also $q_2(n; m_1, \ldots, m_n)$ are finite for all n and masses $m_1, \ldots, m_n$.

In recent years, Smale (1970a,b) and his followers have studied central configurations with topological methods via Morse critical point theory. In this section, we shall restrict ourselves to the actual determination of rigid or central configurations by treating it as a problem of polynomial equation solving. For this purpose we shall adopt an inertial coordinate system associated with the rigid or central configuration $\mathcal{C}$ in (1.11) and the following notations will be used:

$$r_i = (x_i, y_i), \quad i \in J_n;$$
$$x_{ij} = x_i - x_j, \quad y_{ij} = y_i - y_j, \quad (i,j) \in J_n^2;$$
$$r_{ij} = \sqrt{x_{ij}^2 + y_{ij}^2}, \quad (i,j) \in J_n^2.$$

For a rigid configuration $\mathcal{C}$ of the form (1.11) in a plane, it is known that during the motion the plane will be fixed to keep the configuration congruent to the original one. Take now planar inertial coordinate system (x, y) with origin O at the center of mass. Then the configuration will move around O with an angular velocity ω such that

$$\omega^2 x_i = \sum_{j \neq i} m_j \frac{x_{ij}}{r_{ij}^3}, \qquad \omega^2 y_i = \sum_{j \neq i} m_j \frac{y_{ij}}{r_{ij}^3}, \tag{1.13}$$

Up to non-zero constant factors, the right-hand side of the equations represents the total Newtonian attractive forces exerted on the masses m_i by all the other masses, while the left-hand side represents the corresponding centrifugal forces. The equations (1.13) show in particular that ω is a constant. Furthermore, the equations are invariant under orientation-preserving similarity transformations with the origin kept fixed. However the constant ω may then be changed.

We have assumed $r_{ij} \neq 0$, $(i,j) \in J_n^2$, which means that collisions are out of consideration. The determination of rigid configurations and hence also central configurations amounts then to the solving of polynomial equations obtained from (1.13) by clearing of fractions. To achieve this we shall introduce some preliminary transformations of (1.13) to make the solving easier.

Let us take the field $\boldsymbol{C}$ of complex numbers as the basic field and introduce complex variables u_i, v_i, z_{ij}, w_{ij} as follows (with $\boldsymbol{i}$ standing for $\sqrt{-1}$):

$$u_i = x_i + \boldsymbol{i} y_i, \quad v_i = x_i - \boldsymbol{i} y_i, \quad i \in J_n,$$
$$z_{ij} = \frac{1}{(u_i - u_j) r_{ij}},$$
$$w_{ij} = \frac{1}{(v_i - v_j) r_{ij}}, \quad (i,j) \in J_n^2,$$
$$m_0 = -\omega^2.$$

From (1.13) we have the following set of polynomial equations:

$$m_0 u_i + \sum_{j \neq i} m_j w_{ij} = 0,$$

$$m_0 v_i + \sum_{j \neq i} m_j z_{ij} = 0.$$

$$(u_i - u_j) z_{ij} r_{ij} - 1 = 0,$$
$$(v_i - v_j) w_{ij} r_{ij} - 1 = 0,$$
$$r_{ij}^2 - (u_i - u_j)(v_i - v_j) = 0.$$
$$r_{ij} - r_{ji} = 0, \; z_{ij} + z_{ji} = 0, \; w_{ij} + w_{ji} = 0, \quad (i, j) \in J_n^2.$$

To these we may also add

$$\sum_i m_i u_i = 0, \qquad \sum_i m_i v_i = 0,$$

which follows from (1.12) or (1.13).

For a rigid or central configuration $\mathcal{C}$ of the form (1.11) the *reality conditions* are:

(1) $m_0, m_1, \ldots, m_n$ are all real numbers, with $m_0 < 0, m_i > 0, i \in J_n$;

(2) all r_{ij}, $(i, j) \in J_n^2$, are real and positive;

(3) u_i, v_i for $i \in J_n$ are conjugate complex numbers;

(4) $z_{ij} \neq 0, w_{ij} \neq 0$ for $(i, j) \in J_n^2$.

Now the problem of finding central configurations is transformed into a problem of polynomial equation solving. With the method described in this chapter, we have found many interesting results for this problem. The determination of rigid or central configurations for $n = 2$ is trivial. Let us proceed to the case $n = 3$ when we have found all the possible central configurations.

Theorem 1.23: (Wu 1995a)

In the case of $n = 3$, the only possible central configurations of three non-vanishing masses are the classical ones due to Euler and Lagrange — there are no others.

For the case of vortex movements in an incompressible and non-viscous fluid extending to infinity, we have proved the following result with our methods.

Theorem 1.24: (Wu 1995a)

The only possible rigid configurations of three parallel vortex filaments in an incompressible and non-viscous fluid extending to infinity for which the strengths are non-zero and verify some hypothesis K are either of Lagrange equilateral type or of Euler rectilinear type — there are no others.

Other interesting results about this problem may be found in Shi and Zhou (1997).

1.4.3 Solving problems involving inequalities

Perhaps the CAD method of Collins (1975) is the most general complete method which has been applied to problems involving inequalities. On the other hand, we have exhibited a method based on characteristic sets for solving problems involving inequalities. This method, though not so general and complete as that of Collins, seems to be quite efficient for certain kinds of problems (Wu 1988, 1992b,c). The method has been used to prove geometric theorems involving inequalities (Wu 1988), to prove inequalities, algebraic or geometric (Wang D.-K. 1993), to solve non-linear programming problems (Wu 1994b), to solve optimization problems (Wu 1992c), etc. A summary of these applications can be found in Wu (1999).

Denote by $\boldsymbol{R}^n(\boldsymbol{x})$ the *real* Euclidean space of dimension n in the coordinates $\boldsymbol{x} = (x_1, \ldots, x_n)$,

- O an open domain in $\boldsymbol{R}^n(\boldsymbol{x})$,
- D a domain in $\boldsymbol{R}^n(\boldsymbol{x})$, open or not, and let
- $HS = \{h_1, \ldots, h_m\} \subset \boldsymbol{R}[\boldsymbol{x}]$,
- $N = \{1, \ldots, n\}$ and $M = \{1, \ldots, m\}$.

Let T be the set of m-tuples $(i_1, \ldots, i_m)$, with $i_k \in N$ and $1 \leq i_1 < \cdots < i_m \leq n$,

$$\boldsymbol{x}_t := (x_{i_1}, \ldots, x_{i_m}) \quad \text{for} \quad t = (i_1, \ldots, i_m) \in T,$$

and f, g be polynomials in $\boldsymbol{R}[\boldsymbol{x}]$.

A point $\boldsymbol{x}^0 \in \boldsymbol{R}^n(\boldsymbol{x})$ is called an *extremal point* or *extremal zero* of f in D, or f attains its *local extremum* at $\boldsymbol{x}^0$ in D under the *constraint conditions* $HS = 0$ and $g \neq 0$ if $\boldsymbol{x}^0 \in D \cap \mathrm{Zero}_{\boldsymbol{R}}(HS/g)$ and there is a neighborhood V of $\boldsymbol{x}^0$ in $\boldsymbol{R}^n(\boldsymbol{x}) \setminus \mathrm{Zero}_{\boldsymbol{R}}(g)$ such that for $V_D = V \cap D$, and either

- $f(\boldsymbol{x}') \geq f(\boldsymbol{x}^0)$ for any $\boldsymbol{x}' \in V_D \cap \mathrm{Zero}_{\boldsymbol{R}}(HS/g)$, or

- $f(\boldsymbol{x}') \leq f(\boldsymbol{x}^0)$ for any $\boldsymbol{x}' \in V_D \cap \mathrm{Zero}_{\boldsymbol{R}}(HS/g)$.

A real number r is an *extremal value* of f in D under the *constraint conditions* $HS = 0$ and $g \neq 0$ if $r = f(\boldsymbol{x}^0)$ for some extremal point $\boldsymbol{x}^0$ of f in D under the constraint conditions $HS = 0$ and $g \neq 0$.

We introduce the following notations.

- $\mathrm{EZero}_{\boldsymbol{R}}\{f, g, D, HS\}$:= the set of extremal points of f in D under the conditions $HS = 0$ and $g \neq 0$.

- $\mathrm{EVal}_{\boldsymbol{R}}\{f, g, D, HS\}$:= the set of extremal values of f in D under the conditions $HS = 0$ and $g \neq 0$.

- $\mathrm{EZero}_{\boldsymbol{R}}\{f, D, HS\}$, or $\mathrm{EZero}_{\boldsymbol{R}}\{f, g, D\}$, or $\mathrm{EZero}_{\boldsymbol{R}}\{f, D\}$ is $\mathrm{EZero}_{\boldsymbol{R}}\{f, g, D, HS\}$ in case g, or HS, or both are absent.

- $\mathrm{EVal}_{\boldsymbol{R}}\{f, D, HS\}$, or $\mathrm{EVal}_{\boldsymbol{R}}\{f, g, D\}$, or $\mathrm{EVal}_{\boldsymbol{R}}\{f, D\}$ is $\mathrm{EVal}_{\boldsymbol{R}}\{f, g, D, HS\}$ in case g, or HS, or both are absent.

Then, we need to solve the following problem.

Extremal Problem E:

Determine $\mathrm{EZero}_{\boldsymbol{R}}\{f, g, D, HS\}$ *and* $\mathrm{EVal}_{\boldsymbol{R}}\{f, g, D, HS\}$ *for given* f, g, D, HS.

To solve this problem, we first assume that D is an *open* domain O in $\boldsymbol{R}^n(\boldsymbol{x})$. The *Lagrangian polynomial* with *Lagrangian multipliers* $\lambda_j, j \in M$ is

$$L = f + \sum_{j \in M} \lambda_j h_j.$$

The *Lagrangian polynomial set* is

$$LS = \left\{ \frac{\partial L}{\partial x_i}, h_j \mid i \in N, j \in M \right\}.$$

The *Jacobian* for $t = (i_1, \ldots, i_m) \in T$ is the determinant

$$J_t = \frac{\partial(h_1, \ldots, h_m)}{\partial(x_{i_1}, \ldots, x_{i_m})} = \left| \frac{\partial h_j}{\partial x_{i_k}} \right|$$

with entry $\partial h_j / \partial x_{i_k}$ in the jth row and kth column, for the m-tuple $t = (i_1, \ldots, i_m) \in T$. The *Jacobian polynomial set* is the polynomial set

$$JS = \{J_t, h_j \mid t \in T, j \in M\}.$$

Let $\Lambda = (\lambda_1, \ldots, \lambda_m)$ be the set of Lagrangian multipliers and $\tilde{O} = O \times \boldsymbol{R}^m(\Lambda) \subset \boldsymbol{R}^{n+m}(\boldsymbol{x}, \Lambda)$. For $\Lambda = (\lambda_1, \ldots, \lambda_m)$, the *Lagrangian projection* is defined as the natural map

$$\text{LProj}: \ \boldsymbol{R}^{n+m}(\boldsymbol{x}, \Lambda) \longrightarrow \boldsymbol{R}^n(\boldsymbol{x}), \quad \text{with LProj}(\boldsymbol{x}, \Lambda) = \boldsymbol{x}.$$

The following theorem is classical.

Theorem 1.25:

For Problem E with domain $D = O$ open, we have

$$\text{EZero}_{\boldsymbol{R}}\{f, g, O, HS\} \subset \text{LProj } \tilde{O}\text{Zero}_{\boldsymbol{R}}(LS/g) \cup O\text{Zero}_{\boldsymbol{R}}(JS/g).$$

Theorem 1.25 furnishes only a necessary condition for a point $\boldsymbol{x}^0$ to be an extremal one in that it should satisfy the equations $LS = 0$ and $JS = 0$ with some suitable values of Lagrangian multipliers. Thus, even if we are able to solve these equations, there is in general no means to decide whether such solutions will give really extremal values or not. These difficulties will, however, be resolved in the case of f, g, h_i being polynomials as shown below. We first introduce several notations.

- $\boldsymbol{x}^+ := (x_0, x_1, \ldots, x_n)$.
- $\boldsymbol{R}^{n+1} := \boldsymbol{R}^{n+1}(\boldsymbol{x}^+)$.
- $O^+ :=$ the open domain in $\boldsymbol{R}^{n+1}(\boldsymbol{x}^+)$ defined by $\boldsymbol{x} \in O$ while x_0 is arbitrary.
- $\Lambda^+ := (\lambda_0, \lambda_1, \ldots, \lambda_m)$.
- $\tilde{O}^+ := O^+ \times \boldsymbol{R}^{m+1}(\Lambda^+)$.

Let $h_0(\boldsymbol{x}^+) = f(\boldsymbol{x}) - x_0$ be a new polynomial, $HS^+ = HS \cup \{h_0\}$ a new polynomial set, g^+ the polynomial defined by $g^+(\boldsymbol{x}^+) = g(\boldsymbol{x})$, and Proj_0 the natural projection of $\boldsymbol{R}^{n+1}(\boldsymbol{x}^+)$ to $\boldsymbol{R}(x_0)$.

Theorem 1.26: (Finite kernel theorem for open domain)

Let HS be an arbitrary polynomial set and f an arbitrary polynomial in $\boldsymbol{R}[\boldsymbol{x}]$. Then one can construct a finite set K of real values such that

$$\text{EVal}_{\boldsymbol{R}}\{f, O, HS\} \subset K \subset \text{Proj}_0\tilde{O}^+\text{Zero}_{\boldsymbol{R}}(HS^+).$$

It is not difficult to solve the extremal problem for open domains using this theorem. A finite kernel theorem for *closed* domain can also be established (see Wu 1999).

Remark 1.4:

The Lagrange's multiplier method is applicable for arbitrary *differentiable* functions f, h_j. However, the *finiteness* property implied in Theorem 1.6 is no longer true as seen from the simple example of $f = x \sin \frac{1}{x}$, $O = \{x > 0\} \subset \boldsymbol{R}^1(x)$. Thus, the *finiteness* of extremal-value set is characteristic for f, h_j being *polynomials.*

The above theorem can be used to solve the following inequality problem (Wu 1999, Chap. 8).

Problem Ineq:

Let $\boldsymbol{R}^n(\boldsymbol{x})$ denote the real Euclidean space of dimension n in the coordinates $\boldsymbol{x} = (x_1, \ldots, x_n)$ and D be a domain in $\boldsymbol{R}^n$. Let f, h_i $(i \in I = \{1, \ldots, r\}, r < n)$ and g be all real polynomials in $\boldsymbol{R}^n[\boldsymbol{x}]$ over the domain D. Determine for what real value c we shall always have

$$f \geq c \quad \text{or} \quad f > c \quad \text{or} \quad f \leq c \quad \text{or} \quad f < c$$

under the conditions

$$HS = 0 \quad \text{and} \quad g \neq 0, \quad \text{where} \quad HS = \{h_i \mid i \in I\}.$$

We use two examples to illustrate the method.

Problems of non-linear programming. The non-linear programming problems for which the objective function and the constraint conditions are all *polynomial* ones can clearly be treated by our method. The following example is one of the simplest which serves for illustration. More examples may be found in the paper by Wu (1994).

Example 1.5:

Minimize

$$f = 1000 - x_1^2 - 2x_2^2 - x_3^2 - x_1x_2 - x_1x_3$$

under the constraints

$$h_1 = x_1^2 + x_2^2 + x_3^2 - 25 = 0, \qquad h_2 = 8x_1 + 14x_2 + 7x_3 - 56 = 0,$$

and $x_i \geq 0, i = 1, 2, 3.$

First of all let us remark that the least value of f necessarily exists owing to the condition $h_1 = 0$. Let us now split the domain defined by $x_i \geq 0, i = 1, 2, 3$ into 8 open domains in taking each x_i either > 0 or $= 0$. With the zero decomposition theorem, it is readily found that the only extremal zero in the open domain D_1 with all $x_i > 0$ is

$$Z_1 = (3.512, 0.217, 3.552).$$

For the other open domains: there are only two other extremal zeros from the open domain D_2 and D_3 defined by $x_1 = 0,\ x_2 > 0,\ x_3 > 0$ and $x_1 > 0, x_2 > 0, x_3 > 0$, viz.

$$Z_2 = \left(0, \frac{32 - 2\sqrt{61}}{10}, \frac{8 + 2\sqrt{61}}{5}\right), \qquad Z_3 = \left(\frac{63}{13}, \frac{16}{13}, 0\right).$$

The corresponding extremal values are respectively

$$f(Z_1) \approx 961, \qquad f(Z_2) \approx 972, \qquad \text{and} \quad f(Z_3) = 967.5.$$

We see thus that the least value of f under the stated conditions is approximately 961 taking place at the point Z_1.

A collision problem. Now we consider a problem which has been studied by Collins and Hong (1991) in some special case.

Example 1.6:

Suppose that two cars of given form are moving respectively along the X-axis and the Y-axis in positive directions with known velocities $v_1 > 0$ and $v_2 > 0$. Decide whether the cars will collide or not, and determine in the colliding case the time and place of first collision.

Let the starting positions of the two cars be given respectively by

$$f_1(x, y) = 0 \quad \text{and} \quad f_2(x, y) = 0.$$

Then the collision problem is seen to be a Problem Ineq for which

$$D = \{t > 0\} \subset \boldsymbol{R}^3(t, x, y), \qquad f = t, \quad g = 1,$$
$$h_1 = f_1(x - v_1 t, y), \qquad h_2 = f_2(x, y - v_2 t).$$

In particular, let us consider the case in which the two cars are both of elliptical form given by

$$f_1(x, y) = b_1^2(x - c_1)^2 + a_1^2 y^2 - a_1^2 b_1^2, \qquad f_2(x, y) = b_2^2 x^2 + a_2^2 (y - c_2)^2 - a_2^2 b_2^2$$

(where $c_1 < 0$ and $c_2 < 0$). Then, in the case of having *generic* values for a_i, b_i, c_i and v_i, our method will give rise to an irreducible polynomial equation of degree 8 in t with 696 terms. Leaving aside some uninteresting cases, the two cars will collide if and only if this equation has a positive root. The time and place of first collision can be easily determined if numerical values of a_i, b_i, c_i and v_i (meeting the genericity condition) are substituted. Consider the further special case in which the two elliptical forms are similar, viz.

$$a_1^2 b_2^2 = a_2^2 b_1^2,$$

which occurs in particular when both of the cars are circular in form. Then we will arrive at a set of linear and quadratic equations in t. The conditions for possible collision as well as time and place of first collision are all expressible by explicit equations in a_i, b_i, c_i and v_i. For details see the paper by Wu (1995).

Note. This chapter was prepared by Xiao-Shan Gao in cooperation with Dongming Wang on the basis of the material taken from the author's recent book (Wu 1999). Both the submitted version and the proof-pages have been read by the author.

Chapter 2

SOME ALGORITHMS FOR ZERO DECOMPOSITION OF POLYNOMIAL SYSTEMS

Dongming Wang

As alternatives to Wu–Ritt's method, this chapter presents some algorithms for decomposing arbitrary systems of multivariate polynomials into triangular systems of different kinds. These algorithms utilize top-down elimination, splitting, and subresultant regular subchain as basic algorithmic and computational strategies. Most of them are suggested by the author on the basis of some early ideas and work of A. Seidenberg and J. M. Thomas, and they are among some of the most efficient elimination algorithms available by this time. We shall explain various triangular systems (regular, simple, irreducible, or with projection property), their relationships, and how to compute them from any given polynomial system. Some theoretical results will be summarized, and several selected applications of the algorithms will be reviewed.

2.1 Triangular systems and zero decomposition

Let $\boldsymbol{K}$ be a computable field of characteristic 0 and $x_1, \ldots, x_n$, abbreviated $\boldsymbol{x}$, be n indeterminates. In this chapter $\tilde{\boldsymbol{K}}$ always denotes some extension field of $\boldsymbol{K}$, and $\boldsymbol{K}[\boldsymbol{x}]$ the ring of polynomials in $\boldsymbol{x}$ with coefficients in $\boldsymbol{K}$. A polynomial set is a finite set of non-zero polynomials in $\boldsymbol{K}[\boldsymbol{x}]$. For two polynomial sets PS and QS, $\mathrm{Zero}(PS/QS)$ denotes the set of all common zeros (in $\tilde{\boldsymbol{K}}$) of the polynomials in PS which are not zeros of any polynomial in QS. Symbolically,

$$\mathrm{Zero}(PS/QS) := \{\bar{\boldsymbol{x}} \in \tilde{\boldsymbol{K}}^n \mid P(\bar{\boldsymbol{x}}) = 0, Q(\bar{\boldsymbol{x}}) \neq 0, \ \forall P \in PS, Q \in QS\}.$$

Here := reads "is defined to be." We write $\mathrm{Zero}(PS)$ for $\mathrm{Zero}(PS/QS)$ when $QS \subset \boldsymbol{K} \setminus \{0\}$, $\mathrm{Zero}(PS/Q)$ for $\mathrm{Zero}(PS/\{Q\})$, and $\mathrm{Zero}(P/QS)$ for $\mathrm{Zero}(\{P\}/QS)$.

By a *polynomial system* in $\boldsymbol{K}[\boldsymbol{x}]$ we mean a pair $[PS, QS]$ of polynomial sets with which $\mathrm{Zero}(PS/QS)$ is of concern. In other words, we are concerned with

MATHEMATICS MECHANIZATION AND APPLICATIONS
ISBN 0-12-734760-7

the solutions of the following system of simultaneous polynomial equations and inequations

$$P_1 = 0, \ldots, P_s = 0, \qquad Q_1 \neq 0, \ldots, Q_t \neq 0, \tag{2.1}$$

where $PS = \{P_1, \ldots, P_s\}$ and $QS = \{Q_1, \ldots, Q_t\}$. System (2.1) is often written as $PS = 0, QS \neq 0$ for simplicity.

For any polynomial $P \in \boldsymbol{K}[\boldsymbol{x}]$ and variable x_k, the *degree* of P in x_k is denoted $\deg(P, x_k)$. When $P \notin \boldsymbol{K}$, the biggest index p such that $\deg(P, x_p) > 0$ is called the *class*, x_p the *leading variable*, and $\deg(P, x_p)$ the *leading degree* of P; they are denoted by $\mathrm{cls}(P)$, $\mathrm{lv}(P)$, and $\mathrm{ldeg}(P)$ respectively. For any $P \in \boldsymbol{K}$ and $P \neq 0$, we define the *class*, the *leading variable*, and the *leading degree* of P to be 0, x_0, and 0 respectively, where x_0 is a new variable.

Let P be a polynomial of class $p > 0$. Any other polynomial $Q \in \boldsymbol{K}[\boldsymbol{x}]$ is said to be *reduced* with respect to P if $\deg(Q, x_p) < \mathrm{ldeg}(P)$. The leading coefficient $\mathrm{lc}(P, x_p)$ of P in x_p is called the *initial* of P, denoted by $\mathrm{ini}(P)$. The *initial* of any $P \in \boldsymbol{K}$ is defined to be itself. The polynomial $P - \mathrm{ini}(P)\mathrm{lv}(P)^{\mathrm{ldeg}(P)}$ is called the *reductum* of P, denoted by $\mathrm{red}(P)$. For any polynomial set PS, we define

$$\mathrm{ini}(PS) := \{\mathrm{ini}(P) \mid P \in PS\}.$$

The *pseudo-remainder* and *pseudo-quotient* of Q by P with respect to x_k are denoted by $\mathrm{prem}(Q, P, x_k)$ and $\mathrm{pquo}(Q, P, x_k)$ respectively. Moreover, $\mathrm{prem}(Q, P)$ stands for $\mathrm{prem}(Q, P, \mathrm{lv}(P))$ and $\mathrm{prem}(QS, P)$ for $\{\mathrm{prem}(Q, P) \mid Q \in QS\}$, where QS is an arbitrary polynomial set.

2.1.1 Triangular systems

Let $\mathbb{N}$ denote the set of natural numbers and $\mathbb{N}_k = \{1, \ldots, k\}$. The number of elements of a finite set TS is denoted by $|TS|$. An *ordered set* is written by enclosing its elements in a pair of square brackets. For two non-empty ordered sets $TS = [T_1, \ldots, T_r]$ and $US = [U_1, \ldots, U_s]$ which have no intersection, we have

$$TS \cup US := [T_1, \ldots, T_r, U_1, \ldots, U_s].$$

$TS \cup US$ and $US \cup TS$ are distinguished when they are considered as ordered sets. In other words, the ordering is preserved for union of non-intersecting ordered sets. If one or both of TS and US are usual sets, then so is $TS \cup US = US \cup TS$. The ordering is also preserved for difference of ordered sets in the natural way. For example, $[a, b, c, d] \setminus [a, c] = [b, d]$.

Definition 2.1:

A finite non-empty ordered set of polynomials in $\boldsymbol{K}[\boldsymbol{x}]$

$$TS = [T_1, T_2, \ldots, T_r]$$

is called a triangular set *if* $0 < \mathrm{cls}(T_1) < \mathrm{cls}(T_2) < \cdots < \mathrm{cls}(T_r)$.

Let TS be a triangular set as in Definition 2.1 and P any polynomial. P is said to be *reduced* with respect to TS if P is reduced with respect to every T_i. The polynomial

$$R = \mathrm{prem}(\cdots \mathrm{prem}(P, T_r), \ldots, T_1),$$

denoted simply by $\mathrm{prem}(P, TS)$, is called the *pseudo-remainder* of P with respect to TS. It is easy to deduce the following *pseudo-remainder formula*

$$I_1^{q_1} \cdots I_r^{q_r} P = \sum_{i=1}^{r} Q_i T_i + R,$$

where each q_i is a non-negative integer and $I_i = \mathrm{ini}(T_i), Q_i \in \boldsymbol{K}[\boldsymbol{x}]$ for $i \in \mathbb{N}_r$. For any polynomial set PS, $\mathrm{prem}(PS, TS)$ stands for $\{\mathrm{prem}(P, TS) \mid P \in PS\}$.

We define

$$PS^{(i)} := PS \cap \boldsymbol{K}[x_1, \ldots, x_i], \qquad \mathfrak{P}^{(i)} := [PS^{(i)}, QS^{(i)}],$$
$$PS^{\langle i \rangle} := PS^{(i)} \setminus PS^{(i-1)}, \qquad PS^{[i]} := PS \setminus PS^{(i)},$$
$$\mathrm{Zero}(\mathfrak{P}) := \mathrm{Zero}(PS/QS)$$

for any $i \in \mathbb{N}_n$, polynomial set $PS \subset \boldsymbol{K}[\boldsymbol{x}]$ and polynomial system $\mathfrak{P} = [PS, QS]$.

Definition 2.2:

A polynomial system $[TS, US]$ *in* $\boldsymbol{K}[\boldsymbol{x}]$ *is called a* triangular system *if* TS *is a triangular set and* $I(\bar{\boldsymbol{x}}) \neq 0$ *for any* $I \in \mathrm{ini}(TS)$ *of class* i *and* $\bar{\boldsymbol{x}} \in \mathrm{Zero}(TS^{(i)}/US)$.

A triangular system $\mathfrak{T} = [TS, US]$ *is said to be* fine *if* $0 \notin \mathrm{prem}(US, TS)$. *It is said to be* perfect *over* $\tilde{\boldsymbol{K}}$ *if* $\mathrm{Zero}(\mathfrak{T}) \neq \varnothing$ *in* $\tilde{\boldsymbol{K}}$.

Lemma 2.1: (Wang 1993b; Wu 1984a)

For any triangular system $[TS, US]$ *and polynomial* P *in* $\boldsymbol{K}[\boldsymbol{x}]$, *if* $\mathrm{prem}(P, TS) = 0$ *then* $\mathrm{Zero}(TS/US) \subset \mathrm{Zero}(P)$.

Definition 2.3:

A triangular set $TS \subset \boldsymbol{K}[\boldsymbol{x}]$ *is said to be* fine *or* perfect *over* $\tilde{\boldsymbol{K}}$ *if* $[TS, \mathrm{ini}(TS)]$ *is fine or perfect over* $\tilde{\boldsymbol{K}}$, *respectively.*

A triangular set or system in $\boldsymbol{K}[\boldsymbol{x}]$ *is said to be* perfect *(without reference to any specific field) if it is perfect over some suitable extension field of* $\boldsymbol{K}$.

Consider for instance $TS = [x_1^d - 1, (x_1 - 1)x_2 - 1]$, where d is a positive integer. By definition TS is a triangular set: it is not fine and thus not perfect if $d = 1$; it is both fine and perfect if $d > 1$. Let $US = \{x_1 + 1, x_1^2 - 2x_1 + 1\}$. Then $[TS, US]$ is a triangular system: it is fine for $d > 1$, but not so for $d = 1$; it is perfect for $d > 2$, but not perfect for $d = 1, 2$.

The main objective of this section is to present two algorithms that decompose any given polynomial system $\mathfrak{P}$ into finitely many fine triangular systems $\mathfrak{T}_1, \ldots, \mathfrak{T}_e$ such that

$$\mathrm{Zero}(\mathfrak{P}) = \bigcup_{i=1}^{e} \mathrm{Zero}(\mathfrak{T}_i). \tag{2.2}$$

The algorithms employ a strategy of elimination top-down from x_n to x_1 with splitting.

Definition 2.4:

A finite set or sequence Ψ *of (fine) triangular systems* $\mathfrak{T}_1, \ldots, \mathfrak{T}_e$ *in* $\boldsymbol{K}[\boldsymbol{x}]$ *is called a* (fine) triangular series. *It is called a* (fine) triangular series *of a polynomial system* $\mathfrak{P}$ *in* $\boldsymbol{K}[\boldsymbol{x}]$ *if* (2.2) *holds.*

When $\Psi = \varnothing$ *or* $e = 0$, *it is understood that* $\mathrm{Zero}(\mathfrak{P}) = \varnothing$. *A (fine) triangular series of* $[PS, \varnothing]$ *is also called a* (fine) triangular series *of the polynomial set* PS.

2.1.2 Seidenberg's algorithm refined

The algorithm presented in this subsection proceeds by splitting polynomial systems whenever pseudo-division is performed. It follows the elimination idea of Seidenberg (1956a,b) and has several advantages (see Wang 1993a,b). For this algorithm and that in the next subsection, the following simple lemma is used.

Lemma 2.2: (Wang 1993b)

Let P *be a polynomial of positive class with* $\mathrm{ini}(P) = I$ *and* $[PS, QS]$ *a polynomial system in* $\boldsymbol{K}[\boldsymbol{x}]$, *and* $RS = \mathrm{prem}(PS, P) \setminus \{0\}$. *Then*

$$\mathrm{Zero}(PS \cup \{P\}/QS) = \mathrm{Zero}(RS \cup \{P\}/QS \cup \{I\}) \cup \mathrm{Zero}(PS \cup \{I, \mathrm{red}(P)\}/QS).$$

The *level* of a polynomial set $PS \subset \boldsymbol{K}[\boldsymbol{x}]$, denoted $\mathrm{level}(PS)$, is the smallest integer l such that $PS \subset \boldsymbol{K}[x_1, \ldots, x_l]$. The level of PS is also called the *level* of the polynomial system $[PS, QS]$ for any QS. A *triplet* of level $i \in \mathbb{N}_n$ is a list $[PS, QS, TS]$ of three elements, where

(a) $[PS, QS]$ is a polynomial system of level i in $\boldsymbol{K}[\boldsymbol{x}]$;

(b) TS, if non-empty, is a triangular set in $\boldsymbol{K}[\boldsymbol{x}]$ with $TS^{(i)} = \varnothing$.

Our elimination procedure will start with a triplet $[PS, QS, TS]$ with $TS = \varnothing$. The variables x_i are eliminated and the obtained triangularized polynomials are adjoined to TS successively for $i = n, n-1, \ldots, 1$.

Let the polynomial system $[PS, QS]$ be of level $i > 0$. Clearly, $FS = PS^{\langle i \rangle}$ is not empty and every polynomial in FS has class i. We want to eliminate the variable x_i for the polynomials in FS, so that afterwards only one polynomial has class i. For this purpose, let us take one polynomial T from FS which has minimal degree in x_i and pseudo-divide all the other polynomials in FS by T with respect to x_i. Meanwhile, $\mathrm{ini}(T)$ is assumed to be non-zero and the case in which $\mathrm{ini}(T)$ happens to be 0 is considered disjunctively by replacing T with $\mathrm{ini}(T)$ and $\mathrm{red}(T)$. Then, we reset FS to be $\{T\} \cup \mathrm{prem}(FS, T) \setminus \{0\}$ (see Lemma 2.2). Repeating this process, we shall finally get an FS that contains exactly one polynomial T of class i, and a set of other polynomial systems of level $\leq i$.

Iterating the above elimination procedure top-down for each x_i ($i = n, n-1, \ldots, 1$), we can compute a (*principal*) fine triangular system $[TS, US]$ from any polynomial system $[PS, QS]$ according to the following algorithm.

Algorithm P: (Principal triangular system of a polynomial system)

Given a polynomial system $[PS, QS]$ in $\boldsymbol{K}[\boldsymbol{x}]$, this algorithm computes another polynomial system $[TS, US]$ and a set Ω of triplets such that

$$\mathrm{Zero}(PS/QS) = \mathrm{Zero}(TS/US) \cup \bigcup_{[PS^*, QS^*, TS^*] \in \Omega} \mathrm{Zero}(PS^* \cup TS^*/QS^*),$$

and either $[TS, US]$ is a fine triangular system or $TS = [1]$; in the latter case $\mathrm{Zero}(TS/US) = \varnothing$.

P1. [Initialization.] Set $TS \leftarrow \varnothing$, $FS \leftarrow PS$, $US \leftarrow QS$, $\Omega \leftarrow \varnothing$.

P2. [Elimination for x_i.] For $i = \mathrm{level}(PS), \ldots, 1$ do:

P2.1. [Trivial case.] If $FS \cap \boldsymbol{K} \setminus \{0\} \neq \varnothing$ then set $TS \leftarrow [1]$ and the algorithm terminates. If $\mathrm{level}(FS) < i$ then go to P2 for next i. Otherwise, set $T \leftarrow 0$.

P2.2. While $FS^{\langle i \rangle} \neq \{T\}$ do:

P2.2.1. Let T be an element of $FS^{\langle i \rangle}$ with minimal degree in x_i.

P2.2.2. [Splitting according to Lemma 2.2.] Set

$$\Omega \leftarrow \Omega \cup \{[FS \setminus \{T\} \cup \{\mathrm{red}(T), \mathrm{ini}(T)\}, US, TS]\},$$
$$US \leftarrow US \cup \{\mathrm{ini}(T)\}.$$

P2.2.3. [Reduction.] Compute $FS \leftarrow \{T\} \cup \mathrm{prem}(FS, T) \setminus \{0\}$.

P2.3. Set $FS \leftarrow FS \setminus \{T\}$ and compute $US \leftarrow \mathrm{prem}(US, T)$.

P2.4. If $0 \in US$ then set $TS \leftarrow [1]$ and the algorithm terminates else set $TS \leftarrow [T] \cup TS$. ■

The polynomial system $[PS, QS]$ is decomposed into fine triangular systems by applying Algorithm P to all the computed triplets recursively. More precisely, we have the following algorithm (Wang 1993b).

Algorithm T: (Triangular series of a polynomial system)
Given a polynomial system $[PS, QS]$ in $\boldsymbol{K}[\boldsymbol{x}]$, this algorithm computes a fine triangular series Ψ of $[PS, QS]$.

T1. Set $\Psi \leftarrow \varnothing$, $\Phi \leftarrow \{[PS, QS, \varnothing]\}$.

T2. While $\Phi \neq \varnothing$ do:

T2.1. Let $[FS, GS, TS']$ be an element of Φ and set $\Phi \leftarrow \Phi \setminus \{[FS, GS, TS']\}$.

T2.2. Compute a polynomial system $[TS, US]$ and a set Ω of triplets from $[FS, GS]$ by Algorithm P.

T2.3. Set

$$\Phi \leftarrow \Phi \cup \{[FS^*, GS^*, TS^* \cup TS'] \mid [FS^*, GS^*, TS^*] \in \Omega\}.$$

If $TS \neq [1]$ then set $\Psi \leftarrow \Psi \cup \{[TS \cup TS', US]\}$. ■

2.1.3 Subresultant-based algorithm

The decomposition algorithm T* presented in this section has the same functionality and employs the same strategies of splitting and top-down elimination as Algorithm T. The difference is that Algorithm T* makes use of the key concept and computation of *subresultant regular subchains* (SRS).

The reader is referred to Collins (1967), Brown and Traub (1971), Loos (1982), Mishra (1993) and Chap. 12 of this book for the theory of subresultants and the definition and computation of subresultant chains of polynomials in $\mathcal{R}[x]$, where $\mathcal{R}$ is a domain.

Definition 2.5:

Let $S_{\mu+1}$ *and* S_μ *be two polynomials in* $\mathcal{R}[x]$ *with* $\deg(S_{\mu+1}, x) \geq \deg(S_\mu, x) > 0$ *and*

$$S_{\mu+1}, S_\mu, \ldots, S_0$$

be the subresultant chain of $S_{\mu+1}$ *and* S_μ *with respect to* x*.* S_j *is said to be* defective *if* $\deg(S_j, x) < j$*, and* regular *otherwise.*

Let $d_2, \ldots, d_r$ *be a finite sequence of steadily decreasing non-negative integers with* $d_2 \geq \mu$ *such that each* S_{d_i} *is regular for* $2 \leq i \leq r$*, and* S_j *is defective for any* $0 \leq j \leq \mu$ *and* $j \notin \{d_2, \ldots, d_r\}$*. The sequence of regular subresultants*

$$S_{d_2}, \ldots, S_{d_r}$$

is called the subresultant regular subchain (SRS) *of* $S_{\mu+1}$ *and* S_μ *with respect to* x*.*

We rename the regular subresultants $S_{d_2}, \ldots, S_{d_r}$ in the above definition $H_2, \ldots, H_r$ and set $P_1 = S_{\mu+1}, P_2 = S_\mu$.

Lemma 2.3: (Wang 1998a)

Let P_1 *and* P_2 *be two polynomials in* $\boldsymbol{K}[x_1, \ldots, x_k]$ *with* $\deg(P_1, x_k) \geq \deg(P_2, x_k) > 0$*,* $H_2, \ldots, H_r$ *be the SRS of* P_1 *and* P_2 *with respect to* x_k*,* $I = \mathrm{lc}(P_2, x_k)$*, and* $I_i = \mathrm{lc}(H_i, x_k)$ *for* $i = 2, \ldots, r$*. Then*

$$\mathrm{Zero}(\{P_1, P_2\}/I) = \bigcup_{i=2}^{r} \mathrm{Zero}(\{H_i, I_{i+1}, \ldots, I_r\}/II_i).$$

By Lemma 2.2 and as in Algorithm T, we may assume that $I \neq 0$ and deal with the case $I = 0$ separately. Lemma 2.3 exhibits the splitting of $[\{P_1, P_2\}, \{I\}]$ into finitely many subsystems $[\{H_i, I_{i+1}, \ldots, I_r\}, \{I, I_i\}]$. Its significance lies in the fact that each $\{H_i, I_{i+1}, \ldots, I_r\}$ contains at most one polynomial of class k, whereas P_1 and P_2 are both of class k. With this splitting strategy and a top-down elimination for x_k ($k = n, n-1, \ldots, 1$), any polynomial system $[PS, QS]$ in $\boldsymbol{K}[\boldsymbol{x}]$ can be decomposed into triangular systems by the following algorithm (Wang 2000a,b).

Algorithm T*: (Triangular series of a polynomial system)

Given a polynomial system $[PS, QS]$ in $\boldsymbol{K}[\boldsymbol{x}]$, this algorithm computes a fine triangular series Ψ of $[PS, QS]$.

T1. [Initialization.] Set $\Phi \leftarrow \{[PS, QS, n]\}$, $\Psi \leftarrow \varnothing$.

T2. While $\Phi \neq \varnothing$ do:

T2.1. Let $[TS, US, \ell]$ be an element of Φ and set $\Phi \leftarrow \Phi \setminus \{[TS, US, \ell]\}$.

T2.2. [Elimination for x_k.] For $k = \ell, \ldots, 1$ do:

T2.2.1. [Trivial case.] If $TS \cap \boldsymbol{K} \setminus \{0\} \neq \varnothing$ or $0 \in US$ then go to T2. If $TS^{\langle k \rangle} = \varnothing$ then go to T2.2 for next k.

T2.2.2. Repeat:

T2.2.2.1. [Splitting according to Lemma 2.2.] Let P_2 be an element of $TS^{\langle k \rangle}$ with minimal degree in x_k and set

$$\Phi \leftarrow \Phi \cup \{[TS \setminus \{P_2\} \cup \{\mathrm{ini}(P_2), \mathrm{red}(P_2)\}, US, k]\},$$
$$US \leftarrow US \cup \{\mathrm{ini}(P_2)\}.$$

If $|TS^{\langle k \rangle}| = 1$ then go to T2.2.3. Otherwise, let P_1 be an element of $TS^{\langle k \rangle} \setminus \{P_2\}$.

T2.2.2.2. [SRS computation.] Compute the SRS $H_2, \ldots, H_r$ of P_1 and P_2 with respect to x_k and set $I_i \leftarrow \mathrm{lc}(H_i, x_k)$ for $2 \leq i \leq r$. If $\mathrm{cls}(H_r) < k$ then set $\bar{r} \leftarrow r - 1$ else set $\bar{r} \leftarrow r$.

T2.2.2.3. [Splitting according to Lemma 2.3.] Set

$$\Phi \leftarrow \Phi \cup \{[TS \setminus \{P_1, P_2\} \cup \{H_i, I_{i+1}, \ldots, I_r\}, US \cup \{I_i\}, k] \mid 2 \leq i \leq \bar{r} - 1\},$$
$$TS \leftarrow TS \setminus \{P_1, P_2\} \cup \{H_r, H_{\bar{r}}\},$$
$$US \leftarrow US \cup \{I_{\bar{r}}\}.$$

T2.2.3. [Reduction.] Compute $US \leftarrow \mathrm{prem}(US, P_2)$.

T2.3. Set $\Psi \leftarrow \Psi \cup \{[TS, US]\}$, with TS ordered as a triangular set. ■

Two slightly different data structures are adopted for Algorithms T and T*. It is possible to use the data structure of one algorithm for the other. When the implementation of Algorithms T and T* is concerned, several details have to be taken into account for the sake of efficiency. The interested reader may consult some relevant papers by this author.

2.2 Regular systems and simple systems

The fine triangular systems computed by the algorithms T and T* are not necessarily perfect. This can be seen from the simple example that $[[x_1^2, x_1x_2 + 1], \{x_1\}]$ is an imperfect triangular system. We show how to impose different kinds of conditions to rule out those trivial triangular systems that have no zero.

2.2.1 Decomposition into regular systems

For any $k \in \mathbb{N}_n$ we write $\boldsymbol{x}_k$ for $x_1, \ldots, x_k$ or $(x_1, \ldots, x_k)$ with $\boldsymbol{x} = \boldsymbol{x}_n$, and similarly $\bar{\boldsymbol{x}}_k = (\bar{x}_1, \ldots, \bar{x}_k)$.

Definition 2.6:

A triangular system $[\mathit{TS}, \mathit{US}]$ *in* $\boldsymbol{K}[\boldsymbol{x}]$ *is said to be* regular *or called a* regular system *if for any* $k \in \mathbb{N}_n$*:*

(*a*) *either* $\mathit{TS}^{\langle k \rangle} = \varnothing$ *or* $\mathit{US}^{\langle k \rangle} = \varnothing$;

(*b*) $I(\bar{\boldsymbol{x}}_{k-1}) \neq 0$ *for any* $I \in \mathrm{ini}(\mathit{US}^{\langle k \rangle})$ *and* $\bar{\boldsymbol{x}}_{k-1} \in \mathrm{Zero}(\mathit{TS}^{(k-1)}/\mathit{US}^{(k-1)})$.

A triangular set TS is said to be regular *or called a* regular set *if there exists a polynomial set US such that* $[\mathit{TS}, \mathit{US}]$ *is a regular system.*

A triangular series Ψ *is called a* regular series *if every* $\mathfrak{T} \in \Psi$ *is a regular system.*

In the above definition, condition (b) is satisfied also for every $I \in \mathrm{ini}(\mathit{TS}^{\langle k \rangle})$ as $[\mathit{TS}, \mathit{US}]$ is a triangular system.

For example, let $T = (x_1 - 1)x_2 - 1$; $[T]$ is a regular set because $[[T], \{x_1 - 1\}]$ is a regular system. However, $\mathit{TS} = [x_1^d - 1, T]$ (as already given below Definition 2.3) is not a regular set for any integer $d > 0$. For $[\mathit{TS}, \varnothing]$ is not a triangular system by definition, while $\mathit{US} = \varnothing$ is the only possible set such that condition (a) holds.

Lemma 2.4: (Wang 2000a,b)

Let P_1 *and* P_2 *be two polynomials in* $\boldsymbol{K}[\boldsymbol{x}_k]$ *with* $\deg(P_1, x_k) \geq \deg(P_2, x_k) > 0$, $H_2, \ldots, H_r$ *be the SRS of* P_1 *and* P_2 *with respect to* x_k *and*

$$I = \mathrm{lc}(P_2, x_k), \qquad I_i = \mathrm{lc}(H_i, x_k), \quad 2 \leq i \leq r.$$

Let $\mathit{PS}, \mathit{QS} \subset \boldsymbol{K}[\boldsymbol{x}_{k-1}]$ *be two polynomial sets and assume that* $I(\bar{\boldsymbol{x}}_{k-1}) \neq 0$ *for any* $\bar{\boldsymbol{x}}_{k-1} \in \mathrm{Zero}(\mathit{PS}/\mathit{QS})$*. Then*

$$\mathrm{Zero}(\mathit{PS} \cup \{P_2\}/\mathit{QS} \cup \{P_1\}) = \bigcup_{i=2}^{r} \mathrm{Zero}(\mathit{PS} \cup \mathit{PS}_i/\mathit{QS} \cup \{P_1, I_i\}),$$

where $\mathit{PS}_i = \{\mathrm{pquo}(P_2, H_i, x_k), I_{i+1}, \ldots, I_r\}$ *for each* i*. If* $\mathrm{cls}(H_r) < k$*, then* $I_r = H_r$ *and*

$$\mathrm{Zero}(\mathit{PS} \cup \mathit{PS}_r/\mathit{QS} \cup \{P_1, I_r\}) = \mathrm{Zero}(\mathit{PS} \cup \{P_2\}/\mathit{QS} \cup \{I_r\}).$$

The algorithm T* may be extended for decomposing any polynomial system into finitely many regular systems (Wang 2000a,b). The elimination strategy for equation-polynomials remains almost the same. The main new ingredient is to use the only polynomial P_2 of class k in FS, when obtained, to eliminate all the polynomials of class k from GS, where $[FS, GS]$ corresponding to $[TS, US]$ in Algorithm T* is the polynomial system under consideration and $\mathrm{ini}(P_2)$ does not vanish on $\mathrm{Zero}(FS/GS)$. If $FS^{\langle k\rangle} = \varnothing$, one simply adds $\mathrm{ini}(GS^{\langle k\rangle})$ to GS and meanwhile takes into account the systems generated from $[FS, GS]$ with $I = 0$ for $I \in \mathrm{ini}(GS^{\langle k\rangle})$.

The elimination of polynomials from $GS^{\langle k\rangle}$ by P_2 is performed with further splitting according to Lemma 2.4 roughly as follows.

Assume that GS contains a polynomial, say P_1, of class k. Compute the SRS $H_2, \ldots, H_r$ of P_1 and P_2 with respect to x_k, and let $I_i = \mathrm{lc}(H_i, x_k)$ for $2 \le i \le r$. Then, it follows from Lemma 2.4 that

$$\mathrm{Zero}(FS/GS) = \bigcup_{i=2}^{r} \mathrm{Zero}(FS \setminus \{P_2\} \cup \{Q_i, I_{i+1}, \ldots, I_r\}/GS \cup \{I_i\}),$$

where $Q_i = \mathrm{pquo}(P_2, H_i, x_k)$. Observe that, when H_i is of class k, Q_i has smaller degree than P_2 in x_k. It is always the case for $2 \le i \le r-1$. If H_r has class smaller than k, then by Lemma 2.4

$$\mathrm{Zero}(FS \setminus \{P_2\} \cup \{Q_r\}/GS \cup \{I_r\}) = \mathrm{Zero}(FS/GS \setminus \{P_1\} \cup \{I_r\}),$$

i.e., P_1 may be eliminated from GS. Otherwise, the process may continue, for example, by computing the SRS of $\mathrm{pquo}(P_2, H_i, x_k)$ and P_1 with respect to x_k for each i.

The above procedure can be repeated for each split system and will terminate eventually so that all the polynomials of class k in the second set of the system are eliminated. In this way, we shall obtain from $[FS, GS]$ finitely many new systems $[FS_i, GS_i]$ such that each FS_i contains either no or exactly one polynomial of class k, while in the latter case GS_i does not contain any polynomial of class k.

Besides the algorithm using SRS described in Wang (2000b), there is an algorithm proposed by Kalkbrener (1993) for computing a regular series of any given polynomial system; see also Lazard (1991) and Yang and Zhang (1994).

2.2.2 Decomposition into simple systems

For any $P \in \boldsymbol{K}[\boldsymbol{x}_k]$ and $\bar{\boldsymbol{x}}_{k-1} \in \tilde{\boldsymbol{K}}^{k-1}$, the polynomial $P(\bar{\boldsymbol{x}}_{k-1}, x_k)$ is said to be *squarefree* with respect to x_k if

$$\gcd\left(P(\bar{\boldsymbol{x}}_{k-1}, x_k), \frac{\partial P}{\partial x_k}(\bar{\boldsymbol{x}}_{k-1}, x_k), x_k\right) \in \tilde{\boldsymbol{K}}.$$

Definition 2.7:

A pair $[\mathit{TS}, \widetilde{\mathit{TS}}]$ *of triangular or empty sets in* $\boldsymbol{K}[\boldsymbol{x}]$ *is called a* simple system *if*

(*a*) $\mathit{TS} \cap \widetilde{\mathit{TS}} = \varnothing$ *and* $\mathit{TS} \cup \widetilde{\mathit{TS}}$ *can be reordered as a triangular set;*

(*b*) *for every* $P \in \mathit{TS} \cup \widetilde{\mathit{TS}}$ *of class* p *and any* $\bar{\boldsymbol{x}}_{p-1} \in \mathrm{Zero}(\mathit{TS}^{(p-1)}/\widetilde{\mathit{TS}}^{(p-1)})$,

$$\mathrm{ini}(P)(\bar{\boldsymbol{x}}_{p-1}) \neq 0 \quad \textit{and} \quad P(\bar{\boldsymbol{x}}_{p-1}, x_p) \textit{ is squarefree with respect to } x_p.$$

A triangular system is said to be simple *if it is a simple system.*

A triangular set $\mathit{TS} \subset \boldsymbol{K}[\boldsymbol{x}]$ *is said to be* simple *or called a* simple set *if there exists another triangular or empty set* $\widetilde{\mathit{TS}}$ *such that* $[\mathit{TS}, \widetilde{\mathit{TS}}]$ *is a simple system.*

A triangular series Ψ *is called a* simple series *if every triangular system in* Ψ *is simple.*

The concept of simple systems is due to Thomas (1937, Chap. VI), while what he called a simple system is stronger than the one defined above.

Lemma 2.5: (Wang 1998a)

Let $P_1, P_2, I, \mathit{PS}, \mathit{QS}, H_i, I_i, \mathit{PS}_i$ $(2 \leq i \leq r)$ *be as in Lemma 2.4 and assume that* $P_2(\bar{\boldsymbol{x}}_{k-1}, x_k)$ *is squarefree with respect to* x_k *for any* $\bar{\boldsymbol{x}}_{k-1} \in \mathrm{Zero}(\mathit{PS}/\mathit{QS})$. *Then*

$$\mathrm{Zero}(\mathit{PS} \cup \{P_2\}/\mathit{QS} \cup \{P_1\}) = \bigcup_{i=2}^{r} \mathrm{Zero}(\mathit{PS} \cup \mathit{PS}_i/\mathit{QS} \cup \{I_i\}). \tag{2.3}$$

Observe that on the right-hand side of (2.3), P_1 does not appear and the only polynomial of class k is $\mathrm{pquo}(P_2, H_i, x_k)$ for each i. In this sense, the polynomial P_1 is eliminated by means of splitting. The purpose of splitting in the following lemma is to make an arbitrary polynomial squarefree.

Lemma 2.6: (Wang 1998a)

Let P be a polynomial in $\boldsymbol{K}[\boldsymbol{x}_k]$ with $\deg(P, x_k) > 1$ *and* $I = \mathrm{lc}(P, x_k)$*, $H_2, \ldots, H_r$ be the SRS of P and its derivative $\partial P/\partial x_k$ with respect to x_k, and*

$$H_2^* = H_2, \qquad H_i^* = \frac{H_i}{I}, \quad 3 \le i \le r; \qquad I_i = \mathrm{lc}(H_i^*, x_k), \quad 2 \le i \le r.$$

Then

$$\mathrm{Zero}(P/I) = \bigcup_{i=2}^{r} \mathrm{Zero}(\{Q_i, I_{i+1}, \ldots, I_r\}/II_i), \tag{2.4}$$

$$\mathrm{Zero}(\varnothing/PI) = \bigcup_{i=2}^{r} \mathrm{Zero}(\{I_{i+1}, \ldots, I_r\}/Q_iII_i), \tag{2.5}$$

where $Q_i = \mathrm{pquo}(P, H_i^*, x_k)$ *for each i. Moreover, $Q_i(\bar{\boldsymbol{x}}_{k-1}, x_k)$ is squarefree with respect to x_k for any $2 \le i \le r$ and* $\bar{\boldsymbol{x}}_{k-1} \in \mathrm{Zero}(\{I_{i+1}, \ldots, I_r\}/II_i)$.

An algorithm has been devised in Wang (1998a) for computing a simple series of any given polynomial system. This is done by modifying Algorithm T* roughly as follows.

As in Algorithm T*, a single polynomial P_2 of class k may be produced from TS when $TS^{\langle k \rangle} \neq \varnothing$ for each k. Now we incorporate the following three major steps into the algorithm.

A. Make the polynomial P_2 squarefree with respect to x_k according to Lemma 2.6: splitting is performed as in (2.4), P_2 is reset to be Q_r, I_r is added to US, and the other split systems are added to Φ.

B. Eliminate the polynomials from $US^{\langle k \rangle}$, if there are any, by P_2 according to Lemma 2.5: splitting is performed as in (2.3), P_2 is reset to be $\mathrm{pquo}(P_2, H_r, x_k)$, I_r is added to US, and the other split systems are added to Φ.

C. Let P be the product of all the polynomials in $US^{\langle k \rangle}$, if there are any, and make P squarefree with respect to x_k according to Lemma 2.6: splitting is performed as in (2.5), Q_r and I_r are added to US, and the other split systems are added to Φ.

Computing simple series is expensive in general, mainly because of the high price that has to be borne to make polynomials squarefree and to eliminate inequation-polynomials. In practice, it is even preferable to compute irreducible triangular series instead (see Sect. 2.3.2), making use of powerful routines available for polynomial factorization.

2.3 Projection and irreducible triangular systems

2.3.1 Zero decomposition with projection

Let a polynomial system $\mathfrak{P}$ in $\boldsymbol{K}[\boldsymbol{x}]$ be given. We want to eliminate the variables $x_n, \ldots, x_{k+1}$ $(0 \leq k < n)$ to obtain finitely many other polynomial systems $\mathfrak{P}_1, \ldots, \mathfrak{P}_e$ in $\boldsymbol{K}[\boldsymbol{x}_k]$ such that

$$\mathrm{Zero}(\mathfrak{P}) \neq \varnothing \iff \bigcup_{i=1}^{e} \mathrm{Zero}(\mathfrak{P}_i) \neq \varnothing,$$

and to establish some zero relations between $\mathfrak{P}$ and the eliminated (triangular) systems.

Definition 2.8:

For any polynomial system $\mathfrak{P}$ in $\boldsymbol{K}[\boldsymbol{x}]$ and $i \in \mathbb{N}_{n-1}$, the projection *of* Zero($\mathfrak{P}$) *onto $\boldsymbol{x}_i$ is defined as*

$$\mathrm{Proj}_{\boldsymbol{x}_i}\mathrm{Zero}(\mathfrak{P}) := \{\bar{\boldsymbol{x}}_i \in \tilde{\boldsymbol{K}}^i \mid \exists \bar{x}_{i+1}, \ldots, \bar{x}_n \in \tilde{\boldsymbol{K}} \textit{ such that } \bar{\boldsymbol{x}} \in \mathrm{Zero}(\mathfrak{P})\}.$$

Moreover, we define

$$\mathrm{Proj}_{\boldsymbol{x}}\mathrm{Zero}(\mathfrak{P}) := \mathrm{Zero}(\mathfrak{P}), \qquad \mathrm{Proj}\mathrm{Zero}(\mathfrak{P}) := \begin{cases} \varnothing & \textit{if } \mathrm{Zero}(\mathfrak{P}) = \varnothing, \\ \{0\} & \textit{otherwise} \end{cases}$$

for the two extreme cases $i = n$ and $i = 0$.

It is easy to see that

$$\mathrm{Proj}_{\boldsymbol{x}_i}\mathrm{Zero}(\mathfrak{P}) \neq \varnothing \iff \mathrm{Zero}(\mathfrak{P}) \neq \varnothing.$$

There are two kinds of projection as shown in the following two lemmas.

Lemma 2.7: (Jacobson 1974; Seidenberg 1956a,b; Wang 1993b; Wu 1990a)
Let $[PS, QS]$ be a polynomial system of level $\leq i$ in $\boldsymbol{K}[\boldsymbol{x}]$. Suppose that $QS^{[i]} \neq \varnothing$ and let $H = \prod_{Q \in QS^{[i]}} Q$. Denote by $H_1, \ldots, H_m$ all the non-zero coefficients of the monomials in H with respect to those variables which are $\succ x_i$. Then

$$\mathrm{Proj}_{\boldsymbol{x}_i}\mathrm{Zero}(PS/QS) = \bigcup_{j=1}^{m} \mathrm{Zero}(PS/QS^{(i)} \cup \{H_j\}), \tag{2.6}$$

$$\mathrm{Zero}(PS/QS) = \bigcup_{j=1}^{m} \mathrm{Zero}(PS/QS \cup \{H_j\}). \tag{2.7}$$

With practical considerations, the zero relations (2.6) and (2.7) in Lemma 2.7 may be made more complicated (see Wang 1993b).

Lemma 2.8: (Jacobson 1974; Seidenberg 1956a,b; Wang 1993b; Wu 1990a) *Let P be a polynomial in $\boldsymbol{K}[\boldsymbol{x}]$ with*

$$\mathrm{cls}(P) = i > 0, \qquad \mathrm{ini}(P) = I, \qquad \mathrm{ldeg}(P) = d,$$

and $[\mathit{PS}, \mathit{QS}]$ a polynomial system of level $\ell \le i-1$ with $\mathrm{level}(\mathit{QS}) \le i$.

(*a*) *If $\mathit{QS}^{\langle i \rangle} = \varnothing$, then for any $\ell \le j \le i-1$*

$$\mathrm{Proj}_{\boldsymbol{x}_j}\mathrm{Zero}(\mathit{PS} \cup \{P\}/\mathit{QS} \cup \{I\}) = \mathrm{Proj}_{\boldsymbol{x}_j}\mathrm{Zero}(\mathit{PS}/\mathit{QS} \cup \{I\}).$$

(*b*) *Suppose that $\mathit{QS}^{\langle i \rangle} \neq \varnothing$ and let*

$$H = \prod_{Q \in \mathit{QS}^{\langle i \rangle}} Q, \qquad R = \mathrm{prem}(H^d, P), \qquad \mathit{QS}' = \mathit{QS}^{(i-1)} \cup \{I, R\}.$$

Then, for any $\ell \le j \le i-1$,

$$\mathrm{Proj}_{\boldsymbol{x}_j}\mathrm{Zero}(\mathit{PS} \cup \{P\}/\mathit{QS} \cup \{I\}) = \mathrm{Proj}_{\boldsymbol{x}_j}\mathrm{Zero}(\mathit{PS}/\mathit{QS}'),$$
$$\mathrm{Zero}(\mathit{PS} \cup \{P\}/\mathit{QS} \cup \{I\}) = \mathrm{Zero}(\mathit{PS} \cup \{P\}/\mathit{QS}').$$

Now consider a fine triangular system $[\mathit{TS}, \mathit{US}]$ with $\mathit{TS} = [T_1, \ldots, T_r]$ and $\mathrm{cls}(T_i) = p_i$ for each i. We can carry out projection for each x_{p_i} top-down ($i = r, r-1, \ldots, 1$) according to

- Lemma 2.7 when $p_i < n$ and $\mathit{US}^{[p_i]} \neq \varnothing$,
- Lemma 2.8 (*a*) when $\mathit{US}^{\langle p_i \rangle} = \varnothing$, and
- Lemma 2.8 (*b*) when $\mathit{US}^{\langle p_i \rangle} \neq \varnothing$.

In the first case of projection, the triangular system $[\mathit{TS}, \mathit{US}]$ may be split into several triangular systems. So the last two cases should actually be considered for all of them. In particular, when projection is needed only for $x_n, \ldots, x_k$, let l be such that $p_l < k+1 \le p_{l+1}$. Then, the projection may be performed first for $x_{p_r}, \ldots, x_{p_{l+1}}$, and finally for x_k in addition.

In this way of projection, either all the split triangular systems are removed and thus $\mathrm{Zero}(TS/US) = \varnothing$, or a finite sequence of polynomial sets $US_1, \ldots, US_s$ are finally obtained such that

$$\mathrm{Zero}(TS/US) = \bigcup_{i=1}^{s} \mathrm{Zero}(TS/US_i),$$

$$\mathrm{Proj}_{\boldsymbol{x}_k}\mathrm{Zero}(TS/US) = \bigcup_{i=1}^{s} \mathrm{Zero}(TS^{(k)}/US_i^{(k)}).$$

Definition 2.9:

Let $\mathfrak{T} = [TS, US]$ be a fine triangular system in $\boldsymbol{K}[\boldsymbol{x}]$ and k a non-negative integer. $\mathfrak{T}$ is said to possess the projection property *of dimension k if*

$$\mathrm{Zero}(\mathfrak{T}^{(i)}) \subset \mathrm{Proj}_{\boldsymbol{x}_i}\mathrm{Zero}(\mathfrak{T}) \tag{2.8}$$

holds for $i = k$ and all $i \in \{\mathrm{cls}(T) \mid T \in TS, \mathrm{cls}(T) > k\}$, and to possess the strong projection property *if* (2.8) *holds for all $0 \le i < n$.*

Lemmas 2.7 and 2.8 ensure that the above-mentioned triangular systems $[TS, US_j]$ ($j \in \mathbb{N}_s$) all possess the projection property of dimension k.

Note that the projection according to Lemma 2.7 is performed once for all the variables $x_{p_{i+1}}, \ldots, x_{p_i+1}$. This is mainly for practical considerations. One can project for one variable each time as well.

For an arbitrary polynomial system $\mathfrak{P}$, one can first compute a fine triangular series Ψ of $\mathfrak{P}$ and then perform projection for each triangular system in Ψ (see Wu 1990a; Gao and Chou 1993b). A more effective way is to embed the projection process into the algorithm for computing triangular series.

Take for example the algorithm T. The reduction step P2.3 in Algorithm P can be replaced by two projection steps for x_i, first according to Lemma 2.7 when there are polynomials of classes $> i$ to be "projected," and then according to Lemma 2.8 when there are polynomials of class i but no polynomial of class $> i$ to be "projected." If projection is needed only for $x_n, \ldots, x_k$, then an additional step is inserted for x_k according to Lemma 2.7 when there are polynomials of classes $> k$ to be "projected."

Such an algorithm with projection modified from Algorithm T is proposed in Wang (1993b). It may either detect that $\mathrm{Zero}(\mathfrak{P}) = \varnothing$ or compute a zero decomposition of the form (2.2) such that

$$\mathrm{Proj}_{\boldsymbol{x}_k}\mathrm{Zero}(\mathfrak{P}) = \bigcup_{i=1}^{e} \mathrm{Zero}(\mathfrak{T}_i^{(k)})$$

and each $\mathfrak{T}_i$ is a fine triangular system possessing the projection property of dimension k. The algorithm also provides a quantifier elimination procedure and thus a decision procedure for the existential theory of algebraically closed fields. One of its consequences is the projection theorem of elimination theory in the affine case (Jacobson 1974, Sect. 5.4, pp. 305–306; Seidenberg 1956a,b; Wu 1990a). See Wang (1993b, 2000a) for further details of projection and Sect. 2.5 for its applications.

2.3.2 Decomposition into irreducible triangular systems

Polynomial factorization is not required theoretically for the algorithms discussed in the previous sections. Here we explain how to decompose triangular systems further, making use of factorization in order to compute zero decompositions possessing better properties.

A triangular set $\mathit{TS} \subset \boldsymbol{K}[\boldsymbol{x}]$ is said to be *quasi-irreducible* if every polynomial in TS is irreducible over the ground field $\boldsymbol{K}$. A triangular system $[\mathit{TS}, \mathit{US}]$ in $\boldsymbol{K}[\boldsymbol{x}]$ is said to be *quasi-irreducible* if TS is quasi-irreducible. Using polynomial factorization over $\boldsymbol{K}$, one has no difficulty computing zero decompositions of the form (2.2) with all triangular systems $\mathfrak{T}_i$ quasi-irreducible.

Let $\mathit{TS} = [T_1, \ldots, T_r]$ be a triangular set with $\mathrm{cls}(T_i) = p_i$ for each i, and the leading variables $x_{p_1}, \ldots, x_{p_r}$ be renamed $y_1, \ldots, y_r$. Denote all the x_i other than $x_{p_1}, \ldots, x_{p_r}$ by $u_1, \ldots, u_d$, abbreviated to $\boldsymbol{u}$ $(d + r = n)$. We call $\boldsymbol{u}$ the *parameters* of TS. Clearly, TS can be written as

$$\mathit{TS} = [T_1(\boldsymbol{u}, y_1), \ldots, T_r(\boldsymbol{u}, y_1, \ldots, y_r)]. \tag{2.9}$$

Let $\boldsymbol{K}_0 = \boldsymbol{K}(\boldsymbol{u})$, the transcendental extension field of $\boldsymbol{K}$ acquired by adjoining $\boldsymbol{u}$.

For any triangular set $\mathit{TS} = [T_1, \ldots, T_r]$ and $i \in \mathbb{N}_r$,

$$\mathit{TS}^{\{i\}} := [T_1, \ldots, T_i].$$

Definition 2.10:

A fine triangular set TS of the form (2.9) *is said to be* irreducible *if for each $i \in \mathbb{N}_r$ the polynomial*

$$T_i(\boldsymbol{u}, \eta_1, \ldots, \eta_{i-1}, y_i) \in \boldsymbol{K}_{i-1}[y_i]$$

is irreducible over $\boldsymbol{K}_{i-1}$, where

$$(\boldsymbol{u}, \eta_1, \ldots, \eta_{i-1}) \in \mathrm{Zero}(\mathit{TS}^{\{i-1\}}) \quad \textit{and} \quad \boldsymbol{K}_{i-1} = \boldsymbol{K}_0(\eta_1, \ldots, \eta_{i-1})$$

(the algebraic extension field of $\boldsymbol{K}_0$ acquired by adjoining $\eta_1, \ldots, \eta_{i-1}$) when $i > 1$.

A fine triangular system $[TS, US]$ is said to be irreducible *if TS is irreducible.*

Lemma 2.9: (Wang 1993b; Wu 1984a)

Let $TS \subset \boldsymbol{K}[\boldsymbol{u}, y_1, \ldots, y_r]$ be any fine triangular set of the form (2.9). There is an algorithm which determines whether TS is irreducible or not, and if not, computes an integer $k \in \mathbb{N}_r$ and an irreducible factorization of T_k of the form

$$DT_k \doteq F_1 \cdots F_t \tag{2.10}$$

over $\boldsymbol{K}_{k-1}$, where $\boldsymbol{K}_{k-1}$ is as in Definition 2.10, the polynomials

$$D \in \boldsymbol{K}[\boldsymbol{u}, y_1, \ldots, y_{k-1}], \qquad F_i \in \boldsymbol{K}[\boldsymbol{u}, y_1, \ldots, y_k], \quad i \in \mathbb{N}_t,$$

are all reduced with respect to $TS^{\{k-1\}}$ and the dotted equality means that $\operatorname{prem}(DT_k - F_1 \cdots F_t, TS^{\{k-1\}}) = 0$.

See Chap. 13 for a review of such algorithms of algebraic factorization.

Lemma 2.10: (Wang 1993b)

Let $[TS, US]$ be a fine triangular system in $\boldsymbol{K}[\boldsymbol{x}]$. Assume that TS is reducible, so there exists an integer $k > 0$ such that the kth term T_k of TS has an irreducible factorization into polynomials $F_1, \ldots, F_t$ as in (2.10), and $TS^{\{k-1\}}$ is irreducible when $k > 1$. Then the following zero decomposition holds

$$\operatorname{Zero}(TS/US) = \bigcup_{i=1}^{t} \operatorname{Zero}(TS_i/US \cup \{D\}) \cup \operatorname{Zero}(\{D\} \cup TS/US), \tag{2.11}$$

where $TS_i = TS \setminus \{T_k\} \cup \{F_i\}$ for each $i \in \mathbb{N}_t$.

If, in particular, $D \in \boldsymbol{K}$ or the irreducible $TS^{\{k-1\}}$ contains no parameter (i.e., its *dimension* is 0), then (2.11) may be simplified to

$$\operatorname{Zero}(TS/US) = \bigcup_{i=1}^{t} \operatorname{Zero}(TS_i/US).$$

This is trivial for $D \in \boldsymbol{K}$. The case in which $TS^{\{k-1\}}$ is of dimension 0 follows from Proposition 2.30.

Let the notations be as in Lemma 2.10 and $US_i = \operatorname{prem}(US \cup \{D\}, TS_i)$. If $0 \in US_i$ for some i, then the corresponding component in (2.11) can be simply removed. For those components in which US_i does not contain 0, it is easy to see that $[TS_i, US_i]$ is still a fine triangular system and, in particular, $TS_i^{\{k\}}$ is irreducible for each i. Moreover, all the TS_i have the same set of parameters as TS. One can go on examining whether each TS_i is irreducible, and if not, decompose $[TS_i, US_i]$ further. In this way, $[TS, US]$ will be decomposed eventually into finitely many irreducible triangular systems $[TS_i, US_i]$ with each TS_i having the same set of parameters as TS, unless $\mathrm{Zero}(TS/US) = \varnothing$ is detected.

Note that $[\{D\} \cup TS, US]$ may no longer be a triangular system, yet it can be further triangularized and decomposed by applying Algorithm T or T*. The obtained triangular systems may of course be decomposed into irreducible ones as well.

A triangular series Ψ is said to be *irreducible* if every triangular system in Ψ is irreducible. Combining Algorithm T or T* and the decomposition process with polynomial factorization discussed above, we are able to compute an irreducible triangular series from any given polynomial system. See the algorithms Decom and IrrTriSer in Wang (1993b) for technical details.

In addition to the method of Ritt (1950) and Wu (1984a, 1986a; Chap. 1) used extensively in this book and those presented in this and the previous sections, there are several other related methods including those developed by Lazard (1991), Kalkbrener (1993, 1994), Yang et al. (1995) and their co-workers for computing various zero decompositions. For implementations and practical performance of the algorithms discussed in this chapter, the interested reader may refer to Aubry and Moreno Maza (1999), Wang (1993b, 1996a, 2000b), and the comparisons therein.

2.4 Properties of triangular sets and systems

All polynomials, polynomial sets and systems, triangular, regular or simple sets and systems mentioned in this section are in $\boldsymbol{K}[\boldsymbol{x}]$ unless stated otherwise. Denote by $\bar{\boldsymbol{K}}$ the *algebraic closure* and by $\tilde{\boldsymbol{K}}$ some extension of the ground field $\boldsymbol{K}$. For any polynomial set or system $\mathfrak{P}$, by $\mathrm{Zero}(\mathfrak{P}) = \varnothing$ and $\mathrm{Zero}(\mathfrak{P}) \neq \varnothing$ we mean that $\mathfrak{P}$ has no zero in any extension field of $\boldsymbol{K}$ and $\mathfrak{P}$ has a zero in some suitable (algebraic) extension field of $\boldsymbol{K}$, respectively. Proofs of the results stated in this section can be found from the indicated references.

Proposition 2.11: (Wang 2000a)

Any simple set is a regular set, any simple system is a regular system, any simple series is a regular series, and any irreducible triangular set is a simple set.

Note that an irreducible triangular system (or series) is not necessarily a simple system (or series), nor necessarily a regular system (or series).

Theorem 2.12: (Kalkbrener 1993; Wang 2000b; Wu 1984a)

Any regular system or irreducible triangular system is perfect over $\bar{\boldsymbol{K}}$*, and thus is perfect. Any regular set is perfect over* $\bar{\boldsymbol{K}}$*, and thus is also perfect.*

Corollary 2.13:

Let $\mathfrak{P}$ *be any polynomial system and* Ψ *be a regular series, or a simple series, or an irreducible triangular series of* $\mathfrak{P}$*. Then*

$$\mathrm{Zero}(\mathfrak{P}) = \varnothing \iff \Psi = \varnothing.$$

Corollary 2.14:

Let $[PS, QS]$ *be any polynomial system and* P *a polynomial, and let* Ψ^* *be a regular series, or a simple series, or an irreducible triangular series of* $[PS, QS \cup \{P\}]$*. Then*

$$\mathrm{Zero}(PS/QS) \subset \mathrm{Zero}(P) \iff \Psi^* = \varnothing.$$

Theorem 2.15: (Wang 2000b)

For any regular system $[TS, US]$ *and polynomial* P*,*

$$\mathrm{Zero}(TS/US) \subset \mathrm{Zero}(P) \iff \exists d \in \mathbb{N} \textit{ such that } \mathrm{prem}(P^d, TS) = 0.$$

Theorem 2.16: (Wang 1993b, 1998a)

For any simple system or irreducible triangular system $[TS, US]$ *and polynomial* P*,*

$$\mathrm{Zero}(TS/US) \subset \mathrm{Zero}(P) \iff \mathrm{prem}(P, TS) = 0.$$

As corollaries of the above theorems, we have the following results.

Corollary 2.17:

For any regular set TS *and polynomial* P*,*

$$\mathrm{Zero}(TS/\mathrm{ini}(TS)) \subset \mathrm{Zero}(P) \iff \exists d \in \mathbb{N} \textit{ such that } \mathrm{prem}(P^d, TS) = 0.$$

Corollary 2.18:

For any simple set TS *and polynomial* P*,*

$$\mathrm{Zero}(TS/\mathrm{ini}(TS)) \subset \mathrm{Zero}(P) \iff \mathrm{prem}(P, TS) = 0.$$

Theorem 2.19: (Wang 2000b)

Let $[PS, QS]$ *be a polynomial system and* Ψ *a regular series of* $[PS, QS]$. *Then:*

(*a*) $\exists d \in \mathbb{N}$ *such that* $\text{prem}(P^d, TS) = 0$, $\forall P \in PS$ *and* $[TS, US] \in \Psi$;

(*b*) $\forall m \in \mathbb{N}$, $[TS, US] \in \Psi$ *and* $Q \in QS$, $\text{prem}(Q^m, TS) \neq 0$.

Theorem 2.20: (Wang 1993b, 1998a)

Let $[PS, QS]$ *be a polynomial system and* Ψ *a simple series or an irreducible triangular series of* $[PS, QS]$. *Then*

$$\text{prem}(PS, TS) = \{0\} \text{ and } 0 \notin \text{prem}(QS, TS), \quad \forall\, [TS, US] \in \Psi.$$

The above two theorems can be used to prove the following remarkable property.

Theorem 2.21: (Wang 1993b, 1998a, 2000b)

Let $[PS, QS]$ *be a polynomial system and* Ψ *a regular series, or a simple series, or an irreducible triangular series of* $[PS, QS]$. *Then*

$$\text{Zero}(PS/QS) = \bigcup_{[TS,US]\in\Psi} \text{Zero}(TS/\text{ini}(TS) \cup QS). \tag{2.12}$$

In view of Theorems 2.21 and 2.40, it is proper to call $TS_1, \ldots, TS_e$ a *regular series*, or a *simple series*, or an *irreducible triangular series* of PS when there exist polynomial sets US_i such that $[TS_1, US_1], \ldots, [TS_e, US_e]$ is a regular series, or a simple series, or an irreducible triangular series of PS, respectively.

Definition 2.11:

Let $\mathfrak{T} = [TS, US]$ *be an arbitrary triangular system. A zero* $(\xi_1, \ldots, \xi_n)$ *of* $\mathfrak{T}$ *is said to be* regular *if either* $\xi_i = x_i$, *or* x_i *is the leading variable of some polynomial in TS for any* $i \in \mathbb{N}_n$.

When $\mathfrak{T}$ *is regular, any regular zero of* $\mathfrak{T}$ *is also called a* regular zero *of TS. The set of all regular zeros of* $\mathfrak{T}$ *or TS is denoted* $\text{RegZero}(\mathfrak{T})$ *or* $\text{RegZero}(TS)$.

In the case TS is regular, the definition of regular zeros of TS here accords with that given in Kalkbrener (1993).

Let $TS = [T_1, \ldots, T_r]$ be a regular set with $d_i = \text{ldeg}(T_i)$ and $d = d_1 \cdots d_r$; by Theorem 2.12 TS is perfect. If TS is irreducible, then it has d distinct regular zeros which are also called *generic zeros* of TS and generate one and the same extension field of $\boldsymbol{K}$. If TS is simple and reducible, then it has d distinct regular zeros which generate more than one extension field of $\boldsymbol{K}$ of the same transcendence degree. If TS is reducible but not simple, then it has less than d distinct regular zeros which generate one or more extension fields of $\boldsymbol{K}$ of the same transcendence degree.

We write $\boldsymbol{\xi}_i$ for $(\boldsymbol{u}, \eta_1, \ldots, \eta_i)$ with $\boldsymbol{\xi} = \boldsymbol{\xi}_r$.

Proposition 2.22: (Wang 2000b)

Let TS as in (2.9) be a regular set. Then

$$\mathrm{ini}(T_{i+1})(\boldsymbol{\xi}_i) \neq 0, \quad \forall i \in \mathbb{N}_{r-1}, \quad \boldsymbol{\xi}_i \in \mathrm{RegZero}(TS^{\{i\}}).$$

Definition 2.12:

Let P be any polynomial and $TS = [T_1, \ldots, T_r]$ a triangular set. The polynomial

$$\mathrm{res}(P, TS) := \mathrm{res}(\cdots \mathrm{res}(P, T_r, \mathrm{lv}(T_r)), \ldots, T_1, \mathrm{lv}(T_1))$$

is called the resultant *of P with respect to TS, where* $\mathrm{res}(P, Q, x_k)$ *denotes the* resultant *of P and Q with respect to x_k.*

Proposition 2.23: (Wang 2000b; Yang and Zhang 1994)

For any regular set TS and polynomial P,

$$\mathrm{res}(P, TS) \neq 0 \iff P(\boldsymbol{\xi}) \neq 0, \quad \forall \boldsymbol{\xi} \in \mathrm{RegZero}(TS).$$

Lemma 2.24: (Wang 2000b; Yang and Zhang 1994)

Let $TS = [T_1, \ldots, T_r]$ be a triangular set. Then

$$TS \textit{ is regular} \iff \mathrm{res}(\mathrm{ini}(T_{i+1}), TS^{\{i\}}) \neq 0, \quad \forall\, i \in \mathbb{N}_{r-1}.$$

It follows from this lemma and Proposition 2.23 that a triangular set TS is regular if and only if for any $k \in \mathbb{N}_{n-1}$, $TS^{(k)}$ is regular and

$$I(\boldsymbol{\xi}_k) \neq 0 \text{ for } I \in \mathrm{ini}(TS^{\langle k+1 \rangle}) \text{ and all } \boldsymbol{\xi}_k \in \mathrm{RegZero}(TS^{(k)}).$$

These results show the equivalence of the concepts of *regular chains* in Kalkbrener (1993), *proper ascending chains* in Yang and Zhang (1994) and regular sets defined in Wang (2000b) and Sect. 2.2.1.

Theorem 2.25: (Thomas 1937; Wang 1998a)

Let $\mathfrak{S}$ be a simple system. Then for any $1 < k \leq n$ and

$$\bar{\boldsymbol{x}}_{k-1} \in \mathrm{Zero}(\mathfrak{S}^{(k-1)})$$

there exist $\bar{x}_k, \ldots, \bar{x}_l \in \tilde{\boldsymbol{K}}$ such that $\bar{\boldsymbol{x}}_l \in \mathrm{Zero}(\mathfrak{S}^{(l)})$ for all $k \leq l \leq n$. In particular, $\mathfrak{S}$ is perfect over $\bar{\boldsymbol{K}}$.

Corollary 2.26:

Every simple system possesses the strong projection property.

Theorem 2.27: (Wang 1998a, 2000b)

Let $\mathfrak{P}$ be any polynomial system and Ψ a regular or simple series of $\mathfrak{P}$. Then

$$\mathrm{Zero}(\mathfrak{P}) \textit{ is finite} \iff |TS| = n \textit{ and } US = \varnothing, \quad \forall\,[TS, US] \in \Psi.$$

Lemma 2.28: (Wu 1984a)

Let TS be an irreducible triangular set with a generic zero $\boldsymbol{\xi}$. Then for any polynomial P

$$\mathrm{prem}(P, TS) = 0 \iff P(\boldsymbol{\xi}) = 0.$$

Lemma 2.29: (Wang 2000a; Wu 1984a,b)

Let $TS = [T_1, \ldots, T_r]$ be a triangular set of the form (2.9) *and P a polynomial in $\boldsymbol{K}[\boldsymbol{u}, y_1, \ldots, y_r]$, and $R = \mathrm{res}(P, TS)$. Then in $\boldsymbol{K}[\boldsymbol{u}, y_1, \ldots, y_r]$ one can determine polynomials Q and $Q_1, \ldots, Q_r$ such that*

$$QP = Q_1 T_1 + \cdots + Q_r T_r + R.$$

If TS is irreducible with a generic zero $\boldsymbol{\xi} = (\boldsymbol{u}, \eta_1, \ldots, \eta_r)$ and $\mathrm{prem}(P, TS) \neq 0$, *then*

$$R(\boldsymbol{u}) \neq 0, \qquad Q(\boldsymbol{\xi}) \neq 0.$$

Proposition 2.30: (Wu 1984a; Wang 1993b)

Let TS be an irreducible triangular set and P a polynomial with $\mathrm{prem}(P, TS) \neq 0$. *If $|TS| = n$, then*

$$\mathrm{Zero}(\{P\} \cup TS) = \varnothing, \qquad \mathrm{Zero}(TS/IS) = \mathrm{Zero}(TS),$$

where $IS = \mathrm{ini}(TS)$. In this case, the corresponding $\mathrm{Zero}(TS/IS \cup QS)$ *in* (2.12) *can be simplified to* $\mathrm{Zero}(TS/QS)$.

It is also true that $\mathrm{Zero}(TS/\mathrm{ini}(TS)) = \mathrm{Zero}(TS)$ for any regular set TS with $|TS| = n$ (see Aubry 1999, Proposition 4.3.8).

For any non-empty polynomial set PS, $\mathrm{Ideal}(PS)$ denotes the ideal generated by (the polynomials in) PS.

Definition 2.13:

Let $\mathfrak{I}$ be an ideal and P a polynomial in $\boldsymbol{K}[\boldsymbol{x}]$. The saturation *of $\mathfrak{I}$ with respect to P is the ideal*

$$\mathfrak{I} : P^\infty := \{F \in \boldsymbol{K}[\boldsymbol{x}] \mid P^q F \in \mathfrak{I} \textit{ for some } q \in \mathbb{N}\}.$$

For any triangular set TS, the saturation *of TS is the ideal*

$$\mathrm{sat}(\mathit{TS}) := \mathrm{Ideal}(\mathit{TS}) : \Big(\prod\nolimits_{T \in \mathit{TS}} \mathrm{ini}(T)\Big)^\infty.$$

Theorem 2.31: (Aubry et al. 1999; Wang 2000b)

Let TS be any triangular set. Then

$$\mathit{TS} \textit{ is regular} \iff \mathrm{sat}(\mathit{TS}) = \{P \in \boldsymbol{K}[\boldsymbol{x}] \mid \mathrm{prem}(P, \mathit{TS}) = 0\}.$$

Lemma 2.32: (Wang 2000b)

Let TS be a regular set and P any polynomial. If $\mathrm{res}(P, \mathit{TS}) \neq 0$, *then* $\mathrm{sat}(\mathit{TS}) : P^\infty = \mathrm{sat}(\mathit{TS})$.

Proposition 2.33: (Wang 2000b)

Let $[\mathit{TS}, \mathit{US}]$ be a regular system. Then

$$\mathrm{Ideal}(\mathit{TS}) : (\prod\nolimits_{U \in \mathit{US}} U)^\infty = \mathrm{sat}(\mathit{TS}).$$

An *algebraic variety* is a geometric figure defined by the zeros of a polynomial set. It is said to be *unmixed* or *equidimensional* if all its irredundant irreducible components have the same dimension.

Theorem 2.34: (Gao and Chou 1993a)

Let TS be any triangular set. If TS is not perfect then $\mathrm{sat}(\mathit{TS}) = \boldsymbol{K}[\boldsymbol{x}]$*; if TS is perfect then* $\mathrm{Zero}(\mathrm{sat}(\mathit{TS}))$ *is an unmixed variety of dimension $n - |\mathit{TS}|$.*

The result of equidimensionality in Theorem 2.34 has been generalized by Aubry (1999, Theorem 4.1.4) for the ideal $\mathrm{sat}(\mathit{TS})$ instead of the corresponding variety.

Theorem 2.35: (Wang 2000a)

For any simple set TS, the ideal $\mathrm{sat}(\mathit{TS})$ *is radical.*

An ideal $\mathfrak{I} \subset \boldsymbol{K}[\boldsymbol{x}]$ is said to be *prime* if whenever $F, G \in \boldsymbol{K}[\boldsymbol{x}]$ and $FG \in \mathfrak{I}$, either $F \in \mathfrak{I}$ or $G \in \mathfrak{I}$.

Theorem 2.36: (Wang 2000a; Wu 1984a)

For any irreducible triangular set TS, *the ideal* $\mathrm{sat}(TS)$ *is prime.*

Proposition 2.37: (Wang 2000a)

Let TS_1 *and* TS_2 *be two simple sets which have the same set of regular zeros. Then* $\mathrm{sat}(TS_1) = \mathrm{sat}(TS_2)$.

Proposition 2.38: (Chou and Gao 1990b; Wang 2000a)

Let TS_1 *and* TS_2 *be two triangular sets which have the same set of parameters, and* TS_2 *be irreducible. If* $\mathrm{prem}(TS_2, TS_1) = \{0\}$, *then* TS_1 *is also irreducible and has the same set of generic zeros as* TS_2; *thus* $\mathrm{sat}(TS_1) = \mathrm{sat}(TS_2)$.

Proposition 2.39: (Chou and Gao 1990b)

Let TS_1 *and* TS_2 *be two triangular sets, of which* TS_1 *is irreducible. If* $\mathrm{prem}(TS_2, TS_1) = \{0\}$ *and* $0 \notin \mathrm{prem}(\mathrm{ini}(TS_2), TS_1)$, *then* $\mathrm{sat}(TS_2) \subset \mathrm{sat}(TS_1)$.

Other properties of triangular sets and systems of different kinds may be found in Aubry et al. (1999), Chou and Gao (1990b), Gao and Chou (1993a), Kalkbrener (1993, 1994), Lazard (1991), Wang (1998a, 2000a,b), Wu (1984a, 1994a), and Yang and Zhang (1994).

2.5 Applications

2.5.1 Decomposition of algebraic varieties

When $QS = \varnothing$, the zero decomposition (2.12) provides a representation of the algebraic variety defined by PS in terms of its subvarieties determined by TS_i. However, each $\mathrm{Zero}(TS_i/\mathrm{ini}(TS_i))$ is not necessarily an algebraic variety. The following theorem shows that a corresponding variety decomposition can be obtained as long as a finite basis may be determined for each $\mathrm{sat}(TS_i)$.

Theorem 2.40: (Gao and Chou 1993a; Wang 2000a)

Let PS *be a non-empty polynomial set in* $\boldsymbol{K}[\boldsymbol{x}]$ *and* $TS_1, \ldots, TS_e$ *be a regular series of* PS, *or a triangular series of* PS *with* $\mathrm{prem}(PS, TS_i) = \{0\}$ *for each* i. *Then*

$$\mathrm{Zero}(PS) = \bigcup_{i=1}^{e} \mathrm{Zero}(\mathrm{sat}(TS_i)).$$

In fact, there are several algorithms by which a finite basis for $\mathrm{sat}(\mathit{TS}_i)$ can be determined from TS_i; see Gianni et al. (1988), Chou et al. (1990), Cox et al. (1992), Gao and Chou (1993a) and Wang (1989e, 1992a) using Gröbner bases, Ritt (1950) using resolvents, and Wu (1989d) using Chow bases. For each i let PS_i be a finite basis for $\mathrm{sat}(\mathit{TS}_i)$. If $\mathrm{sat}(\mathit{TS}_i) = \boldsymbol{K}[\boldsymbol{x}]$, then the constant 1 is contained in the Gröbner basis of PS_i. Let us assume that such PS_i are simply removed; this is of course not necessary for regular series in which every regular set TS_i is perfect. Thus, a variety decomposition of the following form is obtained:

$$\mathrm{Zero}(\mathit{PS}) = \bigcup_{i=1}^{e} \mathrm{Zero}(\mathit{PS}_i). \tag{2.13}$$

By Theorem 2.34, each PS_i defines an unmixed algebraic variety.

When the triangular series Ψ is irreducible, each finite basis for $\mathrm{sat}(\mathit{TS}_i)$ defines an irreducible algebraic variety with any generic zero of TS_i as its generic point. In this case, $\mathrm{sat}(\mathit{TS}_i)$ is a prime ideal, its finite basis PS_i is called a *prime basis* of TS_i, and an irreducible decomposition of the variety defined by PS is then accomplished.

More generally, when Ψ is a regular series, any regular zero of TS_i is a generic point of an irreducible component of the variety associated to $\mathrm{sat}(\mathit{TS}_i)$.

In view of the relationship between varieties and ideals, a decomposition of an algebraic variety will lead to one of the radical of the corresponding ideal, and vice versa. For example, when $\mathit{TS}_1, \ldots, \mathit{TS}_e$ is a simple series of PS, each $\mathfrak{I}_i = \mathrm{Ideal}(\mathit{PS}_i)$ is radical. This suggests the following ideal decomposition

$$\sqrt{\mathfrak{I}} = \bigcap_{i=1}^{e} \mathfrak{I}_i, \tag{2.14}$$

where $\mathfrak{I} = \mathrm{Ideal}(\mathit{PS})$ and each $\mathfrak{I}_i$ is radical and unmixed (Wang 1998c).

Together with ideal intersection computation, one can thus devise an algorithm to find a generating set of $\sqrt{\mathfrak{I}}$ for any ideal $\mathfrak{I}$ with given generating set (without need of polynomial factorization in theory).

Remark that it is possible to compute radical ideal decompositions of the form (2.14) without having the complete triangular series (see Aubry 1999, Sects. 5.2 and 5.5; Gao and Chou 1993a; Kalkbrener 1993; Wang 1999b). Computing (2.14) directly may avoid the computation of some branches that is redundant for (2.14) yet necessary for obtaining the triangular series. The decompositions of algebraic varieties or radicals of the corresponding ideals may be made *minimal* by removing all the redundant subvarieties.

2.5.2 Radical ideal membership and primary decomposition

One of the fundamental problems in constructive algebraic geometry and polynomial ideal theory is about radical ideal membership, that is to test whether a given polynomial P vanishes on an algebraic variety defined by a given polynomial set $\mathbb{PS}$, or equivalently whether P belongs to the radical of $\mathrm{Ideal}(\mathbb{PS})$. In other words, the problem is to give effective means to verify whether $\mathrm{Zero}(\mathbb{PS}) \subset \mathrm{Zero}(P)$ holds. Corollary 2.14 and Theorem 2.16 together with the algorithms for computing regular series, simple series, and irreducible triangular series provide several different solutions to this problem.

An important application of radical ideal membership test is to automated theorem proving in geometry. This will be discussed in Sect. 2.5.4. The membership test is also needed for examining the containment relationship between algebraic varieties.

Decomposing polynomial ideals into primary components is another classical task in commutative algebra. From an irreducible decomposition of the corresponding algebraic variety, a primary decomposition of any polynomial ideal can be constructed by means of localization and extraction, as suggested by Shimoyama and Yokoyama (1996). The basic idea is explained as follows.

From an irreducible variety decomposition of the form (2.13), one immediately gets the decomposition (2.14) for the radical of $\mathrm{Ideal}(\mathbb{PS})$ with each $\mathfrak{I}_i = \mathrm{Ideal}(\mathbb{PS}_i)$ prime. For each $i \in \mathbb{N}_e$, a pseudo-primary ideal $\mathfrak{J}_i$ may be constructed such that $\mathfrak{I}_i$ is the prime ideal associated with $\mathfrak{J}_i$. Here an ideal is said to be *pseudo-primary* if its radical is prime. Meanwhile, an additional ideal $\mathfrak{I}^*$ is constructed, so that the following decomposition is reached

$$\mathfrak{I} = \bigcap_{i=1}^{e} \mathfrak{J}_i \cap \mathfrak{I}^*.$$

From each pseudo-primary ideal $\mathfrak{J}_i$, one can determine a primary ideal $\bar{\mathfrak{J}}_i$ and a polynomial set $\mathbb{GS}_i$ by extraction such that

$$\mathfrak{J}_i = \bar{\mathfrak{J}}_i \cap \mathrm{Ideal}(\mathbb{GS}_i).$$

One will finally arrive at a primary decomposition of the ideal $\mathfrak{I}$ by applying the above process to the ideal $\mathfrak{I}^*$ and $\mathrm{Ideal}(\mathbb{GS}_i)$ recursively. See Shimoyama and Yokoyama (1996) for details.

2.5.3 Solving polynomial systems

The various zero decompositions presented in the previous sections apply naturally to solving systems of polynomial equations and inequations of the form

(2.1). According to the projection property and Corollary 2.13, the solvability of such systems can be determined by means of computing their triangular series with projections, regular series or simple series (without polynomial factorization), or irreducible triangular series (with factorization).

Now let $\boldsymbol{K} = \boldsymbol{Q}(u_1, \ldots, u_d)$ be specified, where $\boldsymbol{Q}$ is the field of rational numbers.

Proposition 2.41:

Let $[TS, US]$ *be a triangular system in* $\boldsymbol{K}[\boldsymbol{x}]$ *with* $|TS| = n$. *Then*

$$TS = 0, \qquad US \neq 0 \tag{2.15}$$

has at most finitely many solutions in any extension field of $\boldsymbol{K}$. *All the solutions of* (2.15) *in* $\boldsymbol{K}$ *can be exactly computed.*

If, in particular, $d = 0$, *then all the solutions of* (2.15) *in the field of real numbers and in the field* $\boldsymbol{C}$ *of complex numbers can be approximately computed.*

Proposition 2.42:

Let $[TS, US]$ *be a regular system, or a simple system, or an irreducible triangular system, or a triangular system possessing the projection property of dimension* 0 *in* $\boldsymbol{K}[\boldsymbol{x}]$. *Then system* (2.15) *must have solutions in some extension field of* $\boldsymbol{K}$. *If the number of solutions is finite, then* $|TS| = n$.

Theorem 2.43:

Let Ψ *be a regular series, or simple series, or irreducible triangular series of any polynomial system* $[PS, QS]$ *in* $\boldsymbol{K}[\boldsymbol{x}]$, *or a triangular series of* $[PS, QS]$ *computed with projection. Then:*

(*a*) (2.1) *has no solution in any extension field of* $\boldsymbol{K}$ *if and only if* $\Psi = \varnothing$;

(*b*) (2.1) *has at most finitely many solutions if and only if* $|TS| = n$ *for every* $[TS, US] \in \Psi$. *In this case, the solutions of* (2.1) *may be found by means of computing the solutions of* $TS = 0, US \neq 0$ *for all* $[TS, US] \in \Psi$.

When Ψ is a simple series and (2.1) has only finitely many solutions, the exact number of solutions can be counted according to the leading degrees of the polynomials in TS for each $[TS, US] \in \Psi$.

Consider systems of polynomial equations and inequations of the form (2.1), with coefficients in $\boldsymbol{Q}$ and $\boldsymbol{u} = (u_1, \ldots, u_d)$ as *parameters*. Theorem 2.44 below permits us to identify the parametric values for which the considered system has solutions for the unknowns x_i over some extension field of $\boldsymbol{Q}$ and thus to solve any parametric polynomial system (Gao and Chou 1992c; Wang 2000a): by computing simple systems or triangular systems with projection, one knows for what values of the parameters $\boldsymbol{u}$ the system $PS = 0, QS \neq 0$ has solutions for the unknowns $\boldsymbol{x}$.

Theorem 2.44:

Let Ψ be a simple series of a polynomial system $\mathfrak{P}$ in $\boldsymbol{Q}[\boldsymbol{u}, \boldsymbol{x}]$, or a triangular series of $\mathfrak{P}$ computed with projection for $x_n, \ldots, x_1$ and assume that $\Psi \neq \varnothing$. Then

(*a*) *for any $[TS, US] \in \Psi$ and $\bar{\boldsymbol{u}} \in \tilde{\boldsymbol{Q}}^d$ (where $\tilde{\boldsymbol{Q}} \supset \boldsymbol{Q}$), the system*

$$(TS \setminus \boldsymbol{Q}[\boldsymbol{u}])|_{\boldsymbol{u}=\bar{\boldsymbol{u}}} = 0, \qquad (US \setminus \boldsymbol{Q}[\boldsymbol{u}])|_{\boldsymbol{u}=\bar{\boldsymbol{u}}} \neq 0$$

has solutions for $\boldsymbol{x}$ in $\boldsymbol{C}$ if and only if $\boldsymbol{u} = \bar{\boldsymbol{u}}$ is a solution of

$$TS \cap \boldsymbol{Q}[\boldsymbol{u}] = 0, \qquad US \cap \boldsymbol{Q}[\boldsymbol{u}] \neq 0;$$

(*b*)

$$\mathrm{Proj}_{\boldsymbol{u}}\mathrm{Zero}(\mathfrak{P}) = \bigcup_{\mathfrak{T} \in \Psi} \mathrm{Proj}_{\boldsymbol{u}}\mathrm{Zero}(\mathfrak{T}) = \bigcup_{[TS,US] \in \Psi} \mathrm{Zero}(TS \cap \boldsymbol{Q}[\boldsymbol{u}]/US \cap \boldsymbol{Q}[\boldsymbol{u}]).$$

For any given parametric values $\bar{\boldsymbol{u}}$, the solutions for $\boldsymbol{x}$ may be computed from or represented by the simple or triangular systems

$$[(TS \setminus \boldsymbol{Q}[\boldsymbol{u}])|_{\boldsymbol{u}=\bar{\boldsymbol{u}}}, (US \setminus \boldsymbol{Q}[\boldsymbol{u}])|_{\boldsymbol{u}=\bar{\boldsymbol{u}}}], \quad [TS, US] \in \Psi,$$

where Ψ is as in Theorem 2.44.

2.5.4 Automated theorem proving in geometry

Proving geometric theorems automatically has become an active area of research since the work of Wu (1978). The reader may refer to Wang (1996c), Wu (1994a) and Chap. 5 for more information and an introduction to the basic ideas, theory and methods.

The first step of proving geometric theorems using algebraic methods is to algebraize the geometric problems in question. This is done by choosing a coordinate system and denoting the coordinates of points as well as other involved geometric entities by indeterminates $\boldsymbol{x} = (x_1, \ldots, x_n)$. The hypothesis of a theorem $\mathbb{T}$ of *equality type* may be expressed then as a finite system of polynomial equations and inequations

$$H_1(\boldsymbol{x}) = 0, \ldots, H_s(\boldsymbol{x}) = 0, \qquad D_1(\boldsymbol{x}) \neq 0, \ldots, D_t(\boldsymbol{x}) \neq 0$$

(where each $D_i = 0$ usually corresponds to a predetermined degenerate case), and the conclusion expressed as a single polynomial equation $G(\boldsymbol{x}) = 0$. All the polynomials are in $\boldsymbol{x}$ with coefficients in the geometry-associated field $\boldsymbol{K}$. Let

$$\mathbb{PS} = \{H_1, \ldots, H_s\}, \qquad \mathbb{QS} = \{D_1, \ldots, D_t\}.$$

The problem of proving the theorem amounts to

(a) deciding whether $\mathrm{Zero}(\mathbb{PS}/\mathbb{QS}) = \varnothing$; and if not,

(b) deciding whether $\mathrm{Zero}(\mathbb{PS}/\mathbb{QS}) \subset \mathrm{Zero}(G)$; and if not,

(c) finding an appropriate set $\mathbb{DS}$ of polynomials so that $\mathrm{Zero}(\mathbb{PS}/\mathbb{QS} \cup \mathbb{DS}) \subset \mathrm{Zero}(G)$.

The additional inequations $D(\boldsymbol{x}) \neq 0$ for $D \in \mathbb{DS}$ are determined to ensure that the configuration of the geometric hypotheses is in a generic position.

The determination of (a) and (b) can be done by computing regular series, simple series, or irreducible triangular series with pseudo-division according to Corollaries 2.13 and 2.14 and Theorem 2.16. For (c), one may compute first a principal triangular system $[\mathbb{TS}, \mathbb{US}]$ of $[\mathbb{PS}, \mathbb{QS}]$, for instance, and then the pseudo-remainder $R = \mathrm{prem}(G, \mathbb{TS})$. The theorem is proved to be true under the so-called *subsidiary/non-degeneracy* conditions $\mathrm{ini}(\mathbb{TS}) \neq 0$ when $R \equiv 0$. This elementary method is very efficient for confirming many geometric theorems.

However, it is necessary to go further if $R \not\equiv 0$ or one wants to check whether the theorem is also true in the degenerate cases. Moreover, some ambiguities corresponding to the reducibility of geometric configurations may occur when geometric statements are transformed into polynomial equations and inequations (no inequalities with $>$ and $<$). So one may still meet difficulties even though non-degeneracy conditions have been taken into account. It may happen that the conclusion of the theorem holds true only for some components of the configuration variety $\mathcal{V}$ defined by the hypothesis-equations. Those components for which the theorem is false have to be excluded either as degenerate cases or as the unwanted cases that have been included due to the ambiguities indistinguishable in the algebraic formulation.

Although there are special techniques dealing with reducibility (see Wu 1987a and Wang and Gao 1987 for instance), a complete and systematic treatment of the problem is to determine (a) above and

(d) on which components of Zero(PS/QS) the polynomial G vanishes (and thus $\mathbb{T}$ is true)

by decomposing $\mathcal{V}$ into irreducible components. Subsidiary conditions may be provided by excluding those components of Zero(PS/QS) for which $\mathbb{T}$ is *generically* false. See Chou and Gao (1990b), Wang (1995b, 2000a), Wu (1984a, 1994a) and Chap. 5 for further details.

The algorithms of zero decomposition may also be used to automatically derive unknown algebraic relations among some geometric entities (*geometric formulas*), where an adequate description of the geometric hypotheses among the geometric entities is given (see Chou 1987; Chou and Gao 1990a; Wang 1995a; Wu 1986c,d; Chap. 5). The idea is to first formulate the geometric hypotheses as a system of polynomial equations and inequations, then compute a triangular set or series from the corresponding polynomial system using an appropriate variable ordering, and finally get the desired relations from the triangular sets.

2.5.5 Other applications

Implicitization of parametric objects. Geometric objects like curves and surfaces may be represented algebraically by implicit equations or parametric equations. In geometric modeling, one often needs to convert one representation into the other. The problem of implicitization amounts to finding the implicit equations which define the same geometric object as the parametrized representation does. This can be done by using the zero decomposition algorithm with projection (see Li 1989, Wang 1995a and Chap. 14).

Automatic derivation of locus equations. Similar to the automatic derivation of geometric formulas, the zero decomposition method with projection may be used to derive the locus equations of a motion whose geometric description is given (Chou and Gao 1990b; Wang 1995a).

Here by locus equations is meant a system or the disjunction of several systems of polynomial equations and inequations with parameters such that not only is the system a formal consequence of the geometric hypotheses, but also for any point on the locus there is at least one configuration which satisfies the geometric hypotheses. This is why projection is required (see Wang 1995a).

Existence conditions and detection of singularities. Sufficient and necessary conditions for parametric algebraic hypersurfaces to have singularities of an arbitrary multiplicity may be established by using zero decomposition with projection (see Wang 1995a). The structure of the singular varieties may be depicted by computing their irreducible decomposition, or all the singular points when they are finite. The latter is actually identical to the problem of solving polynomial systems.

The algorithms of zero decomposition may also be used to factorize polynomials over successive algebraic extension fields (see Hu and Wang 1986; Wang 1992d; Chap. 13) and to study the qualitative properties of certain differential systems such as establishing center conditions, distinguishing between center and focus, searching for higher order foci and constructing limit cycles (see Wang 1989c, 1991a, 1999a and Chap. 15). In fact, they are applicable to almost all the problems that Wu–Ritt's method of characteristic sets can deal with.

Note and acknowledgments. The final version of this chapter was prepared at LIP6, Université Paris VI. The author wishes to thank Philippe Aubry for his helpful comments on the early manuscript. A French version of the chapter is available as `résumé_étendu.ps.gz` on the web page

`http://www-leibniz.imag.fr/ATINF/Dongming.Wang/dhdr.html`.

Chapter 3

THE EIGENVALUE APPROACH TO POLYNOMIAL SYSTEM SOLVING

Shugong Zhang and Guochen Feng

It is well known that solving a univariate polynomial equation is equivalent to finding the eigenvalues of the related Frobenius matrix. We prove that for a system of multivariate polynomial equations, one can build a joint eigenvalue problem such that finding the solutions of the system of equations is equivalent to finding all eigenvalues and eigenvectors of a matrix pencil.

3.1 Introduction

In Chap. 1, Wu reviewed the history, methods, and applications of polynomial equation solving and gave a detailed introduction to the characteristic set method. It is known that numerical methods for solving algebraic systems are restricted in practice. For instance, the famous Newtonian method and homotopy method need good initial approximation and can find only isolated solutions. On the other hand, symbolic manipulation methods need large amounts of computing effort and hence are also restricted for practical use. Hence it becomes more and more important to develop a method that combines numerical and symbolic computing to meet the practical needs. The eigenvalue method is such a hybrid approach.

The thought of solving algebraic systems by means of an eigenvalue approach is due to H. J. Stetter. In 1988, he first constructed a square matrix eigenvalue problem based on the given algebraic system by means of the U-resultant of the system. Then he showed that the eigenvalues of the square matrix are zeros of the algebraic system under some restricted conditions (Auzinger and Stetter 1988) and thereafter did a lot of work in this direction. After that, Huang and Wu (1989) discussed this problem more carefully. In the early 1990s, Feng and others systematically studied various aspects of the eigenvalue approach to algebraic system solving. They first construct a joint eigenvalue

MATHEMATICS MECHANIZATION AND APPLICATIONS
ISBN 0-12-734760-7

problem from the given algebraic system and then prove that the solutions of the algebraic system are joint eigenvalues of the joint eigenvalue problem, and vice versa (Feng et al. 1993c). They also showed that the projection of the variety determined by the given system onto some axis is a finite set of points if and only if the corresponding matrix in the eigenvalue problem is square, and under certain conditions the solutions of the algebraic system can be read off from the eigenvectors. Thus, if the corresponding ideal is zero-dimensional, all the solutions of the given system can be obtained from the eigenvectors and the multiplicity of each solution can be computed from the Jordan form of the matrix (Feng and Zhang 1992). If the projection of some irreducible component of the variety is a finite set of points, it must consist of a single point and its value is a regular eigenvalue of the corresponding matrix pencil, and vice versa (Feng et al. 1993a,b). In this case, the variety contains manifold components and some manifold components can sometimes be found from the eigenvalue problem.

In this chapter, we will mainly review the results of our research groups at Jilin University and Beijing Municipal Computing Center. Those who are interested in this problem and want to learn more may find relevant materials from the indicated references.

3.2 Equivalence theorems and algorithm

3.2.1 Equivalence theorems

It is well known that the equation

$$x^m + a_1 x^{m-1} + \cdots + a_{m-1} x + a_m = 0 \tag{3.1}$$

is equivalent to the eigenvalue problem of the Frobenius matrix

$$\begin{pmatrix} -a_1 & -a_2 & \cdots & -a_{m-1} & -a_m \\ 1 & 0 & \cdots & 0 & 0 \\ & & \cdots & & \\ 0 & 0 & \cdots & 1 & 0 \end{pmatrix} \tag{3.2}$$

That is, x^* is a solution of (3.1) iff it is an eigenvalue of (3.2). For the multivariate case, we can also build up an eigenvalue problem based on the given system and show that similar results hold true. In what follows we describe how to do this.

Let $\boldsymbol{K}$ be an algebraically closed field of characteristic zero. We denote the ring of multivariate polynomials in $x_1, \ldots, x_n$ over the field $\boldsymbol{K}$ by $\boldsymbol{K}[\boldsymbol{x}]$. Given a finite set $FS = \{F_1, \ldots, F_m\} \subset \boldsymbol{K}[\boldsymbol{x}]$, set

$$F_\mu = \sum_j a_j^{(\mu)} x^j, \quad \mu = 1, \ldots, m,$$

where

$$j = (j_1, \ldots, j_n), \qquad x^j = x_1^{j_1} \cdots x_n^{j_n}.$$

We will arrange the power product x^i with respect to the total degree ordering. Denote the total degree of F_μ by $d_\mu = \deg(F_\mu)$ and $|i| = \sum_{\nu=1}^n i_\nu$ for $i = (i_1, \ldots, i_n)$. Without loss of generality, we assume that $d_1 \geq d_2 \geq \cdots \geq d_m$. Select some integer $d \geq d_1$ and introduce the auxiliary polynomials $F_{m+1}^{(\nu)} = x_\nu - \lambda_\nu$. Consider the system

$$\begin{cases} x^i F_\mu(\boldsymbol{x}) = \sum_j a_j^{(\mu)} x^i x^j = 0, \; |i| \leq d - d_\mu, \quad \mu = 1, \ldots, m, \\ x^i F_{m+1}^{(\nu)} = x^i (x_\nu - \lambda_\nu), \; |i| \leq d - 1, \quad \nu = 1, \ldots, n. \end{cases}$$

Let

$$\overline{Z}_0 = \overline{Z}_0(x_1, \ldots, x_n) = \{x^i \mid |i| < d, \; x^i \text{ arranged decreasingly}\},$$
$$\overline{Z}_1 = \overline{Z}_1(x_1, \ldots, x_n) = \{x^i \mid |i| = d, \; x^i \text{ arranged decreasingly}\}.$$

The above equations can be written in matrix form as

$$\begin{pmatrix} A_{00} & A_{01} \\ A_{10}^{(\nu)} & A_{11}^{(\nu)} - \lambda_\nu I \end{pmatrix} \begin{pmatrix} \overline{Z}_1 \\ \overline{Z}_0 \end{pmatrix} = 0 \mod \text{Ideal}(FS), \quad \nu = 1, \ldots, n,$$

in which the matrices A_{00} and A_{01} have the coefficients of F_ν and 0 as their entries, $A_{10}^{(\nu)}$ and $A_{11}^{(\nu)}$ have only 0 and 1 as their entries with only one non-zero entry in each row of the matrices, I is a unit matrix of the same dimension as that of $\overline{Z}_0$, and Ideal(FS) denotes the *ideal* generated by the polynomials in FS. Sometimes, we also write Ideal($F_1, \ldots, F_m$) for Ideal(FS).

By means of row elimination, the above equations can be equivalently simplified to the following form

$$(B_\nu \; A_\nu - \lambda_\nu I) \begin{pmatrix} Z_1 \\ Z_0 \end{pmatrix} = 0 \mod \text{Ideal}(FS), \quad \nu = 1, \ldots, n, \tag{3.3}$$

where Z_1, Z_0 are term-entry vectors similar to $\overline{Z}_1, \overline{Z}_0$. It is easy to see that if $(x_1^*, \ldots, x_n^*)$ is a zero of FS, then

$$(B_\nu \; A_\nu - \lambda_\nu^* I) \begin{pmatrix} U_1^* \\ U_0^* \end{pmatrix} = 0, \quad \nu = 1, \ldots, n, \tag{3.4}$$

where $\lambda_\nu^* = x_\nu^*$, $\nu = 1, \ldots, n$, $U_1^* = Z_1(x_1^*, \ldots, x_n^*)$ and $U_0^* = Z_0(x_1^*, \ldots, x_n^*)$.

On the other hand, one can prove that if there exist $\lambda_1^*, \ldots, \lambda_n^*$ and $U^* = (U_1^* \; U_0^*)^T$ satisfying (3.4) then $(\lambda_1^*, \ldots, \lambda_n^*)$ is a zero of *FS*. Equation (3.4) presents a *joint eigenvalue problem*. $\lambda_1^*, \ldots, \lambda_n^*$ are called *joint eigenvalues* and $U^* = (U_1^* \; U_0^*)^T$ the *joint eigenvector* of (3.4). For a fixed ν, we call the single equation a *matrix pencil eigenvalue problem*. If $B_\nu = 0$ for all ν, we get a family of square matrix eigenvalue problems.

The following is the main theorem we have proved.

Theorem 3.1: (Feng et al. 1993c)

$\xi_1, \ldots, \xi_n \in \boldsymbol{K}$ are joint eigenvalues of (3.4) if and only if $(\xi_1, \ldots, \xi_n)$ is a zero of FS.

According to this theorem, to solve an algebraic system, one can construct an eigenvalue problem in the above manner and then solve the resulting joint eigenvalue problem. But it is not a pleasant task to solve such a joint eigenvalue problem constituted of n matrix pencil eigenvalue problems. Instead we want to solve a single one, which is, in fact, possible.

Notice that in (3.3) the entries of (Z_1, Z_0) are terms. Hence each term entry x^i of (Z_1, Z_0) in (3.3) corresponds to an entry u_i of (U_1, U_0) in (3.4). Relating one matrix pencil eigenvalue problem, we have the following.

Theorem 3.2: (Feng et al. 1993c)

Suppose that ξ_{ν_0} is an eigenvalue of (3.4) and $U(\xi_{\nu_0})$ is a joint eigenvector with 1 as the last entry. If there exists a solution to the system of equations

$$x^i = u_i(\xi_{\nu_0}), \quad |i| \le d, \tag{3.5}$$

where $u_i(\xi_{\nu_0})$ is the entry of $U(\xi_{\nu_0})$ corresponding to x^i, then this solution must be a zero of FS.

Such a system is easy to solve. Firstly, one can get

$$x_1 = u_{(1,0,\ldots,0)}, \; x_2 = u_{(0,1,\ldots,0)}, \; \ldots, \; x_n = u_{(0,0,\ldots,1)}.$$

Then one can get the value of x^i by products of x_ν, $\nu = 1, \ldots, n$. Finally one needs only to check if $x^i = u_i(\xi_{\nu_0})$.

Theorem 3.2 shows that once the equation (3.5) is solved, one can find the solution of *FS* from the eigenvector (U_1, U_0).

Example 3.1:

Let

$$F_1 = x_1^3 + x_1x_2^2 - x_1^2 - x_2^2 - x_1 + 1 = 0,$$
$$F_2 = x_1^2x_2 + x_2^3 + x_1^2 + x_2^2 - x_2 - 1 = 0.$$

Take the total degree ordering with $x_1 \succ x_2$. Consider the system

$$F_1 = 0,$$
$$F_2 = 0,$$
$$x^i(x_1 - \lambda) = 0, \quad |i| < 3.$$

Writing the above equations in matrix form and applying Gaussian elimination, we get a new system. Taking an equivalent subsystem, we have

$$(B_1 \; A_1 - \lambda I)Z = 0 \quad \text{mod Ideal}(F_1, F_2),$$

where

$$B_1 = \begin{pmatrix} -1 & 0 \\ 0 & -1 \\ 0 & 0 \\ 0 & 0 \\ 0 & 0 \\ 0 & 0 \end{pmatrix}, \qquad A_1 = \begin{pmatrix} 1 & 0 & 1 & 1 & 0 & -1 \\ -1 & 0 & -1 & 0 & 1 & 1 \\ 0 & 0 & 0 & 0 & 0 & 0 \\ 1 & 0 & 0 & 0 & 0 & 0 \\ 0 & 1 & 0 & 0 & 0 & 0 \\ 0 & 0 & 0 & 1 & 0 & 0 \end{pmatrix},$$

$$Z = (Z_1 \; Z_0)^T, \qquad Z_1 = (x_1x_2^2 \; x_2^3), \qquad Z_0 = (x_1^2 \; x_1x_2 \; x_2^2 \; x_1 \; x_2 \; 1).$$

It is easy to check that $\lambda = 1$ and $U = (1 \; -1 \; 1 \; -1 \; 1 \; 1 \; -1 \; 1)^T$ fit for the matrix pencil eigenvalue problem $(B_1 \; A_1 - \lambda I)U = 0$. From the correspondence between the entries of Z and U, we can verify Theorem 3.2. Then a zero $(1, -1)$ of the given system can be read off from U.

Example 3.2:

Now we consider

$$\begin{cases} F_1 = x_1^2 + 4x_1 - 4x_2 + 3x_3 + 21 = 0, \\ F_2 = x_1x_2 - 3x_1 + 5x_2 - 15 = 0, \\ F_3 = x_1x_3 + 2x_1 - 2x_3 - 4 = 0. \end{cases}$$

Let $d = 2$. Similar to the above example, we get a joint eigenvalue problem. Taking the first matrix pencil eigenvalue problem, we have

$$\left[\begin{pmatrix} -4 & 4 & -3 & -21 \\ 3 & -5 & 0 & 15 \\ -2 & 0 & 2 & 4 \\ 1 & 0 & 0 & 0 \end{pmatrix} - \lambda_1 I\right]\begin{pmatrix} x_1 \\ x_2 \\ x_3 \\ 1 \end{pmatrix} = 0.$$

The four eigenvalues of the square matrix are

$$\lambda_1^{(1)} = -5, \qquad \lambda_1^{(2)} = -3, \qquad \lambda_1^{(3)} = -1, \qquad \lambda_1^{(4)} = 2.$$

From the associated eigenvectors we can read off all of the solutions of the given system as

$$\begin{pmatrix} x_1 \\ x_2 \\ x_3 \end{pmatrix} = \begin{pmatrix} -5 \\ 5 \\ -2 \end{pmatrix}, \begin{pmatrix} -3 \\ 3 \\ -2 \end{pmatrix}, \begin{pmatrix} -1 \\ 3 \\ -2 \end{pmatrix}, \begin{pmatrix} 2 \\ 3 \\ -7 \end{pmatrix}.$$

3.2.2 Algorithm

To describe the algorithm, we first introduce some notation. For any polynomial F, we denote the *leading power* product of F by $\mathrm{lp}(F)$. Let FS be a given polynomial set. We define some polynomial sets as follows

$$\begin{aligned} &\mathcal{T}(FS) = \{\mathrm{lp}(F) \mid F \in FS\}, \\ &\mathcal{K}(FS) = \{F \in FS \mid \nexists G \in FS \setminus \{F\} \text{ such that } \mathrm{lp}(G) \mid \mathrm{lp}(F)\}, \\ &\mathrm{lcm}(\mathcal{K}(FS)) = \mathrm{lcm}\{\mathrm{lp}(F) \mid F \in \mathcal{K}(FS)\}. \end{aligned}$$

Again we define two numbers

$$\mathcal{N}(FS) = \#\left\{F \in \mathcal{T}(FS) \,\middle|\, \begin{array}{l} \exists \nu \text{ such that } \deg(F, x_\nu) > 0 \\ \text{while } \deg(F, x_\mu) = 0 \text{ for } \mu \neq \nu \end{array} \right\}$$

and

$$t(PS) = \begin{cases} \deg(\mathrm{lcm}(\mathcal{K}(FS))) & \text{if } \mathcal{N}(FS) < 2, \\ \deg(\mathrm{lcm}(\mathcal{K}(FS))) - \mathcal{N}(FS) + 1 & \text{otherwise.} \end{cases}$$

With the given FS and the determined $t(FS)$, construct a new polynomial set $W(FS, d, t(FS))$ which has the same zero set as that of FS and possesses the property: if there exist $F, G \in W(FS, d, t)$ and i, j such that $\deg(x^i F), \deg(x^j G) \leq t$ then $x^i F - x^j G$ can be linearly represented by polynomials in $W(FS, d, t)$, in which d is a controlling parameter. Such a polynomial set $W(FS, d, t)$ is in fact a total degree ordering version of Wu's well-arranged basis of FS (see Wu 1991a).

In the following algorithm, there are two possible outputs. No matter whether a square matrix eigenvalue problem or a matrix pencil eigenvalue problem is obtained, it can be solved by using one of the rich numerical packages. Then we can either get all of the solutions of the given system or divide the given system into some subsystems.

Figure 3.1 is a sketch of the algorithmic procedure.

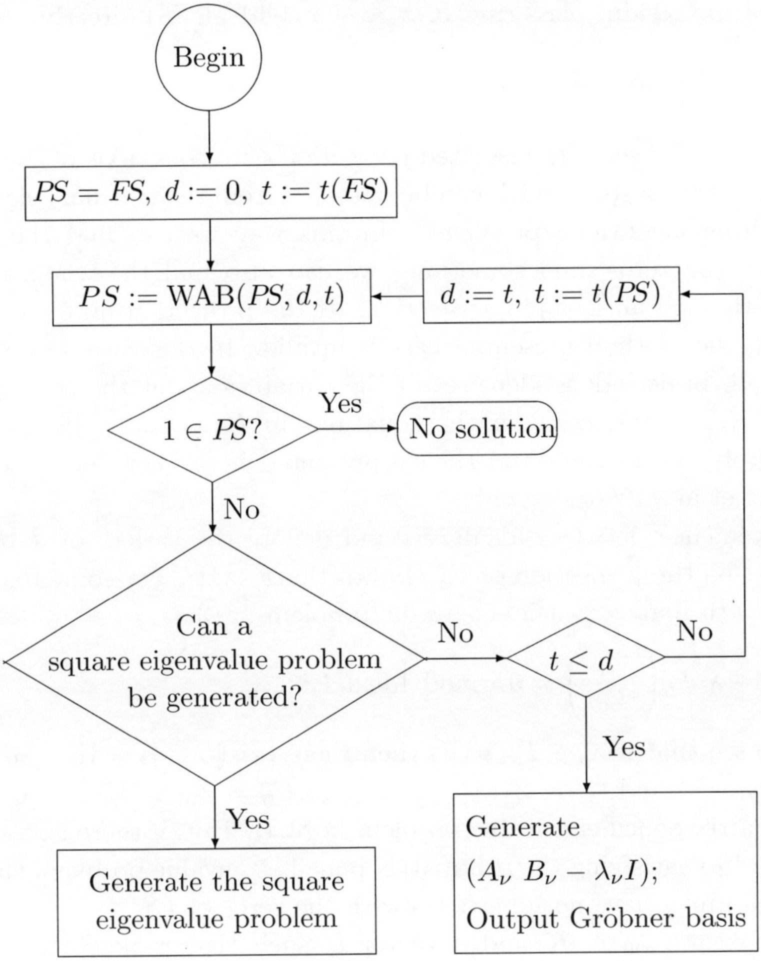

Figure 3.1

Remark 3.1:

- WAB(PS, d, t) is a sub-algorithm which produces a well-arranged basis of Ideal(FS). It is practically an elimination.
- The algorithm gives either an ordinary eigenvalue problem or a matrix pencil eigenvalue problem in a finite number of steps. In the latter case, a reduced Gröbner basis of Ideal(FS) with respect to the total degree ordering is given (Feng et al. 1995a).
- If the Gröbner basis of Ideal(FS) with respect to the total degree ordering

is given, we can construct the eigenvalue problems from the multiplication table of the residue class ring $\boldsymbol{K}[x_1,\ldots,x_n]/\text{Ideal}(FS)$ directly.

3.3 The projection of zeros

In this section, we discuss the isolated projection onto some axis of the zeros of the given algebraic system which can be obtained from the (regular) eigenvalues of the resulting eigenvalue problems. In this way we can find the isolated projection of the zeros, and sometimes we can also find the whole manifold solutions with such an isolated projection. If the projection onto each axis is not isolated, the method presented here is invalid. In this case, one may find a maximal algebraic independent set of the variables about the corresponding ideal, say $\{x_1,\ldots,x_\sigma\}$, consider the equations in $\boldsymbol{K}(x_1,\ldots,x_\sigma)[x_{\sigma+1},\ldots,x_n]$ and then apply the method and theory presented here. For details, one may refer to Feng et al. (1996).

For convenience, let $J = \text{Ideal}(FS)$ and denote the variety of J by $V(J)$. Let $P_{x_\nu}V(J)$ be the projection of $V(J)$ onto the x_ν-axis. For some fixed ν, we consider the νth matrix pencil eigenvalue problem in (3.3)

$$(B_\nu\ A_\nu - \lambda_\nu I)\begin{pmatrix} Z_1 \\ Z_0 \end{pmatrix} = 0 \mod \text{Ideal}(FS).$$

It is easy to see that if $\lambda_\nu^* \in P_{x_\nu}V(J)$ then there exist λ_μ^*, $\mu = 1,\ldots,n$, $\mu \neq \nu$, such that $\lambda_\nu = \lambda_\nu^*$ and $U_1^* = Z_1(\lambda_1^*,\lambda_2^*,\ldots,\lambda_n^*), U_0^* = Z_0(\lambda_1^*,\lambda_2^*,\ldots,\lambda_n^*)$ satisfy the single matrix pencil eigenvalue problem in (3.4). But, if there exist a λ and a vector (U_1,U_0) satisfying the νth matrix pencil eigenvalue problem, then such λ and (U_1,U_0) may have nothing to do with the zeros of FS.

If there exist a $\lambda_0 \in \boldsymbol{K}$ and a vector U such that $\text{rank}(B\ A - \lambda_0 I) < \max_{\lambda\in\boldsymbol{K}} \text{rank}(B\ A - \lambda I)$ and $(B\ A - \lambda_0 I)U = 0$, then λ_0 is called a *regular eigenvalue* and U a *right eigenvector* of $(B\ A - \lambda I)$.

We will show that the regular eigenvalues of the above matrix pencil eigenvalue problem and the projection $P_{x_\nu}V(J)$ are closely related. The following result can be proved.

Theorem 3.3: (Feng et al. 1993a,b)

If V_0 is an irreducible component of $V(J)$ and $P_{x_\nu}V_0$ is a finite set of points, then $P_{x_\nu}V_0$ must contain only one point, i.e., there is a $\lambda \in \boldsymbol{K}$ such that $P_{x_\nu}V_0 = \{\lambda\}$ and λ must be a regular eigenvalue of the νth matrix pencil in (3.3). Conversely, if d is large enough and λ is a regular eigenvalue of the νth matrix pencil in (3.3), then λ must be the νth component of some point in $V(J)$.

This theorem shows that for a given irreducible component V_0 of $V(J)$, if it is an isolated point or its projection onto some axis, say the x_ν-axis, is isolated, then we can find its νth component through the regular eigenvalues of the νth matrix pencil eigenvalue problem provided that its scale is large enough. But, if the projection of V_0 is not isolated, we cannot find such a projection in this manner.

In the special case where the projection onto the x_ν-axis of the whole $V(J)$ is a finite set of points, all the νth components of zeros of FS can certainly be obtained in this way. In this case these components can also be found by solving a square matrix eigenvalue problem.

Theorem 3.4: (Feng and Zhang 1992)

$P_{x_\nu}V(J)$ is a finite set of points iff there exist finitely many linearly independent $H_1 = 1, H_2, \ldots, H_q$ in the residue class ring $\boldsymbol{K}[x_1, \ldots, x_n]/J$ and a square matrix A_ν such that

$$x_\nu(H_q, , \ldots, H_2, H_1)^T = A_\nu(H_q, , \ldots, H_2, H_1)^T \mod J.$$

In this case $P_{x_\nu}V(J) = \sigma(A_\nu)$, where $\sigma(A_\nu)$ is the spectrum of A_ν.

Corollary 3.5: (Feng et al. 1993c)

If J is zero-dimensional, then determining $V(J)$ can be realized by solving a family of ordinary eigenvalue problems.

In Theorem 3.4, suppose that $\lambda_\nu^{(1)}, \ldots, \lambda_\nu^{(\gamma)}$ are eigenvalues of A_ν. Then:

- System FS can be divided into subsystems $(F_1, \ldots, F_m,\ x_\nu - \lambda_\nu^{(\delta)})$, $\delta = 1, \ldots, \gamma$.

- If λ_ν^* is an eigenvalue of A_ν, then there exist λ_μ^*, $\mu \neq \nu$, $\nu = 1, \ldots, n$, such that $(\lambda_1^*, \ldots, \lambda_n^*)$ is a zero of J and $Z(\lambda_1^*, \ldots, \lambda_n^*)$ is the associated eigenvector. In this case, the zero set of $(F_1, \ldots, F_m, x_\nu - \lambda_\nu)$ is a single point if $\dim\ker(A_\nu - \lambda_\nu^* I) = 1$. All the components can be obtained from the associated eigenvector.

- If λ_ν is an eigenvalue of A_ν, from $x_\nu Z = A_\nu Z \bmod J$ we can get $(x_\nu - \lambda_\nu)Z = (A_\nu - \lambda_\nu I)Z \bmod J$. Therefore

$$(A_\nu - \lambda_\nu I)Z = 0 \mod (J + \mathrm{Ideal}(x_\nu - \lambda_\nu)).$$

Sometimes one may find zero manifolds of FS from the above equation.

Example 3.3:

Consider the polynomial system

$$\begin{aligned}
F_1 &= 2x_1^2 + x_1x_2 + 3x_1x_3 - x_2^2 + 3x_2x_3 + 3x_2 + 2x_1^2 - 8x_3 = 0,\\
F_2 &= 3x_1^2 + 3x_1x_2 + x_1x_3 + x_1 + x_2x_3 + 4x_2 + x_3^2 - 5x_3 = 0,\\
F_3 &= x_1^2 + x_1x_3 - x_2^2 + 4x_2 - 3 = 0.
\end{aligned}$$

For variable x_3, we can generate a square matrix eigenvalue problem

$$x_3Z = \begin{pmatrix}
\frac{-917}{35} & \frac{260}{35} & \frac{-621}{35} & \frac{1353}{35} & \frac{596}{35} & \frac{-1611}{35}\\
\frac{1071}{35} & \frac{-40}{35} & \frac{648}{35} & \frac{-1494}{35} & \frac{-1013}{35} & \frac{1818}{35}\\
\frac{-8}{5} & -1 & \frac{1}{5} & \frac{7}{5} & \frac{19}{5} & \frac{-9}{5}\\
1 & 0 & 0 & 0 & 0 & 0\\
0 & 1 & 0 & 0 & 0 & 0\\
0 & 0 & 0 & 0 & 1 & 0
\end{pmatrix} Z \mod J,$$

in which $J = \text{Ideal}(F_1, F_2, F_3)$ and $Z = (x_2x_3\ x_3^2\ x_1\ x_2\ x_3\ 1)^T$. We can deduce that the projection onto x_3-axis of $V(J)$ is a finite set of points. By computation, 2 is an eigenvalue of the square matrix and hence a projection value. From $(A_3 - 2I)Z = 0$, we can get a zero manifold of $V(J)$

$$x_3 - 2 = 0, \qquad x_1 + x_2 - 1 = 0.$$

Example 3.4:

We still consider Example 3.2. The four eigenvalues of the square matrix are

$$\lambda_1^{(1)} = -5, \qquad \lambda_1^{(2)} = -3, \qquad \lambda_1^{(3)} = -1, \qquad \lambda_1^{(4)} = 2.$$

It is easy to check that

$$\dim\ker(A_1 - \lambda_1^{(\mu)} I) = 1, \quad \mu = 1, \ldots, 4.$$

Thus, from the associated eigenvectors we can read off all the solutions of the given system.

3.4 Further results

In the univariate case, the multiplicity of a zero of the polynomial in (3.1) is equal to the order of the corresponding (unique) Jordan block in the Frobenius matrix (3.2). For the multivariate case, especially in the case of $\dim(J) = 0$, similar results hold true. In this section, we discuss the relation of algebraic invariants between the matrix and/or matrix pencil in the eigenvalue problem and the ideal generated by the given polynomial set.

3.4.1 Zero-dimensional case

In the case of $\dim(J) = 0$, what we are interested in is the condition under which (3.6) holds true since in this case we can find all components of the zeros of FS from the eigenvectors.

Proposition 3.6: (Feng et al. 1995b)

Suppose that $\lambda_\nu^{(\mu_0)}$ *is an eigenvalue of* A_ν *satisfying*

$$\dim\ker(A_\nu - \lambda_\nu^{(\mu_0)} I) = 1. \tag{3.6}$$

Then $\mathrm{Ideal}(J, x_\nu - \lambda_\nu^{(\mu_0)})$ *is maximal and each component of the points in* $V(J, x_\nu - \lambda_\nu^{(\mu_0)})$ *can be obtained from the eigenvector associated with* $\lambda_\nu^{(\mu_0)}$.

Definition 3.1:

If $\boldsymbol{K}[x_\nu] \cap J \neq \{0\}$ *and* $\Psi(x_\nu) \in J$ *has the lowest degree in* $\boldsymbol{K}[x_\nu] \cap J$, *then* Ψ *is called the* minimal polynomial *of* x_ν *with respect to* J.

It is easy to see that the minimal polynomial of x_ν is unique up to a constant scale and if J is zero-dimensional then there exists a minimal polynomial for each x_ν, $\nu = 1, \ldots, n$.

Theorem 3.7: (Feng et al. 1995b)

Suppose that J *is zero-dimensional and* $\Psi(x_\nu) = \prod_{\mu=1}^{q}(x_\nu - \lambda_\nu^{(\mu)})^{d_\mu}$ *is the minimal polynomial of* x_ν, *in which the* $\lambda_\nu^{(\mu)}$ *are pairwise different and* $d_\mu \geq 1$. *If*

$$\begin{aligned} J + \mathrm{Ideal}(x_\nu - \lambda_\nu^{(\mu_0)})^{d_{\mu_0}}) &= \sum_{\mu\neq\nu,\, \mu=1}^{n} \mathrm{Ideal}(x_\mu - \beta_\mu) \\ &\quad + \mathrm{Ideal}((x_\nu - \lambda_\nu^{(\mu_0)})^{d_{\mu_0}}), \end{aligned}$$

then $\dim\ker(A_\nu - \lambda_\nu^{(\mu_0)} I) = 1$.

Example 3.5:

Consider the equations in Example 3.3. Constructing the eigenvalue problem, we can get a matrix A_1 of order 4. The minimal polynomial of x_1 is

$$\Psi(x_1) = (x_1 + 5)(x_1 + 3)(x_1 + 1)(x_1 - 2)$$

and

$$\begin{aligned}
I_1 &= \mathrm{Ideal}(J, x_1 + 5) = \mathrm{Ideal}(x_1 + 5, x_2 - 5, x_3 + 2), \\
I_2 &= \mathrm{Ideal}(J, x_1 + 3) = \mathrm{Ideal}(x_1 + 3, x_2 - 3, x_3 + 2), \\
I_3 &= \mathrm{Ideal}(J, x_1 + 1) = \mathrm{Ideal}(x_1 + 1, x_2 - 3, x_3 + 2), \\
I_4 &= \mathrm{Ideal}(J, x_1 - 2) = \mathrm{Ideal}(x_1 - 2, x_2 - 3, x_3 - 7).
\end{aligned}$$

Thus,

$$\dim \ker(A_1 - \lambda_1^{(\mu)} I) = 1 \quad \text{for} \quad \mu = 1, \ldots, 4$$

and all the components of zeros of the given equations can be gotten from the corresponding eigenvectors.

Example 3.6:

Let $J = \mathrm{Ideal}(F_1, F_2, F_3)$, in which

$$F_1 = 2x_1^2 - x_2, \qquad F_2 = x_1 x_2 - x_2, \qquad F_3 = x_2^2 - 2x_2.$$

In the corresponding eigenvalue problem, the matrix associated with x_1 is

$$A_1 = \begin{pmatrix} 1 & 0 & 0 \\ \frac{1}{2} & 0 & 0 \\ 0 & 1 & 0 \end{pmatrix}.$$

The minimal polynomial of x_1 is $\Psi(x_1) = x_1^2(x_1 - 1)$. Since

$$I_1 = \mathrm{Ideal}(J, x_1^2) = \mathrm{Ideal}(x_1^2, x_2),$$

by Theorem 3.7 we have

$$\dim \ker(A_1 - 0I) = 1.$$

In fact, $\ker(A_1 - 0I) = \mathrm{Span}\{v\}$, where $v = (0\ 0\ 1)^T$.

Suppose that $J \subset \boldsymbol{K}[\boldsymbol{x}]$ is zero-dimensional, the minimal polynomial of x_ν is

$$\Psi(x_\nu) = \prod_{\mu=1}^{\gamma} (x_\nu - \lambda_\nu^{(\mu)})^{d_\mu}$$

and A_ν is the νth matrix. Then in the Jordan form of A_ν, the greatest order of the Jordan block with $\lambda_\nu^{(\mu)}$ as an eigenvalue equals the multiplicity of the factor $x_\nu - \lambda_\nu^{(\mu)}$ in $\Psi(x_\nu)$.

Theorem 3.8: (Feng et al. 1995b)

Suppose that $\Psi(x_\nu) = \prod_{\mu=1}^{\gamma}(x_\nu - \lambda_\mu^{(\nu)})^{d_\mu}$ *is the minimal polynomial of* x_ν, *and* l_{μ_0} *is the greatest order of the Jordan block with* $\lambda_\nu^{(\mu_0)}$ *as an eigenvalue in the Jordan form of* A_ν. *Then* $l_{\mu_0} = d_{\mu_0}$.

Suppose that $\boldsymbol{K}[x_1, \ldots, x_n]/J$ is a $\boldsymbol{K}$-vector space of dimension p. Then for each ν, $\nu = 1, \ldots, n$, we have an eigenvalue problem. Möller and Stetter (1995) discussed the problem of solving polynomial systems with multiple zeros. Here we discuss the relation between the structure of the Jordan form of A_ν and the multiplicity of zeros in $V(J)$.

Definition 3.2:

Let $\boldsymbol{K}[\boldsymbol{x}]^*$ *be the vector space composed of all the* $\boldsymbol{K}$*-functionals over* $\boldsymbol{K}[\boldsymbol{x}]$ *and* $V \subset \boldsymbol{K}[\boldsymbol{x}]$ *a linear vector subspace. If* $\{L_1, \ldots, L_t\}$ *is a basis of the vector space*

$$V^\perp := \{L \in \boldsymbol{K}[\boldsymbol{x}]^* \mid L(F) = 0, F \in V\},$$

then $\{L_1, \ldots, L_t\}$ *is called a dual-basis of* V.

Lemma 3.9: (Marinari et al. 1991)

If J *is zero-dimensional, then*

$$\dim J^\perp = \dim(\boldsymbol{K}[\boldsymbol{x}]/J).$$

If $\{L_1, \ldots, L_t\}$ *is a* dual-basis *of* J, *then*

$$F = G \mod J \Longleftrightarrow L_\mu(F) = L_\mu(G),\ \mu = 1, \ldots, t.$$

For univariate polynomial $F(x) = (x-\lambda_1)^{d_1} \cdots (x-\lambda_m)^{d_m}$, where $\lambda_1, \ldots, \lambda_m$ are pairwise different, the functionals

$$L_l^{(\mu)} :\longrightarrow \frac{F^{(l-1)}(\lambda_\mu)}{(l-1)!}, \quad \mu = 1, \ldots, m,\ l = 1, \ldots, d_m,$$

constitute a dual-basis of Ideal(F). They describe the multiplicity of the zeros. We generalize this fact to the multivariate case. Denote the term set by T.

Definition 3.3:

For a term $t = x_1^{i_1} \cdots x_n^{i_n}$, *we define the differentials*

$$\begin{aligned} \partial_t : \boldsymbol{K}[\boldsymbol{x}] &\longrightarrow \boldsymbol{K}[\boldsymbol{x}] \\ F &\mapsto \frac{1}{i_1! \cdots i_n!} \frac{\partial^{i_1+\cdots+i_n}}{\partial x_1^{i_1} \cdots \partial x_n^{i_n}} F. \end{aligned}$$

The shift σ_u *(where* u *is a term) is defined for all* $D = \sum c_t \partial_t$ *by* $\sigma_u(D) = \sum c_t \partial_{\frac{t}{u}}$ *(where* $\frac{t}{u}$ *is a term if* u *divides* t*; otherwise* $\partial_{\frac{t}{u}} = 0$*).*

A linear vector subspace W *of* Span$\{\partial_t \mid t \in T\}$ *is called a* closed vector space *(of differential conditions), if* $W \neq \varnothing$*, its dimension is finite, and*

$$D = \sum c_t \partial_t \in W \Longrightarrow \sigma_{x_\nu}(D) \in W, \ \nu = 1, \ldots, n.$$

Lemma 3.10: (Möller and Stetter 1995)

If J *is zero-dimensional, that is,* $V(J) = \{\xi_1, \ldots, \xi_r\}$ $(\subset \boldsymbol{K}^n)$*, then there exist closed vector spaces* $\{D_1^{(\nu)}, \ldots, D_{l_\nu}^{(\nu)}\}$*,* $\nu = 1, \ldots, r$*, of differential condition with* $W_1, \ldots, W_r$ *as bases such that*

$$F \in J \Longleftrightarrow D_\mu^{(\nu)}(F)(\xi_\nu) = 0, \quad \mu = 1, \ldots, l_\nu, \ \nu = 1, \ldots, r.$$

The $\sum_{\nu=1}^r l_\nu$ *functionals*

$$L_\mu^{(\nu)} : \ F \mapsto D_\mu^{(\nu)} F(\xi_\nu), \qquad \boldsymbol{K}[\boldsymbol{x}] \longrightarrow \boldsymbol{K} \tag{3.7}$$

are linearly independent. In other words, the $\sum_{\nu=1}^r l_\nu$ *functionals constitute a dual-basis of* J*.*

Definition 3.4:

Let J *be zero-dimensional. The multiplicity of a zero* ξ $(\in \boldsymbol{K}^n)$ *of* J *is defined as*

$$\text{mult}(\xi, J) = \dim_{\boldsymbol{K}} \{D \mid \ D \in \text{Span}_{\boldsymbol{K}}\{\partial_t \mid \ t \in T\}, \ D(F)(\xi) = 0, \ \forall F \in J\}.$$

By Lemma 3.10, we can see that the multiplicity of $\xi_\nu \in V(J)$ equals l_ν, the number of linearly independent functionals in (3.7) with respect to ξ_ν.

Theorem 3.11:

Let $\Psi(x_\nu) = \prod_{\mu=1}^{\gamma}(x_\nu - \lambda_\mu^{(\nu)})^{d_\mu}$ *be the minimal polynomial of* x_ν*,*

$$I_\mu = \text{Ideal}(J, (x_\nu - \lambda_\mu^{(\nu)})^{d_\mu}),$$

and A_ν *the matrix of* (3.4)*. If* $V(I_\mu) = \{\xi_1, \ldots, \xi_{s_\mu}\} \subset \boldsymbol{K}^n$*,* $\mu = 1, \ldots, \gamma$*, then*

$$\begin{aligned} \textstyle\sum_{\delta=1}^{s_\mu} \text{mult}(\xi_\delta, J) &= \dim_{\boldsymbol{K}} \{\boldsymbol{K}[\boldsymbol{x}]/I_\mu\} \\ &= \dim \ker (A_\nu - \lambda_\mu^{(\nu)} I)^{d_\mu}. \end{aligned}$$

Definition 3.5:

A ideal $J \subset \boldsymbol{K}[\boldsymbol{x}]$ *is said to be at a* regular position *with respect to* x_ν *if the* x_ν*-components of zeros in* $\boldsymbol{K}^n$ *of* J *are pairwise different.*

Corollary 3.12:

If J is zero-dimensional and at a regular position with respect to x_ν, then for an arbitrary $\xi \in V(J)$,

$$\mathrm{mult}(\xi, J) = \dim\ker\left(A_\nu - \lambda_\mu^{(\nu)} I\right)^{d_\mu},$$

where $\lambda_\mu^{(\nu)}$ is the νth component of ξ.

Example 3.7:

Consider the univariate case and let

$$\Psi(x) = \prod_{\mu=1}^{\gamma} (x - \lambda_\mu)^{d_\mu} = \sum_{\delta=0}^{m} c_\delta x^\delta,$$

where $c_m = 1$ and

$$J = \mathrm{Ideal}(\Psi(x)) \subset \boldsymbol{K}[x].$$

Then, $\boldsymbol{K}[x]/J = \mathrm{Span}\{1, x, \ldots, x^{m-1}\}$ and

$$x \begin{pmatrix} x^{m-1} \\ x^{m-2} \\ \vdots \\ x \\ 1 \end{pmatrix} = \begin{pmatrix} -c_{m-1} & -c_{m-2} & \cdots & -c_1 & -c_0 \\ 1 & 0 & \cdots & 0 & 0 \\ \cdots & \cdots & \cdots & \cdots & \cdots \\ 0 & 0 & \cdots & 0 & 0 \\ 0 & 0 & \cdots & 1 & 0 \end{pmatrix} \begin{pmatrix} x^{m-1} \\ x^{m-2} \\ \vdots \\ x \\ 1 \end{pmatrix} \mod J$$

or simply

$$xZ = AZ \mod J.$$

For zero λ_μ, we have

$$\begin{aligned} \mathrm{mult}(\lambda_\mu, J) &= \dim_{\boldsymbol{K}} \{\boldsymbol{K}[x]/(x - \lambda_\mu)^{d_\mu}\} \\ &= d_\mu \\ &= \dim\ker(A - \lambda_\mu I)^{d_\mu}. \end{aligned}$$

By Theorem 3.8, the order of the greatest Jordan block with λ_μ as an eigenvalue is d_μ. Therefore, in the Jordan form of A there is only one Jordan block associated with each λ_μ. In fact, the closed vector space of differential condition with respect to λ_μ is $\mathrm{Span}\{\frac{d^{d_\mu - \nu}}{dx^{d_\mu - \nu}} \mid \nu = 1, \ldots, d_\mu\}$.

Example 3.8:

Consider the system of equations

$$\begin{cases} F_1 = x_1^2 + 2x_2^2 - 2x_2 = 0, \\ F_2 = x_1 x_2^2 - x_1 x_2 = 0, \\ F_3 = x_2^3 - 2x_2^2 + x_2 = 0. \end{cases}$$

A basis of the residue class ring is $\{1, x_2, x_1, x_2^2, x_1 x_2\}$. The matrix associated with x_2 is

$$A_2 = \begin{pmatrix} 1 & 0 & 0 & 0 & 0 \\ 0 & 2 & 0 & -1 & 0 \\ 1 & 0 & 0 & 0 & 0 \\ 0 & 1 & 0 & 0 & 0 \\ 0 & 0 & 0 & 1 & 0 \end{pmatrix}.$$

The Jordan form of A_2 is

$$J_{\text{ordan}} = \begin{pmatrix} 0 & & & & \\ & 0 & & & \\ & & 1 & & \\ & & & 1 & 1 \\ & & & & 1 \end{pmatrix}.$$

The minimal polynomial of x_2 is $\Psi(x_2) = x_2(x_2-1)^2 = F_3$. Two components of J are

$$I_1 = \mathrm{Ideal}(J, x_2) = \mathrm{Ideal}(x_1^2, x_2),$$
$$I_2 = \mathrm{Ideal}(J, (x_2-1)^2) = \mathrm{Ideal}(x_1^2 + 2x_2 - 2, x_1 x_2 - x_1, x_2^2 - 2x_2 + 1).$$

The zero set of the given system is $V(J) = \{(0,0), (0,1)\}$. From the Jordan form of A_2, we see that

$$\mathrm{mult}((0,0), J) = 2, \qquad \mathrm{mult}((0,1), J) = 3.$$

The closed subspaces of differential condition are $W_1 = \mathrm{Span}\{\partial_1, \partial_{x_1}\}$ at point $(0,0)$ and $W_2 = \mathrm{Span}\{\partial_1, \partial_{x_1}, \partial_{x_1^2} - \frac{1}{2}\partial_{x_2}\}$ at point $(0,1)$.

3.4.2 Positive-dimensional cases

In this part we assume that the ideal J is positive-dimensional and the basis of the residue class ring $\boldsymbol{K}[x_1, \ldots, x_n]/J$ has been computed. $\boldsymbol{K}[x_1, \ldots, x_n]/J$ as a $\boldsymbol{K}$-vector space is infinite-dimensional. Let d be large enough, $BT_{d-1}(J)$ the set of term bases with total degree less than d, Z_1 the vector with entries of terms of total degree d in $BT_d(J)$ arranged in decreasing order and Z_0 the vector with entries of terms in $BT_{d-1}(J)$ arranged in decreasing order. By computing the Gröbner basis of J or the multiplication table of $\boldsymbol{K}[x_1, \ldots, x_n]/J$, we can get

$$(B_\nu \;\; A_\nu - \lambda_\nu I) \begin{pmatrix} Z_1 \\ Z_0 \end{pmatrix} = 0, \quad \nu = 1, \ldots, n. \tag{3.8}$$

For some fixed ν, $1 \leq \nu \leq n$, according to Kronecker's theory (Gantmacher 1953), we can prove the following.

Proposition 3.13: (Feng et al. 1995b)

Let $(B_\nu \;\; A_\nu - \lambda I)$ be the matrix pencil in (3.8). Then its Kronecker form has the form

$$K_{\text{ron}} = \text{Diag}(L_{\epsilon_1}, \ldots, J_{\theta_1}, \ldots),$$

where the triangle block has the form

$$L_{\epsilon_l} = \begin{pmatrix} 1 & -\lambda & & & \\ & 1 & -\lambda & & \\ & & \ddots & \ddots & \\ & & & 1 & -\lambda \end{pmatrix} \in \boldsymbol{K}^{\epsilon_l \times (\epsilon_l + 1)}, \quad \epsilon_l \geq 0,$$

$$J_{\theta_k} = \begin{pmatrix} \lambda_k - \lambda & 1 & & \\ & \lambda_k - \lambda & \ddots & \\ & & \ddots & 1 \\ & & & \lambda_k - \lambda \end{pmatrix} \in \boldsymbol{K}^{\theta_k \times \theta_k}, \quad \theta_k > 0.$$

Without loss of generality, we assume that the Kronecker form of the matrix pencil $(B_\nu \;\; A_\nu - \lambda_\nu I)$ is

$$K_{\text{ron}} = \begin{pmatrix} L_{\epsilon_1} & & & & & & 0 \\ & \ddots & & & & & \vdots \\ & & L_{\epsilon_p} & & & & 0 \\ & & & J_{\theta_1} & & & 0 \\ & & & & \ddots & & \vdots \\ & & & & & J_{\theta_q} & 0 \end{pmatrix},$$

in which $\epsilon_1 \geq \cdots \geq \epsilon_p > 0$. We call L_{ϵ_l} a *singular block* and J_{θ_k} a *regular block*.

According to Kronecker's theory, one may find the right eigenvector $Z_\rho(\lambda)$ of $(B_\nu \ A_\nu - \lambda I)$, whose components are polynomials in λ such that

$$(B_\nu \ A_\nu - \lambda I) \sum_\rho c_\rho Z_\rho(\lambda) = 0.$$

If one can determine the c_ρ such that Theorem 3.2 holds, then $x_\nu = x_\nu(\lambda)$, $\nu = 1, \ldots, n$, give a zero manifold of dimension 1 of $V(J)$.

Definition 3.6:

If there exists a sequence of vectors $V_1, \ldots, V_r$ such that

$$V_l^T(B, A) = \lambda(0, V_l^T) + (0, V_{l+1}^T), \quad 1 \le l < r,$$
$$V_r^T(B, A) = \lambda(0, V_r^T),$$

and there is no V_0 such that $V_0^T(B, A) = \lambda(0, V_0^T) + (0, V_1^T)$, then we say that the matrix pencil $(B \ A - \lambda I)$ has a regular vector sequence of length *r with respect to λ and V_r^T is the* left main vector *of the matrix pencil.*

Now we discuss the relation between the regular blocks in the Kronecker form of the matrix pencil $(B \ A - \lambda I)$ and the decomposition of polynomials in the ideal J. We can prove the following result.

Proposition 3.14: (Feng et al. 1995b)

In equation (3.8) the matrix pencil $(B_\nu \ A_\nu - \lambda I)$ has a regular vector sequence $V_1, \ldots, V_\theta$ of length θ with respect to λ iff there exists a polynomial sequence $T_1, \ldots, T_\theta \in \mathrm{Span}_{\boldsymbol{K}} BT_{d-1}(J)$ such that

$$(x_\nu - \lambda)T_l = T_{l+1} \ne 0 \mod J, \quad l = 1, \ldots, \theta - 1,$$
$$(x_\nu - \lambda)T_\theta = 0 \mod J$$

and $T_1 \ne (x_\nu - \lambda)S \bmod J$ for an arbitrary $S \in \mathrm{Span}_{\boldsymbol{K}} BT_{d-1}(J)$.

It is easy to see that for each regular block of the Kronecker form, there exists a regular vector sequence of the matrix pencil. Thus we have the following theorem.

Theorem 3.15:

In the Kronecker form K_{ron}, there exists a regular block of order θ with respect to λ iff there exists a polynomial $T \in \mathrm{Span}_{\boldsymbol{K}} BT_{d-1}(J)$ such that

$$(x_\nu - \lambda)^l T \ne 0 \mod J, \quad l = 1, \ldots, \theta - 1,$$
$$(x_\nu - \lambda)^\theta T = 0 \mod J$$

and

$$T \neq (x_\nu - \lambda)S \mod J$$

for an arbitrary $S \in \mathrm{Span}_{\boldsymbol{K}} BT_{d-1}(J)$.

Let

$$J = (\bigcap_{l=1}^{\beta} Q_l) \cap (\bigcap_{k=1}^{\gamma} F_k)$$

be a minimal decomposition of J and define

$$d_\infty = \min\{s \mid \ J : (x_\nu - \lambda)^\infty = J : (x_\nu - \lambda)^s\}.$$

Let s_k, $k = 1, \ldots \gamma$, be such that

$$(x_\nu - \lambda)^{s_k} \in F_k, \qquad (x_\nu - \lambda)^{s_k - 1} \notin F_k.$$

Set $d_M = \max\{s_k \mid \ k = 1, \ldots, \gamma\}$. It is easy to see the results stated below.

Corollary 3.16:

Let d_∞ *and* d_M *be defined as above. Then* $d_\infty \leq d_M$. *If there is at least one primary component in* $\{F_k \mid \ s_k = d_M\}$ *which is isolated, then* $d_\infty = d_M$.

Theorem 3.17: (Feng et al. 1995b)

Suppose that the ideal J *has a minimal primary decomposition*

$$J = (\bigcap_{l=1}^{\beta} Q_l) \cap (\bigcap_{k=1}^{\gamma} F_k)$$

and $d_M(\lambda)$ *is the greatest order of the regular block with* λ *as an eigenvalue in the Kronecker form* K_{ron}. *Then*

$$d_M(\lambda) \leq d_M.$$

The equality holds if the component F_{k_0} *satisfying* $s_{k_0} = d_M$ *is isolated.*

Chapter 4

SOLVING THE YANG–BAXTER EQUATION BY WU'S METHOD

Shi-kun Wang and Ke Wu

The Yang–Baxter equation (YBE) is an important matrix equation from mathematical physics that may be written as a system of non-linear polynomial function equations. This chapter shows how Wu's method may be used effectively to solve YBE. All the eight-vertex type solutions — comprising three sets of gauge solutions, three sets of degenerate-form solutions and two sets of trivial solutions up to five solution transformations — of YBE with spectral and color parameters are constructed. Moreover, all the non-trivial solutions are classified into two types, the Baxter type and the Free-Fermion type. Some other types of solutions of YBE found by using Wu's method are mentioned briefly.

4.1 Introduction

Motivated by the many-body problem in one dimension with repulsive delta-function interaction and the statistical model with six vertices, Yang (1967, 1968; a Nobel prize winner) and Baxter (1972; one of the most famous physicists in statistical physics) independently proposed a matrix equation. From some solutions of the equation they could obtain exact analytic results of the physical systems concerned. From that time on it has been found that the matrix equation also applies widely in many branches of physics and mathematics. These include statistics physics, quantum field theory, quantum integrable system, quantum groups, lower dimension topology and knot theory (Jimbo 1989; Zamolodchikov 1979; Drinfel'd 1987; Turaev 1988; Frenkel and Reshetikhin 1992). The equation is called the *Yang–Baxter equation* (YBE) nowadays.

YBE is a very important equation in mathematical physics. The construction of its solutions has been a basic problem. A few solutions were found in the

MATHEMATICS MECHANIZATION AND APPLICATIONS
ISBN 0-12-734760-7

investigation of some statistical models (Ma 1993). However, until the beginning of the 1990s there was no systematic method for solving YBE exactly (Ma 1993). Due to the important role the matrix equation plays in many branches of physics and mathematics, finding more general solutions of YBE (Ruiz-Altaba 1992) has been a hot research topic that attracts attention from both physicists and mathematicians.

The YBE becomes a system of non-linear polynomial function equations if it is represented by the entries of a matrix. So we actually need to solve a system of non-linear polynomial equations. Inspired by some creative ideas from ancient Chinese mathematics, Wu (1984a,b, 1986a, 1987b) established a theory on decomposing the zero set of any system of non-linear polynomial equations and thus a systematic method for solving such systems. The theory and method, which are now called *Wu's theory and method*, provide a basis for automated proving of theorems in geometry and a mechanical device for solving systems of polynomial equations. In particular, they provide us with an efficient way of solving the YBE.

In this chapter we discuss the type of YBE with color and spectral parameters (CSYBE). Only two solutions of the equation are known (Murakami 1992; Bazhanov and Stroganov 1985). Based on Wu's method, we introduce a systematic method to find all solutions of CSYBE. The chapter contains eight sections. In Sect. 4.2, CSYBE is introduced and is given as a system of 28 function equations. In Sect. 4.3 a symmetric method for solution transformation is described. By using this method the 28 function equations can be simplified to 12. In Sect. 4.4 we introduce four concepts: the initial condition, Hamiltonian coefficient, symmetric operation and gauge solution of CSYBE. Using Wu's method we prove that the 12 function equations have three sets of solutions for the initial condition. Two of the three sets are found to be trivial solutions. The symmetric operation and another initial condition are important and are applied frequently in our finding of solutions for CSYBE. Section 4.5 is a core, in which we present the system of polynomial equations satisfied by CSYBE. Moreover, Wu's method is used to obtain two theorems that are the basis of finding classified solutions of CSYBE. However, we omit the details of the heavy calculations. In Sect. 4.6 Wu's method is applied again to obtain a differential equation from which one can calculate the solutions of CSYBE. In Sect. 4.7, all gauge solutions are found and are classified into two types. In Sect. 4.8 we provide all solutions of CSYBE, which cover Murakami's solution and Bazhanov's solution. Some concluding remarks are given in the last section.

4.2 CSYBE

By the *Yang–Baxter equation with color and spectral parameters* (CSYBE) we mean the following matrix equation:

$$\begin{aligned}&\check{R}_{12}(u,\xi,\eta)\check{R}_{23}(u+v,\xi,\lambda)\check{R}_{12}(v,\eta,\lambda)\\&\qquad=\check{R}_{23}(v,\eta,\lambda)\check{R}_{12}(u+v,\xi,\lambda)\check{R}_{23}(u,\xi,\eta),\\&\check{R}_{12}(u,\xi,\eta)=\check{R}(u,\xi,\eta)\otimes E,\qquad \check{R}_{23}(u,\xi,\eta)=E\otimes\check{R}(u,\xi,\eta),\end{aligned}\tag{4.1}$$

where $\check{R}(u,\xi,\eta)$ is a function matrix of dimension N^2, E is the unit matrix of dimension N and $\otimes$ is tensor product of two matrices. u, v are called *spectral parameters* and ξ, η *color parameters*. This type of YBE is from the statistical model related to free Freemason (Fan and Wu 1970). If the matrix is independent of color parameters, then (4.1) is the usual Yang–Baxter equation. If it is independent of spectral parameters, then (4.1) is reduced to the Yang–Baxter equation depending on colored parameters only.

For CSYBE (4.1), the main interest in physics is to discuss the solutions of the following form:

$$\check{R}(u,\xi,\eta)=\begin{pmatrix}a_1(u,\xi,\eta)&0&0&a_7(u,\xi,\eta)\\0&a_2(u,\xi,\eta)&a_5(u,\xi,\eta)&0\\&a_6(u,\xi,\eta)&a_3(u,\xi,\eta)&0\\a_8(u,\xi,\eta)&0&0&a_4(u,\xi,\eta)\end{pmatrix}.\tag{4.2}$$

Following physical references, we call the eight entries in the matrix *weight functions* and the solution of (4.2) an *eight-vertex type solution* if it satisfies in addition $a_i(u,\xi,\eta)\not\equiv 0$ $(i=1,2,\ldots,8)$. In this chapter we consider only weight functions which are meromorphic functions of u and ξ,η.

For eight-vertex type solutions, the matrix equation (4.1) is equivalent to the following 28 equations:

$$\begin{cases}u_7w_3v_8-u_8w_2v_7=0,\\u_7w_8v_3-u_8w_7v_2=0,\\u_2w_3v_2-u_3w_2v_3=0,\\u_2w_8v_7-u_3w_7v_8=0,\end{cases}\tag{4.3}$$

$$\begin{cases}u_1w_5v_2+u_7w_8v_6-v_2w_1u_5-v_5w_2u_3=0,\\u_1w_1v_7+u_7w_3v_4-v_7w_5u_5-v_1w_7u_3=0,\\u_2w_6v_1+u_5w_7v_8-v_6w_1u_2-v_3w_2u_6=0,\\u_1w_2v_1+u_7w_4v_8-v_2w_1u_2-v_5w_2u_6=0,\\u_1w_7v_5+u_7w_6v_3-v_7w_5u_2-v_1w_7u_6=0,\\u_1w_7v_2+u_7w_6v_6-v_1w_1u_7-v_7w_2u_4=0,\end{cases}\tag{4.4}$$

$$\begin{cases} u_4w_6v_2 + u_7w_8v_5 - v_2w_4u_6 - v_6w_2u_3 = 0, \\ u_4w_4v_7 + u_7w_3v_1 - v_7w_6u_6 - v_4w_7u_3 = 0, \\ u_2w_5v_4 + u_6w_7v_8 - v_5w_4u_2 - v_3w_2u_5 = 0, \\ u_4w_2v_4 + u_7w_1v_8 - v_2w_4u_2 - v_6w_2u_5 = 0, \\ u_4w_7v_6 + u_7w_5v_3 - v_7w_6u_2 - v_4w_7u_5 = 0, \\ u_4w_7v_2 + u_7w_5v_5 - v_4w_4u_7 - v_7w_2u_1 = 0, \end{cases} \tag{4.5}$$

$$\begin{cases} u_1w_5v_3 + u_8w_7v_6 - v_3w_1u_5 - v_5w_3u_2 = 0, \\ u_1w_1v_8 + u_8w_2v_4 - v_8w_5u_5 - v_1w_8u_2 = 0, \\ u_3w_6v_1 + u_5w_8v_7 - v_6w_1u_3 - v_2w_3u_6 = 0, \\ u_1w_3v_1 + u_8w_4v_7 - v_3w_1u_3 - v_5w_3u_6 = 0, \\ u_1w_8v_5 + u_8w_6v_2 - v_8w_5u_3 - v_1w_8u_6 = 0, \\ u_1w_8v_3 + u_8w_6v_6 - v_1w_1u_8 - v_8w_3u_4 = 0, \end{cases} \tag{4.6}$$

$$\begin{cases} u_4w_6v_3 + u_8w_7v_5 - v_3w_4u_6 - v_6w_3u_2 = 0, \\ u_4w_4v_8 + u_8w_2v_1 - v_8w_6u_6 - v_4w_8u_2 = 0, \\ u_3w_5v_4 + u_6w_8v_7 - v_5w_4u_3 - v_2w_3u_5 = 0, \\ u_4w_3v_4 + u_8w_1v_7 - v_3w_4u_3 - v_6w_3u_5 = 0, \\ u_4w_8v_6 + u_8w_5v_2 - v_8w_6u_3 - v_4w_8u_5 = 0, \\ u_4w_8v_3 + u_8w_5v_5 - v_4w_4u_8 - v_8w_3u_1 = 0. \end{cases} \tag{4.7}$$

Here and throughout this chapter,

$$u_i = a_i(u, \xi, \eta), \quad v_i = a_i(v, \eta, \lambda), \quad w_i = a_i(u + v, \xi, \lambda), \quad i = 1, \ldots, 8,$$

for notational simplicity.

Solving CSYBE is equivalent to solving the 28 polynomial function equations. It is clearly difficult to solve these equations directly. How to solve the system of non-linear polynomial function equations (4.3)–(4.7) alternatively? The main ideas of ours are as follows.

(1) First find symmetries or transformation properties for the solutions of the system of equations (4.3)–(4.7) so that the system can be simplified.

(2) Find initial conditions for the solutions of CSYBE.

(3) Generate the algebraic curves or surfaces that the weight functions should satisfy and calculate the genus for the curves.

(4) Generate differential equations that the weight functions should satisfy.

(5) Apply the theories of algebraic curves and surfaces to get solutions of (4.3)–(4.7).

4.3 Solution transformations

Assume that $\check{R}(u, \xi, \eta)$ is a solution of (4.1). Having carefully studied the system of equations (4.3)–(4.7), we find that there are five symmetries for eight-vertex type solutions of (4.1).

A. Symmetry of interchanging indices. The system of equations (4.3)–(4.7) is invariant if one interchanges the two sub-indices 2 and 3 as well as the two sub-indices 7 and 8 or the sub-indices 1 and 4 as well as the two sub-indices 5 and 6.

B. The scaling symmetry. The solution $\check{R}(u, \xi, \eta)$ multiplied by an arbitrary function $g(u, \xi, \eta)$ is still a solution of CSYBE (4.1).

C. Symmetry of weight functions. If the weight functions $a_2(u, \xi, \eta), a_3(u, \xi, \eta)$, $a_7(u, \xi, \eta)$ and $a_8(u, \xi, \eta)$ are replaced by the new weight functions

$$\bar{a}_2(u, \xi, \eta) = \frac{N(\xi)}{N(\eta)} a_2(u, \xi, \eta), \qquad \bar{a}_3(u, \xi, \eta) = \frac{N(\eta)}{N(\xi)} a_3(u, \xi, \eta),$$
$$\bar{a}_7(u, \xi, \eta) = s\, N(\eta) N(\xi) a_7(u, \xi, \eta), \qquad \bar{a}_8(u, \xi, \eta) = \frac{1}{s\, N(\eta) N(\xi)} a_8(u, \xi, \eta)$$

respectively, or $a_5(u, \xi, \eta)$ and $a_6(u, \xi, \eta)$ by $-a_5(u, \xi, \eta)$ and $-a_6(u, \xi, \eta)$, where $N(\xi)$ is an arbitrary function of color parameter and s is a complex constant, then the new matrix $\check{R}(u, \xi, \eta)$ is still a solution of CSYBE (4.1).

D. Symmetry of spectral parameter. If one takes a new spectral parameter $\bar{u} = \mu u$, where μ is a complex constant independent of spectral and the color parameters, then the new matrix $\check{R}(\bar{u}, \xi, \eta)$ is still a solution of CSYBE (4.1).

E. Symmetry of color parameters. If one takes new color parameters $\zeta = f(\xi)$ and $\theta = f(\eta)$, where $f(\xi)$ is an arbitrary function, then the new matrix $\check{R}(u, \zeta, \theta)$ is also a solution of CSYBE (4.1).

The five symmetries A–E are called *solution transformations* of the eight-vertex type solutions of CSYBE.

To reduce the system of equations by using these symmetries, we perform the following elementary calculations.

Dividing both sides of the third equation of (4.3) by $a_2(u, \xi, \eta) a_2(u + v, \xi, \lambda)$ $a_2(v, \eta, \lambda)$, we get

$$f(u + v, \xi, \lambda) = f(u, \xi, \eta) f(v, \eta, \lambda), \tag{4.8}$$

where $f(u, \xi, \eta) = a_3(u, \xi, \eta)/a_2(u, \xi, \eta)$. Putting $u = v = \eta = 0$ in (4.8) we have

$$f(0, \xi, \lambda) = f(0, \xi, 0) f(0, 0, \lambda).$$

Substituting this formula into the one obtained by taking $u = v = \xi = 0$ in (4.8), we get

$$f(0,0,\lambda) = f(0,0,\eta)f(0,\eta,\lambda) = f(0,0,\eta)f(0,\eta,0)f(0,0,\lambda).$$

This means that

$$f(0,\eta,0)f(0,0,\eta) = 1.$$

Otherwise, it is easy to show that $f(u,\xi,\eta) = 0$, i.e. $a_3(u,\xi,\eta) = 0$. Therefore

$$f(0,\xi,\eta) = \frac{M(\xi)}{M(\eta)},$$

where $M(\xi) = f(0,\xi,0)$. On the other hand, if we differentiate both sides of (4.8) with respect to the spectral variable v and then set $v = 0$ and $\lambda = \eta$, then

$$f'(u,\xi,\eta) = f'(0,\eta,\eta)f(u,\xi,\eta) \tag{4.9}$$

holds, where the prime means derivative with respect to u and the simple formula

$$\left.\frac{dG(u+v)}{dv}\right|_{v=0} = \frac{dG(u)}{du}$$

for any function $G(u)$ is used. Similarly, one has

$$f'(v,\xi,\lambda) = f'(0,\xi,\xi)f(v,\xi,\lambda) \tag{4.10}$$

if one differentiates both sides of (4.8) with respect to u and then sets $u = 0$ and $\eta = \xi$. The two formulas (4.9) and (4.10) above imply that $f'(0,\xi,\xi)$ is a constant independent of the color parameter ξ. Hence

$$f(u,\xi,\eta) = \frac{M(\xi)}{M(\eta)}\exp(\nu u), \tag{4.11}$$

where ν is a complex constant.

From the set of equations (4.3) we have

$$f(u,\xi,\eta)h(v,\eta,\lambda) = h(u+v,\xi,\lambda), \tag{4.12}$$

$$f(u,\xi,\eta)f(v,\eta,\lambda) = f(u+v,\xi,\lambda), \tag{4.13}$$

$$h(u,\xi,\eta) = h(u+v,\xi,\lambda)f(v,\eta,\lambda), \tag{4.14}$$

$$h(u,\xi,\eta) = f(u+v,\xi,\lambda)h(v,\eta,\lambda), \tag{4.15}$$

where $f(u,\xi,\eta) = a_3(u,\xi,\eta)/a_2(u,\xi,\eta)$ and $h(u,\xi,\eta) = a_8(u,\xi,\eta)/a_7(u,\xi,\eta)$. If we let $v = 0$ and $\lambda = 0$ in (4.15), then

$$h(u,\xi,\eta) = f(u,\xi,0)h(0,\eta,0).$$

Substituting this into (4.14) and using (4.13), one obtains

$$f(u,\xi,0)h(0,\eta,0) = f(u,\xi,0)f(v,0,0)h(0,\lambda,0)f(v,\eta,\lambda)$$

or

$$h(0,\eta,0) = f(v,0,0)h(0,\lambda,0)f(v,\eta,\lambda).$$

This formula implies that $\nu = 0$ in (4.11) and thus one can obtain

$$f(u,\xi,\eta) = \frac{M(\xi)}{M(\eta)}, \qquad h(u,\xi,\eta) = l\, M(\xi)M(\eta),$$

where l is a complex constant independent of the spectral and color parameters. So, up to the solution transformations B and C, one can assume that

$$a_3(u,\xi,\eta) = a_2(u,\xi,\eta) = 1, \qquad a_8(u,\xi,\eta) = a_7(u,\xi,\eta)$$

without any loss of generality. Therefore, (4.3)–(4.7) can be simplified to the following 12 equations:

$$\begin{aligned}
&a_5(v,\eta,\lambda) + a_5(u,\xi,\eta)a_1(u+v,\xi,\lambda) - a_1(u,\xi,\eta)a_5(u+v,\xi,\lambda)\\
&\quad - a_7(u,\xi,\eta)a_7(u+v,\xi,\lambda)a_6(v,\eta,\lambda) = 0,\\
&a_7(u+v,\xi,\lambda)a_1(v,\eta,\lambda) + a_5(u,\xi,\eta)a_5(u+v,\xi,\lambda)a_7(v,\eta,\lambda)\\
&\quad - a_1(u,\xi,\eta)a_1(u+v,\xi,\lambda)a_7(v,\eta,\lambda) - a_7(u,\xi,\eta)a_4(v,\eta,\lambda) = 0,\\
&a_6(u,\xi,\eta) + a_1(u+v,\xi,\lambda)a_6(v,\eta,\lambda) - a_6(u+v,\xi,\lambda)a_1(v,\eta,\lambda)\\
&\quad - a_5(u,\xi,\eta)a_7(u+v,\xi,\lambda)a_7(v,\eta,\lambda) = 0,\\
&a_6(u,\xi,\eta)a_5(v,\eta,\lambda) + a_1(u+v,\xi,\lambda) - a_1(u,\xi,\eta)a_1(v,\eta,\lambda)\\
&\quad - a_7(u,\xi,\eta)a_4(u+v,\xi,\lambda)a_7(v,\eta,\lambda) = 0,\\
&a_6(u,\xi,\eta)a_7(u+v,\xi,\lambda)a_1(v,\eta,\lambda) + a_5(u+v,\xi,\lambda)a_7(v,\eta,\lambda)\\
&\quad - a_1(u,\xi,\eta)a_7(u+v,\xi,\lambda)a_5(v,\eta,\lambda)\\
&\quad - a_7(u,\xi,\eta)a_6(u+v,\xi,\lambda) = 0,\\
&a_7(u,\xi,\eta)a_1(u+v,\xi,\lambda)a_1(v,\eta,\lambda) + a_4(u,\xi,\eta)a_7(v,\eta,\lambda)\\
&\quad - a_1(u,\xi,\eta)a_7(u+v,\xi,\lambda)\\
&\quad - a_7(u,\xi,\eta)a_6(u+v,\xi,\lambda)a_6(v,\eta,\lambda) = 0
\end{aligned} \tag{4.16}$$

plus six equations obtained by interchanging the sub-indices 1 and 4 as well as 5 and 6 in each of the equations (4.16). We call these last six equations the *counterpart* of (4.16).

Denote the set of all solutions of (4.16) and its counterpart by $\mathcal{S}$. From the argument above we know the set of all eight-vertex solutions of CSYBE is $\mathcal{S}$ up to the five solution transformations A–E.

4.4 Initial condition and trivial solutions

Taking $u = 0$ and $\eta = \xi$ in (4.16) and its counterpart, we have

$$\begin{aligned}
&x_7x_1 + x_5x_5x_7 - x_1x_1x_7 - x_7x_4 = 0,\\
&x_7x_4 + x_6x_6x_7 - x_4x_4x_7 - x_7x_1 = 0,\\
&x_6x_7x_1 + x_5x_7 - x_1x_7x_5 - x_7x_6 = 0,\\
&x_5x_7x_4 + x_6x_7 - x_4x_7x_6 - x_7x_5 = 0,\\
&x_5x_6 + x_4 - x_4x_4 - x_7x_1x_7 = 0,\\
&x_6x_5 + x_1 - x_1x_1 - x_7x_4x_7 = 0,\\
&x_7x_1x_1 + x_4x_7 - x_1x_7 - x_7x_6x_6 = 0,\\
&x_7x_4x_4 + x_1x_7 - x_4x_7 - x_7x_5x_5 = 0,\\
&x_5 - x_7x_7x_6 = 0,\\
&x_6 - x_5x_7x_7 = 0,
\end{aligned} \tag{4.17}$$

where $x_i = a_i(0, \xi, \xi)$.

Now we solve the system of equations (4.17) for the variables $a_1(0, \xi, \xi)$, $a_4(0, \xi, \xi), \ldots, a_7(0, \xi, \xi)$. By Wu's method it is easy to obtain three sets of solutions of (4.17). From two of them we can obtain two sets of trivial solutions of CSYBE. So we have the following proposition.

Proposition 4.1:

For one set of solutions of (4.16) *and its counterpart, the weight functions satisfy*

$$\begin{aligned}
&a_1(0, \xi, \xi) = a_4(0, \xi, \xi) = 1,\\
&a_5(0, \xi, \xi) = a_6(0, \xi, \xi) = a_7(0, \xi, \xi) = a_8(0, \xi, \xi) = 0.
\end{aligned} \tag{4.18}$$

Up to the solution transformations A–E, CSYBE (4.1) *has two sets of trivial solutions. The first set is*

$$\begin{aligned}
&a_1(u, \xi, \eta) = a_4(u, \xi, \eta) = a_5(u, \xi, \eta) = a_6(u, \xi, \eta) = H(u, \xi, \eta),\\
&a_2(u, \xi, \eta) = a_3(u, \xi, \eta) = a_7(u, \xi, \eta) = a_8(u, \xi, \eta) = 1,
\end{aligned} \tag{4.19}$$

where $H(u, \xi, \eta)$ is an arbitrary function of the spectral parameter u and the colored parameters ξ and η, and the second set of trivial solutions is

$$\begin{aligned}
&a_1(u, \xi, \eta) = a_4(u, \xi, \eta) = a_5(u, \xi, \eta) = -a_6(u, \xi, \eta) = \frac{F(\xi)}{F(\eta)} \exp(u),\\
&a_1(u, \xi, \eta) = a_2(u, \xi, \eta) = 1, \qquad a_7(u, \xi, \eta) = a_8(u, \xi, \eta) = \boldsymbol{i},
\end{aligned} \tag{4.20}$$

where $\boldsymbol{i}^2 = -1$ and $F(\xi)$ is an arbitrary function of the color parameter ξ.

Definition 4.1:

A gauge solution of CSYBE (4.1) *is meant the solution whose weight functions satisfy* $a_2(u,\xi,\eta) = a_3(u,\xi,\eta) = 1$, $a_7(u,\xi,\eta) = a_8(u,\xi,\eta)$ *and the condition* (4.18).

The condition (4.18) *is called the* initial condition *of the gauge solution of CSYBE.*

Let $\mathcal{S}_G$ denote the set of all gauge solutions of CSYBE.

Proposition 4.2:

$\mathcal{S} = \mathcal{S}_G \cup \{$*the trivial solutions* (4.19) *and* (4.20)$\}$.

Differentiating both sides of all the equations in (4.16) and their counterparts with respect to the variable v and letting $v = 0$ and $\lambda = \eta$, by virtue of the initial condition (4.18) one immediately obtains

$$\begin{aligned}
&m_5(\eta) + a_5(u,\xi,\eta)\,a_1'(u,\xi,\eta) - a_1(u,\xi,\eta)\,a_5'(u,\xi,\eta) \\
&\quad - m_6(\eta)\,a_7(u,\xi,\eta)^2 = 0, \\
&a_7'(u,\xi,\eta) + [m_1(\eta) - m_4(\eta)]\,a_7(u,\xi,\eta) \\
&\quad + m_7(\eta)\,[a_5(u,\xi,\eta)^2 - a_1(u,\xi,\eta)^2] = 0, \\
&a_6'(u,\xi,\eta) - m_6(\eta)\,a_1(u,\xi,\eta) + m_1(\eta)\,a_6(u,\xi,\eta) \\
&\quad + m_7(\eta)\,a_5(u,\xi,\eta)\,a_7(u,\xi,\eta) = 0, \\
&a_1'(u,\xi,\eta) - m_1(\eta)\,a_1(u,\xi,\eta) + m_5(\eta)\,a_6(u,\xi,\eta) \\
&\quad - m_7(\eta)\,a_4(u,\xi,\eta)\,a_7(u,\xi,\eta) = 0, \\
&a_6(u,\xi,\eta)\,a_7'(u,\xi,\eta) - a_7(u,\xi,\eta)\,a_6'(u,\xi,\eta) \\
&\quad + m_1(\eta)\,a_6(u,\xi,\eta)\,a_7(u,\xi,\eta) - m_5(\eta)\,a_1(u,\xi,\eta)\,a_7(u,\xi,\eta) \\
&\quad + m_7(\eta)\,a_5(u,\xi,\eta) = 0, \\
&a_7(u,\xi,\eta)\,a_1'(u,\xi,\eta) - a_1(u,\xi,\eta)\,a_7'(u,\xi,\eta) \\
&\quad + m_1(\eta)\,a_1(u,\xi,\eta)\,a_7(u,\xi,\eta) - m_6(\eta)\,a_6(u,\xi,\eta)\,a_7(u,\xi,\eta) \\
&\quad + m_7(\eta)\,a_4(u,\xi,\eta) = 0
\end{aligned} \tag{4.21}$$

and their counterparts. Here and throughout the chapter,

$$a_i'(u,\xi,\eta) = \frac{\partial}{\partial u} a_i(u,\xi,\eta), \qquad m_i(\xi) = a_i'(u,\xi,\eta)|_{u=0,\eta=\xi},$$

for $i = 1, 4, \ldots, 7$. The following result is clear.

Proposition 4.3:

$\mathcal{S}_G \subseteq \mathcal{S}_{DI}$, where $\mathcal{S}_{DI}$ denotes the set of all solutions of (4.21) *and its counterpart.*

Definition 4.2:

$m_i(\eta)$ $(i = 1, 4, \ldots, 7)$ are called the Hamiltonian coefficients *or simply* coefficients *of the weight functions with respect to a spectral parameter.*

If we differentiate (4.16) with respect to u, let $u = 0$, $\eta = \xi$ and then replace the variables v and λ by u and η, we get

$$
\begin{aligned}
& m_6(\xi) + a_6(u,\xi,\eta)\, a_1'(u,\xi,\eta) - a_1(u,\xi,\eta)\, a_6'(u,\xi,\eta) \\
& \qquad - m_5(\xi)\, a_7(u,\xi,\eta)^2 = 0, \\
& a_7'(u,\xi,\eta) + [m_1(\xi) - m_4(\xi)]\, a_7(u,\xi,\eta) \\
& \qquad + m_7(\xi)\, [a_6(u,\xi,\eta)^2 - a_1(u,\xi,\eta)^2] = 0, \\
& a_5'(u,\xi,\eta) - m_5(\xi)\, a_1(u,\xi,\eta) + m_1(\xi)\, a_5(u,\xi,\eta) \\
& \qquad + m_7(\xi)\, a_6(u,\xi,\eta)\, a_7(u,\xi,\eta) = 0, \\
& a_1'(u,\xi,\eta) - m_1(\xi)\, a_1(u,\xi,\eta) + m_6(\xi)\, a_5(u,\xi,\eta) \\
& \qquad - m_7(\xi)\, a_4(u,\xi,\eta)\, a_7(u,\xi,\eta) = 0, \\
& a_5(u,\xi,\eta)\, a_7'(u,\xi,\eta) - a_7(u,\xi,\eta)\, a_5'(u,\xi,\eta) \\
& \qquad + m_1(\xi)\, a_5(u,\xi,\eta)\, a_7(u,\xi,\eta) - m_6(\xi)\, a_1(u,\xi,\eta)\, a_7(u,\xi,\eta) \\
& \qquad + m_7(\xi)\, a_6(u,\xi,\eta) = 0, \\
& a_7(u,\xi,\eta)\, a_1'(u,\xi,\eta) - a_1(u,\xi,\eta)\, a_7'(u,\xi,\eta) \\
& \qquad + m_1(\xi)\, a_1(u,\xi,\eta)\, a_7(u,\xi,\eta) - m_5(\xi)\, a_5(u,\xi,\eta)\, a_7(u,\xi,\eta) \\
& \qquad + m_7(\xi)\, a_4(u,\xi,\eta) = 0
\end{aligned}
\tag{4.22}
$$

and their counterparts. Similar to Proposition 4.3 we have the following.

Proposition 4.4:

$\mathcal{S}_G \subseteq \mathcal{S}_{DII}$, where $\mathcal{S}_{DII}$ denotes the set of all solutions of (4.22) *and its counterpart.*

Remark 4.1:

The following trick will often be employed to obtain equations like (4.21) and (4.22). We first perform the calculation with respect to v and take $\lambda = \eta$ to yield some equations. Then repeat the same operation with u in place of v and take $\eta = \xi$ to yield another equation. Finally we compare the two results to get some new formulas. For example, formula (4.11) is obtained by this method. Comparing (4.21) with (4.22), we find that the trick, in fact, is to interchange the sub-indices 5 and 6 (or 1 and 4) and then replace $m_i(\eta)$ by $m_i(\xi)$ in the original equations (4.21) to get (4.22). We call this trick the *symmetric operation*.

Remark 4.2:

If the variable with respect to which we differentiate (4.16) is not a spectral parameter v but a color parameter λ, then the equations obtained are the same as (4.21) when we still let $v = 0$ and $\lambda = \eta$. Of course, in this case the prime means the derivative with respect to the color parameter η, that is the second color variable in $a_i(u, \xi, \eta)$, and $m_i(\eta) = da_i(v, \eta, \lambda)/d\lambda|_{v=0,\lambda=\eta}$. Similarly, (4.22) also represents the equations obtained by differentiating (4.16) with respect to the first color variable ξ in $a_i(u, \xi, \eta)$. However, in this case the prime means the derivative with respect to ξ and $m_i(\xi) = da_i(u, \xi, \eta)/d\lambda|_{u=0,\eta=\xi}$.

It follows from the second equation of (4.21) and its counterpart that

$$\begin{aligned} &2a_7'(u, \xi, \eta) + m_7(\eta)\,[a_5(u, \xi, \eta)^2 + a_6(u, \xi, \eta)^2 - a_1(u, \xi, \eta)^2 \\ &\quad - a_4(u, \xi, \eta)^2] = 0. \end{aligned} \tag{4.23}$$

If the symmetric operation described in Remark 4.1 is used, then one also has

$$\begin{aligned} &2a_7'(u, \xi, \eta) + m_7(\xi)\,[a_5(u, \xi, \eta)^2 + a_6(u, \xi, \eta)^2 - a_1(u, \xi, \eta)^2 \\ &\quad - a_4(u, \xi, \eta)^2] = 0. \end{aligned} \tag{4.24}$$

We know that $(u_5^2 + u_6^2 - u_1^2 - u_4^2) \not\equiv 0$ due to the initial condition (4.18). Comparing the formulas (4.23) and (4.24), one can claim that

$$m_7(\xi) = m_7(\eta).$$

This means that $m_7(\xi)$ is independent of the color parameter.

Remark 4.3:

We set $m_7(\xi) = \alpha$ throughout the chapter. If $\alpha = 0$, then

$$a_7'(u, \xi, \eta) = 0, \qquad m_1(\eta) = m_4(\eta)$$

due to (4.23) and the second equation of (4.21). This implies that $a_7(u, \xi, \eta)$ is a function of the color variables ξ and η only.

Proposition 4.5:

For a gauge eight-vertex solution, $m_7(\xi)$ is a constant independent of color parameters and $a_7(u,\xi,\eta)$ is independent of the spectral parameter if $m_7(\eta)=0$.

Furthermore, if $\alpha=0$, let $u=0$ and $\eta=\xi$ in the equation

$$a_1'(u,\xi,\eta)=m_6(\eta)\,a_6(u,\xi,\eta)-m_1(\eta)\,a_1(u,\xi,\eta),$$

which arises from the sixth equation in (4.21); then we obtain $m_1(\eta)=0$ owing to the initial condition (4.18). Therefore, $m_4(\xi)=0$. Similarly, we have

$$a_6'(u,\xi,\eta)=m_1(\eta)\,a_6(u,\xi,\eta)-m_5(\eta)\,a_1(u,\xi,\eta)$$

from the fifth equation in (4.21) and thus

$$m_6(\eta)=-m_5(\eta).$$

Substituting $m_6(\eta)=-m_5(\eta)$ and $m_1(\eta)=m_4(\eta)=0$ into the third and fourth equations in (4.21) and their counterparts, we obtain the following differential equations:

$$a_1'(u,\xi,\eta)=-m_5(\eta)a_6(u,\xi,\eta),\qquad a_4'(u,\xi,\eta)=m_5(\eta)a_5(u,\xi,\eta),$$

$$a_6'(u,\xi,\eta)=-m_5(\eta)a_1(u,\xi,\eta),\qquad a_5'(u,\xi,\eta)=m_5(\eta)a_4(u,\xi,\eta).$$

Therefore, if $\alpha=0$ then the weight functions satisfy

$$\frac{d^2}{du^2}a_i(u,\xi,\eta)=m_5(\eta)^2\,a_i(u,\xi,\eta),\quad i=1,4,5,6.$$

Furthermore, using the symmetric operation one sees that

$$\frac{d^2}{du^2}a_i(u,\xi,\eta)=m_5(\xi)^2\,a_i(u,\xi,\eta),\quad i=1,4,5,6$$

hold. Hence $m_5(\eta)$ actually is a constant independent of the color parameters and not identically zero; otherwise the solutions will be independent of the spectral parameter. Thus we may set $m_5(\eta)=\beta$.

The argument presented above enables us to infer the following proposition.

Proposition 4.6:

For a gauge eight-vertex solution, there is at least one of $m_7(\xi)$ and $m_5(\xi)$ (or $m_6(\xi)$) which is not identically zero. Otherwise, the solution will be independent of the spectral parameter.

4.5 Algebraic curves

The simplified system of equations and its counterpart is not a system of algebraic equations. However, we can regard u_i, v_i and w_i $(i = 1, 4, \ldots, 7)$ in (4.16) and its counterpart as indeterminates. By Wu's method we can eliminate $w_1, w_4, \ldots, w_7$ in numerically increasing order from the system of equations (4.16) and its counterpart so as to obtain seven equations which do not contain the indeterminates w_i $(i = 1, 4, \ldots, 7)$. Differentiating these equations with respect to the spectral variable v, letting $v = 0, \lambda = \eta$ and then substituting the initial values (4.18) into the resulting ones, we obtain the following seven polynomial equations:

$$
\begin{aligned}
& m_7 u_1^3 - m_7 u_1 u_5^2 - 3\, m_1 u_1 u_7 + m_4 u_1 u_7 - m_7 u_4 u_7^2 - m_7 u_4 \\
& \quad + m_5 u_6 u_7 + m_6 u_6 u_7 = 0, \\
& -m_6 u_1 u_4 + m_7 u_1 u_6 u_7 + m_7 u_4 u_5 u_7 + m_1 u_4 u_6 + m_4 u_4 u_6 \\
& \quad - m_6 u_5 u_6 - m_5 u_7^2 + m_6 = 0, \\
& m_7 u_1^2 - m_7 u_4^2 - m_7 u_5^2 + m_7 u_6^2 + 2\, m_4 u_7 - 2\, m_1 u_7 = 0, \\
& -m_5 u_1 u_4 + m_1 u_1 u_5 + m_4 u_1 u_5 + m_7 u_1 u_6 u_7 + m_7 u_4 u_5 u_7 \\
& \quad - m_5 u_5 u_6 - m_6 u_7^2 + m_5 = 0, \\
& m_7 u_1^2 u_6 - m_6 u_1 u_7 - m_5 u_1 u_7 - m_7 u_5^2 u_6 + m_7 u_5 u_7^2 + m_7 u_5 \\
& \quad + m_4 u_6 u_7 + m_1 u_6 u_7 = 0, \\
& m_7 u_1^3 u_5 - m_6 u_1 u_4 u_7 - m_7 u_1 u_5^3 + 2\, m_4 u_1 u_5 u_7 - 2\, m_1 u_1 u_5 u_7 \\
& \quad + m_7 u_1 u_6 - m_7 u_4 u_5 u_7^2 + m_5 u_5 u_6 u_7 + m_6 u_7^3 - m_5 u_7 = 0, \\
& -m_7 u_1^2 u_4 + m_7 u_1 u_7^2 + m_7 u_1 + m_7 u_4 u_5^2 + m_4 u_4 u_7 + m_1 u_4 u_7 \\
& \quad - m_5 u_5 u_7 - m_6 u_5 u_7 = 0,
\end{aligned}
\tag{4.25}
$$

where m_i and u_i are used instead of $m_i(\eta)$ and $a_i(u, \xi, \eta)$ respectively for brevity. Since Wu's elimination of indeterminates ensures that the solutions of the system of (4.16) and its counterpart do not get lost, we have the following result.

Proposition 4.7:

$\mathcal{S}_G \subseteq \mathcal{S}_{WG}$, *where* $\mathcal{S}_{WG}$ *denotes the set of all solutions of* (4.25).

Applying Wu's method to (4.25) we have the following theorems.

Theorem 4.8:

The Hamiltonian coefficients of the gauge solutions of CSYBE satisfy

$$m_1^2 - m_4^2 = 0, \qquad m_5^2 - m_6^2 = 0.$$

Theorem 4.9:

The weight functions of the gauge solutions of CSYBE must satisfy the following two systems of equations:

$$\begin{aligned} &a_5(u,\xi,\eta) = a_6(u,\xi,\eta),\\ &a_1(u,\xi,\eta) = a_4(u,\xi,\eta),\\ &\alpha\, a_1(u,\xi,\eta)\, a_5(u,\xi,\eta) - m_6(\eta)\, a_7(u,\xi,\eta) = 0,\\ &\alpha^2\, a_1(u,\xi,\eta)^2 a_5(u,\xi,\eta)^2 - m_5^2(\eta)\, a_5(u,\xi,\eta)^2 - m_5^2(\eta) a_1(u,\xi,\eta)^2\\ &\qquad + 2m_5(\eta)\, m_1(\eta)\, a_1(u,\xi,\eta) a_5(u,\xi,\eta) + m_5^2(\eta) = 0, \end{aligned} \tag{4.26}$$

and

$$\begin{aligned} &1 + a_7(u,\xi,\eta)^2 - a_1(u,\xi,\eta) a_4(u,\xi,\eta) - a_5(u,\xi,\eta) a_6(u,\xi,\eta) = 0,\\ &\alpha\left[a_1(u,\xi,\eta) a_6(u,\xi,\eta) + a_4(u,\xi,\eta) a_5(u,\xi,\eta)\right]\\ &\qquad = \left[m_5(\eta) + m_6(\eta)\right] a_7(u,\xi,\eta),\\ &\alpha\left[a_1(u,\xi,\eta)^2 + a_6(u,\xi,\eta)^2 - a_4(u,\xi,\eta)^2 - a_5(u,\xi,\eta)^2\right]\\ &\qquad = 4m_1(\eta)\, a_7(u,\xi,\eta), \end{aligned} \tag{4.27}$$

where the definition of α appears in Remark 4.3.

Remark 4.4:

Proposition 4.6 is used in the proofs of Theorems 4.8 and 4.9 which are omitted.

Proposition 4.10:

$\mathcal{S}_{WG} = \mathcal{S}_{BAX} \cup \mathcal{S}_{FER}$, where $\mathcal{S}_{BAX}$ and $\mathcal{S}_{FER}$ denote the set of all solutions of the system of equations (4.26) and that of the system of equations (4.27) respectively.

The fourth formula in (4.26) represents the algebraic curve that the two weight functions $a_1(u,\xi,\eta)$ and $a_5(u,\xi,\eta)$ satisfy. It follows from the theory for algebraic curves that its genus is 1 (Griffiths and Harris 1978). Similarly, the genus of the algebraic curve in the second system is also 1. This means that the weight functions $a_1(u,\xi,\eta), a_5(u,\xi,\eta), a_7(u,\xi,\eta)$ are all elliptic functions.

4.6 Differential equations

If the weight functions satisfy (4.26), we make the symmetric operation with respect to (4.26). Thus we have

$$\begin{aligned} &a_5(u,\xi,\eta) = a_6(u,\xi,\eta), \qquad a_1(u,\xi,\eta) = a_4(u,\xi,\eta), \\ &\alpha\, a_1(u,\xi,\eta)\, a_6(u,\xi,\eta) - m_5(\xi)\, a_7(u,\xi,\eta) = 0, \\ &\alpha^2\, a_1(u,\xi,\eta)^2 a_6(u,\xi,\eta)^2 - m_6^2(\xi)\, a_6(u,\xi,\eta)^2 - m_6^2(\xi) a_1(u,\xi,\eta)^2 \\ &\quad + 2m_6(\xi)\, m_1(\xi)\, a_1(u,\xi,\eta) a_6(u,\xi,\eta) + m_6^2(\xi) = 0. \end{aligned} \tag{4.28}$$

Combining the first and the third equations in (4.26) and the third one in (4.28), we have $m_5(\xi) = m_6(\xi)$ independent of the color parameter. Similarly, combining the second and the fourth equations in (4.26) and the fourth in (4.28) we have $m_1(\xi)) = m_4(\xi)$ independent of the color parameter.

Let $m_5(\xi) = m_6(\xi) = \beta$ and $m_1(\xi) = m_4(\xi) = \gamma$.

If we combine (4.26) and (4.21), (4.27) and (4.21), two new systems of equations including some differential equations are obtained. Similarly, we regard $a_1'(u,\xi,\eta)$, $a_4'(u,\xi,\eta), \ldots, a_7'(u,\xi,\eta)$ in the two new systems as indeterminates and then solve the systems by applying Wu's method. In this way, we can get an even stronger result than Theorem 4.9.

Theorem 4.11:

The weight functions of the gauge solutions of CSYBE must satisfy the following two systems of equations:

$$\begin{aligned} &a_5(u,\xi,\eta) = a_6(u,\xi,\eta), \\ &a_1(u,\xi,\eta) = a_4(u,\xi,\eta), \\ &\alpha\, a_1(u,\xi,\eta)\, a_5(u,\xi,\eta) - \beta\, a_7(u,\xi,\eta) = 0, \\ &\alpha^2\, u_1^2 u_5^2 - \beta^2\, u_5^2 - \beta^2\, u_1^2 + 2\beta\,\gamma\, u_1 u_5^2 + \beta^2 = 0, \\ &a_5'(u,\xi,\eta)^2 = \beta^2 - (\beta^2 - \gamma^2 + \alpha^2)\, a_5(u,\xi,\eta)^2 + \alpha^2\, a_5(u,\xi,\eta)^4, \\ &a_1'(u,\xi,\eta)^2 = \beta^2 - (\beta^2 - \gamma^2 + \alpha^2)\, a_1(u,\xi,\eta)^2 + \alpha^2\, a_1(u,\xi,\eta)^4, \end{aligned} \tag{4.29}$$

and

$$\begin{aligned}
&1 + a_7(u,\xi,\eta)^2 - a_1(u,\xi,\eta)a_4(u,\xi,\eta) - a_5(u,\xi,\eta)a_6(u,\xi,\eta) = 0,\\
&\alpha\,[a_1(u,\xi,\eta)a_6(u,\xi,\eta) + a_4(u,\xi,\eta)a_5(u,\xi,\eta)]\\
&\quad = [m_5(\eta) + m_6(\eta)]\,a_7(u,\xi,\eta),\\
&\alpha\,[a_1(u,\xi,\eta)^2 + a_6(u,\xi,\eta)^2 - a_4(u,\xi,\eta)^2 - a_5(u,\xi,\eta)^2]\\
&\quad = 4m_1(\eta)\,a_7(u,\xi,\eta),\\
&a_7'(u,\xi,\eta)^2 = \alpha^2 - \{[m_5(\eta)^2 + m_6(\eta)]^2 - 4\,m_1^2(\eta) - 2\,\alpha^2\}a_7(u,\xi,\eta)^2\\
&\quad + \alpha^2 a_7(u,\xi,\eta)^4.
\end{aligned} \tag{4.30}$$

When making the symmetric operation with respect to the fourth equation in (4.30), we have

$$\begin{aligned}
&a_7'(u,\xi,\eta)^2 = \alpha^2 - \{[m_5(\xi)^2 + m_6(\xi)]^2 - 4\,m_1^2(\xi) - 2\,\alpha^2\}a_7(u,\xi,\eta)^2\\
&\quad + \alpha^2 a_7(u,\xi,\eta)^4.
\end{aligned} \tag{4.31}$$

Comparing the fourth equation in (4.30) with (4.31), we claim that $[m_5(\xi)^2 + m_6(\xi)]^2 - 4\,m_1^2(\xi) - 2\,\alpha^2$ is independent of the color parameter. Let it be Δ. So

$$a_7'(u,\xi,\eta)^2 = \alpha^2 - \Delta a_7(u,\xi,\eta)^2 + \alpha^2 a_7(u,\xi,\eta)^4. \tag{4.32}$$

4.7 Gauge solutions

In this section $\mathrm{sn}(\zeta)$ and $\mathrm{cd}(\zeta) = \mathrm{cn}(\zeta)/\mathrm{dn}(\zeta)$ are the Jacobian elliptic functions.

Based on the equations in Theorem 4.11, all gauge solutions can be given and classified into two types, called *Baxter type* and *Free-Fermion type*. In what follows we explain how to write down the solutions and to classify them.

4.7.1 Baxter type solutions: $m_1 = m_4$

We consider the first case in Theorem 4.11; then $m_1 = m_4$. Since the curve defined by the fourth equation in (4.29) only includes two weight functions $a_1(u,\xi,\eta)$ and $a_5(u,\xi,\eta)$, we can parametrize $a_1(u,\xi,\eta)$ and $a_5(u,\xi,\eta)$ as functions of one variable. If $\beta \pm \alpha \pm \gamma \neq 0$, the fifth and the sixth equations in (4.29) are elliptic differential equations. Therefore, the solutions should be

$$\begin{aligned} a_1(u,\xi,\eta) &= a_4(u,\xi,\eta) = \frac{\operatorname{sn}(\lambda u + F(\xi) - F(\eta) + \mu)}{\operatorname{sn}(\mu)}, \\ a_2(u,\xi,\eta) &= a_3(u,\xi,\eta) = 1, \\ a_5(u,\xi,\eta) &= a_6(u,\xi,\eta) = \pm\frac{\operatorname{sn}(\lambda u + F(\xi) - F(\eta))}{\operatorname{sn}(\mu)}, \\ a_7(u,\xi,\eta) &= \pm k \operatorname{sn}(\lambda u + F(\xi) - F(\eta)) \operatorname{sn}(\lambda u + F(\xi) - F(\eta) + \mu), \end{aligned} \tag{4.33}$$

where k, as the module of Jacobi elliptic functions, is an arbitrary constant.

If $\beta \pm \alpha \pm \gamma = 0$, then the elliptic solutions (4.33) degenerate into trigonometric solutions

$$\begin{aligned} a_1(u,\xi,\eta) &= a_4(u,\xi,\eta) = \frac{\tan(\lambda u + F(\xi) - F(\eta) + \mu)}{\tan(\mu)}, \\ a_2(u,\xi,\eta) &= a_3(u,\xi,\eta) = 1, \\ a_5(u,\xi,\eta) &= a_6(u,\xi,\eta) = \pm\frac{\tan(\lambda u + F(\xi) - F(\eta))}{\tan(\mu)}, \\ a_7(u,\xi,\eta) &= \pm \tan(\lambda u + F(\xi) - F(\eta)) \tan(\lambda u + F(\xi) - F(\eta) + \mu). \end{aligned} \tag{4.34}$$

In (4.33) and (4.34), $\lambda \neq 0$ and $\mu \neq 0$ are two arbitrary constants and $F(\xi)$ is an arbitrary function.

4.7.2 Free-Fermion type solutions: $m_1(\xi) = -m_4(\xi)$.

Now we consider the second case in Theorem 4.11. According to Theorem 4.8 it is divided into two sub-cases: $m_5(\xi) = m_6(\xi)$ and $m_5(\xi) = -m_6(\xi)$.

A. The sub-case $m_5(\xi) = m_6(\xi)$. Assume that $m_5(\xi) \not\equiv 0$ (the case $m_5(\eta) = 0$ will be put into the second sub-case). It is clear that $\alpha \neq 0$ due to the second equation in (4.30). For brevity we let $\alpha = 1$ up to the solution transformation D. When

$$m_5(\xi)^2 - m_1(\xi)^2 \not\equiv 0 \quad \text{and} \quad m_5(\xi)^2 - m_1(\xi)^2 \not\equiv 1,$$

$\Delta \not\equiv 0$ and thus the equation (4.32) is an elliptic differential equation. From the initial condition (4.18), the solution should be

$$a_7(u,\xi,\eta) = k \operatorname{sn}(\lambda u + F(\xi) - F(\eta)) \operatorname{cd}(\lambda u + F(\xi) - F(\eta)), \tag{4.35}$$

where k (as the module of elliptic functions) and λ are two constants with $k\lambda = 1$ and $F(\xi)$ an arbitrary function.

Substituting (4.35) into the second equation in (4.21) and its counterpart as well as the first and second equations in (4.30) and using properties of elliptic functions, we have

$$\mathrm{cd}^2 - \mathrm{sn}^2 + 2m_1(\eta)k\,\mathrm{cd}\,\mathrm{sn} + u_5^2 - u_1^2 = 0, \tag{4.36}$$

$$\mathrm{cd}^2 - \mathrm{sn}^2 - 2m_1(\eta)k\,\mathrm{cd}\,\mathrm{sn} + u_6^2 - u_4^2 = 0, \tag{4.37}$$

$$u_1u_4 + u_5u_6 - \mathrm{cd}^2 - \mathrm{sn}^2 = 0, \tag{4.38}$$

$$u_1u_6 + u_4u_5 - 2m_5(\eta)k\,\mathrm{cd}\,\mathrm{sn} = 0, \tag{4.39}$$

where $m_5(\eta)$ and $m_1(\eta)$ are two functions satisfying

$$m_5(\eta)^2 - m_1(\eta)^2 = \frac{1}{k^2}.$$

In the formulas (4.36)–(4.39) and in what follows, we simply write sn and cd instead of the elliptic functions $\mathrm{sn}(\lambda\,u + F(\xi) - F(\eta))$ and $\mathrm{cd}(\lambda\,u + F(\xi) - F(\eta))$ for brevity. Using the symmetric operation we also have

$$\mathrm{cd}^2 - \mathrm{sn}^2 + 2m_4(\xi)k\,\mathrm{cd}\,\mathrm{sn} + u_5^2 - u_4^2 = 0.$$

Since $m_1(\xi) + m_4(\xi) = 0$, one obtains

$$\mathrm{cd}^2 - \mathrm{sn}^2 - 2m_1(\xi)k\,\mathrm{cd}\,\mathrm{sn} + u_5^2 - u_4^2 = 0. \tag{4.40}$$

From (4.38), (4.39) and (4.36) it follows that

$$-(\mathrm{cd}^2 + \mathrm{sn}^2)u_1 + [\mathrm{cd}^2 - \mathrm{sn}^2 + 2m_1(\eta)k\,\mathrm{sn}\,\mathrm{cd}]u_4 + 2m_5(\eta)k\,\mathrm{sn}\,\mathrm{cd}u_5 = 0. \tag{4.41}$$

If we perform the symmetric operation with respect to the counterpart of (4.41), then

$$-(\mathrm{cd}^2 + \mathrm{sn}^2)u_4 + [\mathrm{cd}^2 - \mathrm{sn}^2 - 2m_1(\xi)k\,\mathrm{sn}\,\mathrm{cd}]u_1 + 2m_5(\xi)k\,\mathrm{sn}\,\mathrm{cd}u_5 = 0. \tag{4.42}$$

Similarly, one has

$$[\mathrm{sn}^2 + 2m_1(\eta)k\,\mathrm{sn}\,\mathrm{cd} - \mathrm{cd}^2]u_5 - (\mathrm{cd}^2 + \mathrm{sn}^2)u_6 + 2m_5(\eta)k\,\mathrm{sn}\,\mathrm{cd}u_4 = 0, \tag{4.43}$$

$$[\mathrm{sn}^2 + 2m_1(\xi)k\,\mathrm{cd}\,\mathrm{sn} - \mathrm{cd}^2]u_6 + (\mathrm{cd}^2 + \mathrm{sn}^2)u_5 + 2m_5(\xi)k\,\mathrm{cd}\,\mathrm{sn}u_4 = 0. \tag{4.44}$$

Solving the equations (4.41)–(4.44) for u_1, u_4, u_5, u_6, we have

$$\frac{a_4(u,\xi,\eta)}{a_1(u,\xi,\eta)} = \frac{H_4}{H_1}, \qquad \frac{a_6(u,\xi,\eta)}{a_5(u,\xi,\eta)} = \frac{H_6}{H_5},$$

where

$$\begin{aligned}
H_1 &= [m_5(\xi) + m_5(\eta)]\,\mathrm{cd} + [m_1(\xi)m_5(\eta) + m_5(\xi)m_1(\eta)]k\,\mathrm{sn},\\
H_4 &= [m_5(\xi) + m_5(\eta)]\,\mathrm{cd} - [m_1(\xi)m_5(\eta) + m_5(\xi)m_1(\eta)]k\,\mathrm{sn},\\
H_5 &= [m_5(\xi) + m_5(\eta)]\,\mathrm{sn} + [m_1(\xi)m_5(\eta) - m_5(\xi)m_1(\eta)]k\,\mathrm{cd},\\
H_6 &= [m_5(\xi) + m_5(\eta)]\,\mathrm{sn} - [m_1(\xi)m_5(\eta) - m_5(\xi)m_1(\eta)]k\,\mathrm{cd}.
\end{aligned}$$

Let $u_1 = H_1X, u_4 = H_4X$ and $u_5 = H_5Y, u_6 = H_6Y$. From (4.36) and (4.40) one obtains

$$\begin{aligned}
(H_1^2 - H_4^2)X^2 &= 4\,[m_5(\xi) + m_5(\eta)]\,[m_5(\eta)m_1(\xi) + m_5(\xi)m_1(\eta)]\,k\,\mathrm{sn}\,\mathrm{cd}X^2\\
&= 2[m_1(\xi) + m_1(\eta)]k\,\mathrm{sn}\,\mathrm{cd},\\
(H_5^2 - H_6^2)Y^2 &= 4\,[m_5(\xi) + m_5(\eta)]\,[m_5(\eta)m_1(\xi) - m_5(\xi)m_1(\eta)]\,k\,\mathrm{sn}\,\mathrm{cd}Y^2\\
&= 2[m_1(\xi) - m_1(\eta)]k\,\mathrm{sn}\,\mathrm{cd}.
\end{aligned}$$

It is easy to show, using $m_5(\xi)^2 - m_1(\xi)^2 = 1/k^2$, that

$$\begin{aligned}
X^2 &= \frac{m_1(\xi) + m_1(\eta)}{2[m_5(\xi) + m_5(\eta)][m_1(\xi)m_5(\eta) + m_5(x)m_1(\eta)]}\\
&= \frac{1 + k^2[m_5(\xi)m_5(\eta) - m_1(\xi)m_1(\eta)]}{2[m_5(\xi) + m_5(\eta)]^2}\\
&= \frac{-1 + k^2[m_5(\xi)m_5(\eta) + m_1(\xi)m_1(\eta)]}{2[m_1(\xi)m_5(\eta) + m_5(\xi)m_1(\eta)]^2k^2},\\
Y^2 &= \frac{m_1(\xi) - m_1(\eta)}{2[m_5(\xi) + m_5(\eta)][m_1(\xi)m_5(\eta) - m_5(x)m_1(\eta)]}\\
&= \frac{1 + k^2[m_5(\xi)m_5(\eta) + m_1(\xi)m_1(\eta)]}{2[m_5(\xi) + m_5(\eta)]^2}\\
&= \frac{-1 + k^2[m_5(\xi)m_5(\eta) - m_1(\xi)m_1(\eta)]}{2[m_1(\xi)m_5(\eta) - m_5(\xi)m_1(\eta)]^2k^2}.
\end{aligned}$$

Hence the gauge solutions of the color Yang–Baxter equation (4.1) should have the following form:

$$\begin{aligned}
a_1(u,\xi,\eta) &= A(\xi,\eta)\,\mathrm{cd} + B(\xi,\eta)\,\mathrm{sn},\\
a_2(u,\xi,\eta) &= a_3(u,\xi,\eta) = 1,\\
a_4(u,\xi,\eta) &= A(\xi,\eta)\,\mathrm{cd} - B(\xi,\eta)\,\mathrm{sn},\\
a_5(u,\xi,\eta) &= C(\xi,\eta)\,\mathrm{sn} + D(\xi,\eta)\,\mathrm{cd},\\
a_6(u,\xi,\eta) &= C(\xi,\eta)\,\mathrm{sn} - D(\xi,\eta)\,\mathrm{cd},\\
a_7(u,\xi,\eta) &= \pm k\,\mathrm{sn}\,\mathrm{cd},
\end{aligned} \tag{4.45}$$

where k, as the module of the elliptic functions, is an arbitrary constant and

$$\begin{aligned}
A(\xi,\eta) &= \sqrt{[1+G(\xi)G(\eta)-H(\xi)H(\eta)]/2},\\
B(\xi,\eta) &= \sqrt{[-1+G(\xi)G(\eta)+H(\xi)H(\eta)]/2},\\
C(\xi,\eta) &= \delta\sqrt{[1+G(\xi)G(\eta)+H(\xi)H(\eta)]/2},\\
D(\xi,\eta) &= \delta\sqrt{[-1+G(\xi)G(\eta)-H(\xi)H(\eta)]/2}\,M,
\end{aligned} \tag{4.46}$$

with

$$\delta^2 = 1, \qquad M = \frac{H(\xi)G(\eta)-G(\xi)H(\eta)}{\sqrt{[H(\xi)G(\eta)-G(\xi)H(\eta)]^2}}$$

and $G(\xi)$, $H(\xi)$ satisfying $G(\xi)^2 - H(\xi)^2 = 1$. If we consider the solution transformation D, then the restrictive condition $k\,\lambda = 1$ can be cancelled; namely, $\lambda \neq 0$ is also an arbitrary constant.

When taking the module $k = 1$, the Jacobian elliptic functions cd and sn should degenerate into 1 and $\tanh(\lambda\,u + F(\xi) - F(\eta))$ respectively. Hence, we have

$$\begin{aligned}
a_1(u,\xi,\eta) &= A(\xi,\eta) + B(\xi,\eta)\tanh(\lambda\,u + F(\xi) - F(\eta)),\\
a_2(u,\xi,\eta) &= a_3(u,\xi,\eta) = 1,\\
a_4(u,\xi,\eta) &= A(\xi,\eta) - B(\xi,\eta)\tanh(\lambda\,u + F(\xi) - F(\eta)),\\
a_5(u,\xi,\eta) &= C(\xi,\eta)\tanh(\lambda\,u + F(\xi) - F(\eta)) + D(\xi,\eta),\\
a_6(u,\xi,\eta) &= C(\xi,\eta)\tanh(\lambda\,u + F(\xi) - F(\eta)) - D(\xi,\eta),\\
a_7(u,\xi,\eta) &= \pm\tanh(\lambda\,u + F(\xi) - F(\eta))[\lambda\,u + F(\xi) - F(\eta)],
\end{aligned} \tag{4.47}$$

where $A(\xi,\eta)$, $B(\xi,\eta)$, $C(\xi,\eta)$ and $D(\xi,\eta)$ are defined by (4.46) and $\lambda \neq 0$ is an arbitrary constant.

If $m_5(\eta)^2 - m_1(\eta)^2 = 0$, then the differential equation (2.29) can be rewritten as

$$a_7'(u,\xi,\eta)^2 = \alpha^2[1 + 2\,a_7(u,\xi,\eta)^2 + a_7(u,\xi,\eta)^4].$$

So, following the calculation of (4.45) we can show that the gauge solutions are

$$
\begin{aligned}
a_1(u,\xi,\eta) &= X\left[\frac{G(\xi)+G(\eta)}{C} + 2G(\xi)G(\eta)S\right],\\
a_2(u,\xi,\eta) &= a_3(u,\xi,\eta) = 1,\\
a_4(u,\xi,\eta) &= X\left[\frac{G(\xi)+G(\eta)}{C} - 2G(\xi)G(\eta)S\right],\\
a_5(u,\xi,\eta) &= Y\left[\frac{G(\xi)-G(\eta)}{C} + 2G(\xi)G(\eta)S\right],\\
a_6(u,\xi,\eta) &= Y\left[-\frac{G(\xi)-G(\eta)}{C} + 2G(\xi)G(\eta)S\right],\\
a_7(u,\xi,\eta) &= \pm\tan(\lambda u + F(\xi) - F(\eta)),
\end{aligned}
\tag{4.48}
$$

where

$$
X = \frac{1}{2\sqrt{G(\xi)G(\eta)}}, \qquad Y = \pm X,
$$

$$
C = \cos(\lambda u + F(\xi) - F(\eta)), \qquad S = \sin(\lambda u + F(\xi) - F(\eta))
$$

and $G(\xi)$ is an arbitrary function.

B. The sub-case $m_5(\xi) = -m_6(\xi)$. In the case $m_5(\eta) = -m_6(\eta)$, the weight functions of the gauge solutions are

$$
\begin{aligned}
a_1(u,\xi,\eta) &= a_4(u,\xi,\eta) = \frac{\cosh(\lambda u + F(\xi) - F(\eta))}{\cos(\mu u + G(\xi) - G(\eta))},\\
a_2(u,\xi,\eta) &= a_3(u,\xi,\eta) = 1,\\
a_5(u,\xi,\eta) &= -a_6(u,\xi,\eta) = \pm\frac{\sinh(\lambda u + F(\xi) - F(\eta))}{\cos(\mu u + G(\xi) - G(\eta))},\\
a_7(u,\xi,\eta) &= \pm\tan(\mu u + G(\xi) - G(\eta)),
\end{aligned}
\tag{4.49}
$$

where λ and μ are two arbitrary constants, not simultaneously zero, and $F(\xi)$, $G(\xi)$ are two arbitrary functions.

To prove it we first consider the sub-case $\alpha \neq 0$. It follows from the fourth equation in (4.30) that

$$
a_1(u,\xi,\eta)a_6(u,\xi,\eta) + a_4(u,\xi,\eta)a_5(u,\xi,\eta) = 0
$$

and

$$
a_1(u,\xi,\eta)a_5(u,\xi,\eta) + a_4(u,\xi,\eta)a_6(u,\xi,\eta) = 0
$$

by performing the symmetric operation. Since $a_1(u,\xi,\eta) \neq -a_4(u,\xi,\eta)$ owing to the initial condition (4.18), one can get

$$\begin{aligned} &a_1(u,\xi,\eta) = a_4(u,\xi,\eta),\\ &a_5(u,\xi,\eta) = -a_6(u,\xi,\eta),\\ &a_1(u,\xi,\eta)^2 - a_5(u,\xi,\eta)^2 = 1 + a_7(u,\xi,\eta)^2, \end{aligned} \tag{4.50}$$

and thus $m_1(\eta) = 0$ because of the first equation in (4.50) and the condition $m_1(\xi) + m_4(\xi) = 0$. Hence (4.32) has solution

$$a_7(u,\xi,\eta) = \tan(\mu\, u + G(\xi) - G(\eta))$$

and (4.49) is true.

Now we consider the sub-case $\alpha = 0$. It follows from the argument of Proposition 4.6 in Sect. 4.4 that the weight functions are

$$\begin{aligned} &a_1(u,\xi,\eta) = A_1(\xi,\eta)\cosh(u) - A_6(\xi,\eta)\sinh(u),\\ &a_4(u,\xi,\eta) = A_4(\xi,\eta)\cosh(u) + A_5(\xi,\eta)\sinh(u),\\ &a_5(u,\xi,\eta) = A_5(\xi,\eta)\cosh(u) + A_4(\xi,\eta)\sinh(u),\\ &a_6(u,\xi,\eta) = A_6(\xi,\eta)\cosh(u) - A_1(\xi,\eta)\sinh(u),\\ &a_7(u,\xi,\eta) = A_7(\xi,\eta), \end{aligned} \tag{4.51}$$

where $A_i(\xi,\eta) = a_i(0,\xi,\eta)$ $(i = 1,4,\ldots,7)$ are functions with respect to the color parameters ξ,η considered as indeterminates. It is clear that $A_i(\xi,\eta)$ $(i = 1,4,\ldots,7)$ satisfy the Yang–Baxter equation with color parameters only.

If we substitute the initial condition (4.18) into the system obtained from the system (4.16) by putting $v = -u$ and $\lambda = \eta$ and combine the resulting equations with the first equation in (4.30), which is called the *Free-Fermion condition* (Fan and Wu 1970), one can show that

$$\begin{aligned} &a_5(u,\xi,\eta) = -a_5(-u,\eta,\xi), \qquad a_6(u,\xi,\eta) = -a_6(-u,\eta,\xi),\\ &a_7(u,\xi,\eta) = -a_7(-u,\eta,\xi), \qquad a_4(u,\xi,\eta) = a_1(-u,\eta,\xi). \end{aligned} \tag{4.52}$$

By (4.52) it is easily shown that

$$A_4(\xi,\eta) = A_1(\xi,\eta), \qquad A_6(\xi,\eta) = -A_5(\xi,\eta). \tag{4.53}$$

As mentioned in Remark 4.2, all formulas obtained in this and other sections, except those obtained by using symmetric operation, should hold for the YBE with only color parameters. So we still have

$$\begin{aligned} &1 + A_7(\xi,\eta)^2 - A_1(\xi,\eta)A_4(\xi,\eta) - A_5(\xi,\eta)A_6(\xi,\eta) = 0,\\ &l_7(\eta)(A_1(\xi,\eta)A_6(\xi,\eta) + A_4(\xi,\eta)A_5(\xi,\eta)) = (l_5(\eta) + l_6(\eta))A_7(\xi,\eta),\\ &l_7(\eta)(A_1(\xi,\eta)^2 + A_6(\xi,\eta)^2 - A_4(\xi,\eta)^2 - A_5(\xi,\eta)^2 = 4\,l_1(\eta)A_7(\xi,\eta), \end{aligned} \tag{4.54}$$

where

$$l_i(\eta) = \frac{\partial}{\partial\eta} A_i(\xi,\eta)\Big|_{\xi=\eta} \qquad (i = 1, 4, \ldots, 7)$$

as mentioned in Remark 4.2. Substituting (4.53) into (4.54), we see that

$$l_5(\eta) + l_6(\eta) = 0, \qquad l_1(\eta) = 0.$$

Therefore, as done for getting (4.32) we also have

$$\left[\frac{\partial}{\partial\eta} A_7(\xi,\eta)\right]^2 = l_7(\eta)[1 + 2\,a_7(u,\xi,\eta)^2 + a_7(u,\xi,\eta)^4],$$

and thus the solution for $A_i(\xi,\eta)$ is

$$\begin{aligned} A_1(\xi,\eta) = A_4(\xi,\eta) &= \frac{\cosh(F(\xi) - F(\eta))}{\cos(G(\xi) - G(\eta))}, \\ A_5(\xi,\eta) = -A_6(\xi,\eta) &= \frac{\sinh(F(\xi) - F(\eta))}{\cos(G(\xi) - G(\eta))}, \\ A_7(\xi,\eta) &= \tan(G(\xi) - G(\eta)), \end{aligned} \tag{4.55}$$

where F and G are two arbitrary functions of a single variable. Substituting (4.55) into (4.51), one can find that (4.49) is also true for the case $\alpha = 0$, only if $\mu = 0$.

The discussions above tell us that, for the case of Hamiltonian coefficients $m_5(\eta) = -m_6(\eta)$, the weight functions of a gauge eight-vertex type solution must be (4.49).

Finally, straightforward calculations show that (4.33), (4.45), and (4.49) and their degenerate forms (4.34), (4.47), and (4.48) are indeed solutions of CSYBE. Based on Propositions 4.2, 4.3, 4.4, 4.7, and 4.10, we have the main theorem of this chapter.

Theorem 4.12:

The gauge eight-vertex solutions of CSYBE (4.1) *comprise* (4.33), (4.45), *and* (4.49) *and their degenerate forms* (4.34), (4.47), *and* (4.48).

The solutions (4.33) and (4.34) are called *Baxter type solutions*. They are just the solutions for "zero field" eight-vertex model by Baxter (1982). The solutions (4.45), (4.49) and their degenerate forms (4.47), (4.48) satisfy the Free-Fermion condition (Fan and Wu 1970) and are called *Free-Fermion type solutions*.

If we take $\lambda = 1$, $G(\xi) = \cosh(2\xi)$, $H(\xi) = \sinh(2\xi)$ and $F(\xi) = 0$ in the solution (4.45), then (4.45) reduces to the following solution:

$$\begin{aligned}
a_1(u,\xi,\eta) &= \cosh(\xi-\eta)\mathrm{cd}(u) + \sinh(\xi+\eta)\mathrm{sn}(u),\\
a_4(u,\xi,\eta) &= \cosh(\xi-\eta)\mathrm{cd}(u) - \sinh(\xi+\eta)\mathrm{sn}(u),\\
a_5(u,\xi,\eta) &= \cosh(\xi+\eta)\mathrm{sn}(u) - \sinh(\xi-\eta)\mathrm{cd}(u),\\
a_6(u,\xi,\eta) &= \cosh(\xi+\eta)\mathrm{sn}(u) + \sinh(\xi-\eta)\mathrm{cd}(u),\\
a_7(u,\xi,\eta) &= k\,\mathrm{sn}(u)\,\mathrm{cd}(u),
\end{aligned}$$

which is given in Murakami (1992).

4.8 General solutions

From Theorem 4.12 and the argument in the second section, we obtain the following result.

Theorem 4.13:

Up to the five solution transformations A–E the gauge solutions (4.33), (4.45), (4.49) *and their degenerate forms* (4.34), (4.47), (4.48) *as well as the two trivial solutions* (4.19) *and* (4.20) *give all eight-vertex type solutions of CSYBE.*

We take

$$G(\xi) = \frac{1}{\mathrm{sn}(\xi)}, \qquad H(\xi) = \frac{\mathrm{cn}(\xi)}{\mathrm{sn}(\xi)}, \qquad F(\xi) = 0, \qquad \lambda = \frac{1}{2}$$

in (4.45) and (4.46) and the solution transformation B with

$$g(u,\xi,\eta) = \sqrt{e(\xi)e(\eta)\mathrm{sn}(\xi)\mathrm{sn}(\eta)}\,\frac{[1-e(u)]}{\mathrm{sn}(u/2)},$$

where

$$e(\zeta) = \mathrm{cn}(\zeta) + \boldsymbol{i}\,\mathrm{sn}(\zeta)$$

is the elliptic exponential. After some detailed calculation by using additional theorems for elliptic functions $\mathrm{sn}(\zeta), \mathrm{cn}(\zeta)$ and $\mathrm{dn}(\zeta)$, we can obtain the following solution given by Bazhanov and Stroganov (1985):

$$\begin{aligned}
a_1(u,\xi,\eta) &= 1 - e(u)\, e(\xi)\, e(\eta),\\
a_2(u,\xi,\eta) &= a_3(u,\xi,\eta) = \sqrt{e(\xi)e(\eta)\mathrm{sn}(\xi)\mathrm{sn}(\eta)}\ \frac{[1-e(u)]}{\mathrm{sn}(u/2)},\\
a_4(u,\xi,\eta) &= e(u) - e(\xi)\, e(\eta),\\
a_5(u,\xi,\eta) &= e(\xi) - e(u)\, e(\eta),\\
a_6(u,\xi,\eta) &= e(\eta) - e(u)\, e(\xi),\\
a_7(u,\xi,\eta) &= a_8(u,\xi,\eta) = -\boldsymbol{i}\, k \sqrt{e(\xi)e(\eta)\mathrm{sn}(\xi)\mathrm{sn}(\eta)}\ [1-e(u)]\,\mathrm{sn}(u/2).
\end{aligned}$$

We omit the details for the involved calculations.

Moreover all non-trivial general solutions can also be classified into two types. The first is Baxter type of solutions if they can be obtained via gauge Baxter solutions together with some solution transformations. The second is Free-Fermion type of solutions if they can be obtained via gauge Free-Fermion solutions with some solution transformations.

According to the standard method of Baxter (1982), for a given R-matrix solution of YBE the Hamiltonian for spin-chain models is generally of the following form:

$$H = \sum_{j=1}^{N} [J_x\, \sigma_j^x \sigma_{j+1}^x + J_y\, \sigma_j^y \sigma_{j+1}^y + J_z\, \sigma_j^z \sigma_{j+1}^z + \frac{1}{2} h\, (\sigma_j^z + \sigma_{j+1}^z)],$$

where σ^x, σ^y, σ^z are Pauli matrices and J_x, J_y, J_z, h are the coupling constants. The relations between them and the Hamiltonian coefficients m_I are

$$J_x = \frac{1}{4}(m_5 + m_6 + m_7 + m_8), \qquad J_y = \frac{1}{4}(m_5 + m_6 - m_7 - m_8),$$

$$J_z = \frac{1}{4}(m_1 - m_3 + m_4 - m_2), \qquad h = \frac{1}{4}(m_1 - m_3 - m_4 + m_2).$$

In this work we have proved that the Hamiltonian coefficients of a gauge solution must satisfy

$$m_1^2 = m_4^2, \qquad m_5^2 = m_6^2.$$

It follows from the solution transformations B and D that

$$(m_1 - m_3)^2 = (m_4 - m_2)^2, \qquad m_5^2 = m_6^2$$

for general solutions. This clearly describes the relations between classification of eight-vertex type solutions and spin-chain Hamiltonians. For example, if

$J_x + J_y = h$, $J_z = 0$, i.e., a special Free-Fermion model in a magnetic field, then $m_5 = m_1 - m_3 = -m_4 + m_2$. The corresponding solution of the color YBE should be (4.48). Thus the transfer matrix is of trigonometric function type.

4.9 Conclusions

This chapter describes how to construct solutions of YBE using Wu's method. As an example, all the eight-vertex type solutions of CSYBE dependent on spectral as well as color parameters are given. It is proved that they comprise three sets of gauge solutions, three sets of degenerate forms and two sets of trivial solutions up to five solution transformations. Moreover, all the non-trivial solutions can be classified into two types, the Baxter type and the Free-Fermion type. By a similar method, Shi (1992, 1993) obtained all the solutions of YBE with simple type and spin 1. Sun and Wang (1995), and Sun et al. (1995a,b, 1996) gave all the solutions of YBE of the following four types:

(1) YBE of six-vertex type with spectral parameter,

(2) YBE of six-vertex type with color parameter,

(3) YBE of six-vertex type with spectral and color parameter,

(4) YBE of eight-vertex type with spectral parameter.

Combining Wu's method and a method from algebraic geometry, Wang (1996) obtained all the eight-vertex solutions of YBE with color as well as spectral parameters.

Symbolic methods like Wu's have a lot of advantages in solving YBE. We know that only isolated solutions of YBE may be obtained by some other ways of solving; in this case no information about all the solutions is provided. However, it is guaranteed by Wu's theory that the solutions found by Wu's method cover all possible ones. For example, from the general solutions and some solution transformations given by us (Wang 1996, Sect. 7) it is easy to derive the solution by Murakami (1992) and Bazhanov and Stroganov (1985). Furthermore, the classification of solutions of YBE is also very important. This is because the classification of statistical models can be treated by using the classification for solutions of YBE. Some important formulas, including information about classifying the solutions, can be gotten by using Wu's method. For example, Theorem 4.8 obtained in this chapter also gives the information that the solutions can classified into two types, the Baxter type and the *Free-Fermion type*.

Applying Wu's method to eliminate indeterminates, we can obtain some algebraic curves or surfaces that weight functions of YBE satisfy. Thus we know how many parameters the solutions depend on. For example, there are only two variables and two independent parameters in the curve given by the fourth equation of (4.29). So there is only one independent variable in the Baxter type solution. This means that there is no difference between the Baxter type solutions with spectral parameter and those with color parameter. However, for Fermion type solutions there are three variables. Hence there is a big difference between the Fermion type solutions with spectral parameter and those with color parameter. This information may have relevance in some physical problems.

In principle and subject to the power of computers, one can solve the YBE of any type using Wu's method. On the other hand, the YBE provides a challenging example for Wu's theory and method to show their applicability and capability.

Part II

Automated Geometric Reasoning

Chapter 5

WU'S METHOD FOR AUTOMATED GEOMETRY THEOREM PROVING AND DISCOVERING

Shang-Ching Chou and Dongdai Lin

Starting with a brief review of automated geometry theorem proving (AGTP), this chapter concentrates on Wu's method of AGTP, the first method that can be used to prove hundreds of difficult geometric theorems efficiently. We give a short introduction to the general ideas, principles and algorithms underlying Wu's method for automated theorem proving in elementary geometry. Variants and improved versions of the method for proving and discovering different classes of theorems are discussed. In particular, we explain how theorems of constructive type, of equality type, in Euclidean geometry, non-Euclidean geometry, finite geometry, and ordered geometry can be proved or even discovered effectively and automatically.

5.1 A brief review of AGTP history

By *automated geometry theorem proving* (AGTP) we mean to take some class of geometric theorems as a whole and to establish a single uniform method for proving the theorems in the whole class, such that one may simply follow the instructions of the method, step by step and in a mechanical way, to reach the final conclusion of each geometric theorem in question. Proving a single theorem by a particularly designed method is not what we are concerned about here.

Generally, there are two approaches to AGTP: the logic approach and the algebraic computation approach. The earliest work on AGTP by computer programs was done by Gelernter (1959) and his collaborators. It was based on the human simulation approach and has been considered a landmark in the area of artificial intelligence for its time. L. A. Wos and his collaborators used their powerful general-purpose resolution theorem prover (based on the logic approach) to experiment with proving theorems in Tarski's axioms for elementary geometry (McCharen et al. 1976). In spite of the success with these

MATHEMATICS MECHANIZATION AND APPLICATIONS
ISBN 0-12-734760-7

techniques, the results did not lead to the development of an automated *and powerful* geometry theorem prover. For more discussion about and some current developments in this approach, see Chap. 10.

The original thought of algebraic approaches can be dated back to G. W. Leibniz and R. Descartes in the 17th century, and formulated by Hilbert at the beginning of the 20th century. In his classic book, Hilbert (1971) outlined a decision method for constructive statements in affine geometry. As pointed out by Wang (1960), the essence of mechanical theorem proving is to convert the qualitative difficulties inherent in usual mathematical proofs into quantitative complexities of calculation by standardizing the proof procedure in an algebraic manner. But because such quantitative complexities of calculation are far beyond the reach of human abilities, the idea of mechanization did not attract the proper attention until the middle of the 20th century.

Tarski (1951) proved that Euclidean geometry is mechanizable and presented a decision method for it. The method of Tarski is too complicated to be efficient, so that actually few theorems of geometrical interest have ever been proved by means of large modern computer. In spite of the subsequent improvements by Seidenberg (1954) and others, for years variations of Tarski's method have remained impractical for proving non-trivial theorems in geometry.

Along the Tarski line, Collins (1975) made an important contribution. His cylindrical algebraic decomposition (CAD) algorithm is currently one of the best algorithms of the Tarski type. This algorithm was implemented by D. Arnon, and several difficult algebra-geometry related problems were solved by Arnon's program (Arnon 1988; Arnon and Mignotte 1988). More recently, another major practical improvement of Collins' method has been made by Collins and Hong (1991). The program of Hong has been used to solve a number of quite difficult problems.

In fact, P. J. Davis was the first to explore the algebraic approach for proving geometric theorems using computer (Davis and Cerutti 1969). His approach for the computer proof of Pappus' theorem is essentially the one described by Hilbert, but he did not provide a unifying mechanical way to do it.

Restricting himself to a class of geometric statements of equality type, Wu (1978) introduced an algebraic method for proving theorems by computer. This method has been used to prove many difficult geometric theorems efficiently, and even to discover several new ones (Chou 1988; Wang 1989a,b; Wu 1984a,b). Wu's work became known outside China mainly through the two papers by Wu (1984c) and Chou (1984), and the quite encouraging fact that over 130 theorems were proved by the method in Chou (1984). It was clarified in Wu

(1984a) that the algebraic tools needed in Wu's approach can be developed from Ritt's work (Ritt 1938, 1950). The algebraic aspect of this approach is now known as the *characteristic set* (CS) *method* (see Chap. 1). It is now the case that hundreds of theorems in Euclidean and non-Euclidean geometries can be proved automatically by computer programs with Wu's method with various modifications (Chou 1988; Chou and Gao 1990b; Ko 1988; Kapur and Wan 1990; Lin and Liu 1992; Wang and Gao 1987; Wang and Zhi 1998).

Encouraged by the success of Wu's work, many authors have proposed methods for geometric theorem proving. In particular, the application of the Gröbner basis (GB) method to the same class of geometric theorems that Wu's method addresses has been investigated. In 1985–1986, three groups (Chou and Schelter 1986; Kapur 1986a,b; Kutzler and Stifter 1986a,b) reported practical successes; see Wang (1998b) for a recent review of AGTP based on GB. Other algebraic approaches of AGTP include the resultant approach (Kalkbrener 1995; Yang 1991; Yang et al. 1992; Kapur et al. 1994), and the approach based on Seidenberg's elimination method (Wang 1995b). Another interesting approach is to prove geometric theorems by checking single or multiple numerical examples (Hong 1986b,c; Wang 1988; Zhang et al. 1990), for which the details can be found in Chap. 9.

At the same time, automated derivation of geometric locus equations and other geometric formulas by using the CS methods were investigated (Wu 1986c,d; Chou 1987; Chou and Gao 1990a; Wang 1991c, 1995a). About 120 problems in geometry were solved by a program based on the CS method (Chou and Gao 1989a). For the automated derivation of formulas in differential case, see Chap. 6.

Here we would like to remind the reader that all the methods mentioned in the two preceding paragraphs can only deal with theorems involving equalities, but not inequalities. By modifying his method, Wu proposed a method that can be used to solve inequality and optimization problems (Wu 1992c,d, 1994b; see also Chap. 1). In Chou et al. (1992), a different approach based on Wu's method was proposed and about 40 geometric theorems involving inequalities were proved. In McPhee et al. (1994), Wu's method and Collins' method are combined to solve a number of difficult problems including the so-called 8_3 configuration problem. Yang (1998) has proposed a method that can be used to prove a class of geometric inequalities efficiently; nearly one thousand inequalities have been proved by the program based on this method. For more details on proving theorems with inequalities, see Chou et al. (1992), Wang (1991c), Wu (1988, 1992c,d, 1994b) and Yang (1998).

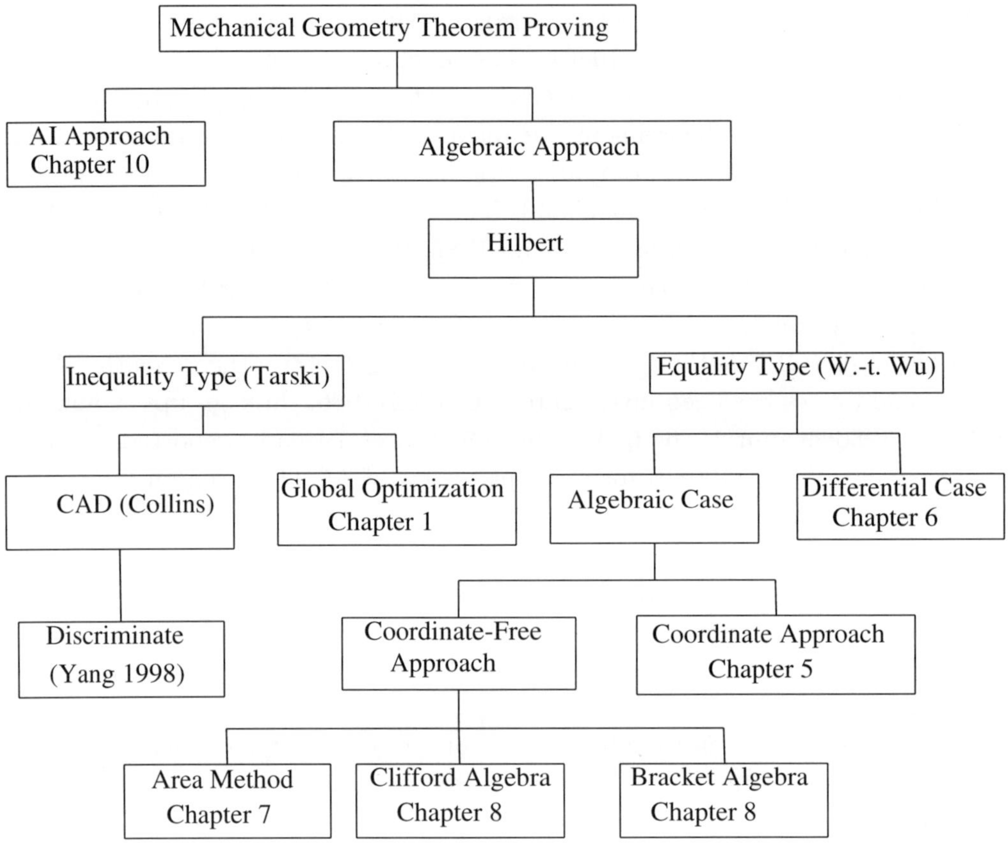

Figure 5.1 Methods of AGTP

In the early 1990s, the area method for proving geometric theorems was introduced (Zhang et al. 1992; Chou et al. 1993a, 1994a). This method is a coordinate-free approach and is based on a few geometric invariants, such as area and Pythagorean difference. The main advantage of this new method is that the proofs generated by it are generally short and human-readable. Detailed discussion of this topic can be found in Chap. 7. By combining Clifford algebraic operations and Wu's method, Li and Cheng (1997) proposed a coordinate-free AGTP method that can be used to prove theorems from both ordinary geometry and differential geometry. Details of this method and other coordinate free approaches of AGTP can be found in Chap. 8.

The above-mentioned work is concerned mainly with elementary geometries in Wu's sense, i.e., geometries in which no differentiation is involved. In fact,

Wu's method is also applicable to differential geometry (Wu 1979, 1980); refer to Chap. 6 for details.

In the rest of this chapter, we will focus on AGTP by Wu's method. Variants and improved versions of the method for automatically proving and discovering different classes of theorems, in particular, theorems of constructive type, of equality type, in Euclidean geometry, non-Euclidean geometry, finite geometry, and ordered geometry are presented. Illustrative examples are provided to show the high power of this method.

5.2 Wu's method of AGTP

AGTP is the most successful application of the general characteristic set method of Wu. Using Wu's method, more than 600 theorems have been proved by computer (Chou 1988; Wang and Gao 1987; Wu 1984a,b); some of them are even new.

5.2.1 A simple version of Wu's method

The procedure of proving geometric theorems by Wu's method consists of the following steps.

W1. [Algebraization.] Choose an appropriate coordinate system and assign coordinates to the points involved; then the geometric hypotheses can be translated to a set of polynomial equations

$$\begin{aligned} &H_1(y_1, \ldots, y_m) = 0, \\ &H_2(y_1, \ldots, y_m) = 0, \\ &\quad \cdots\cdots \\ &H_r(y_1, \ldots, y_m) = 0, \end{aligned}$$

and so can the conclusion to a polynomial equation $C(y_1, \ldots, y_m) = 0$, where the H_i's and C are polynomials with coefficients in a *base field* $\boldsymbol{K}$. Usually, as we shall assume in this chapter, $\boldsymbol{K} = \boldsymbol{Q}$ (the field of rational numbers). Thus, proving the theorem is converted to deciding whether one can derive the conclusion equation from the hypothesis equations.

W2. [Triangulation.] Transform the set of hypothesis polynomials into a triangular form $CS = \{F_1, \ldots, F_p\}$ by Wu's well-ordering principle (see Chap. 1).

W3. [Successive division.] Divide the conclusion polynomial C by the polynomials in the triangular form CS successively to obtain the final remainder $R = \text{prem}(C, CS)$. If $R = 0$, then the theorem is valid, or valid under the subsidiary conditions $I_k \neq 0$, where I_k is the initial of F_k for each k.

See Chap. 1 for the definition of *prem* and *initial*. Now let us give an example to illustrate how this method works.

Example 5.1: (Desargues' theorem)

Let $A_1B_1C_1$ and $A_2B_2C_2$ be two triangles such that the corresponding sides are parallel to each other (Fig. 5.2). Suppose that A_1A_2 and B_1B_2 intersect at a point O. Then C_1C_2 also passes through O.

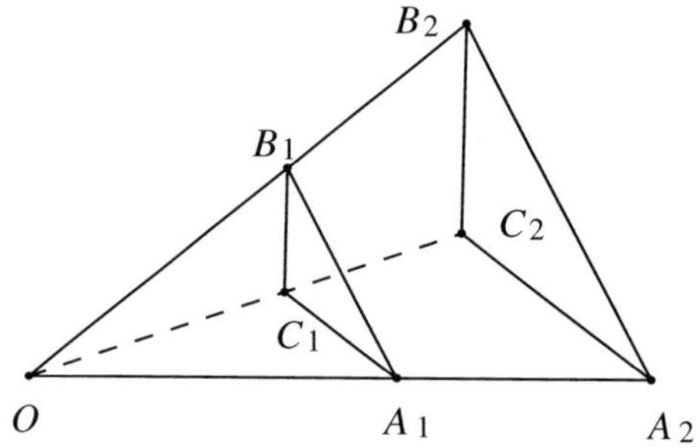

Figure 5.2 Desargues' theorem

Step 1. To convert the statement into algebraic form, we let $O = (0,0), A_1 = (x_1, 0), A_2 = (x_2, 0), B_1 = (x_3, x_4), B_2 = (x_5, x_6), C_1 = (x_7, x_8)$ and $C_2 = (x_9, x_{10})$. The hypotheses of the theorem can be translated into the following equations:

$$\begin{aligned}
H_1 &= x_4(x_5 - x_2) - x_6(x_3 - x_1) = 0, && (A_1B_1//A_2B_2)\\
H_2 &= x_8(x_9 - x_2) - x_{10}(x_7 - x_1) = 0, && (A_1C_1//A_2C_2)\\
H_3 &= (x_4 - x_8)(x_5 - x_9) - (x_6 - x_{10})(x_3 - x_7) = 0, && (B_1C_1//B_2C_2)\\
H_4 &= x_6x_3 - x_5x_4 = 0, && (O = A_1A_2 \wedge B_1B_2)
\end{aligned}$$

and the conclusion to one polynomial equation

$$C = x_9x_8 - x_{10}x_7 = 0. \qquad (C_1C_2 \text{ pass through } O)$$

Step 2. Triangularizing $HS = \{H_1, \ldots, H_4\}$ by Wu's well-ordering principle, we get $CS = \{F_1, \ldots, F_4\}$ with

$$\begin{aligned}
F_1 &= x_4(-x_3x_2 + x_5x_1),\\
F_2 &= x_3(x_6x_1 - x_4x_2),\\
F_3 &= x_4x_3(x_4x_1x_7x_9 - x_3x_9x_8x_1 - x_4x_1^2x_9 - x_1x_2x_7x_8 + x_1^2x_9x_8\\
&\quad + x_3x_2x_7x_8 - x_4x_2x_7^2 + x_4x_2x_7x_1),\\
F_4 &= x_4x_3(-2x_8x_4x_2x_7x_1 - x_4x_1x_7^2x_{10} + 2x_4x_1^2x_7x_{10} + x_1x_8^2x_3x_2\\
&\quad + x_1x_8x_3x_{10}x_7 - x_1^2x_8x_3x_{10} + x_8x_4x_2x_1^2 - x_4x_1^3x_{10} - x_1^2x_8^2x_2\\
&\quad - x_1^2x_8x_{10}x_7 + x_1^3x_8x_{10} + x_1x_7x_8^2x_2 - x_3x_2x_7x_8^2 + x_8x_4x_2x_7^2).
\end{aligned}$$

Step 3. Dividing the conclusion polynomial C by $F_4, \ldots, F_1$ successively, we get $R = \mathrm{prem}(C, CS) = 0$. Hence the theorem is true under the subsidiary conditions $D_i \neq 0$, where

$$DS = \{D_1, \ldots, D_5\} = \{x_1, x_3, x_4, -x_7 + x_1, x_4x_7 - x_3x_8 - x_4x_1 + x_1x\}.$$

In the above example, we have in fact proved that the zeros of the hypothesis-polynomial set are also zeros of the conclusion polynomial if they do not make $D_k = 0$, i.e.

$$\mathrm{Zero}(HS/DS) \subset \mathrm{Zero}(C),$$

where $\mathrm{Zero}(HS/DS)$ denotes the set of all common zeros of the polynomials in HS which are not zeros of any polynomial of DS.

The simple AGTP method explained above was introduced by Wu (1978). It is powerful enough to prove hundreds of non-trivial geometric theorems. Several new results including theorems about Pascal conics (Wu 1984a,b; Chou and Gao 1995) and Gauss lines (Wang 1989b) were discovered using the method; see also Wu (1982, 1983, 1984c, 1986b). The provers reported in Chou (1984, 1988) and Wang and Gao (1987) are based on this version of Wu's method. Table 5.1 contains the running times (in seconds) for 25 geometric theorems on a Symbolic Machine.

Theorem	Sources	Time
Parallelogram	[Chou and Schelter 1986]	0.1
Centroid	E1 [Chou and Schelter 1986]	0.35
Altitudes	E133 [Chou 1988]	0.83
Ceva's Th	E334 [Chou 1988]	0.81
Simson's Th	E2 [Chou and Schelter 1986]	1.2
Brahmagupta	[Chou and Schelter 1986]	1.1
Pappus' Th	E6 [Chou and Schelter 1986]	1.3
Pappus Dual	E17 [Chou 1988]	1.7
Butterfly	E5 [Chou and Schelter 1986]	1.5
Simson Dual	E21 [Chou 1988]	4.8
Pascal's Th	[Chou and Gao 1990b]	3.9
Brianchon	E19 [Chou 1988]	9.6
Pappus Point	E7 [Chou and Schelter 1986]	5.6
Isosceles	E8 [Chou and Schelter 1986]	2.6
Gauss' Th	E9 [Chou and Schelter 1986]	1.5
Gauss Point	E10 [Chou and Schelter 1986]	3.7
Gauss Conic	E17 [Chou 1988]	1511
Feuerbach	E204 [Chou 1988]	2.8
Miquel Point	E308 [Chou 1988]	40.2
Miquel Circle	E309 [Chou 1988]	412.3
Steiner's Th	E13 [Chou 1988]	10.2
Pas-Conic 1	E9 [Chou 1988]	11.8
Morley's Th	[Chou and Gao 1990b]	20.2
V. Pratt's Th	E80 [Chou 1988]	275
Coxeter's Th	E40 [Chou 1988]	218

Table 5.1 Time statistics for Wu's method of AGTP

The reader might have noticed that the conditions $D_i \neq 0$ are not superfluous. In the above example, if $D_5 = 0$, then the two triangles $A_1B_1C_1$ and $A_2B_2C_2$ degenerate to two lines. In such case, line C_1C_2 can be drawn freely, and the conclusion is not true. The possibly appearing subsidiary conditions $D_k \neq 0$ are called *non-degeneracy conditions* of the theorems. Proving a theorem by Wu's method not only decides the validity of the theorem but also finds out all the non-degeneracy conditions. If one wants to decide whether the theorem is still valid in some degenerate cases, one needs Wu–Ritt's zero decomposition theorem.

5.2.2 AGTP based on Wu–Ritt's zero decomposition theorem

Although the simple version of Wu's method described in Sect. 5.2.1 is powerful enough for proving hundreds of non-trivial geometric theorems, its application is restrictive. First, the non-degeneracy conditions $D_i \neq 0$ depend on the choices of the coordinates, which may cause problems. See the 8_3 configuration problem in Chou et al. (1989).

Let HS be the hypothesis-polynomial set and D a polynomial which is assumed to be non-zero. According to Wu–Ritt's zero decomposition theorems (see Theorems 1.5 and 1.11 in Chap. 1), we have

$$\mathrm{Zero}(HS/D) = \bigcup_i \mathrm{Zero}(AS_i/DJ_i), \tag{5.1}$$

$$\mathrm{Zero}(HS/D) = \bigcup_i \mathrm{Zero}(\mathrm{QD}(AS_i)/D), \tag{5.2}$$

where $\mathrm{QD}(AS_i)$ is defined in Chap. 1 and J_i is the initial-product of the ascending set AS_i. Thus, proving the geometric theorem in question is equivalent to checking the validity of the theorem on each component of the decomposition.

Using (5.2) allows us to remove those redundant components $\mathrm{Zero}(\mathrm{QD}(AS_k))$ for which $|AS_k| > |HS|$ according to the affine dimension theorem. Therefore, we need only to check the conclusion on much fewer components.

The following theorem is a direct consequence of the remainder formula and the property of irreducible ascending sets (see Chap. 1).

Theorem 5.1:

Using the notations introduced above:

(1) *A conclusion* $C = 0$ *is true on the component* $\mathrm{Zero}(AS_i/DJ_i)$ *or* $\mathrm{Zero}(\mathrm{QD}(AS_i)/D)$ *if* $\mathrm{prem}(C, AS_i) = 0$.

(2) *If* AS_i *is irreducible, then* $C = 0$ *is true on the component* $\mathrm{Zero}(AS_i/DJ_i)$ *or* $\mathrm{Zero}(\mathrm{QD}(AS_i)/D)$ *if and only if* $\mathrm{prem}(C, AS_i) = 0$.

As pointed out by Wu, one may check whether the conclusion is valid on each component and hence give a complete analysis of the correctness of the theorem.

Example 5.2:

Let ABC be a triangle, and $ABDE$ and $CAFG$ be two squares with AB and CA as their sides. Then EC is perpendicular to BF.

Let $B = (0,0), C = (x_1, 0), A = (x_2, x_3), E = (x_4, x_5), F = (x_6, x_7)$. Then the hypotheses and conclusion of the theorem can be translated into polynomial equations.

Hypotheses:

$$\begin{aligned}
H_1 &= x_5^2 - 2x_3x_5 + x_4^2 - 2x_2x_4 = 0, && (AB = AE)\\
H_2 &= x_3x_5 + x_2x_4 - x_3^2 - x_2^2 = 0, && (AB \perp AE)\\
H_3 &= x_7^2 + x_6^2 - 2x_2x_6 + 2x_1x_2 - x_1^2 = 0, && (AC = AF)\\
H_4 &= x_3x_7 + (x_2 - x_1)x_6 - x_3^2 - x_2^2 + x_1x_2 = 0. && (AC \perp AF)
\end{aligned}$$

Conclusion:

$$C = x_7x_5 + (x_4 - x_1)x_6 = 0. \qquad (EC \perp BF)$$

non-degeneracy conditions:

$$\begin{aligned}
D_1 &= x_2^2 + x_3^2 \neq 0, && (A \neq B)\\
D_2 &= (x_2 - x_1)^2 + x_3^2 \neq 0, && (A \neq C)\\
D_3 &= x_1 \neq 0. && (B \neq C)
\end{aligned}$$

Let $HS = \{H_1, \ldots, H_4\}$ and $DS = \{D_1, D_2, D_3\}$. By Wu's decomposition principle, we can get

$$\mathrm{Zero}(HS/DS) = \mathrm{Zero}(AS_1/J_1) \cup \cdots \cup \mathrm{Zero}(AS_4/J_4),$$

where

$$\begin{aligned}
AS_1 = \{&x_4 + x_3 - x_2, x_3x_5 - x_3^2 - x_2x_3, x_6 - x_3 - x_2,\\
&x_3x_7 - x_3^2 + (x_2 - x_1)x_3\},\\
AS_2 = \{&x_4 - x_3 - x_2, x_3x_5 + x_2x_4 - x_3^2 - x_2^2, x_6 + x_3 - x_2,\\
&x_3x_7 - x_3^2 - (x_2 - x_1)x_3\},\\
AS_3 = \{&x_4 + x_3 - x_2, x_3x_5 - x_3^2 - x_2x_3, x_6 + x_3 - x_2,\\
&x_3x_7 - x_3^2 + (x_2 - x_1)x_3\},\\
AS_4 = \{&x_4 - x_3 - x_2, x_3x_5 + x_2x_4 - x_3^2 - x_2^2, x_6 - x_3 - x_2,\\
&x_3x_7 + (x_2 - x_1)x_6 - x_3^2 - x_2^2 + x_1x_2\}.
\end{aligned}$$

It is easy to see that the above four branches correspond respectively to the four diagrams in Fig. 5.3.

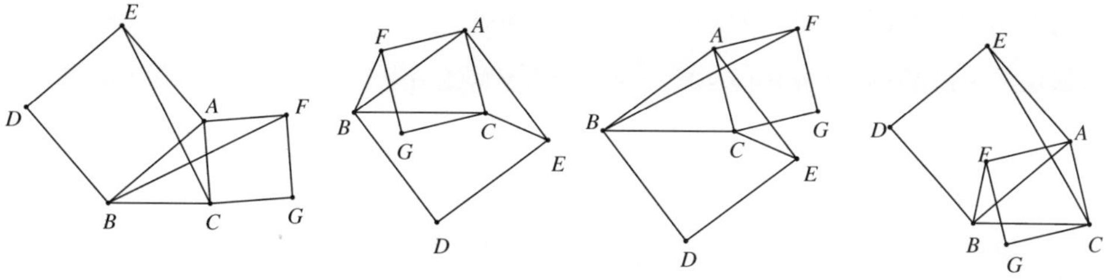

Figure 5.3 A reducible geometric problem

Computations show that $\text{prem}(C, AS_1) = \text{prem}(C, AS_2) = 0$, $\text{prem}(C, AS_3) \neq 0$ and $\text{prem}(C, AS_4) \neq 0$. This means that under the conditions $A \neq B, A \neq C$ and $B \neq C$, the theorem is true for the first two figures in Fig. 5.3, and is generically false for the other two.

Theorem 5.1 combined with Wu–Ritt's zero decomposition theorems provides a complete way of analyzing a geometric theorem. Below, we will introduce two formulations studied extensively in the literature.

Formulation I. Generally, by identifying some of the variables among $y_1, \ldots, y_m$ in the hypothesis polynomials as parameters, the decomposition (5.2) can be refined as one into irreducible components (in an algebraically closed field) as follows:

$$\text{Zero}(HS) = V_1^* \cup \cdots V_d^* \cup V_1^{\text{dege}} \cup \cdots \cup V_l^{\text{dege}},$$

where $V_1^*, \ldots, V_d^*$ are all those components on which the parameters are algebraically independent and the non-parameters are algebraically dependent on the parameters. These components are for the non-degenerate cases. The others correspond to degenerate cases. We say that the geometric statement is *generally* (*generically*) *true* if the conclusion is valid in all the non-degenerate cases, and *generally false* if the conclusion is not valid in any of the non-degenerate cases. For most problems in practice, there is only one non-degenerate case (i.e., $d = 1$); then we always have a clear-cut answer: it is either generally true or generally false. If the conclusion is valid in some, but not all, of the non-degenerate cases, then the statement needs further investigation. The concept of *generic true* was introduced by Wu (1978). The above definition is taken from Chou (1988).

The generically-true formulation is used to prove theorems in Wu (1978, 1984a,b), Chou (1988), Wang and Gao (1987), Chou and Schelter (1986) and Kutzler and Stifter (1986a).

Example 5.3: (Desargues' theorem continued)

According to Wu's decomposition theorem (Chap. 1, Theorem 1.5), we have

$$\mathrm{Zero}(HS) = \mathrm{Zero}(\mathrm{QD}(AS_1)) \cup \cdots \cup \mathrm{Zero}(\mathrm{QD}(AS_4)),$$

where

$$\begin{aligned}
AS_1 = \{&x_1x_5 - x_2x_3, (x_3 - x_1)x_6 - x_4x_5 + x_2x_4,\\
&[(x_3 - x_1)x_8 - x_4x_7 + x_1x_4]x_9\\
&+ [(-x_5 + x_2)x_7 + x_1x_5 - x_2x_3]x_8 + x_6x_7^2\\
&+ [(-x_3 - x_1)x_6 + x_4x_5]x_7 + x_1x_3x_6 - x_1x_4x_5,\\
&(x_7 - x_1)x_{10} - x_8x_9 + x_2x_8\},\\
AS_2 = \{&x_4, x_6, x_8, (x_7 - x_1)x_{10}\},\\
AS_3 = \{&x_4, (x_3 - x_1)x_6,\\
&[(x_3 - x_1)x_8]x_9 + [(-x_5 + x_2)x_7 + x_1x_5 - x_2x_3]x_8\\
&+ x_6x_7^2 + [(-x_3 - x_1)x_6]x_7 + x_1x_3x_6,\\
&(x_7 - x_1)x_{10} - x_8x_9 + x_2x_8\},\\
AS_4 = \{&x_1x_5 - x_2x_3, (x_3 - x_1)x_6 - x_4x_5 + x_2x_4,\\
&(x_3 - x_1)x_8 - x_4x_7 + x_1x_4, (x_7 - x_1)x_{10} - x_8x_9 + x_2x_8\}.
\end{aligned}$$

If $x_1, \ldots, x_4$ are treated as parameters, the component $\mathrm{Zero}(\mathrm{QD}(AS_1))$ corresponds to the non-degenerate case; $\mathrm{Zero}(\mathrm{QD}(AS_2))$, $\mathrm{Zero}(\mathrm{QD}(AS_3))$ and $\mathrm{Zero}(\mathrm{QD}(AS_4))$ correspond to the degenerate cases. Since $\mathrm{prem}(C, AS_1) = 0$, the theorem is generically true. It is easy to check that the theorem is valid on the first degenerate component, and fails in the other two degenerate cases.

Notice that if decomposition (5.1) is used instead of (5.2), then the resulting decomposition will consists of twelve components. Eight of them will become redundant when (5.2) is used.

The above formulation of AGTP can help us find the missing non-degeneracy conditions. Furthermore, it addresses the nature of the statement: if a statement is proved to be generally false, it cannot be a theorem no matter how many reasonable additional non-degeneracy conditions are added. However, non-degeneracy conditions are usually in algebraic form and it is hard to transform them into geometric form. In the next subsection a method will be introduced for dealing with non-degeneracy conditions in a geometric way.

Formulation II. Given a set of polynomial equations

$$H_1(y_1, \ldots, y_m) = 0, \ldots, H_r(y_1, \ldots, y_m) = 0$$

and non-degeneracy conditions $D_1 \neq 0, \ldots, D_q \neq 0$ as the hypotheses of a geometric theorem, decide whether an assertion $C(y_1, \ldots, y_m) = 0$ as the conclusion of the theorem is valid under the hypotheses without adding any subsidiary conditions.

This formulation has been used to prove theorems in Chou and Gao (1990b), Chou and Yang (1989), Kapur (1986a), Ko (1988) and Kutzler and Stifter (1986a).

In this formulation of AGTP, the geometric statement is exactly specified, and the user can select non-degeneracy conditions which are considered suitable. However, if the statement is proved to be false, one does not know the nature of the statement: whether it is generally false or the proof failed due to missing non-degeneracy conditions. Practical experience shows that non-degeneracy conditions are often implicit in statements from geometry textbooks and identifying them is sometimes very hard and subtle, even in Euclidean geometry. Note that the equality part of the hypotheses (i.e., $H_i = 0$) remains *the same* for a given specification of a geometric statement regardless of how non-degeneracy conditions are added or understood.

In view of the fact that formulation I does not allow us to exclude any obvious degenerate cases or unwanted components by inequations while formulation II requires excluding all the degenerate cases predetermined, Wang (1991c, 1995b) formulated the AGTP problem by allowing both polynomial equations and inequations in the hypotheses and by decomposing the quasi-variety into irreducible components in order to find possible missing conditions. Actually, Example 5.2 has been given according to this formulation. There are also discussions about *conditional true* in Wang (1995b), for which the identification of parameters is not obligate.

5.2.3 Proving theorems of constructive type

As we discussed in the last section, non-degeneracy conditions represent a very important feature of geometric theorems. However, the non-degeneracy conditions of a concrete theorem are not so simple to identify. For a given geometric statement, a non-degeneracy condition that is obvious to one person might not be obvious to another, and a third person might refuse to accept the condition as relevant or appropriate. It is now accepted by most researchers that how to treat non-degeneracy conditions is an essential part of an AGTP method. In this section, we show that for a class of constructive geometric statements, the problem of non-degeneracy conditions can be handled perfectly.

The great success of Wu's method is closely connected to some special geometric statements: the class of constructive statements (Wu 1984a; Wang and Hu 1987; Wang 1989b; Chou 1988; Chou and Gao 1992b). By *constructive* we mean that each geometric entity is either arbitrary or can be constructed as the intersection of lines and/or circles. About 400 among the 512 geometric theorems proved in Chou (1988) are of this kind.

The statements considered by Hilbert are constructive ones, but they have a rather restricted character: they are about those configurations formed by straight lines in a constructive way. Wang and Hu (1987) pointed out that the applicability of Hilbert's idea of explicit constructibility may be much extended beyond the geometrical constructions of a linear character, e.g., to perpendicularity (see also Wang 1989b). Chou (1984, 1988) proposed a complete algorithm for geometric statements of constructive type by using a special kind of polynomial factorization. Chou and Gao (1992b) showed that coordinate-independent non-degeneracy conditions in geometric form can be obtained for the class of constructive geometric statements, and the non-degeneracy conditions thus obtained are *sufficient* for a statement to be valid. A constructive geometric statement is said to be *irreducible* if it has only one non-degenerate component and the ascending set for this component is irreducible. The statements that can be handled by Hilbert's algorithm are irreducible ones. About 70% of the 512 theorems in Chou (1988) are irreducible. The following result has been proved.

Theorem 5.2: (Chou and Gao 1992b)

There is a method which generates, for any constructive geometric statement $\mathcal{S}$, a set DS of coordinate-independent non-degeneracy conditions in geometric form. If $\mathcal{S}$ is irreducible, then the non-degeneracy conditions generated by the method are sufficient for $\mathcal{S}$ to be true over the field $\boldsymbol{C}$ of complex numbers. More precisely, let HS be the equational hypotheses and C the conclusion of $\mathcal{S}$. If

$$HS \quad \text{and} \quad DS \Rightarrow C$$

is not valid over $\boldsymbol{C}$, then the above statement cannot be made valid by adding more non-degeneracy conditions to DS unless the statement becomes a conflict one (i.e., its hypothesis is false).

Example 5.4:

Below is a constructive description of Desargues' theorem.

- Take free points O, A_1, B_1 and C_1.
- Take a free point A_2 on line OA_1.
- Take the intersection B_2 of line OB_1 and PLINE(A_2, A_1, B_1) (the line passing through A_2 and parallel to A_1B_1).
- Take the intersection C_2 of PLINE(A_2, A_1, C_1) and PLINE(B_2, B_1, C_1).
- The conclusion is that points O, C_1 and C_2 are collinear.

The non-degeneracy conditions generated are:

- $O \neq A_1$ (from the second construction),
- O, A_1, B_1 are not collinear (from the third construction), and
- A_1, B_1, C_1 are not collinear (from the fourth construction).

This is an irreducible statement since it is a theorem of Hilbert type.

5.3 Proving theorems in non-Euclidean, finite, and ordered geometries

5.3.1 Theorem proving in non-Euclidean geometries

In the previous sections, we have mainly discussed AGTP by Wu's method in elementary geometry. Apart from Euclidean geometry, there is a variety of other geometries, e.g., projective geometries, line geometry, non-Euclidean geometries and circle geometry, in which Wu's method can be used for theorem proving. In Wu (1984b) it is shown how these geometries may be defined based on an axiomatic system and how a number system and then a coordinate system may be introduced as in the case of ordinary Euclidean geometry. It has also been shown that the number systems intrinsically associated to such geometries are actually fields of characteristic 0 so far as some axiom analogous to the Pascal theorem in Hilbert's terminology is adjoined to the defining axiomatic system of the relevant geometry. For such geometries one may then introduce coordinate systems and treat geometry theorem proving in the Cartesian fashion without further reference to its origin — the underlying axiomatic system.

Automated theorem proving in non-Euclidean geometry with algebraic methods has been reported in a few papers. A method for non-Euclidean geometries was introduced by Wu (1984b) as a direct consequence of his general algebraic method. Transformation theorems among Euclidean, Minkowskian, hyperbolic and elliptic geometries were studied in Chou and Ko (1986), Chou and Gao (1994) and Gao (1990). In Yang (1989), properties of trigonometric functions were used to derive new properties of non-Euclidean geometries. The polynomials involved in the proofs of theorems in non-Euclidean geometries are generally larger than those in Euclidean geometry.

5.3.2 Theorem proving in geometry over finite fields

To prove theorems in geometry over finite fields, we have to deal with polynomial ring over finite fields. The characteristic set method for polynomials over finite fields has been discussed in Lin and Liu (1992). Unlike in ordinary geometry where the theorems are discussed over a fixed field such as $\mathcal{R}$ or $\boldsymbol{C}$, in geometry over a finite field, when we talk about a geometric statement, we do not clearly indicate which finite field we are discussing over, not even its characteristic. In other words, the finite field may be $\mathbf{F}_2$, $\mathbf{F}_3$, or $\mathbf{F}_{2^6}$. Here is an example.

Example 5.5:

Let $\mathbf{F}_q$ be the finite field with q elements. If q is odd, then in the finite projective plane over $\mathbf{F}_q$, any three distinct tangents of a conic are not concurrent.

Without loss of generality, we may suppose that the equation of the conic is $X^2 + YZ = 0$. Then the theorem is equivalent to the following statement: for any three points A, B and C on the conic, if the three tangents at A, B and C meet at a point O, then two of the three points A, B and C must coincide.

Let $A = (x_1, x_2, x_3)$, $B = (x_4, x_5, x_6)$ and $C = (x_7, x_8, x_9)$. Then the equations of the tangents at A, B and C are

$$\begin{aligned} &2x_1X + x_3Y + x_2Z = 0, && \text{(tangent } l_1 \text{ at } A)\\ &2x_4X + x_6Y + x_5Z = 0, && \text{(tangent } l_2 \text{ at } B)\\ &2x_7X + x_9Y + x_8Z = 0. && \text{(tangent } l_3 \text{ at } C)\end{aligned}$$

So the hypotheses of the theorem can be expressed as

$$\begin{aligned}
P_1 &= x_3x_2 + x_1^2 = 0, && (A \text{ on the conic})\\
P_2 &= x_6x_5 + x_4^2 = 0, && (B \text{ on the conic})\\
P_3 &= x_9x_8 + x_7^2 = 0, && (C \text{ on the conic})\\
P_4 &= -2(x_9x_5x_1 - x_9x_4x_2 - x_8x_6x_1 + x_8x_4x_3\\
&\quad + x_7x_6x_2 - x_7x_5x_3) = 0; && (l_1, l_2, l_3 \text{ are concurrent})
\end{aligned}$$

the conclusion-polynomial equations are

$$\begin{aligned}
&P_5 = x_5x_1 - x_4x_2 = 0, \quad P_6 = x_6x_1 - x_4x_3 = 0, \quad P_7 = x_2x_6 - x_5x_3 = 0, && (A = B)\\
&P_8 = x_8x_1 - x_7x_2 = 0, \quad P_9 = x_9x_1 - x_7x_3 = 0, \quad P_{10} = x_2x_9 - x_3x_8 = 0, && (A = C)\\
&P_{11} = x_8x_4 - x_7x_5 = 0, \quad P_{12} = x_9x_4 - x_7x_6 = 0, \quad P_{13} = x_5x_9 - x_8x_6 = 0. && (B = C)
\end{aligned}$$

Working with $\mathbb{Z}$, we get a characteristic set $\{P_{14}, \ldots, P_{17}\}$ of $\{P_1, \ldots, P_4\}$ with

$$\begin{aligned}
P_{14} &= x_9x_8 + x_7^2,\\
P_{15} &= 2(x_8^2x_5x_4x_1^2 - x_8^2x_4^2x_2x_1 - x_8x_7x_5^2x_1^2 + x_8x_7x_4^2x_2^2 + x_7^2x_5^2x_2x_1\\
&\quad - x_7^2x_5x_4x_2^2),\\
P_{16} &= x_6x_5 + x_4^2, \qquad P_{17} = x_3x_2 + x_1^2.
\end{aligned}$$

Factorize P_{15} as

$$P_{15} = 2(x_8x_4 - x_7x_5)(x_8x_1 - x_7x_2)(x_5x_1 - x_4x_2).$$

It is easy to check that the conclusion is true under the condition that

$$2 \not| q \quad \text{and} \quad J = x_8x_5x_4x_2x_1(x_5x_1 - x_4x_2) \neq 0.$$

In fact, provided that $2 \not| q$, the conclusion is also true even if $J = 0$.

5.3.3 Proving theorems with inequalities

In elementary geometry, there are also many theorems involving inequalities. The CAD method of Collins (1975) is one of the most general and complete methods for problems involving inequalities. On the other hand, as Wu's method works so well for problems of equality type, one might also expect that a revision of Wu's method could be efficient for solving certain kinds of problems involving inequalities. In Chap. 1, an approach of Wu has already been reported. In Chou et al. (1992), a different approach based on Wu's method is proposed and about

40 geometric theorems involving inequalities have been proved. The following example is one of them.

Example 5.6:

Let O be a point, $D = AB \cap OC$, $E = AC \cap OB$, and $F = BC \cap OA$. If D is between B and A, and E is between C and A, then F is between B and C.

Let

$$\begin{array}{llll} A = (0,0), & B = (u_1, 0), & C = (u_2, u_3), & D = (x_1, 0), \\ E = (x_2, x_3), & O = (x_4, x_5), & F = (x_6, x_7). & \end{array}$$

We have seven equations $H_i = 0$ for the hypotheses and one inequality $G < 0$ for the conclusion as follows

$$\begin{array}{lr} H_1 = (y_1^2 + 1)x_1 - u_1 = 0, & (D \text{ is between } A \text{ and } B) \\ H_2 = (y_2^2 + 1)x_2 - u_2 = 0, & (E \text{ is between } A \text{ and } C) \\ H_3 = u_2 x_3 - u_3 x_2 = 0, & (E, A, C \text{ are collinear}) \\ H_4 = (x_2 - u_1)x_5 - x_3 x_4 + u_1 u_3 = 0, & (O \text{ is on } BE) \\ H_5 = (x_2 - u_2)x_5 + u_3 x_4 - u_3 x_1 = 0, & (O \text{ is on } CD) \\ H_6 = x_4 x_7 - x_5 x_6 = 0, & (F \text{ is on } AO) \\ H_7 = (-u_2 + u_1)x_7 + u_3 x_6 - u_1 u_3 = 0, & (F \text{ is on } BC) \\ G = x_6^2 - (u_2 + u_1)x_6 + u_1 u_2 < 0. & (F \text{ is between } B \text{ and } C) \end{array}$$

Let $HS = \{H_1, \ldots, H_7\}$. Using the Wu–Ritt zero decomposition algorithm, one may find that Zero(HS) has only one non-degenerate component; let the corresponding ascending set and its initial-product be denoted by AS and J, respectively. Then on Zero(AS/J), G is evaluated to be

$$-\frac{y_2^2 y_1^2 (u_2 - u_1)^2}{(y_1^2 + y_2^2)^2},$$

which is negative definite if $u_2 - u_1 \neq 0$.

A general scheme based on a combination of Wu's method and Collins' method has been proposed, and used to solve the following problems automatically (Chou et al. 1989).

Example 5.7:

A 8_3 configuration consists of eight points and eight lines such that each line passes through three points and each point is on three lines in the Euclidean plane.

(1) Show that there does not exist an 8_3 configuration in the Euclidean plane such that the points and lines are all distinct.

(2) Find all the possible 8_3 configurations in the Euclidean plane.

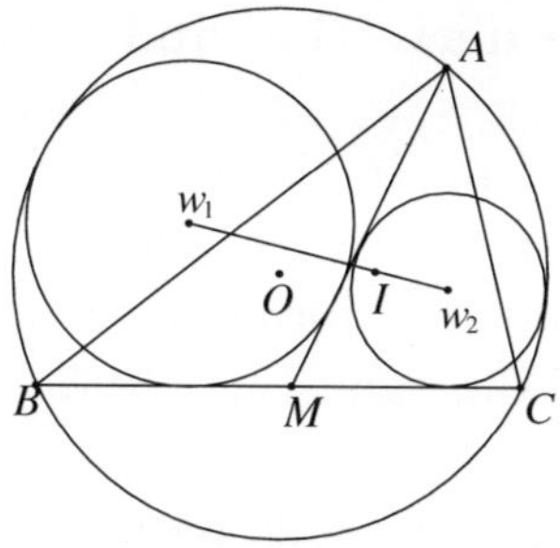

Figure 5.4 Thébault's theorem

These two problems are also solved in Conti and Traverso (1995) with a different method. Problem (1) is solved in Wang and Zhi (1998) with the assistance of the user when dealing with inequalities.

In some geometric theorems of equation type, difficulties related to reducibility are caused by order relations among points and/or lines. In Wu (1987a), the concept of ordered lines and circles are used to reduce some reducible cases to irreducible ones. Wang (1994) showed, with several examples, how algebraic factorization may be used effectively to deal with the reducibility problem. In Li (1997a), the same problem is studied from the geometric viewpoint. Let us consider the following example.

Example 5.8: (Thébault's theorem)

Let I and O be the inner center and circumcenter of an arbitrary triangle ABC respectively (Fig. 5.4). Let M be in the line segment BC, and W_1 and W_2 be the centers of two circles tangent to circle O. Furthermore, circle W_1 is tangent to the two sides of $\angle AMB$ and circle W_2 is tangent to the two sides of $\angle AMC$. Then the points W_1, W_2, I are collinear.

This theorem was proved by Chou (1988) with polynomial factorization and the largest polynomial in the proof consists of 674 927 terms. With his new technique, Wu (1987a) could prove the theorem without using polynomial factorization; the largest polynomial in the proof has 4885 terms. Li (1997a) gave a simpler proof for the theorem by introducing new orders and equations. Other machine proofs of the theorem may be found in Yang and Zhang (1994) and Wang (1994, 1995b).

5.4 Mechanical formula derivation and theorem discovering

There are two kinds of problems in elementary geometry other than theorem proving: one is finding locus equations, and the other is deriving geometric formulas. These problems and their formulation were addressed in Chou (1984, 1985, 1987), Wu (1986c,d), Wang and Gao (1987), Chou and Gao (1990a), and Wang (1991c, 1995a).

The following formulation of formula derivation is due to Wu (1986c). Let *HS* be a polynomial set whose vanishing expresses the underlying geometrical conditions of a problem. We aim to find one or several unknown relations from *HS*. Rearrange the variables x_i occurring in *HS* so that the first few x_i are just the geometrical (or algebraic) entities among which the connecting relations are to be sought for. Let *CS* be a characteristic set of *HS* with C_i as its polynomials. By the well-ordering principle of Wu (Chap. 1), we have

$$\mathrm{Zero}(HS) \subset \mathrm{Zero}(CS).$$

Hence, as a consequence of the geometrical conditions we have $C_1 = 0$, which is in general just the relation to be derived.

We use Qin–Heron's formula to illustrate this type of problem.

Example 5.9:

Find a formula for the area of a triangle ABC in terms of its three sides.

Let $x_1, \ldots, x_4$ denote the three sides $|AB|, |AC|, |BC|$ and the area of triangle ABC, respectively, among which a yet unknown relation is to be sought for. Choose coordinates so that

$$A = (0, 0), \qquad B = (x_5, 0), \qquad C = (x_6, x_7).$$

Remark that the variables x_i occurring in the coordinates of A, B, C have been so chosen to succeed $x_1, \ldots, x_4$ in the natural order. The geometrical conditions are now in the form of the following equations:

$$\begin{aligned}
H_1 &= 2x_4 - x_5 x_7 = 0,\\
H_2 &= x_5 - x_1 = 0,\\
H_3 &= x_2^2 - x_6^2 - x_7^2 = 0,\\
H_4 &= x_3^2 - (x_6 - x_5)^2 - x_7^2 = 0.
\end{aligned}$$

Let $HS = \{H_1, \ldots, H_4\}$. A characteristic set of *HS* is readily found to consist of four polynomials, of which the first is given by

$$C_1 = 2x_1^2x_2^2 + 2x_1^2x_3^2 + 2x_2^2x_3^2 - x_1^4 - x_2^4 - x_3^4 + 16x_4^2.$$

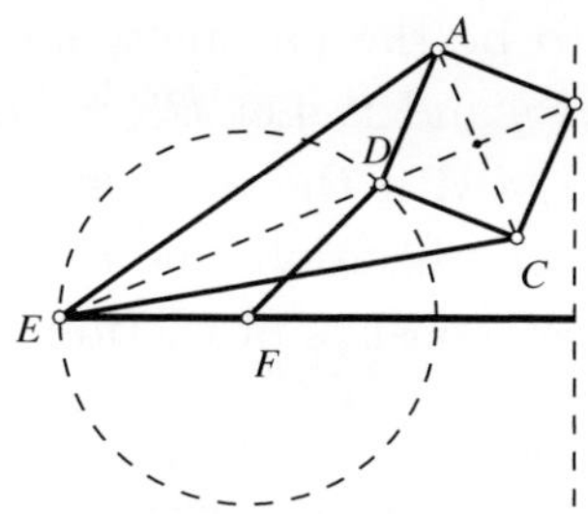

Figure 5.5 Peaucellier's linkage

The relation $C_1 = 0$ is just the Qin–Heron formula to be derived.

In Chou and Gao (1990a) and Wang (1995a), Wu–Ritt's zero decomposition theorem is used to derive formulas automatically. The following is one of the 120 problems solved automatically in Chou and Gao (1989a), and is a more interesting example.

Example 5.10: (Peaucellier's linkage)

Let the links AD, AB, DC and BC have equal length, as do the links EA and EC (Fig. 5.5). The length of FD is equal to the distance from E to F. The locations of the joints E and F are fixed points on the plane, but the linkage is allowed to rotate about these points. As it does, what is the locus of the joint B?

Let

$$F = (0,0), \quad E = (r,0), \quad C = (x_2, y_2), \quad D = (x_1, y_1), \quad B = (x, y),$$

n and m be the lengths of the projections of AD on the direction of line EDB (note that points E, D and B are always collinear) and on the direction perpendicular to it. Then the geometric conditions can be expressed by the following set of polynomial equations:

$$\begin{aligned}
H_1 &= y_1^2 + x_1^2 - r^2 = 0, && (r = FD)\\
H_2 &= y_2^2 - 2y_1y_2 + x_2^2 - 2x_1x_2 + y_1^2 + x_1^2 - n^2 - m^2 = 0, && (CD = n^2 + m^2)\\
H_3 &= y_2^2 - 2yy_2 + x_2^2 - 2xx_2 + x^2 + y^2 - n^2 - m^2 = 0, && (CB = n^2 + m^2)\\
H_4 &= y_2^2 + x_2^2 - 2rx_2 - n^2 - 4rn - m^2 - 3r^2 = 0, && (EC = (n + 2r)^2 + m^2)\\
H_5 &= (x - r)y_1 - yx_1 + ry = 0, && (E \text{ is on } DB)
\end{aligned}$$

together with the polynomial inequation

$$D_1 = x_1 - x \neq 0. \qquad (B \neq D)$$

Selecting m, n, r and y to be the parameters of the problem, we want to find a relation among m, n, r, y and x. Let $HS = \{H_1, \ldots, H_5\}$. Using Method 4.6 in Chou and Gao (1990a) (with the ordering $x \prec x_1 \prec y_1 \prec x_2 \prec y_2$), we have found that Zero(HS/D_1) has only one non-degenerate component; the corresponding ascending set AS consists of the following polynomials:

$$
\begin{aligned}
&x + 2n + r,\\
&(x^2 - 2rx + y^2 + r^2)x_1 + rx^2 - 2r^2x - ry^2 + r^3,\\
&(x - r)y_1 - yx_1 + ry,\\
&c_2x_2^2 + c_1x_2 + c_0,\\
&2yy_2 + (2x - 2r)x_2 - x^2 - y^2 - 4rn - 3r^2,
\end{aligned}
$$

where the c_i are polynomials of r, n, x, x_1, y_1. The first polynomial gives the relation $x + 2n + r = 0$, which tells us that the locus is a line parallel to the y-axis.

It is pointed out in Chou (1990) and Wang (1991c, 1995a) that the problem of deriving locus equations can also be formulated as finding a quantifier-free formula $F(\boldsymbol{u}, x_1)$ such that

$$
\begin{aligned}
F(\boldsymbol{u}, x_1) \iff \exists x_2 \cdots x_p[H_1(\boldsymbol{u}, \boldsymbol{x}) = 0 \wedge \cdots \wedge H_r(\boldsymbol{u}, \boldsymbol{x}) = 0 \wedge\\
D_1(\boldsymbol{u}, \boldsymbol{x}) \neq 0 \wedge \cdots \wedge D_s(\boldsymbol{u}, \boldsymbol{x}) \neq 0],
\end{aligned}
$$

where $\boldsymbol{u}$ stands for $u_1, \ldots, u_q$ and $\boldsymbol{x}$ for $x_1, \ldots, x_p$. Formulated in this way, the problem is actually to calculate the projection of an algebraic set in the affine space $\boldsymbol{E}^{q+p}$ into the subspace $\boldsymbol{E}^{q+1}$. If $\boldsymbol{E}$ is algebraically closed, there are methods, for example, the one in Wu (1990a) based on CS and another in Wang (1993b) based on Seidenberg's elimination technique, for computing such projections. If $\boldsymbol{E}$ is a real closed field, the CAD method of Collins (1975) gives a solution to the problem. This has been discussed briefly in Wang (1991c). It would be interesting to explore the above formulation with more examples and to examine the connection between the real and the algebraically closed cases. For example, we expect Collins' method to produce the following form of $F(\boldsymbol{u}, x_1)$ for Peaucellier's linkage:

$$
(x + 2n + r = 0) \wedge (-d \leq y \leq d) \wedge (\text{other non-degeneracy conditions}),
$$

where d is from the 4th (quadratic) polynomial of the above AS.

Chapter 6

MECHANICAL THEOREM PROVING IN DIFFERENTIAL GEOMETRY

Hongbo Li

This chapter presents the development of mechanical theorem proving in differential geometry initiated by W.-t. Wu in the later 1970s. Wu's method is capable of proving and discovering theorems in differential geometry and mechanics mechanically and efficiently; for example, the theorems of Bertrand, Mannheim, and Schell may be proved or even discovered automatically, and so may Newton's laws be derived from Kepler's laws using an implementation of the method. Following Wu's work, there have been extensive investigations carried out by S.-C. Chou and X.-S. Gao on theorem proving and differential parametric systems with improved algorithms, by D. Wang on zero decomposition using Seidenberg's elimination theory, by Z. Li on theorem proving and discovering in the theory of local surfaces, by S. Li in control theory, and by H. Li and M. Cheng on theorem proving with Clifford algebraic representation, to name a few.

6.1 Overview

Proving theorems in differential geometry mechanically was initiated by Wu (1979, 1980). He extended the characteristic set method, a method developed by J. F. Ritt in algebraic geometry and differential algebra, to a well-ordering principle that can be used for mechanical theorem proving and discovering in differential geometry and mechanics. Hundreds of theorems in the theory of space curves and mechanics have been proved or discovered automatically using Wu's method; among them are the remarkable theorems that are associated with the names of Bertrand, Mannheim, and Schell (Wu 1987d). As a more significant example, Newton's laws of gravitation may be derived automatically from Kepler's laws of planet motion by the same method of Wu (1987c, 1991c).

Wu's initiative has stimulated a new wave of research on automated theorem

MATHEMATICS MECHANIZATION AND APPLICATIONS
ISBN 0-12-734760-7

proving in differential geometry. Chou and Gao (1993a) proposed an improved version of Wu–Ritt's well-ordering principle in the one-variable case, which is capable of reducing the size of polynomials and the branches in computing characteristic sets. They have made extensive investigations on various aspects of Wu's method, from algebraic formulation of geometric statements to automated discovery of new relations to parametric differential systems (Chou and Gao 1992c, 1993b,c; Gao and Chou 1993b). Their experiments on mechanical theorem proving in space curves remain the most extensive up to now.

Inspired by the elimination theory of Seidenberg (1956b), Wang (1995c, 1996d) proposed a zero decomposition technique that is capable of improving the performance of Wu's method for mechanical theorem proving in the theory of space curves. On the other hand, Z. Li (1995b) made investigations on theorem proving in the theory of space surfaces and found a new theorem by computing a characteristic set of the fundamental equations of a surface. S. Li (1993a,b, 1994, 1998) applied Wu's method in control theory and made improvements in that area as well.

The need for integrating vector algebra, differential forms and Cartan's method of moving frames with Wu's method for automated reasoning in differential geometry led Li and Cheng to introduce a Clifford algebraic reduction method (Li H. 1994, 1995, 1997a; Li and Cheng 1998a,b). This method is based on triangulation in Clifford algebraic formalism and can significantly enhance the performance of Wu's method. A prover based on the method may produce proofs that are much the same as those in the textbooks when it comes to space curves, and simpler than those in the textbooks when it comes to space surfaces.

The great success of Wu's method and its variants for mechanical theorem proving in geometry has been a surprise — no other method, either logical or algebraic, has been able to compete with this method. In the following sections we shall review Wu's method and various improvements on it for the differential case; the method for proving theorems in elementary geometries is presented in Chap. 5.

6.2 One-variable case: the basic principle

In this section we give a brief introduction to the basic principle of Wu's method in the one-variable case (Wu 1991c).

A *differential field* is a field equipped with differentiation. A *constant* is an element in the field whose derivatives are zero. All constants in a differential field form a subfield, called the *field of constants* of the differential field.

Let $\mathcal{F}$ be a differential field (*d-field* for short). An *indeterminate* is a symbol having infinitely many derivatives. The ith derivative of an indeterminate x is denoted by $\Delta_i x$. A *differential polynomial* (d-polynomial) in the indeterminates $x_1, \ldots, x_m$ is a polynomial in $\Delta_i x_j$ with coefficients in $\mathcal{F}$. Let P be a d-polynomial.

- The *class* of P is the greatest subscript c for which x_c actually occurs in P.
- The *order* of P is the highest order r for which $\Delta_r x_c$ actually occurs in P, where c is the class of P.
- The *lead* of P is $\Delta_r x_c$, where r is the order of P.
- The *degree* of P is the highest degree d of $\Delta_r x_c$ which actually occurs in P.
- The *initial* of P is the coefficient of $(\Delta_r x_c)^d$.
- The *separant* of P is the formal partial of P with respect to $\Delta_r x_c$.

See Chaps. 1 and 2 for some of these concepts in the algebraic case.

A partial order is defined among d-polynomials as follows: let P_1, P_2 be two d-polynomials, with classes c_1, c_2, orders r_1, r_2, degrees d_1, d_2, respectively. Then $P_1 \prec P_2$ if one of the following holds:

(a) $c_1 < c_2$;

(b) $c_1 = c_2, \quad r_1 < r_2$;

(c) $c_1 = c_2, \quad r_1 = r_2, \quad d_1 < d_2$.

We say that P_1 is *d-reduced* with respect to P_2 if either $c_1 < c_2$, or all the derivatives of x_{c_2} in P_1 have orders less than r_2, or $\Delta_{r_2} x_{c_2}$ is the highest derivative of x_{c_2} in P_1 and the degree of P_1 with respect to x_{c_2} is less than d_2.

Let $\{P_1, \ldots, P_s\}$ be an ordered set of d-polynomials. It is called a (non-trivial) *d-ascending set* if

(a) the class of P_1 is greater than zero,

(b) for any $j > i$, the class of P_j is greater than that of P_i,

(c) the initial of P_j is d-reduced with respect to P_i.

A d-polynomial is said to be *d-reduced* with respect to a d-ascending set if it is d-reduced with respect to every d-polynomial in the d-ascending set.

Let $\{P_1, \ldots, P_s\}$ and $\{Q_1, \ldots, Q_t\}$ be two d-ascending sets. We order $\{P_1, \ldots, P_s\} \prec \{Q_1, \ldots, Q_t\}$ if one of the following holds:

(a) there exists $k \leq s, t$ such that $P_k \prec Q_k$, and for any $i < k$, P_i and Q_i are incomparable in order;

(b) $t < s$, and for any $i \leq t$, P_i and Q_i are incomparable in order.

Let *dPS* be a finite set of d-polynomials. Any subset of *dPS* which is a minimal d-ascending set is called a *d-basic set*.

A d-field $\mathcal{F}'$ is said to be a *d-extension* of a d-field $\mathcal{F}$ if $\mathcal{F}'$ contains $\mathcal{F}$ and the differentiation in $\mathcal{F}$ extends to $\mathcal{F}'$. Let *dPS* be a set of d-polynomials in the indeterminates $x_1, \ldots, x_m$ with coefficients in $\mathcal{F}$. Let $y_1, \ldots, y_m$ be elements in a d-extension of $\mathcal{F}$. If the d-polynomials in *dPS* are all zero when x_i is replaced by y_i (for $i = 1, \ldots, m$), then $(y_1, \ldots, y_m)$ is called a *d-zero* of *dPS*. The set of all the d-zeros of *dPS* is denoted by d-Zero(*dPS*). The set of the d-zeros of *dPS* which are not d-zeros of a d-polynomial G is denoted by d-Zero(*dPS*/G).

For any d-polynomial Q and d-ascending set $dAS = \{A_1, \ldots, A_s\}$, we have the following *pseudo-remainder formula*:

$$\prod_i (I_i^{l_i} S_i^{m_i})\, Q = \sum_{j,k} (Q_{jk}\ \Delta_j A_k) + R,$$

where I_i, S_i are the initial and separant of A_i respectively, l_i, m_i are non-negative integers, and Q_{jk}, R are d-polynomials d-reduced with respect to *dAS*. The d-polynomial R is called a *d-remainder* of Q with respect to *dAS* and denoted by d-prem(Q, *dAS*). For a d-polynomial set *dPS*, the set $\{\text{d-prem}(Q, dAS) \mid Q \in dPS\}$ is called a *d-remainder set* of *dPS* with respect to *dAS* and denoted by d-prem(*dPS*, *dAS*).

Using the pseudo-remainder formula, one can obtain the following scheme from a set of d-polynomials *dPS*:

$$\begin{array}{ccccccccc} dPS & = & dPS_0 & & dPS_1 & & \cdots & & dPS_u \\ & & dBS_0 & \succ & dBS_1 & \succ & \cdots & \succ & dBS_u \\ & & dRS_0 & & dRS_1 & & \cdots & & dRS_u \quad = \quad \varnothing. \end{array} \tag{6.1}$$

Here dBS_i is a d-basic set of dPS_i, dRS_i is a set of non-zero d-remainders of the d-polynomials in dPS_i with respect to dBS_i, u is the smallest integer such that dRS_u is empty, and $dPS_{i+1} = dPS_i \cup dRS_i$. It is easy to prove that the integer u always exists. dBS_u is called a *d-characteristic set* of dPS (Wu 1991c), denoted by dCS.

The above scheme enables us to establish the following *Wu–Ritt well-ordering principle*: Let dCS be a d-characteristic set of dPS, N the product of all the initials and separants of the d-polynomials in dCS, and dQS_k the dPS joined by one of the initials and separants. Then

$$\text{d-Zero}(dPS) = \text{d-Zero}(dCS/N) \cup \left(\bigcup_k \text{d-Zero}(dQS_k) \right).$$

Consider a statement in differential geometry with hypothesis HYP and conclusion CON, both being d-polynomial sets. Let dCS be a d-characteristic set of HYP, and N the product of the initials and separants of the d-polynomials in dCS. Let $R = \text{d-prem}(CON, dCS)$. If $R = \{0\}$, then for any d-zero of HYP which satisfies $N \neq 0$ we have $CON = \{0\}$. Therefore, we have the following (Wu 1991c).

Mechanical theorem proving principle. A geometric theorem with hypothesis HYP and conclusion CON is generically true under certain non-degeneracy conditions if

$$\text{d-prem}(CON, dCS) = \{0\},$$

where dCS is a characteristic set of HYP.

Mechanical theorem discovering principle. For a hypothesis set HYP, any d-polynomial P occurring in dPS_k or dBS_k of the scheme (6.1) will furnish us with a geometric theorem $P = 0$, whenever it can be endowed with an intrinsic geometric meaning.

6.3 Illustrative examples

Theorems on various Bertrand curves are among the most eminent results in the local theory of space curves; the first example below is on curve pairs of Bertrand type. Our second example shows how Newton's gravitational laws can be derived mechanically from Kepler's laws. Both of the examples are taken from Wu (1991c).

6.3.1 Curve pairs of Bertrand type

Let there be two space curves r_1 and r_2, whose Frenet moving frames are $\{r_1; T_1, N_1, B_1\}$ and $\{r_2; T_2, N_2, B_2\}$ respectively. If the two curves are in one-to-one correspondence such that at each pair of corresponding points on r_1 and r_2, one vector in the Frenet frame of curve r_1 coincides with another vector in the Frenet frame of r_2, then the two curves are called a curve pair of Bertrand type.

Let s_1 and s_2 be the arc lengths of the two curves r_1 and r_2 respectively. Let $\Delta_1 s_2$, the derivative of s_2 with respect to s_1, be r. The Frenet frame of r_2 is related to that of r_1 through

$$\begin{cases} r_2 = r_1 + a_1 T_1 + a_2 N_1 + a_3 B_1, \\ T_2 = u_{11} T_1 + u_{12} N_1 + u_{13} B_1, \\ N_2 = u_{21} T_1 + u_{22} N_1 + u_{23} B_1, \\ B_2 = u_{31} T_1 + u_{32} N_1 + u_{33} B_1, \end{cases} \tag{6.2}$$

where the a's and u's are scalars. Let the curvatures and torsions of the two curves be k_1, τ_1 and k_2, τ_2 respectively. By differentiating (6.2), using the Frenet equations of both curves and comparing the coefficients of T_1, N_1, B_1, we get 12 equations: $H_i = 0, i = 1, \ldots, 12$, where

$$\begin{cases} H_1 = r u_{11} - 1 - \Delta_1 a_1 + a_2 k_1, \\ H_2 = r u_{12} - \Delta_1 a_2 - a_1 k_1 + a_3 \tau_1, \\ H_3 = r u_{13} - \Delta_1 a_3 - a_2 \tau_1, \\ H_4 = r k_2 u_{21} - \Delta_1 u_{11} - k_1 u_{12}, \\ H_5 = r k_2 u_{22} - \Delta_1 u_{12} - k_1 u_{11} + \tau_1 u_{13}, \\ H_6 = r k_2 u_{23} - \Delta_1 u_{13} - \tau_1 u_{12}, \\ H_7 = -r k_2 u_{11} + r \tau_2 u_{31} - \Delta_1 u_{21} + k_1 u_{22}, \\ H_8 = -r k_2 u_{12} + r \tau_2 u_{32} - \Delta_1 u_{22} - k_1 u_{21} + \tau_1 u_{23}, \\ H_9 = -r k_2 u_{13} + r \tau_2 u_{33} - \Delta_1 u_{23} - \tau_1 u_{22}, \\ H_{10} = -r \tau_2 u_{21} - \Delta_1 u_{31} + k_1 u_{32}, \\ H_{11} = -r \tau_2 u_{22} - \Delta_1 u_{32} - k_1 u_{31} + \tau_1 u_{33}, \\ H_{12} = -r \tau_2 u_{23} - \Delta_1 u_{33} - \tau_1 u_{32}. \end{cases} \tag{6.3}$$

Consider the classical Bertrand case, i.e., $N_2 = N_1$. This condition together with the orthogonality of the matrix (u_{ij}) provides us with

$$\begin{cases} a_1 = a_3 = 0, \\ u_{12} = u_{21} = u_{23} = u_{32} = 0, \ u_{22} = 1, \\ u_{11} = u_{33}, \ u_{13} = -u_{31}, \ u_{11}^2 + u_{13}^2 = 1. \end{cases} \tag{6.4}$$

Now change the notations as follows:

$$(r, a_1, a_2, a_3, u_{11}, u_{12}, u_{13}, u_{21}, u_{22}, u_{23}, u_{31}, u_{32}, u_{33})$$
$$= (x_5, x_6, x_7, x_8, x_9, x_{10}, x_{11}, x_{12}, x_{13}, x_{14}, x_{15}, x_{16}, x_{17}).$$

We are interested in the relationship between k_2 and τ_2. Let

$$(k_1, \tau_1, k_2, \tau_2) = (x_{25}, x_2, x_{30}, x_4).$$

Substituting (6.4) into (6.3), we obtain 14 d-polynomials, of which 11 are from (6.3) and 3 from the last row of (6.4). These d-polynomials form the hypothesis set *HYP*. A d-characteristic set of *HYP* consists of 10 d-polynomials, of which the first is

$$C_1 = x_4\Delta_1 x_2 + x_2\Delta_1 x_4.$$

$C_1 = 0$ means that $x_2 x_4$ is a constant. Therefore, the following theorem has been discovered automatically.

Schell's Theorem:

The product of the torsions at the corresponding points of a Bertrand pair of curves is a constant.

Suppose that one is interested in the probable constraint on k_1 and τ_1 for the curve r_1 to have a Bertrand companion. For this purpose let

$$(k_1, \tau_1, k_2, \tau_2) = (x_1, x_2, x_{30}, x_{40}).$$

A d-characteristic set *dCS* of the hypothesis set contains the d-polynomial

$$F_1(\Delta_1 x_1\, \Delta_2 x_2 - \Delta_1 x_2\, \Delta_2 x_1),$$

under the non-degeneracy conditions $F_1, \Delta_1 x_1 \neq 0$, etc. Here

$$F_1 = (x_1\, \Delta_1 x_2 - x_2\, \Delta_1 x_1)(x_1\, \Delta_1 x_1 + x_1\, \Delta_1 x_2).$$

$\Delta_1 x_1\, \Delta_2 x_2 - \Delta_1 x_2\, \Delta_2 x_1 = 0$ means that k_1 and τ_1 are linearly dependent over the field of constants. Therefore, we have the following.

Bertrand's Theorem:

If a curve has a Bertrand companion, the curvature and torsion of the curve are in some linear relation with constant coefficients.

Let Z_1 and Z_2 be the centers of curvatures of r_1 and r_2 at the corresponding points, and let c be the cross ratio of (r_1, Z_1, r_2, Z_2). Then we have

$$\text{d-prem}(\Delta_1 G, dCS) = 0,$$

where

$$G = \frac{1}{c} = \frac{1 + a_2 k_2}{1 - a_2 k_1}.$$

The following theorem is thus obtained.

Mannheim's Theorem:

The cross ratio of any two corresponding points and the centers of curvature of a Bertrand pair of curves is a constant.

Similarly, one can discuss other cases of Bertrand type and rediscover various theorems in an automatic manner. Chou and Gao (1991) collected about 100 theorems from differential geometry and mechanics proved or derived automatically with their program. Several new formulas about Bertrand curve pairs in affine space are included in their collection. A list of results for about 18 kinds of Bertrand curves in metric and affine spaces is also contained in Chou and Gao (1993c).

6.3.2 Newton's gravitational laws from Kepler's laws

Kepler's three laws can be formulated as follows.

(K1) The planets move in elliptic orbits with the Sun as a focus.

(K2) The vector from the Sun to a planet sweeps equal areas in equal periods of time.

(K3) The square of the period of a planet motion is proportional to the cube of the major axis of the elliptic orbit.

The following gravitational laws of Newton are also well-known in mechanics.

(N1) The acceleration of a planet is inversely proportional to the square of the distance from the Sun to the planet.

(N2) The acceleration vector of a planet is directed towards the Sun.

(N3) The proportion factor in (N1) is independent of the planets.

In order to apply Wu's method, we first introduce coordinates and translate the above laws into equations. Take the Sun as the origin and the major axis of the elliptic orbit of a planet as the x-axis. Let r be the distance between the Sun and the planet, T be the motion period, $2a, 2b$ be the major and minor axis of the orbit. Also let $hT = 2\pi ab$, $b^2 = ap$, and e be the eccentricity. Then, (K1) corresponds to

$$r = p + ex, \qquad p' = 0, \qquad e' = 0,$$

where the prime denotes derivation with respect to time t, and (K2) corresponds to

$$xy' - x'y = h, \qquad h' = 0.$$

Since $T^2/a^3 = 4\pi^2 p/h^2$, (K3) corresponds to T^2/a^3 being a constant independent of the planet.

Let A be the acceleration. Then, (N1) corresponds to r^2A being constant, i.e.,

$$2r'A + rA' = 0,$$

and (N2) corresponds to

$$xy'' - x''y = 0,$$

which is a direct consequence of (K2). (N3) states that r^2A is a constant independent of the planet. As we have already seen a constant $c = p/h^2$ in (K3) which is independent of the planet, (N3) suggests some probable relationship between the two constants.

Let

$$(r, A, c, p, e, x, y, h) = (x_{11}, x_{12}, x_{15}, x_{21}, x_{22}, x_{31}, x_{32}, x_{51}).$$

Then Kepler's laws may be expressed as 8 d-polynomial equations $P_i = 0$, where

$$\begin{cases} P_1 = x_{11} - x_{31}x_{22} - x_{21}, \\ P_2 = x_{31}^2 + x_{32}^2 - x_{11}^2, \\ P_3 = \Delta_1 x_{21}, \\ P_4 = \Delta_1 x_{22}, \\ P_5 = x_{31}\Delta_1 x_{32} - x_{32}\Delta_1 x_{31} - x_{51}, \\ P_6 = \Delta_1 x_{51}, \\ P_7 = x_{15}x_{51}^2 - x_{21}, \\ P_8 = (\Delta_2 x_{31})^2 + (\Delta_2 x_{32})^2 - x_{12}^2. \end{cases} \tag{6.5}$$

In the process of computing a d-characteristic set of (6.5) according to the scheme (6.1), the following two d-polynomials occur in the d-basic sets:

$$\begin{cases} B_{41} = 2x_{12}\Delta_1 x_{11} + \Delta_1 x_{12} x_{11}, \\ B_{31} = x_{15}^2 x_{12}^2 x_{11}^2 - 1. \end{cases}$$

Observe that $B_{41} = 0$ is just (N1), while $B_{31} = 0$ produces $c = \pm 1/(r^2 A)$, which is identical to (N3). Therefore, Newton's gravitational laws are derived automatically from Kepler's laws.

In Chou and Gao (1993b), a complete discussion of the logical relationship among the laws is presented. Some of the results are:

(1) (K1) and (K2) imply (N1);

(2) (N1) and (N2) imply (K1);

(3) (K1) and (N1) do not imply (K2).

6.4 Later developments

6.4.1 Weak ascending chains

Chou and Gao (1993a) proposed an improved algorithm to control the growth of the size of the d-polynomials and the growth of branches during the computation of d-characteristic sets.

First we need some definitions. Let $\mathcal{F}$ be a d-field, $\{P_1, \ldots, P_s\}$ be an ordered set of d-polynomials. It is called a (non-trivial) *quasi-ascending chain* (*q-ascending chain*) if the class of P_1 is greater than zero and, for any $j > i$, the class of P_j is greater than that of P_i. For a d-polynomial G and a q-ascending chain dAS, G is said to be *quasi-reduced* (*q-reduced*) with respect to dAS if the d-remainders of the initial and the separant of G with respect to dAS are both non-zero. A q-ascending chain $\{P_1, \ldots, P_s\}$ is called a *weak ascending chain* (*w-ascending chain*), if for any $j > 1$, P_j is q-reduced with respect to $\{P_1, \ldots, P_{j-1}\}$. As in the case of d-ascending sets, an order can be defined for the w-ascending chains. For a set of d-polynomials, a w-ascending chain in it which is not higher in order than any other w-ascending chain in the set is called a *weak basic* set (*w-basic set*).

Consider a set dPS of d-polynomials and let d-Ideal(dPS) denote the differential ideal generated by the d-polynomials in dPS. The Wu–Ritt well-ordering principle modified by Chou and Gao is as follows. For dPS, one can find either a non-zero d-polynomial in $\mathcal{F} \cap$ d-Ideal(dPS), or a w-ascending chain dAS and a set of d-polynomials $dPS' \supseteq dPS$ such that

(a) dAS is a w-basic set of dPS';

(b) d-Zero(dPS)=d-Zero(dPS');

(c) $\text{d-Zero}(dPS) = \text{d-Zero}(dAS/HS) \cup \left(\bigcup_i \text{d-Zero}(dPS' \cup \{H_i\}/\{H_1, \ldots, H_{i-1}\})\right);$

(d) $\text{d-Zero}(dPS) = \text{d-Zero}(\text{d-PD}(dAS)) \cup \left(\bigcup_i \text{d-Zero}(dPS' \cup \{H_i\}/\{H_1, \ldots, H_{i-1}\})\right).$

Here $HS = \{H_1, \ldots\}$ is the set of initials and separants of the d-polynomials in dAS, and d-PD(dAS) denotes the set of d-polynomials whose d-remainders with respect to dAS are zero.

Let $dAS = \{A_1, \cdots, A_s\}$ be a q-ascending chain. For a d-polynomial G, we define

$$\text{w-prem}(G, dAS) = \begin{cases} \text{w-prem}(\text{d-prem}(G, A_s), \{A_1, \ldots, A_{s-1}\}), \text{ if } \text{cls}(G) = \text{cls}(A_s); \\ \text{w-prem}(G, \{A_1, \ldots, A_{s-1}\}), \text{ if } \text{cls}(G) < \text{cls}(A_s); \\ G, \text{ if } \text{cls}(G) > \text{cls}(A_s) \text{ and } G \text{ is q-reduced with respect to } dAS; \\ \text{d-prem}(G, dAS), \text{ otherwise.} \end{cases}$$

G is said to be *weak reduced* (*w-reduced*) with respect to dAS if w-prem$(G, dAS) = G$.

A scheme to realize the modified well-ordering principle is as follows. Assume that for dPS one cannot find a non-zero d-polynomial in $\mathcal{F} \cap \text{d-Ideal}(dPS)$; then

$$\begin{array}{ccccccccc} dPS & = & dPS_0 & & dPS_1 & & \cdots & & dPS_u \\ & & dBS_0 & \succ & dBS_1 & \succ & \cdots & \succ & dBS_u \\ & & dRS_0 & & dRS_1 & & \cdots & & dRS_u \quad = \quad \varnothing. \end{array}$$

Here dBS_i is a w-basic set of dPS_i, dRS_i is the result of w-prem applied to dPS_i with respect to dBS_i, and $dPS_{i+1} = dPS_i \cup dRS_i$. Obviously, $dPS' = dPS_u$ and $dAS = dBS_u$.

6.4.2 Zero decomposition techniques

Wang (1995c, 1996d) proposed another decomposition algorithm based on the elimination idea and theory of Seidenberg (1956b). This algorithm can effectively improve the performance of zero decomposition over Wu's method.

We recall some definitions. A pair $[dTS, dUS]$, where dTS and dUS are d-polynomial sets, is called a *d-triangular system*, if dTS is a q-ascending chain such that neither ini(T) (the initial of T) nor sep(T) (the separant of T) vanishes on d-Zero(dTS/dUS) for any $T \in dTS$. A d-triangular system $[dTS, dUS]$ is said to be *fine* if d-prem(dUS, dTS) does not contain zero.

Let dPS be a d-polynomial set; we define

$$\begin{aligned} dPS^{(k)} &= dPS \cap K\{x_1, \ldots, x_k\}, \\ dPS^{\langle k \rangle} &= dPS^{(k)} \backslash dPS^{(k-1)}, \\ dPS^{[k]} &= dPS \backslash dPS^{(k)}. \end{aligned}$$

dPS is said to be of level k, denoted as level$(dPS) = k$, if $dPS^{[k-1]} = dPS^{\langle k \rangle} \neq \varnothing$.

Wang's Algorithm: (Zero decomposition)

The input is $[dPS, dQS]$, where dPS, dQS are d-polynomial sets; the output is an empty set if d-Zero$(dPS/dQS) = \varnothing$ is detected, or otherwise a set Ψ of the form $\{[dTS_1, dUS_1], \ldots, [dTS_e, dUS_e]\}$ such that

$$\text{d-Zero}(dPS/dQS) = \bigcup_{i=1}^{e} \text{d-Zero}(dTS_i/dUS_i),$$

where each $[dTS_i, dUS_i]$ is a fine d-triangular system.

W1. Set $\Psi \leftarrow \varnothing$, $\Phi \leftarrow \{[dPS, dQS, \varnothing]\}$. Repeat steps W2–W6 until $\Phi = \varnothing$.

W2. Let $[dFS, dUS, dTS] \in \Phi$ and set $\Phi \leftarrow \Phi \backslash \{[dFS, dUS, dTS]\}$, $m \leftarrow$ level(dFS).

Do steps W3–W5 for $i = m, \ldots, 1$.

W3. If $dFS \cap K \neq \varnothing$ then go to W2. If $dFS^{\langle i \rangle} \neq \varnothing$ then repeat W4.

W4. Select from $dFS^{\langle i \rangle}$ an element T having minimal rank with respect to x_i. If ini(T) does not occur in $\prod_{U \in dUS} U$ as a factor then set

$$\begin{aligned} &\Phi \leftarrow \Phi \cup \{[dFS \backslash \{T\} \cup \{\text{ini}(T), \text{red}(T)\}, dUS, dTS]\}, \\ &dUS \leftarrow dUS \cup \{\text{ini}(T)\}. \end{aligned}$$

Here red(T) is T minus its leading term. If sep$(T) \neq$ ini(T) then set

$$\begin{aligned} &\Phi \leftarrow \Phi \cup \{[dFS \backslash \{T\} \cup \{\text{d-prem}(T, \text{sep}(T)), \text{sep}(T)\}, dUS, dTS]\}, \\ &dUS \leftarrow dUS \cup \{\text{sep}(T)\}. \end{aligned}$$

If $dFS^{\langle i \rangle}$ has only one element then go to W5; otherwise compute

$$dFS \leftarrow dFS^{(i-1)} \cup \{T\} \cup \text{d-prem}(dFS^{\langle i \rangle}, T) \backslash \{0\}.$$

W5. Compute $dUS \leftarrow$ d-prem(dUS, T). If $0 \in dUS$ then go to W2; otherwise set

$$dFS \leftarrow S \backslash \{T\}, \qquad dTS \leftarrow \{T\} \cup dTS.$$

W6. Set $\Psi \leftarrow \Psi \cup \{[dTS, dUS]\}$. ∎

Using an implementation of this algorithm, Wang (1995c) was able to prove several theorems including results about Bertrand curves and the relations between Kepler's and Newton's laws more efficiently. A detailed presentation of the above algorithm and its variants in the algebraic case for proving theorems in elementary geometry can be found from Wang (1995b).

6.4.3 Elimination of existential quantifiers

Statements in differential geometry often involve existential quantifiers. As an example, one may take the statement that a vector is perpendicular to a fixed direction. Such statements cannot be transformed into equations.

Chou and Gao (1993b) proposed a technique to deal with the problem when the existential quantifier is about the existence of certain constant scalars or vectors. They make use of the *Wronskian* of r functions $y_1, \ldots, y_r$ of the same variable,

$$W(y_1, \ldots, y_r) = \begin{vmatrix} y_1 & y_2 & \cdots & y_r \\ y_1' & y_2' & \cdots & y_r' \\ \vdots & \vdots & \vdots & \vdots \\ y_1^{(r-1)} & y_2^{(r-1)} & \cdots & y_r^{(r-1)} \end{vmatrix},$$

together with the following property of the Wronskian.

Lemma 6.1:

Let $\tau_1, \ldots, \tau_m$ be in a d-field. They satisfy $a_1\tau_1 + \cdots + a_m\tau_m = 0$ for constants $a_1, \ldots, a_m$, which are not all zero, if and only if $W(\tau_1, \ldots, \tau_m) = 0$. In particular, the τ's satisfy $a_1\tau_1 + \cdots + a_m\tau_m = a_0$ for constants $a_0, \ldots, a_m$, which are not all zero, if and only if $W(1, \tau_1, \ldots, \tau_m) = 0$.

Chou–Gao's Algorithm A: (Existential quantifier elimination)

Assume that the variables $\boldsymbol{x} = x_1, \ldots, x_n$ satisfy a d-polynomial equation $P(\boldsymbol{x}) = 0$. We need to know whether the $\boldsymbol{x}$ satisfy a rational polynomial equation $R(\boldsymbol{a}, \boldsymbol{x}) = 0$, where $\boldsymbol{a} = a_1, \ldots, a_m$ are constants; and if there is such a rational polynomial R, we need to know which monomials effectively occur in the relation. Let $W_i = W(1, x_1, \ldots, x_n, \ldots, x_1^i, \ldots, x_n^i)$, and M be a sufficiently large integer.

C1. For $i = 1, \ldots, M$, decide whether d-prem$(W_i, P) = 0$.

Let j be the first i such that d-prem$(W_j, P) = 0$. If such a j does not exist, go to C3.

C2. Output "A polynomial relation of degree $\leq j$ with respect to any of the $\boldsymbol{x}$ exists."

One can further decide which monomials effectively occur in the polynomial relation by computing the Wronskians of functions selected from the following set:

$$\{1, x_1, \ldots, x_n, \ldots, x_1^j, \ldots, x_n^j\}.$$

C3. Output "No polynomial relation of degree lower than M with respect to any of the $\boldsymbol{x}$ exists." ■

Li and Cheng (1998a,b) proposed another technique to eliminate existential quantifiers which are not necessarily about the existence of some constant variables.

Li–Cheng's Algorithm A: (Existential quantifier elimination)

Let *HYP* be a d-polynomial set representing the hypothesis of a theorem and V_{old} the set of variables that occur in the hypothesis. Let the conclusion of the theorem be the existence of variables in $V_{\text{new}} = \{v_1, \ldots, v_m\}$ satisfying a set of properties $P_1 = 0, \ldots, P_n = 0$, where the P's are d-polynomials. Let $CON = \{P_1, \ldots, P_n\}$.

L1. Compute a d-characteristic set *dCS* of *HYP*.

L2. Set $dCS \leftarrow dCS \cup CON$. Compute a d-characteristic set dCS' of *dCS* under the order of variables $V_{\text{new}} \prec V_{\text{old}}$.

L3. Let *dDP* be the set of d-polynomials in dCS' whose leads are derivatives of the variables in V_{new}. Let $dCS'' = dCS' \setminus dDP$.

L4. Do the following for $i = 1, \ldots, m$:

If v_i appears in dCS'' as a lead, then the corresponding d-polynomial can be taken as an algebraic polynomial with v_i as the leading variable. Denote this algebraic polynomial by $P(v_i, u_1, \ldots, u_s)$, where $u_1, \ldots, u_s \prec v_i$. If this polynomial can guarantee the existence[1] of a solution for v_i in the field $\mathcal{F}(u_1, \ldots, u_s)$, the existence of v_i is proved; otherwise the algorithm terminates, with the output that "the conclusion is not proved."

L5. The conclusion *CON* can be proved by dCS''. ∎

The above technique is called *new-variable solving technique*. One example will be given in a later section.

6.4.4 Simplest relation derivation

A technique was introduced by Chou and Gao (1992c) to derive mechanically the simplest relation among certain variables, i.e., an algebraic relation of the derivatives with lowest orders. We present this technique in the following algorithmic form.

Chou–Gao's Algorithm B: (Simplest relation derivation)

Given two sets of d-polynomials: $dPS = \{H_1, \ldots, H_r\}$ in the variables $\boldsymbol{x} = x_1, \ldots, x_n$, and $dQS = \{Q_1(w_1, \boldsymbol{x}), \ldots, Q_s(w_s, \boldsymbol{x})\}$ in the variables $\boldsymbol{w} = w_1, \ldots, w_s$, and $\boldsymbol{x}$, find the relations among the w's with the lowest order of derivation for each w_i, under the condition that no element of $dDS = \{D_1, \ldots, D_t\}$ is equal to zero, where each D_i is a d-polynomial in $\boldsymbol{x}$. Let M be a sufficiently large integer.

C1. Under the order $\boldsymbol{w} \prec \boldsymbol{x}$, use Wu–Ritt's algorithm to compute the following decomposition:

$$\text{d-Zero}(dPS \cup dQS/dDS) = \bigcup_{i=1}^{a} \text{d-Zero}(\text{d-PD}(dAS_i'^{*})/dDS) \cup \left(\bigcup_{j=1}^{b} \text{d-Zero}(\text{d-PD}(dAS_j')/dDS) \right),$$

where all the d-ascending sets $dAS_i'^{*}$ and dAS_j' are irreducible, every $dAS_i'^{*}$ does not contain any d-polynomial of $\boldsymbol{w}$ and every dAS_j' contains at least one such d-polynomial.

If $a \neq 0$, then the w's are algebraically independent and the algorithm terminates.

---

[1] In Li and Cheng's implementation it is required that the polynomial P be linear with respect to v_i.

C2. Use Wu–Ritt's decomposition algorithm once again to get

$$\text{d-Zero}(dPS/dDS) = \bigcup_{i=1}^{u} \text{d-Zero}(dAS_i/dDS \cup JS_i),$$

where JS_i is the set of initials and separants of the d-polynomials in dAS_i.

C3. For each component d-Zero$(dAS_i/dDS \cup JS_i)$, compute

$$dQS_i \leftarrow \text{d-prem}(dQS, dAS_i).$$

Set $dPS_i \leftarrow dQS \cup dQS_i \cup dAS_i$. Treat the d-polynomials in dPS_i as algebraic ones, and use Wu–Ritt's decomposition algorithm for algebraic polynomials to find the relations determined by $[dPS_i, dDS \cup JS_i]$ among the $\boldsymbol{w}$.

If the relations can be found, then the algorithm terminates.

C4. For $j = 1, \ldots, M$ and $i = 1, \ldots, s$ do: add $\Delta_j w_i$ to $\boldsymbol{w}$ and $\Delta_j Q_i$ to dQS, and repeat steps C3 and C4 for the new $\boldsymbol{w}$ and dQS.

C5. Output that "a relation where the derivatives of the W are at most of degree M does not exist." ■

Using this algorithm, Newton's gravitational laws can be derived from Kepler's laws naturally as the simplest relations among certain variables.

6.4.5 Clifford algebra for theorem proving

In differential geometry textbooks, vector algebra and Frenet moving frames are used to solve problems in the local theory of space curves. Li and Cheng (1998a,b) recognized that Clifford algebra is more suitable for symbolic vector equation solving than is vector algebra. Based on Wu's method of characteristic sets, they proposed a method that employs Clifford algebraic representation for geometric entities and constraints. A prover based on this method is capable of producing proofs for theorems and exercises from textbooks in differential geometry that are much the same as those used in the textbooks. In this subsection we introduce this method, which is called a *Clifford algebraic reduction method*. A more comprehensive presentation of methods using Clifford algebra for theorem proving in elementary geometry may be found from Chap. 8.

Li–Cheng's Algorithm B: (Algebraic reduction)

By *algebraic reduction* for a d-polynomial equations set we mean triangulation by taking d-polynomial equations as algebraic ones. Let *HYP* be a set of algebraic equations.

L1. Do substitutions and pseudo-divisions in *HYP* as in Wu's method (see Chap. 5).

L2. Do generations in *HYP*.

By *generation* we mean producing, from the vector equation $\mathbf{A} \times \mathbf{B} = \mathbf{V}$, the following two equations:

$$\mathbf{A} \cdot \mathbf{V} = 0, \qquad \mathbf{B} \cdot \mathbf{V} = 0,$$

where $\mathbf{A}, \mathbf{B}, \mathbf{V}$ are vectors, "$\times$" is the cross product in vector algebra, and $\mathbf{V}, \mathbf{B} \prec \mathbf{A}$.

Repeat steps L1 and L2 until no new generation can be done.

L3. Do vector equation solving in *HYP* according to a set of formulas given in Li and Cheng (1998a).

L4. When no new equation solving is done, the algorithm terminates; otherwise go back to L1. ■

By *differential reduction* for a d-polynomial equations set we mean triangulation by both algebraic reductions and differentiations. The result is called a *reduction set*. Let $v_1 \prec \cdots \prec v_n$ be basic variables, i.e., those variables which are not derivatives of the other variables. For an equation $E = 0$, let v_r be the highest-ordered basic variable that occurs in E; v_r is called the *leading basic variable*. The highest order of derivation of v_r that occurs in E is called the *class* of E.

Li–Cheng's Algorithm C: (Differential reduction)

Let *dPS* be a set of d-polynomial equations.

L1. Do algebraic reduction in *dPS*.

L2. Select equations from *dPS* and form groups:

Let v be a basic variable and dPS_v the set of equations in *dPS* with leading basic variable v; let c_v be the least class in dPS_v. Then equations in dPS_v with class c_v form a group G_v; c_v is called the *class* of G_v.

Do the grouping according to every basic variable and discard empty groups.

L3. [Differentiation.] Let $dQS = dPS \setminus \bigcup_v G_v$, where G_v is non-empty. For any equation $E \in dQS$, if it contains a derivative of v for which $G_v \neq \varnothing$ and the order of the derivative is not lower than the class of G_v, then every member of G_v is differentiated to meet the order of the highest derivative of v in E.

Do the above procedure for every equation in dQS.

L4. When no new differentiation is done, the algorithm terminates; otherwise go back to L1. ■

Li–Cheng's Algorithm D: (Mechanical theorem proving)

For a given geometric theorem, first select one or several moving frames. The motional equations and the orthonormal properties of the frames, together with the hypothesis, form the input set dPS. The conclusion is denoted by CON, whose elements are expressions.

L1. Do differential reduction in dPS. The result is denoted by dRS. The non-degeneracy conditions are denoted by dND_1.

L2. Proving the conclusion by dRS: do substitutions and pseudo-divisions to CON with dRS.

If CON is reduced to $\{0\}$, then the algorithm terminates; in this case the theorem is proved to be true under the non-degeneracy conditions dND_1.

L3. By adding parametric equation solving technique to algebraic reduction, compute the differential reduction of dRS. The result is denoted by $dPRS$. The non-degeneracy conditions are denoted by dND_2.

Parametric equation solving indicates that new parameters, either free or constrained, are allowed in solutions.

L4. Proving the conclusion by $dPRS$. If CON is reduced to $\{0\}$, then the algorithm terminates; in this case the theorem is proved under the non-degeneracy conditions $dND_1 \cup dND_2$.

L5. Introduce coordinates and translate $dPRS$ into d-polynomials. The result is denoted by $dCRS$.

L6. Use Wu's method to compute a d-characteristic set dCS of $dCRS$. The non-degeneracy conditions are denoted by dND_3.

L7. Prove the conclusion by *dCS*. The non-degeneracy conditions are $dND_1 \cap dND_2 \cap dND_3$. ■

Li and Cheng's algorithm is based on Wu's principle of mechanical theorem proving in differential geometry. The correctness and completeness of the algorithm are both guaranteed by the principle. The main difference of the algorithm from Wu's method is that the algebraization procedure of a geometric theorem is intermixed with the proving procedure:

(1) in L1 a high-level algebraic representation of the geometric hypothesis is used, the triangulation is incomplete but is highly geometrically invariant;

(2) if the proof in L2 succeeds, it will be highly efficient and geometrically interpretable;

(3) in L3 some parameters are introduced for "finer" algebraization and hence more complete triangulation — these parameters may limit the geometric interpretabilities, but not generally, and can be considered as "geometric coordinates";

(4) if the proof in L4 succeeds, it will be geometrically interpretable in general;

(5) in L5 coordinates are introduced to all geometric entities, but the hypothesis of the theorem has undergone two processes and is in general much simplified;

(6) because of the previous processes, a characteristic set computed in L6 often has better structure (fewer branches, better geometric interpretation, etc.), which may contribute to simplifying the proving procedure in L7.

Below we present an example, in which the new-variable solving technique is also used.

Example 6.1:

Any Bertrand curve with non-constant k and τ can be written as

$$\mathbf{r} = x \int \mathbf{g}(t)dt + y \int \mathbf{g}(t) \times \mathbf{g}'(t)dt, \tag{6.6}$$

where x, y are non-zero constants and $|\mathbf{g}(t)| = |\mathbf{g}'(t)| = 1$.

The hypotheses are

$$\begin{cases} ak + b\tau = 1, \\ a' = 0, \\ b' = 0, \\ a, b, k, \tau, k', \tau' \neq 0. \end{cases}$$

The new variables are $x, y, \gamma, \mathbf{g}$, where $\gamma = ds/dt$.

Under the order of variables

$$a \prec b \prec \tau \prec k \prec x \prec y \prec \gamma \prec \mathbf{B} \prec N \prec T \prec g,$$

the set of hypothesis d-polynomials is already a reduction set. Now we add the properties of the new variables into the reduction set.

The equation (6.6) is equivalent to

$$\gamma \mathbf{T} = x\mathbf{g} + y\mathbf{g} \times \mathbf{g}'.$$

In fact, from the expression of $\mathbf{r}$ and the definition of γ, it is easy to see that $\gamma^2 = x^2 + y^2$. The new properties are

$$\begin{cases} \gamma \mathbf{T} = x\mathbf{g} + y\mathbf{g} \times \mathbf{g}', \\ \mathbf{g}^2 = 1, \\ \mathbf{g}'^2 = 1, \\ x' = 0, \\ y' = 0, \\ \gamma^2 = x^2 + y^2, \\ x, y \neq 0. \end{cases}$$

A new reduction set may be computed that is composed of the following two branches:

$$\begin{cases} ak = 1 - b\tau, \\ a' = 0, \\ b' = 0, \\ x = -a, \\ y = b, \\ \gamma^2 = a^2 + b^2, \\ \gamma \mathbf{g} = -a\mathbf{T} + b\mathbf{B}, \end{cases} \qquad \begin{cases} ak = 1 - b\tau, \\ a' = 0, \\ b' = 0, \\ x = a, \\ y = b, \\ \gamma^2 = a^2 + b^2, \\ \gamma \mathbf{g} = a\mathbf{T} - b\mathbf{B}. \end{cases}$$

The existence of $x, y, \gamma, \mathbf{g}$ is thus obtained. The conclusions are the new properties of the new variables, which can be easily proved by the new reduction set.

6.5 The two-variable case

Most of the notions in the one-variable case, e.g., d-field, d-ascending set and d-basic set, extend naturally to the present case and we shall keep the same terminology and notations. However, the notion of d-characteristic set needs major revisions (Wu 1991c).

Let the two independent variables be t_1 and t_2. The partial derivation with respect to t_1 i times and with respect to t_2 j times is denoted by Δ_{ij}. The set of partial derivatives of indeterminates $x_1, \ldots, x_n$ will be ordered in some natural way. Let dAS be a d-ascending set, and let F_1 and F_2 be two d-polynomials in dAS with leads $\Delta_{ij}x_c$ and $\Delta_{kl}x_c$ respectively, where $i > k$ and $j < l$. Differentiating F_1 with respect to t_2 $l-j$ times and F_2 with respect to t_1 $i-k$ times, we get two equations of the form

$$S_1\Delta_{0(l-j)}\Delta_{ij}x_c = G_1, \qquad S_2\Delta_{(i-k)0}\Delta_{kl}x_c = G_2,$$

where S_1 and S_2 are the separants of F_1 and F_2 respectively. The equality $S_2G_1 = S_1G_2$ follows as a consequence of the equations from dAS. We call the d-polynomial $S_2G_1 - S_1G_2$ the *integrability d-polynomial* of F_1 and F_2. The scheme (6.1) is modified now as follows. For any i, let dRS_i be the d-remainders of both the d-polynomials in dPS_i and the integrability d-polynomials of the pairs in dBS_i.

6.5.1 An example

We present an example to illustrate mechanical theorem proving and discovering in the two-variable case. This example is on surface pairs in the space.

For two pairs of surfaces S and S' in the space which are in 1-1 correspondence between their points X and X', we may ask for the geometric conditions under which XX' is tangent to both of the surfaces.

Let us attach two moving frames $\mathcal{M} = (X, E_1, E_2, E_3)$ and $\mathcal{M}' = (X', E_1', E_2', E_3')$ to S and S' in such a way that E_3 and E_3' are normal to S and S', and E_1 and E_1' are along the common tangent XX'. Denote the distance between X and X' by R and the angle between E_3 and E_3' by A. Then

$$\begin{cases} X' = X + RE_1, \\ E_1' = E_1, \\ E_2' = UE_2 + VE_3, \\ E_3' = -VE_2 + UE_3. \end{cases} \tag{6.7}$$

Here we have set $U = \cos A$ and $V = \sin A$ so that

$$U^2 + V^2 = 1. \tag{6.8}$$

Using W_i and W_{ij} to denote the Cartan exterior differential forms in parameters t_1 and t_2 on S corresponding to the above $\mathcal{M}$, we have the following Cartan structure equations:

$$\begin{cases} dX = W_1E_1 + W_2E_2, \\ dE_i = \sum\limits_j W_{ij}E_j, \\ W_{ij} = -W_{ji}, \\ W_3 = 0, \\ dW_1 = -W_2 \wedge W_{12}, \\ dW_2 = W_1 \wedge W_{12}, \\ W_1 \wedge W_{13} + W_2 \wedge W_{23} = 0, \\ dW_{12} = -W_{13} \wedge W_{23}, \\ dW_{13} = W_{12} \wedge W_{23}, \\ dW_{23} = -W_{12} \wedge W_{13}, \\ KW_1 \wedge W_2 = W_{13} \wedge W_{23}, \end{cases}$$

where K represents the Gaussian curvature of S at X. Referring the corresponding points on S' to the same parameters, we also have forms W_i' and W_{ij}' for $\mathcal{M}'$ and the corresponding Cartan equations.

To establish the hypothesis set, we do exterior differentiation to the equations in (6.7) and compare the coefficients of E_i. Then we obtain

$$\begin{cases} W_1' = W_1 + dR, \\ UW_2' = W_2 + RW_{12}, \\ VW_2' = RW_{13}, \\ VW_2 + RVW_{12} = RUW_{13}, \\ W_{12}' = UW_{12} + VW_{13}, \\ W_{13}' = -VW_{12} + UW_{13}, \\ W_{23}' = W_{23} + UdV - VdU. \end{cases} \tag{6.9}$$

Consider the case when both R and A are constants. The correspondence between X and X' under this assumption is called a *Bäcklund transformation.*

To study its geometric properties, let us express W_i and W_{ij} in terms of the parameters t_1 and t_2, with $W_1 = 0$ and $W_2 = 0$ as the parametric curves. Then

$$\begin{cases} W_1 = X_{31}dt_1, \\ W_2 = X_{37}dt_2, \\ W_{12} = X_{52}dt_1 + X_{42}dt_2, \\ W_{13} = X_{53}dt_1 + X_{43}dt_2, \\ W_{23} = X_{56}dt_1 + X_{46}dt_2, \\ W_1' = X_{61}dt_1 + X_{66}dt_2, \\ W_2' = X_{62}dt_1 + X_{67}dt_2, \\ W_{12}' = X_{72}dt_1 + X_{82}dt_2, \\ W_{13}' = X_{73}dt_1 + X_{83}dt_2, \\ W_{23}' = X_{76}dt_1 + X_{86}dt_2. \end{cases} \tag{6.10}$$

Let

$$(U, V, R, K, K') = (X_{21}, X_{22}, X_{23}, X_{25}, X_{30}).$$

Now equations (6.8), the Cartan equations, (6.9), and (6.10) give rise to a hypothesis set. In determining its d-characteristic set, it is too complicated to carry out the computations to the full owing to the complexity of the integrability d-polynomials involved. In the third d-basic set there appears the following d-polynomial

$$X_{23}^2 X_{30} + X_{22}^2.$$

The vanishing of this d-polynomial gives us a condition for $X_{30} = K'$ to be satisfied. As the relationship between S and S' is symmetric, the same is true for K. We have thus discovered the following theorem automatically.

Bäcklund Theorem:

Any two surfaces in Bäcklund correspondence have equal constant negative curvatures given by

$$K' = K = -\frac{\sin^2 A}{R^2}.$$

6.5.2 A new discovery

Z. Li (1995b) discovered the following new theorem in his experiment of computing a d-characteristic set of the fundamental equations of a space surface.

Li's Theorem:

If a surface has no umbilici, then the second fundamental form is algebraically determined by the first fundamental form together with the derivatives of the first fundamental form.

Below we show how this theorem may be discovered automatically.

For a surface $r(u_1, u_2)$, let the first fundamental form be

$$I = Edu_1^2 + 2Fdu_1du_2 + Gdu_2^2,$$

and the second fundamental form be

$$II = Ldu_1^2 + 2Mdu_1du_2 + Ndu_2^2.$$

The unit normal vector at r is $n = r_{u_1} \times r_{u_2}/|r_{u_1} \times r_{u_2}|$. The fundamental equations for the frame $\{r; r_{u_1}, r_{u_2}, n\}$ are

$$\begin{cases} \dfrac{\partial r}{\partial u_i} = r_{u_i}, \\ \dfrac{\partial r_{u_i}}{\partial u_j} = \Gamma_{ij}^k r_{u_k} + b_{ij} n, \\ \dfrac{\partial n}{\partial u_i} = -b_i^j r_{u_j}. \end{cases}$$

To compute a characteristic set from the fundamental equations, let

$$(E, F, G, L, M, N, r_{u_1}, r_{u_2}, n) = (X_1, X_2, X_3, X_{24}, X_{25}, X_{26}, X_{32}, X_{33}, X_{34}).$$

An order "$\prec$" on variables $x_1, \ldots, x_m$ and their partial derivatives is said to be *admissible* if

(a) $x_i \prec \Delta_{jk} x_i$;

(b) if $\Delta_{ij} x_p \prec \Delta_{kl} x_p$ then $\Delta_{(i+u)(j+v)} x_p \prec \Delta_{(k+u)(l+v)} x_p$ for any $u, v \geq 0$;

(c) if $\Delta_{ij} x_p \prec \Delta_{kl} x_q$ then $\Delta_{(i+u)(j+v)} x_p \prec \Delta_{(k+u)(l+v)} x_q$ for any $u, v \geq 0$.

For our purpose, an admissible order is made as follows:

(a) $\Delta_{ij} \succ \Delta_{kl}$ if either $i + j > k + l$, or $i + j = k + l$ and $j > l$;

(b) $\Delta_{ij} X_p \succ \Delta_{kl} X_q$ if either $\Delta_{ij} \succ \Delta_{kl}$, or $(i, j) = (k, l)$ and $p > q$;

(c) for $1 \leq p, q, r \leq 9$, $\Delta_{ij} X_p \prec \Delta_{kl} X_{2q} \prec \Delta_{mn} X_{3r}$.

Consider $X_2 = X_{25} = 0$, i.e., there is no umbilicus on the surface. When computing a d-characteristic set from the fundamental equations, Li discovered a d-polynomial of the form $X_{24}P$, where P is a d-polynomial with lead X_{24} and coefficients involving only X_1, X_2, X_3. Therefore, the second fundamental form is algebraically determined by the first fundamental form together with the derivatives of the first fundamental form.

6.6 Differential forms for theorem proving

To overcome the difficulty of integrability d-polynomials in computing d-characteristic sets in the local theory of space surfaces, H. Li (1997b) proposed a simple method to integrate the calculus of differential forms with Wu's method. Surprisingly, a prover based on this method is capable of producing proofs simpler than those in textbooks. Below we introduce this method briefly.

The method is analogous to the Clifford algebraic method proposed by Li and Cheng (1998a). The input consists of (a) the given hypothesis, (b) the motional equations of a first-order field of frames, and (c) the Gauss–Codazzi equations.

The procedure of triangulation for proving a theorem is also composed of three stages, i.e., computing (a) a reduction set, (b) a parametric reduction set, and (c) a d-characteristic set. To compute a reduction set, for scalar equations Wu's method is used; for vector equations, the Clifford algebraic reduction method. For equations of differential forms, the following algorithm realizes the triangulation.

Li's Algorithm: (Triangulating a set of equations on differential forms)

Let dPS be a set of differential forms equations.

L1. Do substitutions in dPS.

L2. Solve differential forms equations in dPS.

Repeat steps L1 and L2 until no new equation solving can be done.

L3. Generate integrability equations. If an exact form, e.g., df, is the leading variable of an equation, then a new equation is generated by substituting this equation into $d(df) = 0$. The newly derived equation is called an *integrability equation*.

If no new integrability equation can be generated, the algorithm terminates; otherwise go to L1. ∎

Example 6.2:

If the principal curvatures of a surface are unequal constants, the surface must be a cylinder.

We choose a second-order field of frames, i.e., a field $\{x; e_1, e_2, e_3\}$ of frames with e_3 normal to the surface at x, and e_1, e_2 along the principal directions. For this second-order field of frames, we have the following motional equations:

$$\begin{cases} dx &= \omega_1 e_1 + \omega_2 e_2, \\ de_1 &= \omega_{12} e_2 + \omega_{13} e_3, \\ de_2 &= -\omega_{12} e_1 + \omega_{23} e_3, \\ de_3 &= -\omega_{13} e_1 - \omega_{23} e_2. \end{cases}$$

We also have the Gauss–Codazzi equations

$$\begin{cases} d\omega_1 &= \omega_{12} \wedge \omega_2, \\ d\omega_2 &= \omega_1 \wedge \omega_{12}, \\ d\omega_{12} &= -ac\omega_1 \wedge \omega_2, \\ d\omega_{13} &= \omega_{12} \wedge \omega_{23}, \\ d\omega_{23} &= \omega_{13} \wedge \omega_{12}, \end{cases}$$

where $a \neq c$ are the principal curvatures. Assume that $c \neq 0$. At last, we have the following representation for ω_{13} and ω_{23}:

$$\omega_{13} = a\omega_1, \qquad \omega_{23} = c\omega_2.$$

The hypothesis for this theorem is

$$\begin{cases} da = 0, \\ dc = 0. \end{cases}$$

The conclusion is: the integral curve along e_1 is a straight line, while the integral curve along e_2 is a circle.

Let the order of variables be $a, c \prec \omega_1, \omega_2 \prec \omega_{12} \prec \omega_{13}, \omega_{23} \prec x, e_1, e_2 \prec e_3$.

Then the triangulation results in

$$\begin{cases} a = 0, \\ dc = 0, \\ d\omega_1 = 0, \\ d\omega_2 = 0, \\ \omega_{12} = 0, \\ \omega_{13} = 0, \\ \omega_{23} = c\omega_2, \\ dx = \omega_1 e_1 + \omega_2 e_2, \\ de_1 = 0, \\ de_2 = c\omega_2 e_3, \\ de_3 = -c\omega_2 e_2. \end{cases}$$

The conclusion is direct from the triangulation result.

Chapter 7

POINTS ELIMINATION METHODS FOR GEOMETRIC PROBLEM SOLVING

Jing-Zhong Zhang

This chapter reports on recent advances in automated production of short and human-readable proofs based on *points elimination methods*. Such methods use high level geometric invariants and may produce a variety of proofs with different styles. They include the area method and the full-angle method for plane Euclidean geometry, the volume method for solid geometry, the argument method for hyperbolic and elliptic geometries, and the vector method for metric geometries.

7.1 Introduction

Since the pioneering work of Wu (1978), highly successful algebraic methods for automated solution of geometric problems have been developed. Up to 1992, computer programs based on these methods were used to prove hundreds of difficult geometric theorems (Chou 1984; Wu 1984b; Kapur 1986a; Chou and Schelter 1986; Wang and Hu 1987; Zhang and Yang 1989; Zhang et al. 1990; Chou and Gao 1990a,b; Chap. 5). Many hard theorems whose traditional proofs need an enormous amount of human intelligence, such as Feuerbach's theorem and Morley's trisector theorem, can be proved by algebraic methods within seconds.

The above-mentioned algebraic methods, which are very different from the traditional proving methods used by geometers since Euclid, generally can only tell whether a statement is true or not. If one wants to know why the statement holds, one usually has to look at the tedious computations of polynomials. The traditional method usually gives elegant, short and readable proofs. Researchers have been studying automated generation of readable proofs using computer programs since the work by Gelernter et al. (1960). Despite the enormous amount of effort and great improvements (Nevins 1975; Reiter 1977; Anderson

MATHEMATICS MECHANIZATION AND APPLICATIONS
ISBN 0-12-734760-7

1981; Coelho and Pereira 1986) over 40 years, the successes in this direction have been very limited in the sense that no computer program has been developed which can produce readable proofs of non-trivial geometric theorems efficiently. A notable exception is the work of Chou et al. (1995c), which will be introduced in Chap. 10.

In 1992, a new idea for automated geometric problem solving was proposed and implemented by Zhang et al. (1992) and Chou et al. (1994a). This ideal has led to the *points elimination method.* The program based on this method can produce short and readable proofs for most non-trivial geometric theorems. The core of the *algebraic methods* consists of various elimination techniques which are in clear contrast to the somewhat random searching methods. The new method adopts the idea of elimination, but instead of eliminating the coordinates of points, it eliminates the points and/or other geometric objects themselves directly. Based on this idea, we can develop various methods for geometric problem solving by choosing different *geometric quantities, constructions* and *formulas* for eliminating points and/or other geometric objects.

In fact, a series of *points elimination methods* for machine solving in geometry has been developed in recent years. These methods can produce *readable proofs* in various geometries.

- The *area method* for Euclidean plane geometry was proposed originally in Zhang et al. (1992) and improved in Chou et al. (1993a, 1994a).

- The *volume method* for solid geometry was also given in Chou et al. (1994a, 1995a).

- The *full-angle method* (Chou et al. 1994b, 1996c), the *vector method* and the *complex number method* (Chou et al. 1993b, 1994a) were then developed. These methods provide more readable solutions for some classes of problems.

- The *argument method* for non-Euclidean geometries was established later in Yang, Gao, Chou and Zhang (1997), by which many new theorems have been discovered.

Using the above methods, about one thousand non-trivial geometric problems were solved by computer programs and most of the solutions are readable. Recently, the points elimination methods have been extended to the classes of geometric problems in which the geometric quantities can be represented as the elements in Clifford algebra (Li 1996a; Wang 1997; Fèvre and Wang 1998a,b;

Yang et al. 1998a; Chap. 8). Using the extended points elimination methods, computer programs have been developed by which one can solve problems involving lines, circles, and conics. The solutions produced are generally short and readable.

Why do we like *readable solving* rather than giving only a *yes/no* solution? There are at least two reasons. Firstly, mathematics is a culture. The developing of mathematics needs understanding and support by most people. Readable solving can make people understand mathematics and close to mathematics. This is important for attracting more people to support mathematics. Secondly, readable solving is more useful in education. It can teach students to learn why the mathematical conclusions are true and how to solve problems in mathematical form. Based on our methods, a software package for readable solving in geometry is being developed. It will be helpful to teachers and students.

This paper is a sketch of various points elimination methods for geometry. In Sect. 7.2, we explain the main idea of the techniques by three examples. Then, in Sect. 7.3, we present the area method for Euclidean plane geometry. This is followed, in Sect. 7.4, by the full-angle method. Section 5 introduces the vector method and the complex numbers method. We present in Sect. 7.6 the volume method for solid geometry and then, in Sect. 7.7, the points elimination method for non-Euclidean geometry. In the last section, we give some comments and propose a few problems for future research and development.

7.2 The main idea of points elimination methods

When faced with a hard geometric problem, the traditional idea for solving it is often to add auxiliary points and lines. The points elimination methods proceed by removing some things from the figure such that the relationship between the geometric elements becomes simpler and simpler. If there exists no relationship between the geometric elements in the figure, then the solution to the problem should express this clearly. We will use two examples to illustrate this idea.

In this chapter, we use capital English letters to denote points. We denote by $\overline{AB}$ the *signed length* of the oriented segment from A to B and its *absolute value* by AB. AB also denotes the line passing through points A and B. The *signed area* S_{ABC} of triangle ABC is the usual area with a sign depending on the order of the three vertices of the triangle. In the area method, the signed area is used as a basic (undefined) geometric quantity which satisfies the following properties.

Proposition 7.1:

(1) $S_{ABC} = S_{CAB} = S_{BCA} = -S_{BAC} = -S_{CBA} = -S_{ACB}$.

(2) *Points* A, B *and* C *are collinear iff* $S_{ABC} = 0$.

(3) *For any four points* A, B, C, *and* D, $S_{ABC} = S_{ABD} + S_{ADC} + S_{DBC}$.

(4) *If points* A, B, C, *and* D *are four collinear points such that* $A \neq B$, *and* P *is any point not on line* AB, *then*

$$\frac{\overline{CD}}{\overline{AB}} = \frac{S_{PCD}}{S_{PAB}}.$$

The *signed area of a quadrilateral* $ABCD$ is defined to be

$$S_{ABCD} = S_{ABC} + S_{ACD}.$$

All the following propositions can be deduced from Proposition 7.1.

Proposition 7.2: (Co-side theorem)

Let M *be the intersection of two non-parallel lines* AB *and* PQ, *and* $Q \neq M$. *Then*

$$\frac{\overline{PM}}{\overline{QM}} = \frac{S_{PAB}}{S_{QAB}}.$$

Proposition 7.3:

Let R *be a point on line* PQ. *Then for any points* A *and* B,

$$S_{RAB} = \frac{\overline{PR}}{\overline{PQ}} S_{QAB} + \frac{\overline{RQ}}{\overline{PQ}} S_{PAB}.$$

We use the notation $AB \parallel CD$ to denote the fact that A, B, C, and D satisfy: (1) lines AB and CD do not intersect; (2) $A = B$ or $C = D$; or (3) A, B, C, and D are on the same line.

Proposition 7.4:

$PQ \parallel AB$ *iff* $S_{PAB} = S_{QAB}$, *i.e., iff* $S_{PAQB} = 0$.

Proposition 7.5:

Let $AB \parallel CD$. *Then for any points* P *and* Q *such that* PQ *is not parallel to* AB, *we have*

$$\frac{\overline{AB}}{\overline{CD}} = \frac{S_{APBQ}}{S_{CPDQ}}.$$

As a consequence, if $ABCD$ *is a parallelogram then* $S_{APBQ} = S_{DPCQ}$.

Example 7.1: (Fig. 7.1)

Let O be the intersection of the two diagonals AC and BD of a parallelogram ABCD. Show that O is the midpoint of AC, or $\frac{\overline{AO}}{\overline{OC}} = 1$.

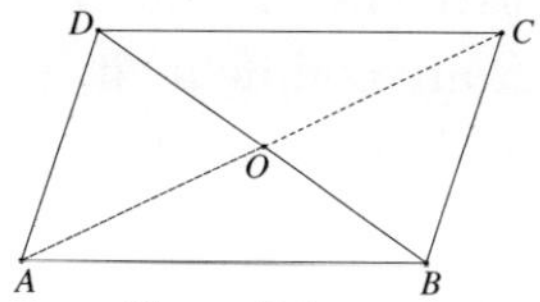

Figure 7.1

Analysis. The task is to show that $\overline{AO} = \overline{OC}$, or equivalently $\frac{\overline{AO}}{\overline{OC}} = 1$. We use the formula

$$\frac{\overline{AO}}{\overline{OC}} = \frac{S_{ABD}}{S_{BCD}}$$

to eliminate point O. Now we need: "Suppose that $ABCD$ is a parallelogram. Show $S_{ABD} = S_{BCD}$." Because $AB \parallel CD$ and $AD \parallel BC$, we have

$$S_{ABD} = S_{ABC}, \qquad S_{BCD} = S_{ABC}.$$

So point D is also eliminated and the problem now becomes a trivial statement: "For any triangle ABC, $S_{ABC} = S_{ABC}$."

It is now possible to get some basic idea about the points elimination method from the above example:

(1) The *geometric quantities* are ratios of signed lengths and signed areas.

(2) Points can be *constructed* in an order: A, B, C, D, O; and be eliminated in the reverse order: O, D, C, B, A.

(3) To eliminate points, *formulas* are needed. The main task in building a proving system is to choose the proper formulas for eliminating points.

The last example in this section will show how to solve non-trivial geometric problems by hand based on the idea of eliminating points.

Example 7.2: (Simson's theorem)

Let D be a point on the circumscribed circle of triangle ABC. From D three perpendiculars are drawn to the three sides BC, AC and AB of triangle ABC with three feet P, Q and R respectively. Show that P, Q and R are collinear.

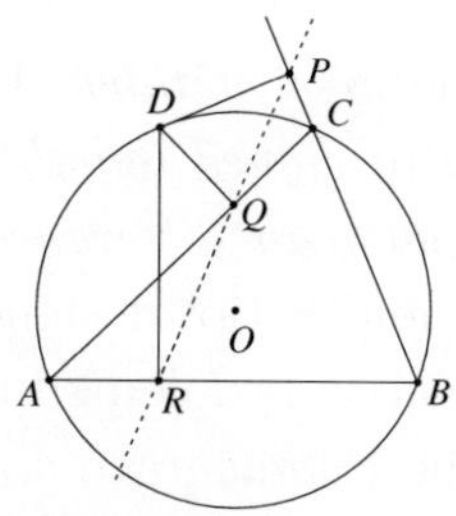

Figure 7.2

Analysis. Let H be the intersection of lines AB and PQ. To show that P, Q and R are collinear, it is enough to prove that R is the same as H. For this, we need only to show that

$$\frac{\overline{AR}}{\overline{RB}} = \frac{\overline{AH}}{\overline{HB}}, \quad \text{or equivalently} \quad \frac{\overline{AR}}{\overline{RB}}\frac{\overline{BH}}{\overline{HA}} = 1.$$

With the above idea, we have the following proof:

$$\frac{\overline{AR}}{\overline{RB}}\frac{\overline{BH}}{\overline{HA}} = \frac{\overline{AR}}{\overline{RB}}\frac{S_{BPQ}}{S_{AQP}}$$

(by the co-side theorem, eliminating H)

$$= \frac{AD\cos(\angle[DAB])}{BD\cos(\angle[DBA])}\frac{S_{BPQ}}{S_{AQP}}$$

(by the cosine function definition, eliminating R)

$$= \frac{AD\cos(\angle[DAB])}{BD\cos(\angle[DBA])}\frac{S_{BPA}}{S_{ACP}}\frac{\overline{QC}}{\overline{AQ}}$$

(by the co-side theorem and Proposition 7.1, $\frac{S_{BPQ}}{S_{BPA}} = \frac{\overline{QC}}{\overline{AC}}, \frac{S_{AQP}}{S_{ACP}} = \frac{\overline{AC}}{\overline{AQ}}$)

$$= \frac{AD\cos(\angle[DAB])}{BD\cos(\angle[DBA])}\frac{S_{BPA}}{S_{ACP}}\frac{DC\cos(\angle[DCA])}{DA\cos(\angle[DAC])}$$

(by the cosine function definition, eliminating Q)

$$= \frac{DC}{BD}\frac{\cos(\angle[DAB])}{\cos(\angle[DAC])}\frac{\cos(\angle[DCA])}{\cos(\angle[DBA])}\frac{BP}{CP}$$

(by Proposition 7.1, $\frac{S_{BPA}}{S_{APC}} = \frac{\overline{BP}}{\overline{CP}}$)

$$= \frac{DC}{BD}\frac{\cos(\angle[DAB])}{\cos(\angle[DAC])}\frac{\cos(\angle[DCA])}{\cos(\angle[DBA])}\frac{-BD\cos(\angle[DBC])}{DC\cos(\angle[DCB])}$$

(by the cosine function definition, eliminating P)

$$= 1.$$

(by the circumference-angle theorem and Fig. 7.2)

This proof depends on Fig. 7.2. In the next section, we will introduce a geometric quantity to make it independent of figures.

7.3 The area method

7.3.1 Basic geometric quantities

In the *area method*, three geometric quantities are used. They are *signed area, ratio of signed lengths* of two collinear or parallel directed segments, and the *Pythagorean difference*. We have introduced some propositions about signed area and ratio of signed lengths. The *Pythagorean difference* is used to treat geometric problems involving perpendicularity or angles. For points A, B and C, the *Pythagorean difference* P_{ABC} is defined to be $P_{ABC} = \overline{AB}^2 + \overline{CB}^2 - \overline{AC}^2$. It is easy to check that

(1) $P_{AAB} = 0$; $P_{ABC} = P_{CBA}$;

(2) $P_{ABA} = 2\overline{AB}^2$;

(3) if A, B, C are collinear, then $P_{ABC} = 2\overline{BA} \cdot \overline{BC}$.

For four points A, B, C and D, the notation $AB \perp CD$ implies that one of the following conditions is satisfied: $A = B$, or $C = D$, or line AB is perpendicular to line CD. The Pythagorean theorem is taken as a basic (unproven) property of the Pythagorean difference.

Proposition 7.6: (Pythagorean theorem)
$AB \perp BC$ iff $P_{ABC} = \overline{AB}^2 + \overline{BC}^2 - \overline{AC}^2 = 0$.

For a quadrilateral $ABCD$, let

$$P_{ABCD} = P_{ABD} - P_{CBD} = \overline{AB}^2 + \overline{CD}^2 - \overline{BC}^2 - \overline{DA}^2.$$

Then, the Pythagorean theorem can be generalized as follows.

Proposition 7.7:
$AC \perp BD$ iff $P_{ABD} = P_{CBD}$ or $P_{ABCD} = 0$.

Proposition 7.8:
For any points P and Q, let R be a point on line PQ with position ratios $r_1 = \frac{\overline{PR}}{\overline{PQ}}$ and $r_2 = \frac{\overline{RQ}}{\overline{PQ}}$. Then for points A and B, we have

(1) $P_{RAB} = r_1 P_{QAB} + r_2 P_{PAB}$, *and*

(2) $P_{ARB} = r_1 P_{AQB} + r_2 P_{APB} - r_1 r_2 P_{PQP}$.

Proposition 7.9:
Let $AB \parallel CD$. Then for any points P and Q such that PQ is not perpendicular to AB, we have

$$\frac{\overline{AB}}{\overline{CD}} = \frac{P_{APBQ}}{P_{CPDQ}}.$$

As a consequence, if $ABCD$ is a parallelogram then $P_{APBQ} = P_{DPCQ}$.

Proposition 7.10:
Let O, A, B be three points such that $OA = OB$, $AO \perp BO$ and $S_{AOB} > 0$. Then

$$S_{OPAQ} = \frac{1}{4} P_{OPBQ} \quad \text{and} \quad P_{OPAQ} = -4 S_{OPBQ}.$$

Example 7.3: (Orthocenter theorem; Fig. 7.3)

Let the two altitudes AF and BE of triangle ABC meet at H. Show that $CH \perp AB$.

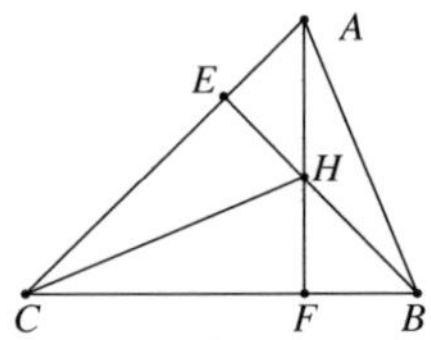

Figure 7.3

Proof. By Proposition 7.7, we need only to show that $P_{ACH} = P_{BCH}$. Since $BH \perp AC$ and $AH \perp BC$, by Proposition 7.7 we can eliminate point H: $P_{ACH} = P_{ACB}$; $P_{BCH} = P_{BCA}$. Thus we have the proof:

$$\frac{P_{ACH}}{P_{BCH}} = \frac{P_{ACB}}{P_{BCA}} = \frac{P_{ACB}}{P_{ACB}} = 1.$$

7.3.2 Constructive geometric statements

We consider three kinds of lines:

LL. (LINE U V) is the line passing through points U and V;

PL. (PLINE W U V) is the line passing through point W and parallel to UV;

TL. (TLINE W U V) is the line passing through point W and perpendicular to (LINE U V);

A *circle* with point O as its center and passing through a point U is denoted by (CIR O U).

We use the following constructions.

C1. (POINTS $Y_1, \ldots, Y_m$). Take arbitrary or free points $Y_1, \ldots, Y_m$ in the plane.

C2. (PRATIO Y R P Q λ). Take a point Y on (PLINE R P Q) such that $\overline{RY} = \lambda \overline{PQ}$, where λ can be a rational number, a variable, or an expression in geometric quantities. The non-degeneracy (ndg) condition is $P \neq Q$.

C3. (TRATIO Y U V r). Take a point Y on (TLINE U U V) such that $r = \frac{4S_{UVY}}{P_{UVU}} (= \frac{UY}{UV})$, where r can be a rational number, a rational expression in geometric quantities, or a variable. The ndg condition is $U \neq V$.

C4. (ON Y l). Take a point Y on a line l.

C5. (INTER Y l_1 l_2). Point Y is the intersection of line l_1 and line l_2. The ndg condition is $l_1 \not\parallel l_2$.

C6. (INTER Y l (CIR O P)). Point Y is the intersection of line l and circle (CIR O P) other than point P. Line l could be (LINE P U), (PLINE P U V), or (TLINE P U V). The ndg conditions are that $O \neq P$, $Y \neq P$, and line l is proper.

C7. (INTER Y (CIR O_1 P) (CIR O_2 P)). Point Y is the intersection of the circle (CIR O_1 P) and the circle (CIR O_2 P) other than point P. The ndg condition is that O_1, O_2, and P are not collinear.

A *constructive geometric statement* is represented by a list

$$S = (C_1, C_2, \ldots, C_k, G),$$

where each construction C_i introduces a new point. From the points introduced by C_j, $j = 1, \ldots, i-1$, the conclusion G of the statement is either a geometric predicate like COLLINEAR or PARALLEL, or an equation $E_1 = E_2$ of polynomials in geometric quantities.

Example 7.4:

The constructive descriptions for the parallelogram theorem (7.1) and the orthocenter theorem (7.3) are

((POINTS A B C)
(INTER D (PLINE A B C) (PLINE C A B))
(INTER O (LINE A C) (LINE B D))
(MIDPOINT O A C))

((POINTS A B C)
(INTER H (TLINE A B C) (TLINE B A C))
(PERPENDICULAR C H A B)).

Constructions C1–C7 are actually some of the constructions used in the area method. For a complete set, see Chou et al. (1994a).

7.3.3 Elimination lemmas

As shown by the above examples, the proof process consists in eliminating the constructed points from geometric invariants. Instead of giving the elimination lemma for each construction and each geometric quantity, what we will do below is to explain the basic ideas behind the method. Each elimination lemma can be divided into two steps:

(1) The first step is to eliminate the point from area and Pythagorean difference.

(2) The second step is to eliminate the point from ratios.

Step 1: Eliminating points from area and Pythagorean difference. This step can be summarized as the the following two basic cases.

Lemma 7.11:

Let Y be the point to be eliminated and there exist points W, U, V such that $YW \parallel UV$.

$$(1) \qquad S_{YNAM} = \begin{cases} S_{WNAM} & \text{if } NM \parallel UV, \\ \frac{\overline{AY}}{\overline{UV}} S_{VNUM} & \text{if } YA \parallel UV, \\ \frac{\overline{UY}}{\overline{UV}} S_{VNAM} + \frac{\overline{YV}}{\overline{UV}} S_{UNAM} & \text{if } W = U, \\ S_{WNAM} + \frac{\overline{WY}}{\overline{UV}} S_{VNUM} & \text{otherwise.} \end{cases}$$

$$(2) \qquad P_{YNAM} = \begin{cases} P_{WNAM} & \text{if } NM \perp UV, \\ \frac{\overline{AY}}{\overline{UV}} P_{VNUM} & \text{if } YA \parallel UV, \\ \frac{\overline{UY}}{\overline{UV}} P_{VNAM} + \frac{\overline{YV}}{\overline{UV}} P_{UNAM} & \text{if } W = U, \\ P_{WNAM} + \frac{\overline{WY}}{\overline{UV}} P_{VNUM} & \text{otherwise.} \end{cases}$$

(3) *If $W = U$, then*

$$P_{YYNM} = \frac{\overline{UY}}{\overline{UV}} P_{VVNM} + \frac{\overline{YV}}{\overline{UV}} P_{UUNM} + \frac{\overline{UY}}{\overline{UV}} \frac{\overline{YV}}{\overline{UV}} P_{NNMM}.$$

Otherwise,

$$P_{YYNM} = P_{WWNM} + \frac{\overline{WY}}{\overline{UV}} (P_{VNUM} - 2P_{WUV} - P_{NNMM}) - \frac{\overline{WY}^2}{\overline{UV}^2} P_{NNMM}.$$

Lemma 7.12:

Let Y be the point to be eliminated and there exist points W, U, V such that $YW \perp UV$.

$$(1) \qquad S_{YNAM} = \begin{cases} S_{WNAM} & \textit{if } NM \perp UV, \\ \frac{\overline{AY}}{\overline{UT}} \frac{P_{VNUM}}{4} & \textit{if } YA \parallel UV, \\ S_{WNAM} + \frac{\overline{WY}}{\overline{UT}} \frac{P_{VNUM}}{4} & \textit{otherwise.} \end{cases}$$

$$(2) \qquad P_{YNAM} = \begin{cases} P_{WNAM} & \textit{if } NM \parallel UV, \\ -4\frac{\overline{AY}}{\overline{UT}} S_{VNUM} & \textit{if } YA \parallel UV, \\ P_{WNAM} - 4\frac{\overline{WY}}{\overline{UT}} S_{VNUM} & \textit{otherwise.} \end{cases}$$

In the above formula, T is a point such that $UT = UV$, $UT \perp UV$, and $S_{TUV} > 0$.

Step 2: Eliminating points from ratios. This step can be summarized as the following cases.

Lemma 7.13:

Let Y be the point to be eliminated and there exist points W, U, V such that $YW \parallel UV$ and point A is not on line WY. Then $\frac{\overline{AY}}{\overline{CD}} = \frac{S_{AUWV}}{S_{CUDV}}$.

Lemma 7.14:

Let Y be the point to be eliminated and there exist points W, U, V such that $YW \perp UV$ and point A is not on line WY. Then $\frac{\overline{AY}}{\overline{CD}} = \frac{P_{AUWV}}{P_{CUDV}}$.

Lemma 7.15:

Let Y be introduced by (PRATIO Y W U V r). If $A \in WY$, then

$$\frac{\overline{AY}}{\overline{CD}} = \frac{\frac{\overline{AW}}{\overline{UV}} + r}{\frac{\overline{CD}}{\overline{UV}}};$$

otherwise, $\frac{\overline{AY}}{\overline{CD}} = \frac{S_{AUWV}}{S_{CUDV}}$.

Lemma 7.16:

Let Y be introduced by (TRATIO Y W U V r). If $A \in WY$, then

$$\frac{\overline{AY}}{\overline{CD}} = \frac{S_{AUWV} - \frac{r}{4} P_{PQQP}}{S_{CUDV}};$$

otherwise $\frac{\overline{AY}}{\overline{CD}} = \frac{P_{AUWV}}{P_{CUDV}}$.

Basically, the area method, using the lemmas given above, eliminates the constructed points in a geometric statement in reverse order in which they were introduced. We omit the detailed description here.

Example 7.5:

Given any triangle ABC with squares $ADFC$ and $BEGC$ described both externally or both internally on sides CA and CB, show that AE and BD intersect on the altitude through C.

Constructive description:

((POINTS A B C)
(TRATIO D A C 1)
(TRATIO E B C -1)
(INTER J (LINE A E) (LINE B D))
(PERPENDICULAR C J A B))

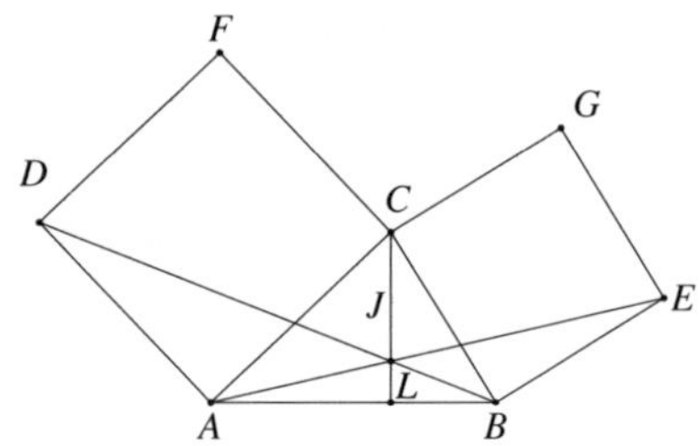

Figure 7.4

The ndg conditions: $A \neq C$, $B \neq C$, $AE \nparallel BD$.

The machine proof:

$$\frac{P_{BAC}}{P_{BAJ}}$$

$$\stackrel{J}{=} \frac{P_{BAC} \cdot S_{ABED}}{P_{BAE} \cdot S_{ABD}}$$

$$\stackrel{E}{=} \frac{P_{BAC} \cdot (\frac{1}{4} P_{CBD} + S_{ABD})}{(P_{ABA} + 4S_{ABC}) \cdot S_{ABD}}$$

$$\stackrel{D}{=} \frac{P_{BAC} \cdot (P_{BAC} + P_{ABC} + 4S_{ABC})}{(4) \cdot (P_{ABA} + 4S_{ABC}) \cdot (\frac{1}{4} P_{BAC})}$$

$$\stackrel{\text{simplify}}{=} \frac{P_{BAC} + P_{ABC} + 4S_{ABC}}{P_{ABA} + 4S_{ABC}}$$

$$= \frac{4P_{ABA} + 16S_{ABC}}{(P_{ABA} + 4S_{ABC}) \cdot (2)^2} \stackrel{\text{simplify}}{=} 1$$

The eliminants:

$$P_{BAJ} \stackrel{J}{=} \frac{P_{BAE} \cdot S_{ABD}}{S_{ABED}}$$

$$P_{BAE} \stackrel{E}{=} P_{ABA} + 4S_{ABC}$$

$$S_{ABED} \stackrel{E}{=} \frac{1}{4}(P_{CBD} + 4S_{ABD})$$

$$S_{ABD} \stackrel{D}{=} \frac{1}{4}(P_{BAC})$$

$$P_{CBD} \stackrel{D}{=} P_{ABC} + 4S_{ABC}$$

$$P_{ABC} = \frac{1}{2}(P_{BCB} - P_{ACA} + P_{ABA})$$

$$P_{BAC} = -\frac{1}{2}(P_{BCB} - P_{ACA} - P_{ABA})$$

The above proof in LaTeX format is *produced automatically* by our program. In the proof, $a \stackrel{J}{=} b$ means that b is the result obtained by eliminating point J from a. The "eliminants" are the results obtained by eliminating points from separate geometric quantities.

7.4 Proving geometric theorems with full-angles

7.4.1 Full-angles

Just as signed area was used instead of the usual concept of area, to mechanize the proof of theorems involving angles we use a new kind of angular concept:

the full-angle. Intuitively, a full-angle $\angle[u, v]$ is the directed angle from line u to line v. Two full-angles $\angle[l, m]$ and $\angle[u, v]$ are equal if there exists a rotation K such that $K(l) \parallel u$ and $K(m) \parallel v$.

For the geometric meaning of the addition of full-angles, let l, m, u and v be four lines and K be a rotation such that $K(l) \parallel v$. Then $\angle[u, v] + \angle[l, m] = \angle[u, K(m)]$. For the application of the full-angle in the coordinate approach, see Wu (1984b) and Chou (1988).

We introduce notations for two special full-angles. For all parallel lines $AB \parallel PQ$, $\angle[0] = \angle[AB, PQ]$ is a constant. For all perpendicular lines $AB \perp PQ$, $\angle[1] = \angle[AB, PQ]$ is a constant. It is clear that $\angle[1] + \angle[1] = \angle[0]$ and $\angle[u, v] + \angle[0] = \angle[u, v]$.

The reason we use full-angle is that it enables many geometric properties to be described concisely. For instance, it has the following basic properties.

Proposition 7.17:

F1. *For any line UV, $\angle[AB, CD] = \angle[AB, UV] + \angle[UV, CD]$.*

F2. *PQ is parallel to UV if there exists a line m such that $\angle[m, PQ] = \angle[m, UV]$. If $PQ \parallel UV$, then $\angle[m, PQ] = \angle[m, UV]$ for any line m.*

F3. *PQ is perpendicular to UV if there exists a line m such that $\angle[m, PQ] = \angle[1] + \angle[m, UV]$. If $PQ \perp UV$, then $\angle[m, PQ] = \angle[1] + \angle[m, UV]$ for any line m.*

F4. *For three non-collinear points X, A and B, $XA = XB$ iff $\angle[AX, AB] = \angle[AB, XB]$.*

F5 (Inscribed angle theorem). *Four non-collinear points A, B, C, and D are cyclic iff $\angle[AD, CD] = \angle[AB, CB]$.*

If ordinary angles are used, then the above propositions are true only under certain additional conditions like "a line is inside an angle" and "two points are on the same side of a line."

7.4.2 Theorem proving with full-angles

Even using the above simple properties of full-angles, we can produce short proofs for many difficult geometric theorems automatically. The concept of eliminating points is now extended to that of *eliminating lines*. For instance, if PQ is perpendicular to UV, by F3 we can eliminate line PQ from full-angle

$\angle[AB, PQ]$: $\angle[AB, PQ] = \angle[1] + \angle[AB, UV]$. We omit the details of the method, which can be found in Chou et al. (1996c). Short proofs for more than 110 geometric theorems have been generated in this way by the program based on full-angles (Chou et al. 1994b).

We illustrate the method with Miquel's six-circle theorem, which is fundamental to the study of the theory of Möbius (inversive) geometry.

Example 7.6: (Miquel's axiom; Fig. 7.5)[1]

If four circles are arranged in sequence, each two successive circles intersecting at a pair of points, and a circle passes through one point in each such pair, then the remaining four intersections lie on another circle.

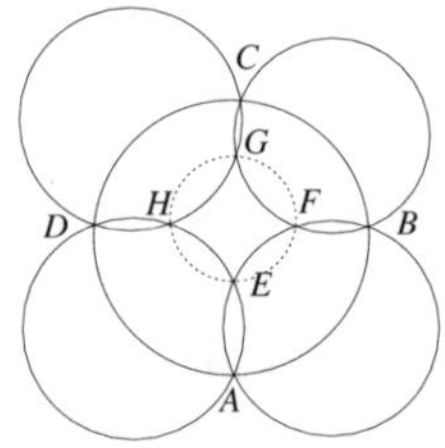

Figure 7.5

Let us assume that the four circles are $ABFE$, $BCGF$, $CDHG$, and $DAEH$. Show that if points A, B, C, and D are cyclic then points E, F, H, and G are also cyclic.

The machine proof:

$\angle[HG,HE] - \angle[FG,FE]$

($\angle[HG,HE] = \angle[HG,HD] + \angle[HD,HE] = \angle[GC,DC] - \angle[EA,DA]$,
because cyclic(G, H, D, C), cyclic(E, H, D, A). (F1, F5))

$= -\angle[FG,FE] + \angle[GC,DC] - \angle[EA,DA]$

($\angle[FG,FE] = \angle[FG,FB] + \angle[FB,FE] = \angle[GC,CB] - \angle[EA,BA]$,
because cyclic(G, F, B, C), cyclic(E, F, B, A). (F1, F5))

$= \angle[GC,DC] - \angle[GC,CB] - \angle[EA,DA] + \angle[EA,BA]$

($\angle[GC,DC] - \angle[GC,CB] = -\angle[DC,CB]$. (F1))

$= -\angle[EA,DA] + \angle[EA,BA] - \angle[DC,CB]$

($-\angle[EA,DA] + \angle[EA,BA] = \angle[DA,BA]$. (F1))

$= -\angle[DC,CB] + \angle[DA,BA]$

($\angle[DC,CB] = \angle[DA,BA]$, because cyclic (C, D, B, A). (F5))

$= \angle[0]$

[1] All previous methods failed to prove this theorem and Example 7.7 automatically due to computer memory limitation.

7.4.3 Automated theorem discovery with full-angles

Instead of using backward chaining (from the conclusion to the hypotheses), we may combine the idea of eliminating points or lines with the forward chaining approach. Then we can not only prove theorems but also discover theorems automatically. We use the following example to explain the process.

Example 7.7: (Fig. 7.6)

$P_0P_1P_2P_3P_4$ is a pentagon, $Q_i = P_{i-1}P_i \cap P_{i+1}P_{i+2}$, and M_i is the intersection of circle $Q_{i-1}P_{i-1}P_i$ and circle $Q_iP_iP_{i+1}$ (where $\cap$ means intersection *and the subscripts are understood to be mod 5). Show that points M_0, M_1, M_2, M_3 and M_4 are cyclic. Let us call this circle the* Miquel circle *associated with the five lines.*

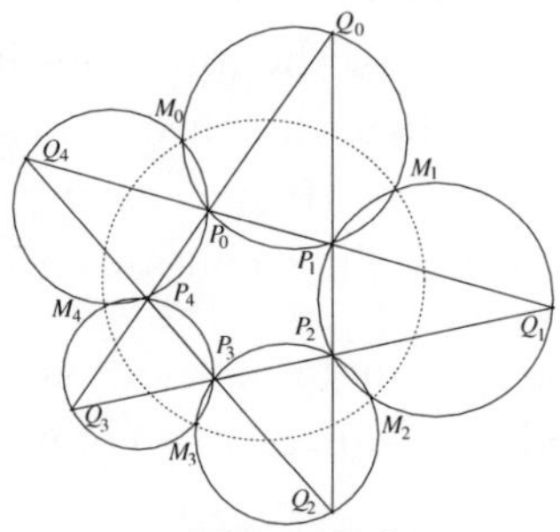

Figure 7.6

The machine proof:

$\angle[M_4M_1,M_4M_0]-\angle[M_2M_1,M_2M_0]$

$(\angle[M_4M_1,M_4M_0]$
$= \angle[M_4M_1,M_4P_0]+\angle[M_4P_0,M_4M_0]$
$= \angle[M_1Q_1,Q_1P_0]-\angle[M_0P_4,P_4P_0]$,
because cyclic(M_1, M_4, P_0, Q_1), cyclic(M_0, M_4, P_0, P_4). (F1, F5))

$= -\angle[M_2M_1,M_2M_0]+\angle[M_1Q_1,Q_1P_0]-\angle[M_0P_4,P_4P_0]$

$(\angle[M_2M_1,M_2M_0]=\angle[M_2M_1,M_2P_1]+\angle[M_2P_1,M_2M_0]=\angle[M_1P_2,P_2P_1]-\angle[M_0Q_4,Q_4P_1]$,
because cyclic(M_1, M_2, P_1, P_2), cyclic(M_0, M_2, P_1, Q_4). (F1, F5))

$= \angle[M_1Q_1,Q_1P_0]-\angle[M_1P_2,P_2P_1]+\angle[M_0Q_4,Q_4P_1]-\angle[M_0P_4,P_4P_0]$

$(\angle[M_1Q_1,Q_1P_0]=\angle[M_1P_2,P_2P_1]$,
because collinear(P_0, P_1, Q_1), cyclic(Q_1, M_1, P_1, P_2). (F2, F5))

$= \angle[M_0Q_4,Q_4P_1]-\angle[M_0P_4,P_4P_0]$

$(\angle[M_0Q_4,Q_4P_1] = \angle[M_0P_4,P_4P_0]$,
because collinear(P_0, P_1, Q_4), cyclic(Q_4, M_0, P_0, P_4). (F2, F5))

$= \angle[0]$

The above proof uses the facts that cyclic(M_1, M_4, P_0, Q_1, Q_3) and cyclic(M_2, M_0, P_1, Q_2, Q_4) which (follow from the Miquel point theorem and) are discovered by the prover using a forward chaining process before proving the theorem. Furthermore, the prover can find the following new fact without knowing it previously.

Example 7.8:

Given five lines in a plane, no two parallel and no three concurrent. Any three of the five lines form a triangle, so there are 10 triangles. For each of the 10 triangles, we draw three lines by connecting each vertex of the triangle with the Miquel point of the four lines in the five by removing the line through other two vertices. Then the three lines are concurrent on the Miquel circle of the five lines. We have 10 such new points on the Miquel circle. Therefore, the Miquel circle for five lines contains 15 notable points.

The full-angle method as reported here is not complete. By using the area method, we can give a complete method for proving theorems involving full-angles.

7.5 The method based on vectors or complex numbers

For each point P in the Euclidean plane, let $\overrightarrow{P}$ be the corresponding vector or complex number, and $\tilde{P}$ the conjugate number of $\overrightarrow{P}$. We denote by $\overrightarrow{AB} = \overrightarrow{B} - \overrightarrow{A}$ the vector from point A to point B. For a point A, $\overrightarrow{A}$ stands for the vector $\overrightarrow{OA}$ where O is the origin of the plane.

The *inner product* of vectors $\overrightarrow{AB}$ and $\overrightarrow{CD}$ is denoted by $\langle \overrightarrow{AB}, \overrightarrow{CD} \rangle$, and the *exterior product* by $[\overrightarrow{AB}, \overrightarrow{CD}]$.

We can define the *signed area of a triangle ABC* as

$$S_{ABC} = \frac{[\overrightarrow{AB}, \overrightarrow{AC}]}{2}.$$

Then the exterior product $[\overrightarrow{AB}, \overrightarrow{CD}]$ is actually the *signed area of the quadrilateral ACBD*, i.e.,

$$S_{ACBD} = S_{ACD} - S_{BCD} = \frac{[\overrightarrow{AB}, \overrightarrow{CD}]}{2}.$$

Thus we have

$$S_{ABC} = \frac{[\overrightarrow{AB}, \overrightarrow{AC}]}{2} = \frac{1}{4\boldsymbol{i}}((\tilde{B} - \tilde{A})(\overrightarrow{C} - \overrightarrow{A}) - (\overrightarrow{B} - \overrightarrow{A})(\tilde{C} - \tilde{A})),$$
$$P_{ABC} = 2\langle \overrightarrow{BA}, \overrightarrow{BC} \rangle = (\overrightarrow{B} - \overrightarrow{A})(\tilde{C} - \tilde{A}) + (\tilde{B} - \tilde{A})(\overrightarrow{C} - \overrightarrow{A}).$$

Therefore, the area method for Euclidean geometry can be translated into the language of vectors or complex numbers. The proofs thus produced are generally longer than those produced by the area method. The reason is that in the area method, the area and Pythagorean difference are treated like one-term variables, while in the vector or complex number approach they are expressions of several terms.

But the vector or complex method is still useful for solving problems in some classes. For this method has two advantages:

(1) The vectors or complex numbers themselves can be eliminated as geometric quantities directly.

(2) The vectors or complex numbers are useful to represent the relationships about angles or similar triangles.

For instance, the fact that two triangles ABC and XYZ are similar with the same around direction can be represented simply by the following equality:

$$\frac{\overrightarrow{C}-\overrightarrow{A}}{\overrightarrow{B}-\overrightarrow{A}}=\frac{\overrightarrow{Z}-\overrightarrow{X}}{\overrightarrow{Y}-\overrightarrow{X}}.$$

So the vector $\overrightarrow{Z}$ can be eliminated easily if points A, B, C, X, Y are given.

In the following, we will introduce the construction that uses the connection between angles and complex numbers. The elimination method for this construction can be used to produce short and readable proofs for theorems involving special angles. Notably, a short and readable machine proof for Morley's trisector theorem is given for the first time.

Here is the construction:

(SAA C A B α β) introduces a point C such that $\angle BAC=\alpha$ and $\angle CBA=\beta$.

Lemma 7.18:

Let C be introduced by (SAA C A B α β). *Then*

$$C=A+(B-A)\frac{1-e^{-2i\beta}}{1-e^{-2i(\alpha+\beta)}}.$$

Proof:

For any angle α, let $f_\alpha=e^{-2i\alpha}$. Then $f_{-\alpha}=1/f_\alpha$. In our program, the expression $P(\alpha,\beta)=\frac{1-e^{-2i\beta}}{1-e^{-2i(\alpha+\beta)}}$ in the lemma is represented as an expression in quantities f_a, f_b, etc., where a, b, ... are the angles in α and β. For instance,

$$P(\alpha,\beta)=\frac{1-f_\beta}{1-f_\alpha f_\beta}\quad\text{and}\quad P(\alpha+\beta,\alpha-\beta)=\frac{1-f_{-\beta}f_\alpha}{(1-f_{2\alpha})}=\frac{f_\beta-f_\alpha}{f_\beta(1-f_\alpha^2)}.$$

Example 7.9: (Morley's theorem; Fig. 7.7)

The three points of intersection of the adjacent *trisectors* of the angles of any triangle form an equilateral triangle.

Constructive description:

((POINTS P Q)
(CONSTANT $f_{\pi/6}^2 - f_{\pi/6} + 1$)
(SAA Z P Q a b)
(SAA R P Q $3a$ $3b$)
(SAA X R P $2PI/6-b-a$ a)
(SAA Y Q R b $2PI/6-b-a$)
($f_{\pi/6}\overrightarrow{Y} - \overrightarrow{X} = (f_{\pi/6}-1)\overrightarrow{Z}$))

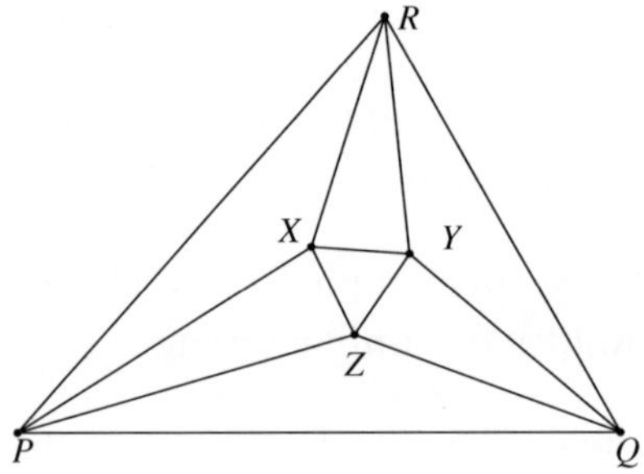

Figure 7.7

In the above description, $PI = \pi$. The construction (CONSTANT $f_{\pi/6}^2 - f_{\pi/6} + 1$) introduces a constant $f_{\pi/6} = e^{-2\boldsymbol{i}(\pi/6)}$ which satisfies the equation $f_{\pi/6}^2 - f_{\pi/6} + 1 = 0$.

In the machine proof below, the equations begin with $\stackrel{\text{cons}}{=}$, which means that $f_{\pi/6}^2$ is replaced by $f_{\pi/6} - 1$. Also, we assume that $P = 0$ and $Q = 1$.

The machine proof:

$$\frac{f_{\pi/6}\overrightarrow{Y} - \overrightarrow{X}}{(f_{\pi/6}-1)\overrightarrow{Z}}$$

$$\stackrel{Y}{=} \frac{(-f_b+f_{\pi/6}^2)\overrightarrow{X}+(f_{\pi/6}f_af_b-f_{\pi/6}^3)\overrightarrow{R}}{((f_{\pi/6}-1)\overrightarrow{Z})\cdot(f_b-f_{\pi/6}^2)}$$

$$\stackrel{\text{cons}}{=} \frac{(-f_b+f_{\pi/6}-1)\overrightarrow{X}+(f_{\pi/6}f_af_b+1)\overrightarrow{R}}{((f_{\pi/6}-1)f_b+f_{\pi/6})\overrightarrow{Z}}$$

$$\stackrel{X}{=} \frac{-(((-f_{\pi/6}f_a^2+(f_{\pi/6}^3+1)f_a)f_b^2-f_{\pi/6}f_af_b+f_{\pi/6}^3-f_{\pi/6}^2)\overrightarrow{R}-f_{\pi/6}^2f_b^2+f_{\pi/6}^3f_b-f_{\pi/6}^3+f_{\pi/6}^2)}{((f_{\pi/6}-1)f_b+f_{\pi/6})\cdot\overrightarrow{Z}\cdot((f_a-f_{\pi/6}^2)f_b)}$$

$$\stackrel{\text{cons}}{=} \frac{(f_{\pi/6}f_a^2f_b^2+f_{\pi/6}f_af_b+f_{\pi/6})\overrightarrow{R}+(f_{\pi/6}-1)f_b^2+f_b-f_{\pi/6}}{(((f_{\pi/6}-1)f_a+f_{\pi/6})f_b^2+(f_{\pi/6}f_a+1)f_b)\overrightarrow{Z}}$$

$$\stackrel{R}{=} \frac{((-f_{\pi/6}+1)f_a^3-f_{\pi/6}f_a^2)f_b^5+(-f_a^3-f_{\pi/6}f_a)f_b^4+(f_{\pi/6}f_a^3-f_{\pi/6})f_b^3+(f_{\pi/6}f_a^2+f_{\pi/6}-1)f_b^2+(f_{\pi/6}f_a+1)f_b}{\overrightarrow{Z}\cdot(((f_{\pi/6}-1)f_a+f_{\pi/6})f_b+f_{\pi/6}f_a+1)\cdot f_b\cdot(-f_a^3f_b^3+1)}$$

$$\stackrel{\text{simplify}}{=} \frac{f_b-1}{(f_af_b-1)\cdot\overrightarrow{Z}} \stackrel{Z}{=} \frac{(f_b-1)\cdot(-f_af_b+1)}{(f_af_b-1)\cdot(-f_b+1)} \stackrel{\text{simplify}}{=} 1$$

The eliminants:

$$\overrightarrow{Y} \stackrel{Y}{=} \frac{(f_a f_b - f_{\pi/6}^2) \cdot \overrightarrow{R}}{f_b - f_{\pi/6}^2}, \quad \overrightarrow{X} \stackrel{X}{=} \frac{(f_a f_b - f_{\pi/6}^2)\overrightarrow{R} - f_{\pi/6}^2 f_b + f_{\pi/6}^2}{(f_a - f_{\pi/6}^2) \cdot f_b},$$

$$\overrightarrow{R} \stackrel{R}{=} \frac{(f_b^2 + f_b + 1) \cdot (f_b - 1)}{(f_a f_b - 1) \cdot (f_a^2 f_b^2 + f_a f_b + 1)}, \quad \overrightarrow{Z} \stackrel{Z}{=} \frac{f_b - 1}{f_a f_b - 1}.$$

The reader can find a review of methods and techniques for proving and discovering geometric theorems using Clifford algebra in Chap. 8.

7.6 The volume method

This section is an extension of the area method to solid geometry. The volume method is complete for the class of constructive geometric statements which covers a large portion of the equational theorems about lines, planes, circles, and spheres. The volume method works not only for Euclidean solid geometry but also for metric solid geometries associated with any number field of characteristic 0.

We have implemented the method and our program[2] has proved more than 80 example theorems from solid geometry. The proofs produced by the program are generally short and readable. The algorithm is also very efficient.

7.6.1 The signed volume

All basic propositions in the *area method* are still used for any set of points on a plane in the space. Besides these, we need more concepts and propositions in solid geometry. For any four points A, B, C, and D in the space, the *signed volume* V_{ABCD} of the tetrahedron $ABCD$ is a real number which satisfies the following properties:

V1. *When two neighboring vertices of the tetrahedron are interchanged, the signed volume of the tetrahedron will change sign, e.g.,* $V_{ABCD} = -V_{ABDC}$.

V2. *Points* A, B, C, *and* D *are coplanar iff* $V_{ABCD} = 0$.

V3. *There exist at least four points* A, B, C, *and* D *such that* $V_{ABCD} \neq 0$.

V4. *For five points* A, B, C, D, *and* O,

$$V_{ABCD} = V_{ABCO} + V_{ABOD} + V_{AOCD} + V_{OBCD}.$$

[2]The prover (euc.tar.Z) is available via ftp at emcity.cs.twsu.edu: pub/geometry/.

V5. *If A, B, C, D, E, and F are six coplanar points and* $S_{ABC} = \lambda S_{DEF}$*, then* $V_{TABC} = \lambda V_{TDEF}$ *for any point T.*

We denote by $PABCQ$ the polyhedron with faces PAB, PBC, PAC, QAB, QBC, and QAC (Fig. 7.8). The volume of $PABCQ$ is defined to be

$$V_{PABCQ} = V_{PABC} - V_{QABC}.$$

We have

$$V_{PABCQ} = V_{PABQ} + V_{PCAQ} + V_{PBCQ}.$$

We now list the basic propositions in solid geometry for the *volume method.* The proofs can be found in Chou et al. (1994a, 1995a).

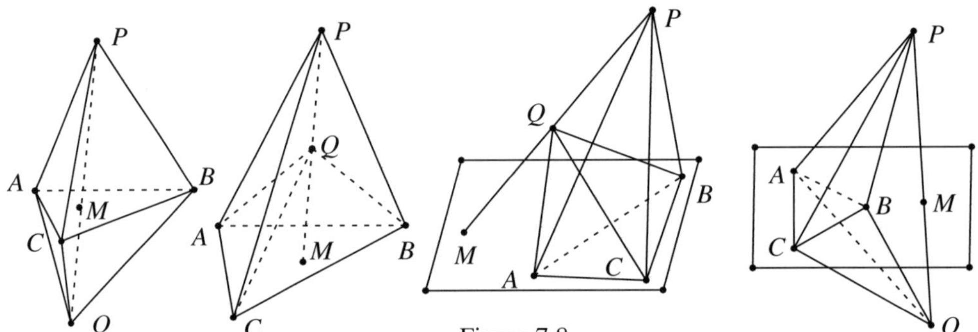

Figure 7.8

Proposition 7.19: (Co-face theorem)

Let a line PQ and a plane ABC meet at M. If $Q \neq M$*, then*

$$\frac{\overline{PM}}{\overline{QM}} = \frac{V_{PABC}}{V_{QABC}}, \qquad \frac{\overline{PM}}{\overline{PQ}} = \frac{V_{PABC}}{V_{PABCQ}}, \qquad \frac{\overline{QM}}{\overline{PQ}} = \frac{V_{QABC}}{V_{PABCQ}}.$$

Proposition 7.20:

Let R be a point on a line PQ and ABC be a triangle. Then

$$V_{RABC} = \frac{\overline{PR}}{\overline{PQ}} V_{QABC} + \frac{\overline{RQ}}{\overline{PQ}} V_{PABC}.$$

Proposition 7.21:

Let R be a point in the plane PQS. Then for three points A, B, and C

$$V_{RABC} = \frac{S_{PQR}}{S_{PQS}} V_{SABC} + \frac{S_{RQS}}{S_{PQS}} V_{PABC} + \frac{S_{PRS}}{S_{PQS}} V_{QABC}.$$

Proposition 7.22:

$PQ \parallel ABC$ iff $V_{PABC} = V_{QABC}$ or equivalently $V_{PABCQ} = 0$.

Proposition 7.23:

$PQR \parallel ABC$ iff $V_{PABC} = V_{QABC} = V_{RABC}$.

Proposition 7.24:

Let $PQTS$ be a parallelogram. Then for three points A, B, and C,

$$V_{PABC} + V_{TABC} = V_{QABC} + V_{SABC} \quad \text{or} \quad V_{PABCQ} = V_{SABCT}.$$

Proposition 7.25:

Let triangle ABC be a parallel translation of triangle DEF. Then for any point P, $V_{PABC} = V_{PDEFA}$.

Proposition 7.26:

Let triangle ABC be a parallel translation of triangle DEF. Then for two points P and Q,

$$V_{PABC} + V_{QDEF} = V_{QABC} + V_{PDEF} \qquad \text{or} \qquad V_{PABCQ} = V_{PDEFQ}.$$

Before presenting the method, we use two examples to show how to use the above properties about volumes to prove theorems. The proofs given below are actually modifications of the proofs produced by our program.

Example 7.10: (Menelaus' theorem; Fig. 7.9)

Let the sides AB, BC, CD, and DA of an arbitrary skew quadrilateral be cut by a plane XYZ at the points E, F, G, and H respectively. Then

$$\frac{\overline{AE}}{\overline{EB}} \cdot \frac{\overline{BF}}{\overline{FC}} \cdot \frac{\overline{CG}}{\overline{GD}} \cdot \frac{\overline{DH}}{\overline{HA}} = 1.$$

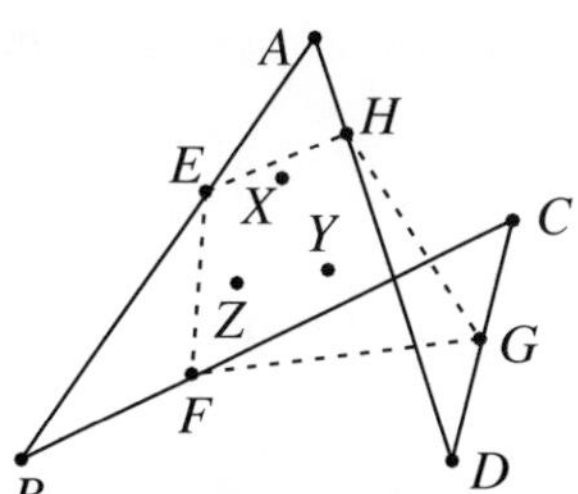

Figure 7.9

Proof. By the co-face theorem,

$$\frac{\overline{DH}}{\overline{AH}} = \frac{V_{DXYZ}}{V_{AXYZ}}, \quad \frac{\overline{CG}}{\overline{DG}} = \frac{V_{CXYZ}}{V_{DXYZ}}, \quad \frac{\overline{BF}}{\overline{CF}} = \frac{V_{BXYZ}}{V_{CXYZ}}, \quad \frac{\overline{AE}}{\overline{BE}} = \frac{V_{AXYZ}}{V_{BXYZ}}.$$

Then it is clear that $\frac{\overline{AE}}{\overline{EB}} \cdot \frac{\overline{BF}}{\overline{FC}} \cdot \frac{\overline{CG}}{\overline{GD}} \cdot \frac{\overline{DH}}{\overline{HA}} = 1.$

Similarly to the area method, a statement from solid geometry may be represented in a constructive form. For instance, Example 7.10 can be described in the following way:

((POINTS A B C D X Y Z)
(INTER E (LINE A B) (PLANE X Y Z))
(INTER F (LINE B C) (PLANE X Y Z))
(INTER G (LINE C D) (PLANE X Y Z))
(INTER H (LINE A D) (PLANE X Y Z))
($\frac{\overline{AE}}{\overline{BE}}\frac{\overline{BF}}{\overline{CF}}\frac{\overline{CG}}{\overline{DG}}\frac{\overline{DH}}{\overline{AH}} = 1$))

The ndg conditions:

$$\begin{array}{llll} AB \not\parallel XYZ, & BC \not\parallel XYZ, & CD \not\parallel XYZ, & AD \not\parallel XYZ, \\ B \neq E, & C \neq F, & D \neq G, & A \neq H. \end{array}$$

7.6.2 Examples for the volume method

Eliminating points from volumes is the basis of the current method. Detailed description of the approach can be found in Chou et al. (1992, 1994a, 1995b). Here, we only give some illustrative examples.

Lemma 7.27:

Let Y be introduced by (INTER Y (LINE U V) (LINE I J)). *Then*

$$V_{ABCY} = \frac{S_{UIJ}}{S_{UIVJ}} V_{ABCV} - \frac{S_{VIJ}}{S_{UIVJ}} V_{ABCU}.$$

Lemma 7.28:

Let Y be introduced by (INTER Y (LINE U V) (PLANE L M N)). *Then*

$$V_{ABCY} = \frac{1}{V_{ULMNV}} (V_{ULMN} V_{ABCV} - V_{VLMN} V_{ABCU}).$$

Lemma 7.29:

Let Y be introduced by (INTER Y (LINE U V) (LINE I J)). *Then*

$$\frac{\overline{ABY}}{\overline{CDE}} = \begin{cases} \frac{S_{UIJ} V_{UABV}}{S_{UIVJ} V_{UCDEA}} & \textit{if one of } U, V, I, J, \textit{ say } U, \textit{ is not in } ABY, \\ \frac{S_{IUV} S_{ABJ} - S_{JUV} S_{ABI}}{S_{CDE} S_{IUJV}} & \textit{if } U, V, I, J, A, B, Y \textit{ are coplanar.} \end{cases}$$

Lemma 7.30:

Let Y be introduced by C =(INTER Y (LINE U V) (PLANE L M N)). *Then*

$$\frac{\overline{ABY}}{\overline{CDE}} = \begin{cases} \frac{V_{ULMN}}{V_{ULMNV}} \frac{V_{UABV}}{V_{UCDEA}} & \textit{if } U \textit{ (or } V\textit{) is not in } ABY, \\ \frac{V_{ULMN} S_{ABV} - V_{VLMN} S_{ABU}}{S_{CDE} V_{ULMNV}} & \textit{if } U, V \textit{ are in } ABY. \end{cases}$$

We can now describe the volume method as follows. For a statement in solid geometry: $S = (C_1, \ldots, C_r, (E_1, E_2))$, let the point introduced by C_i be P_i. Then we can use the above lemmas to eliminate points $P_r, P_{r-1}, \ldots, P_1$ respectively from E_1 and E_2. Finally, we obtain two rational expressions R_1 and R_2 respectively. S is a true geometric statement if R_1 is identical to R_2. For a formal description of the algorithm, see Chou et al. (1995a).

We have implemented the algorithm in Common Lisp on a NeXT workstation. The following is the machine-produced proof for Example 7.10.

Example 7.11:

For an input as in Example 7.10, our program produces the following machine proof (in LaTeX form) automatically.

The machine proof:

$$\frac{\overline{DH}}{\overline{AH}} \cdot \frac{\overline{CG}}{\overline{DG}} \cdot \frac{\overline{BF}}{\overline{CF}} \cdot \frac{\overline{AE}}{\overline{BE}}$$

$$\stackrel{H}{=} \frac{V_{DXYZ}}{V_{AXYZ}} \cdot \frac{\overline{CG}}{\overline{DG}} \cdot \frac{\overline{BF}}{\overline{CF}} \cdot \frac{\overline{AE}}{\overline{BE}}$$

$$\stackrel{G}{=} \frac{V_{DXYZ} \cdot V_{CXYZ}}{V_{AXYZ} \cdot V_{DXYZ}} \cdot \frac{\overline{BF}}{\overline{CF}} \cdot \frac{\overline{AE}}{\overline{BE}}$$

$$\stackrel{\text{simplify}}{=} \frac{V_{CXYZ}}{V_{AXYZ}} \cdot \frac{\overline{BF}}{\overline{CF}} \cdot \frac{\overline{AE}}{\overline{BE}}$$

$$\stackrel{F}{=} \frac{V_{CXYZ} \cdot V_{BXYZ}}{V_{AXYZ} \cdot V_{CXYZ}} \cdot \frac{\overline{AE}}{\overline{BE}}$$

$$\stackrel{\text{simplify}}{=} \frac{V_{BXYZ}}{V_{AXYZ}} \cdot \frac{\overline{AE}}{\overline{BE}}$$

$$\stackrel{E}{=} \frac{V_{BXYZ} \cdot V_{AXYZ}}{V_{AXYZ} \cdot V_{BXYZ}}$$

$$\stackrel{\text{simplify}}{=} 1$$

The eliminants:

$$\frac{\overline{DH}}{\overline{AH}} \stackrel{H}{=} \frac{V_{DXYZ}}{V_{AXYZ}}$$

$$\frac{\overline{CG}}{\overline{DG}} \stackrel{G}{=} \frac{V_{CXYZ}}{V_{DXYZ}}$$

$$\frac{\overline{BF}}{\overline{CF}} \stackrel{F}{=} \frac{V_{BXYZ}}{V_{CXYZ}}$$

$$\frac{\overline{AE}}{\overline{BE}} \stackrel{E}{=} \frac{V_{AXYZ}}{V_{BXYZ}}$$

In the above proof, the symbol $\stackrel{H}{=}$ means point H is eliminated. The *eliminants* are the separate elimination results obtained by using the above lemmas.

7.7 Points elimination method for non-Euclidean geometries

In Yang, Gao, Chou and Zhang (1997), a complete (decision) method for producing short and human-readable proofs for a class of constructive statements in non-Euclidean geometries is presented. The method is a substantial extension of the area method for Euclidean geometry. A computer program based on this method has been used to prove more than 90 theorems including many new ones from non-Euclidean geometries. The proofs produced by the program are generally very short and readable.

The basic geometric quantities used by us in non-Euclidean geometries are the *argument* (roughly speaking, the sine of the area) and the *cosine of distance.* Not all of the properties of these quantities needed in our method are in the classic books of non-Euclidean geometry (Coxeter 1968; Greenberg 1980). Some of the geometric results proved in this way are new.

7.7.1 Outline of the method for hyperbolic geometry

We first use an example to show how the method works.

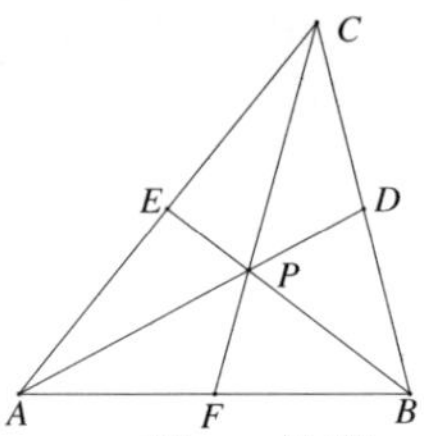

Figure 7.10

Example 7.12: (Ceva's theorem)

In Fig. 7.10, ABC is a triangle and P is any point. D, E, F are the intersection points of lines AP, BP, CP with lines BC, CA, AB respectively. Show that

$$\frac{\sinh(AF)}{\sinh(FB)}\frac{\sinh(BD)}{\sinh(DC)}\frac{\sinh(CE)}{\sinh(EA)} = 1,$$

where $\sinh(AF)$ is the hyperbolic sine of segment AF.

To prove this theorem, we use a geometric invariant or quantity: *the argument.* The absolute value of the argument of a triangle ABC is defined to be

$$|S_{ABC}| = |\sinh(AB) \cdot \sinh(BC) \cdot \sin\angle(AB, BC)|.$$

We also assume that the argument of a clock-wise triangle is positive and

$$S_{ABC} = S_{CAB} = S_{BCA} = -S_{ACB} = -S_{BAC} = -S_{CBA}.$$

It is clear that $S_{ABC} = 0$ iff A, B and C are on the same line. Another quantity is $\frac{\sinh(AB)}{\sinh(PQ)}$, the *ratio of hyperbolic sines* of directed line segments, where A, B, P, Q are collinear. Then it is clear that

$$\frac{\sinh(AB)}{\sinh(PQ)} = -\frac{\sinh(BA)}{\sinh(PQ)} = -\frac{\sinh(AB)}{\sinh(QP)}.$$

The following result is a direct consequence of the definition of the argument.

Proposition 7.31: (Co-side theorem)

Let M be the intersection of two lines AB and PQ and $Q \neq M$. Then

$$\frac{\sinh(PM)}{\sinh(QM)} = \frac{S_{PAB}}{S_{QAB}}.$$

Note that the result is true in all the four cases in Fig. 7.10.

Proof of Example 7.12. In our method, we need to give an order for the points in the geometric statement such that each point can be constructed from the previous points. In this example, a construction order of points is A, B, C, P, D, E, F: first taking four free points A, B, C, D, and then taking intersection points of lines AP, BP, CP with lines BC, CA, AB respectively.

The proof method still proceeds by *eliminating points* from the conclusion. By the co-side theorem, we can eliminate points F, E, and D respectively from the conclusion:

$$\frac{\sinh(AF)}{\sinh(FB)} = \frac{S_{ACP}}{S_{BPC}}, \qquad \frac{\sinh(CE)}{\sinh(EA)} = \frac{S_{BCP}}{S_{BPA}}, \qquad \frac{\sinh(BD)}{\sinh(DC)} = \frac{S_{ABP}}{S_{APC}}$$

Then it is clear that

$$\frac{\sinh(AF)}{\sinh(FB)} \frac{\sinh(BD)}{\sinh(DC)} \frac{\sinh(CE)}{\sinh(EA)} = \frac{S_{ACP}S_{BCP}S_{ABP}}{S_{BPC}S_{BPA}S_{APC}} = 1.$$

We have mentioned that the proof method is for constructive geometric statements which are quite similar to those in Euclidean geometry. But there are some differences. For instance, there is no parallel line in non-Euclidean geometry. As in the area method, the key step of the proof method is to eliminate points from geometric quantities. Some of the elimination lemmas are given below.

Lemma 7.32:

Let point Y be introduced by construction (INTER Y (LINE U V) (LINE P Q)) *and $f = \frac{\sinh(AY)}{\sinh(CD)}$, where points A and Y are on line CD. Then*

$$f = \begin{cases} \frac{S_{AUV}}{S_{CUDV}} & \text{if } A \text{ is not on } UV, \\ \frac{S_{APQ}}{S_{CPDQ}} & \text{otherwise.} \end{cases}$$

Lemma 7.33:

Let point Y be introduced by construction (FOOT Y P U V) *and* $f = \frac{\sinh(AY)}{\sinh(CD)}$, *where points A and Y are on line CD. Then*

$$f = \begin{cases} \frac{\sinh(AV)}{\sinh(CD)} \frac{P_{PAV}}{f_1(P,A,V)} & \text{if } A \in UV \text{ and } A \neq V, \\ \frac{S_{AUV}}{S_{CUDV}} & \text{if } A \notin UV. \end{cases}$$

We have incorporated the method into the Common Lisp prover for the area method (Chou et al. 1994a). Chou, Gao, Yang, and Zhang (1994) is a collection of 90 geometric theorems proved by our prover. Among the 90 theorems, about 40 belong to projective geometry and thus are true in non-Euclidean geometries. The validity of the other 50 theorems is not trivial. Many of the theorems, such as Example 7.14, are new. More detail will be given in the last section.

The program is very fast. More importantly, the average length of the maximal polynomials in the proofs for the 90 theorems is only 2.2. Considering the fact that the 90 theorems are quite non-trivial, we can say that the proofs produced by our method are much shorter than those by the coordinate based methods.

Grünbaum and Shephard (1993) discovered many new results in Euclidean geometry about polygons using numerical search. We may prove the corresponding results in non-Euclidean geometries using the method presented in this section. Furthermore, for any polygon with concrete number of sides, our prover may be used to prove the corresponding theorems automatically.

Example 7.13:

Let $ABCDE$ be a pentagon, $P = AD \cap BE$, $Q = AC \cap BE$, $R = BD \cap AC$, $S = CE \cap BD$, and $T = AD \cap CE$. Then

$$\frac{\sinh(AP)}{\sinh(TD)} \frac{\sinh(DS)}{\sinh(RB)} \frac{\sinh(BQ)}{\sinh(PE)} \frac{\sinh(ET)}{\sinh(SC)} \frac{\sinh(CR)}{\sinh(QA)}$$
$$= \frac{\sinh(AT)}{\sinh(PD)} \frac{\sinh(DR)}{\sinh(SB)} \frac{\sinh(BP)}{\sinh(QE)} \frac{\sinh(ES)}{\sinh(TC)} \frac{\sinh(CQ)}{\sinh(RA)}$$
$$= 1.$$

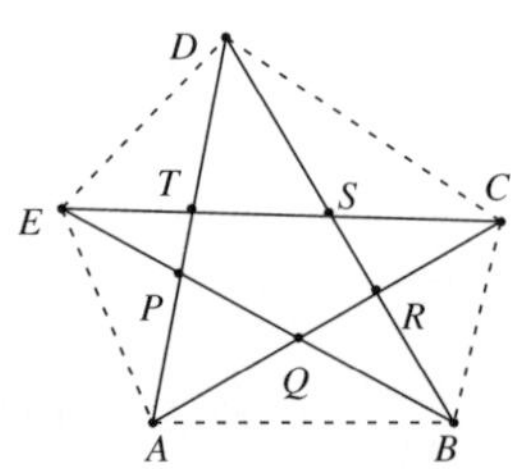

Figure 7.11

In the general case, let $V_1 \cdots V_m$ be a polygon and $1 \leq d \leq \frac{m}{2}$, $1 \leq j \leq \frac{m}{2}$ integers. We denote by $P_{d,j,i}$ the intersection of line V_iV_{i+d} and line $V_{i+j}V_{i+j+d}$, $i = 1, \ldots, m$. Then $P_{d,j,i-j}$ is the intersection of line $V_{i-j}V_{i-j+d}$ and line V_iV_{i+d}.

Let

$$T(m,d,j)=\prod_{i=1}^{m}\frac{\sinh(V_iP_{d,j,i})}{\sinh(P_{d,j,i-j}V_{i+d})},$$
$$S(m,d,j)=\prod_{i=1}^{m}\frac{\sinh(V_iP_{d,j,i-j})}{\sinh(P_{d,j,i}V_{i+d})}.$$

Then Example 7.13 is a special case of the following theorem.

Example 7.14:

(1) $T(m,d,j)=1$ iff $d+2j=m$ or $2d+j=m$.
(2) $S(m,d,j)=1$ iff $d+2j=m$, $2d=j$, or $2j=d$.

All specific cases of Example 7.14 can be proved by our prover. For instance, the proof for the first case of Example 7.13 is as follows.

The machine proof:

$$\frac{-\frac{\sinh(ET)}{\sinh(CE)}\cdot\frac{\sinh(DS)}{\sinh(BD)}\cdot\frac{\sinh(CR)}{\sinh(AC)}\cdot\frac{\sinh(BQ)}{\sinh(BE)}\cdot\frac{\sinh(AP)}{\sinh(AD)}}{\frac{\sinh(EP)}{\sinh(BE)}\cdot\frac{\sinh(DT)}{\sinh(AD)}\cdot\frac{\sinh(CS)}{\sinh(CE)}\cdot\frac{\sinh(BR)}{\sinh(BD)}\cdot\frac{\sinh(AQ)}{\sinh(AC)}}$$

$$\stackrel{T}{=}\frac{S_{ADE}\cdot S_{ACDE}\cdot(-\frac{\sinh(DS)}{\sinh(BD)})\cdot\frac{\sinh(CR)}{\sinh(AC)}\cdot\frac{\sinh(BQ)}{\sinh(BE)}\cdot\frac{\sinh(AP)}{\sinh(AD)}}{(-\frac{\sinh(EP)}{\sinh(BE)}\cdot\frac{\sinh(CS)}{\sinh(CE)}\cdot\frac{\sinh(BR)}{\sinh(BD)}\cdot\frac{\sinh(AQ)}{\sinh(AC)}\cdot S_{CDE})\cdot(-S_{ACDE})}$$

$$\stackrel{\text{simplify}}{=}\frac{-S_{ADE}\cdot\frac{\sinh(AP)}{\sinh(AD)}\cdot\frac{\sinh(BQ)}{\sinh(BE)}\cdot\frac{\sinh(CR)}{\sinh(AC)}\cdot\frac{\sinh(DS)}{\sinh(BD)}}{S_{CDE}\cdot\frac{\sinh(AQ)}{\sinh(AC)}\cdot\frac{\sinh(BR)}{\sinh(BD)}\cdot\frac{\sinh(CS)}{\sinh(CE)}\cdot\frac{\sinh(EP)}{\sinh(BE)}}$$

$$\stackrel{S}{=}\frac{-S_{ADE}\cdot(-S_{CDE})\cdot(-S_{BCDE})}{S_{CDE}\cdot\frac{\sinh(AQ)}{\sinh(AC)}\cdot\frac{\sinh(BR)}{\sinh(BD)}\cdot(-S_{BCD})\cdot\frac{\sinh(EP)}{\sinh(BE)}\cdot S_{BCDE}\cdot\frac{\sinh(AP)}{\sinh(AD)}\cdot\frac{\sinh(BQ)}{\sinh(BE)}\cdot\frac{\sinh(CR)}{\sinh(AC)}}$$

$$\stackrel{\text{simplify}}{=}\frac{S_{ADE}}{\frac{\sinh(EP)}{\sinh(BE)}\cdot S_{BCD}\cdot\frac{\sinh(BR)}{\sinh(BD)}\cdot\frac{\sinh(AQ)}{\sinh(AC)}}\cdot\frac{\sinh(CR)}{\sinh(AC)}\cdot\frac{\sinh(BQ)}{\sinh(BE)}\cdot\frac{\sinh(AP)}{\sinh(AD)}$$

$$\stackrel{R}{=}\frac{(-S_{BCD})\cdot S_{ADE}\cdot(-S_{ABCD})}{\frac{\sinh(EP)}{\sinh(BE)}\cdot S_{BCD}\cdot(-S_{ABC})\cdot\frac{\sinh(AQ)}{\sinh(AC)}\cdot S_{ABCD}}\cdot\frac{\sinh(BQ)}{\sinh(BE)}\cdot\frac{\sinh(AP)}{\sinh(AD)}$$

$$\stackrel{\text{simplify}}{=}\frac{-S_{ADE}}{\frac{\sinh(AQ)}{\sinh(AC)}\cdot S_{ABC}\cdot\frac{\sinh(EP)}{\sinh(BE)}}\cdot\frac{\sinh(AP)}{\sinh(AD)}\cdot\frac{\sinh(BQ)}{\sinh(BE)}$$

$$\stackrel{Q}{=}\frac{-S_{ADE}\cdot(-S_{ABC})\cdot S_{ABCE}}{S_{ABE}\cdot S_{ABC}\cdot\frac{\sinh(EP)}{\sinh(BE)}\cdot(-S_{ABCE})}\cdot\frac{\sinh(AP)}{\sinh(AD)}$$

$$\stackrel{\text{simplify}}{=}\frac{-S_{ADE}}{\frac{\sinh(EP)}{\sinh(BE)}\cdot S_{ABE}}\cdot\frac{\sinh(AP)}{\sinh(AD)}$$

$$\stackrel{P}{=}\frac{-S_{ABE}\cdot S_{ADE}\cdot(-S_{ABDE})}{S_{ADE}\cdot S_{ABE}\cdot S_{ABDE}}$$

$$\stackrel{\text{simplify}}{=}1$$

The elimination lemmas used in the proof:

$\frac{\sinh(DT)}{\sinh(AD)} \stackrel{T}{=} \frac{-S_{CDE}}{S_{ACDE}}$, $\frac{\sinh(ET)}{\sinh(CE)} \stackrel{T}{=} \frac{S_{ADE}}{-S_{ACDE}}$, $\frac{\sinh(CS)}{\sinh(CE)} \stackrel{S}{=} \frac{S_{BCD}}{S_{BCDE}}$, $\frac{\sinh(DS)}{\sinh(BD)} \stackrel{S}{=} \frac{-S_{CDE}}{S_{BCDE}}$, $\frac{\sinh(BR)}{\sinh(BD)} \stackrel{R}{=} \frac{S_{ABC}}{S_{ABCD}}$, $\frac{\sinh(CR)}{\sinh(AC)} \stackrel{R}{=} \frac{-S_{BCD}}{S_{ABCD}}$, $\frac{\sinh(AQ)}{\sinh(AC)} \stackrel{Q}{=} \frac{S_{ABE}}{S_{ABCE}}$, $\frac{\sinh(BQ)}{\sinh(BE)} \stackrel{Q}{=} \frac{S_{ABC}}{S_{ABCE}}$, $\frac{\sinh(EP)}{\sinh(BE)} \stackrel{P}{=} \frac{S_{ADE}}{-S_{ABDE}}$, $\frac{\sinh(AP)}{\sinh(AD)} \stackrel{P}{=} \frac{S_{ABE}}{S_{ABDE}}$.

The proof method for constructive geometric statements in elliptic geometry is quite similar to that in hyperbolic geometry. First, we have the following known result.

Theorem 7.34:

Let points A, B, P, Q, R ... satisfy the same geometric conditions in hyperbolic and elliptic geometries. Then

$$P(\sin(AB), \cos(AB), \sin(\angle PQR), \cos(\angle PQR), \ldots) = 0$$

is a valid algebraic equation in elliptic geometry iff

$$P(\sqrt{-1}\sinh(AB), \cosh(AB), \sin(\angle PQR), \cos(\angle PQR), \ldots) = 0$$

is a valid algebraic equation in hyperbolic geometry.

By the above theorem, all the basic propositions and elimination lemmas proved in this section can be transformed to similar results in elliptic geometry. We may thus have a similar algorithm for theorem proving in elliptic geometry.

Another consequence of Theorem 7.34 is that a constructive geometric statement with conclusion that $P(\sin(AB), \cos(AB), \ldots) = 0$ is true in elliptic geometry iff the statement with same hypotheses and with conclusion

$$P(\sqrt{-1}\sinh(AB), \cosh(AB), \ldots) = 0$$

is true in hyperbolic geometry.

For instance, Ceva's theorem

$$\left(\text{Example 7.12 with conclusion } \frac{\sin(AF)\sin(BD)\sin(CE)}{\sin(FB)\sin(DC)\sin(EA)} = 1\right)$$

is true in elliptic geometry.

7.8 Concluding remarks

We have seen that the presented points elimination methods are complete for constructive geometric theorems, which include most of the commonly used theorems. Experiments show that our program based on these methods can produce short proofs for most of the constructive theorems. In fact, Chou et al. (1992) contains a collection of 400 theorems proved with the area method; Chou et al. (1994b) contains a collection of 110 theorems proved with the full-angle method; Chou et al. (1994c) contains a collection of 135 theorems from the problem section of the American Mathematical Monthly. Most of the proofs are short: they contain only algebraic expressions of a few terms, and can be proved within seconds.

Table 7.1 shows the statistics for 410 theorems proved by vector or complex number methods, where *Maxterm* stands for *maximum number of terms*.

Length of proof		Proving time	
Maxterm	No. of theorems	Time (sec)	Theorems
$m = 1$	61	$t \leq 0.1$	119
$m = 2$	57	$0.1 < t \leq 0.5$	124
$2 < m \leq 5$	137	$0.5 < t \leq 1$	60
$5 < m \leq 10$	58	$1 < t \leq 2$	48
$10 < m \leq 20$	48	$2 < t \leq 5$	36
$20 < m \leq 94$	49	$5 < t \leq 31$	23

Table 7.1 Statistics for 410 problems from plane geometry

Table 7.2 contains the timing and proof-length statistics for 80 theorems in solid geometry proved by the program based on the volume method.

Length of proof		Proving time	
Maxterm	No. of theorems	Time (sec)	Theorems
$m = 1$	18	$t \leq 0.05$	28
$m = 2$	25	$0.05 < t \leq 0.1$	19
$2 < m \leq 5$	24	$0.1 < t \leq 0.5$	27
$5 < m \leq 10$	9	$0.5 < t \leq 1$	3
$10 < m \leq 140$	4	$1 < t \leq 100$	3

Table 7.2 Statistics for 80 problems from solid geometry

Table 7.3 contains the timing and proof-length statistics for 90 geometric problems in non-Euclidean geometry solved by our program. Lm-no is the number of elimination lemmas used to eliminate points from geometric quantities. In other words, lm-no is the number of deduction steps in the proof.

Proving time		Length of proof		Deduction step	
Time	Thms	Maxterm	Thms	Lm-no	Thms
$t \leq 0.05$	46%	$m = 1$	62%	$l \leq 5$	14%
$t \leq 0.1$	62%	$m \leq 2$	82%	$l \leq 10$	46%
$t \leq 1$	94%	$m \leq 5$	91%	$l \leq 20$	77%
$t \leq 5$	97%	$m \leq 10$	96%	$l \leq 30$	90%
$t \leq 6$	100%	$m \leq 15$	100%	$l \leq 69$	100%
0.27	average	2.2	average	15.12	average

Table 7.3 Statistics for 90 problems from non-Euclidean geometry

The method of eliminating points presented here is based on a collection of powerful, high level theorems, such as the co-side theorems about the signed areas and co-face theorems about the signed volumes. This method can be contrasted with the earlier algebraic methods, which also proved astonishingly difficult theorems in geometry, but with lowlevel, mind-numbing polynomial manipulations. For more than eighty percent of 700 theorems proved by the methods of eliminating points, the maximal polynomials in their proofs have fewer than six terms. The Maxterms of the proofs produced using algebraic methods are rarely less than six. On the other hand, the algebraic methods are more general. For example, they can be used to prove theorems involving inequalities and theorems in differential geometry.

There are still many problems unsolved or solved unsatisfactorily by the approach of eliminating points. Although a large portion of the geometric theorems in textbooks can be proved by our prover, there are still cases which are not in the class, e.g., theorems which cannot be described constructively. A moderate goal is to extend the elimination methods to more constructions, such as the ones considered in Chou (1988). The main difficulty is that for more complicated constructions, the elimination formulas needed may become large and obscure. In this case, the goal of producing short and readable proofs may be dampened. Also the current methods are limited to theorems of equational type: geometric inequalities cannot be handled by these methods.

How do we combine the points elimination methods, algebraic methods and AI methods to devise more powerful methods for more readable solving? This is an important and difficult task, but every step towards this goal will be very worthy.

Chapter 8

CLIFFORD ALGEBRA APPROACHES TO MECHANICAL GEOMETRY THEOREM PROVING

Hongbo Li

Stimulated by the need for geometrically invariant representations and computations, H. Li and M. Cheng have developed a method for automated geometry theorem proving (AGTP) based on Clifford algebra since 1994. This method combines Clifford algebraic calculus with Wu's method and is capable of producing automated proofs with enhanced geometric interpretation and improving the efficiency of the proving procedure at the same time. It extends the work of Sturmfels (1993a), Crapo (1991), and others on automated theorem proving in projective geometry using projectively invariant methods, and applies to Euclidean, non-Euclidean and differential geometries. Motivated by the early work of Li and Cheng, Wang (1997) proposed a general framework of combining Clifford algebraic calculus and term rewriting for geometric reasoning. Meanwhile, he investigated the application of non-commutative Gröbner bases in Grassmann algebra to AGTP. The term rewriting approach for simplifying Clifford algebraic expressions has been further studied and refined by Fèvre and Wang (1998a,b) and Boy de la Tour et al. (1999). The question of normalizing Clifford algebraic expressions for AGTP is also addressed in the work of Yang et al. (1998, 1999). This chapter presents most of these developments for AGTP with Clifford algebraic calculus.

8.1 Overview

It has been widely known that algebraic methods for variable elimination, such as Wu's method (Wu 1994a; Chou 1988) and the Gröbner basis method (Kapur 1986b; Kutzler and Stifter 1986b), are very efficient for automatically proving theorems in geometry. In these methods, coordinates are introduced to formulate the geometric theorems in question. Because it is difficult to

MATHEMATICS MECHANIZATION AND APPLICATIONS
ISBN 0-12-734760-7

provide geometric interpretations for algebraic expressions in coordinate form, much of the recent research on automated geometric deduction has focused on developing high-level coordinate-free techniques. One such technique is the bracket algebra (or Grassmann–Cayley algebra) method for automated theorem proving in projective geometry introduced by Sturmfels (1993a), Crapo and Richter-Gebert (1995), and others, which is mainly based on Gröbner basis computation. Another technique is the area method (Chou et al. 1994a; Yang, Gao, Chou and Zhang 1997) for automated theorem proving in both Euclidean and non-Euclidean geometries, which is based on a set of geometric invariants and their rules. Provers based on these methods are capable of producing readable proofs, sometimes even traditional ones.

As pointed out by Hestenes and Sobczyk (1984), Grassmann–Cayley algebra can be realized within Clifford algebra. Clifford algebra (originally motivated by geometric considerations) has a long history, starting with Clifford (1876). It also includes many other algebraic tools for geometry, such as vector algebra and calculus, determinants, differential forms, complex numbers, quaternions and spinors, and can be applied to various branches of geometry, analysis, algebra, mechanics and physics (Chevalley 1954; Hestenes 1966, 1987, 1991; Hestenes and Sobczyk 1984; Hestenes and Ziegler 1991; Doran et al. 1993). The version of Clifford algebra formulated by Hestenes (1966, 1987) and Hestenes and Sobczyk (1984) has significantly enhanced its computational efficiency and is applicable immediately to various geometries, see Havel (1991), Havel and Dress (1992), Mourrain and Stolfi (1995a,b), Bayro-Corrochano and Lasenby (1995), Bayro-Corrochano et al. (1996) and Li (1996b, 1997c). The following is quoted from Hestenes' web page http://modelingnts.la.asu.edu/GC_R&D.html:

> "Though Leibniz articulated the dream of a universal geometric calculus in the seventeenth century, its realization began in 1844 with Hermann Grassmann's great work Die Lineale Ausdehnungslehre. Grassmann's Vision was so far ahead of its time, however, that it took more than a century to be widely appreciated. In the meantime Grassmann penetrated deep into the thinking of such excellent mathematicians as Peano (1888) and Whitehead (1848), but their work failed to advance or promulgate his vision. Many of his ideas were rediscovered and/or further developed anonymously in various branches of mathematics, but without his unifying perspective.
>
> "Grassmann's program to develop a universal geometric calculus reemerged in 1966 with the book Space-Time Algebra by David Hestenes, a refinement of his doctoral dissertation (UCLA, 1963). The idea of geometric algebra (GA) was given its modern form and reinvigorated by more than a century of advances in mathematics and physics since Grassmann. The roles of theoretical physics and the Lecture Notes of Marcel Riesz (1958) in stimulating the initial synthesis are described in the article Clifford Algebra

and the Interpretation of Quantum Mechanics. On balance, applications to physics have done more to advance GA than pure mathematics research.

"The most comprehensive treatment of the mathematical theory is given in the book Clifford Algebra to Geometric Calculus (1984). Except for the last chapter added in 1979, the manuscript was ready for publication by 1976 but did not appear in print until 1984, owing to an unfortunate series of publishing mishaps. With general arguments and many examples, the case is made there and elsewhere that geometric algebra is a more efficient general computational tool than matrix algebra. It is therefore a candidate to supplant (or subsume) matrix algebra in computer software and software designs for scientific computation. Indeed, the use of geometric algebra in Computer-Aided Geometric Design, Computer Vision and robotics is expanding rapidly."

As to AGTP, several methods based on the idea of integrating Clifford algebraic representations and computations with Wu's method have been proposed in Li (1994, 1995, 1997a,b, 2000) and Li and Cheng (1997, 1998a,b), with a general framework for applying Clifford algebra to theorem proving coming in Li (2000). These methods can be applied to Euclidean, projective, spherical, hyperbolic, conformal, and differential geometries. Following Li and Cheng's work, Wang (1997, 1998b) suggested combining Clifford algebraic calculus with term rewriting techniques for geometric reasoning. He also studied noncommutative Gröbner bases in Grassmann algebra for AGTP and applied the method to problems in computer vision. The idea of using rewriting techniques for simplifying Clifford algebraic expressions has been further studied, refined, and implemented by Fèvre and Wang (1998a,b, 1999); see also Boy de la Tour et al. (1999). On the other hand, another method for AGTP by computing normal forms for certain kind of Clifford algebraic expressions is proposed in Yang (1998) and Yang et al. (1999).

Besides their wide applicability, AGTP methods based on Clifford algebra can help in solving problems that are difficult to solve by other algebraic methods (Li 1997a,b,c; Li and Shi 1997), due to the simplification of geometric computations. It is of great potential to further develop Clifford algebraic methods for AGTP and other related branches.

8.2 A brief introduction to Clifford algebra

Clifford algebra can be defined in various ways; see, for example, Riesz (1958), Hestenes and Sobczyk (1984) and Crumeyrolle (1990). A Clifford algebra can be either finite or infinite, either orthogonal or symplectic, either positive definite or not. In this section we present a definition for finite-dimensional positive-definite Clifford algebra, which is the one most often used in Euclidean, affine and projective geometries.

Let $\boldsymbol{R}$ denote the real number field, and let $\boldsymbol{R}^n$ be an n-dimensional real vector space. By defining an anti-commutative and associative product among vectors in $\boldsymbol{R}^n$, called the *outer product*, or *wedge product*, we can generate a *Grassmann algebra* $\mathcal{G}_n$ from $\boldsymbol{R}^n$, which has dimension 2^n and whose elements, called *multivectors*, are graded from 0 up to n:

$$x = \langle x \rangle_0 + \langle x \rangle_1 + \cdots + \langle x \rangle_n, \quad \text{for any } x \in \mathcal{G}_n.$$

Here $\langle x \rangle_i$ is the *i-vector part* of x. A multivector is said to be *homogeneous* if it has only i-vector part for some $0 \leq i \leq n$.

An *i-blade*, or *extensor of step i*, is a wedge product of i vectors in $\boldsymbol{R}^n$. An *i-vector* is a linear combination of *i-blades*. The integer i is called the *grade*, or *step*, of an i-vector.

The Euclidean metric in $\boldsymbol{R}^n$ defines a commutative product, called the *inner product*, or *dot product*, between two vectors in $\boldsymbol{R}^n$. It can be extended to any two multivectors as follows: for any multivectors x, y, z, scalars λ, μ, vectors $\mathbf{A}_i$, $\mathbf{B}_j$,

$$\begin{cases} (\lambda x + \mu y) \cdot z := \lambda x \cdot z + \mu y \cdot z, \\ \lambda \cdot x := 0, \\ (\mathbf{A}_1 \wedge \cdots \wedge \mathbf{A}_p) \cdot (\mathbf{B}_1 \wedge \cdots \wedge \mathbf{B}_q) \\ \qquad := (-1)^{pq - \min(p,q)} (\mathbf{B}_1 \wedge \cdots \wedge \mathbf{B}_q) \cdot (\mathbf{A}_1 \wedge \cdots \wedge \mathbf{A}_p), \\ \mathbf{A}_0 \cdot (\mathbf{A}_1 \wedge \cdots \wedge \mathbf{A}_p) \\ \qquad := \sum_{i=1}^{p} (-1)^{i+1} (\mathbf{A}_0 \cdot \mathbf{A}_i)\ \mathbf{A}_1 \wedge \cdots \wedge \check{\mathbf{A}}_i \wedge \cdots \wedge \mathbf{A}_p, \\ (\mathbf{A}_1 \wedge \cdots \wedge \mathbf{A}_p) \cdot (\mathbf{B}_1 \wedge \cdots \wedge \mathbf{B}_q) \\ \qquad := (\mathbf{A}_1 \wedge \cdots \wedge \mathbf{A}_{p-1}) \cdot [\mathbf{A}_p \cdot (\mathbf{B}_1 \wedge \cdots \wedge \mathbf{B}_q)], \text{ if } p \leq q. \end{cases}$$

Here $\check{\mathbf{A}}_i$ indicates that $\mathbf{A}_i$ does not occur in the product. Two multivectors x and y are said to be *orthogonal* if $x \cdot y = 0$. For an i-blade x, there exist i mutually orthogonal vectors $\mathbf{B}_1, \ldots, \mathbf{B}_i$ such that $x = \mathbf{B}_1 \wedge \cdots \wedge \mathbf{B}_i$.

The *magnitude* of a multivector x is defined as

$$|x| := |\langle x \rangle_0| + \sum_{i=1}^{n} \sqrt{|\langle x \rangle_i \cdot \langle x \rangle_i|}.$$

In particular, for an i-vector x where $i > 0$, $|x| = \sqrt{|x \cdot x|}$.

For two vectors $\mathbf{A}$ and $\mathbf{B}$, we can add their outer product and inner product together, and thus form a new product, called the *Clifford product*, or *geometric product*:

$$\mathbf{AB} := \mathbf{A} \cdot \mathbf{B} + \mathbf{A} \wedge \mathbf{B}.$$

The geometric product can be extended to any finite number of multivectors as follows: let λ and μ be scalars, the x's be multivectors, and the $\mathbf{A}$'s be mutually orthogonal vectors. Then

$$\begin{cases} x_1(x_2x_3) := (x_1x_2)x_3, \\ x_1(\lambda x_2 + \mu x_3)x_4 := \lambda x_1x_2x_4 + \mu x_1x_3x_4, \\ \mathbf{A}x := \mathbf{A} \cdot x + \mathbf{A} \wedge x, \\ (\mathbf{A}_1 \wedge \cdots \wedge \mathbf{A}_j)x := (\mathbf{A}_1 \wedge \cdots \wedge \mathbf{A}_{j-1})(\mathbf{A}_j x). \end{cases}$$

The vector space $\mathcal{G}_n$ equipped with the geometric product, is called a *Clifford algebra* (of positive-definite signature and with dimensions 2^n), or *geometric algebra* (Hestenes and Sobczyk 1984).

8.3 An algebra for linear subspaces

Clifford algebra is essentially an algebra to describe linear subspaces. As we shall see, its applicability goes far beyond this; see Lawson and Michelsohn (1989), Hestenes and Sobczyk (1984), Hestenes (1987) and Li (1997c). First we introduce Grassmann–Cayley algebra, which can be defined within Clifford algebra.

Let x_i be an i-blade, where $i > 0$. The set

$$\{\mathbf{A} \in \boldsymbol{R}^n |\ \mathbf{A} \wedge x_i = 0\}$$

is a vector subspace. It can be represented by x_i uniquely up to a non-zero scalar factor. This space is called the *space* x_i.

For two blades x_i and y_j of grades i and j respectively, if $x_i \wedge y_j \neq 0$, it represents the direct sum of the two spaces x_i and y_j.

To define the intersection of two spaces in Clifford algebra we need the *dual* operator. Let I_m be an m-blade of unit magnitude. The *dual* of a multivector x with respect to the space I_m is defined by

$$x^{\sim} = x^{\sim I_m} := xI_m.$$

Notice that we do not require x to be in the Clifford algebra generated by the space I_m.

Let $x_1, \ldots, x_r$ be multivectors; then their *meet* $x_1 \vee \cdots \vee x_r$ with respect to the space I_m is defined by

$$\begin{aligned} x_1 \vee \cdots \vee x_r &:= x_1 \vee (x_2 \vee \cdots \vee x_r), \\ (x_1 \vee x_2)^{\sim} &:= x_1^{\sim} \wedge x_2^{\sim}. \end{aligned}$$

For two blades x_i and y_j of grades i and j respectively, if $x_i \vee y_j \neq 0$, it represents the intersection of the two spaces x_i and y_j. We have

$$y_i \vee y_j = \begin{cases} 0 & \text{if } i+j<n, \\ y_i^{\sim} \cdot y_j & \text{if } i+j \geq n. \end{cases}$$

The Grassmann algebra $\mathcal{G}_n$ equipped with the meet operator with respect to the space $\boldsymbol{R}^n$, is called a *Grassmann–Cayley algebra*; it has important applications in projective geometry and invariant theory. See Sturmfels (1993a), White (1991) and Whiteley (1991).

Grassmann–Cayley algebra is closely connected to *bracket algebra*. Let x_i be an i-vector. Then for any $(n-i)$-vector y_{n-i},

$$[x_i](y_{n-i}) := (x_i \wedge y_{n-i})^{\sim}$$

defines an element $[x_i]$ in the dual linear space of $(n-i)$-vectors, called a *bracket*[1]. In particular, the bracket of an n-vector equals the dual of the n-vector. In computation, since bracket is scalar-valued, it follows the same rules as a scalar.

We see that the outer product and meet of two blades do not represent the *sum* and *intersection* of their corresponding spaces. Mourrain (1992) defines the two operators in terms of tensor Grassmann algebra.

Mourrain's Algorithm A: (The sum and intersection of two vector subspaces) Let $x_r = \mathbf{A}_1 \wedge \cdots \wedge \mathbf{A}_r$ and $y_s = \mathbf{B}_1 \wedge \cdots \wedge \mathbf{B}_s$ be blades of grade r and s respectively. Their sum and intersection are denoted by $x_r \bigtriangleup y_s$ and $x_r \bigtriangledown y_s$ respectively (different from Mourrain's notations).

M1. For $i = 0, \ldots, r$, do the following:

Set

$$C \longleftarrow \sum_{\substack{\sigma(1)<\cdots<\sigma(r-i), \\ \sigma(r-i+1)<\cdots<\sigma(r)}} \begin{array}{l} \epsilon(\sigma)\mathbf{A}_{\sigma(1)} \wedge \cdots \wedge \mathbf{A}_{\sigma(r-i)} \otimes \\ \mathbf{A}_{\sigma(r-i+1)} \wedge \cdots \wedge \mathbf{A}_{\sigma(r)} \wedge \mathbf{B}_1 \wedge \cdots \wedge \mathbf{B}_s, \end{array}$$

where σ is a permutation of $\{1, \ldots, n\}$ with sign $\epsilon(\sigma)$.

[1] This is the standard definition of a bracket. Alternatively, one can define the bracket $[x_i]$, where $i > 0$, as an element in the dual linear space of i-vectors defined by $[x_i](y_i) := x_i \cdot y_i$ for any i-vector y_i. For $i = 0$, the bracket of a scalar equals itself.

If $C = 0$ then go to M3 else set $D \leftarrow C$, $i \leftarrow i + 1$.

M2. Output $x_r \bigtriangleup y_s \leftarrow C$, $x_r \bigtriangledown y_s \leftarrow \{0\}$, and then exit.

M3. Let C_{r-i+1} be an $(r-i+1)$-blade such that $C_{r-i+1} \cdot D \neq 0$, and let D_{s+i-1} be an $(s+i-1)$-blade such that $D \cdot D_{s+i-1} \neq 0$. Here the inner products of a multivector x in $\mathcal{G}_n$ with a tensor $y \otimes z$ in $\mathcal{G}_n \otimes \mathcal{G}_n$ are defined as

$$x \cdot (y \otimes z) := (x \cdot y) \otimes z, \qquad (y \otimes z) \cdot x := y \otimes (z \cdot x).$$

Output[2]

$$x_r \bigtriangledown y_s \leftarrow C_{r-i+1} \cdot D, \qquad x_r \bigtriangleup y_s \leftarrow D \cdot D_{s+i-1}. \tag{8.1}$$

■

The sum and intersection of finitely many multivectors can be defined by extending the above definition multilinearly and associatively.

This finishes our introduction of Grassmann–Cayley algebra within Clifford algebra. Now we discuss the geometric significance of the inner product to complete the theory of linear subspaces in $\boldsymbol{R}^n$.

Let $x_i = \mathbf{A}_1 \wedge \cdots \wedge \mathbf{A}_i$. The i vectors $\mathbf{A}_1, \ldots, \mathbf{A}_i$ have their common origin at the origin of $\boldsymbol{R}^n$. The ends of these vectors together with the origin span an i-dimensional simplex $\mathcal{S}(x_i)$. The magnitude of x_i equals the volume of this simplex. Let y_j be a j-blade of unit magnitude, where $i \leq j$.

- When $i = j$, the inner product $x_i \cdot y_j$ is a scalar, whose magnitude equals the volume of the i-dimensional simplex which is the orthogonal projection of $\mathcal{S}(x_i)$ onto the space y_j.
- When $i < j$, the inner product $x_i \cdot y_j$ is a $(j-i)$-blade representing the subspace of the space y_j, which is orthogonal to the orthogonal projection of the space x_i onto the space y_j. The scalar $|x_i \cdot y_j|$ equals the volume of the orthogonal projection of $\mathcal{S}(x_i)$ onto the space y_j.
  In particular, when all the $\mathbf{A}$'s are in the space y_j, then $x_i \cdot y_j = x_i^{\sim y_j}$ is the dual of x_i with respect to the space y_j.

[2]The output in Mourrain (1992) is different from (8.1). Mourrain defined the following two operators in $\mathcal{G}_n \otimes \mathcal{G}_n$: let x and y be multivectors in $\mathcal{G}_n$; then

$$[x \otimes y]_1 := [x]y, \qquad [x \otimes y]_2 := [y]x.$$

Then he set

$$x_r \bigtriangledown y_s \leftarrow [D]_2, \qquad x_r \bigtriangleup y_s \leftarrow [D]_1$$

as the output. Here we choose a different output (8.1) for practical computation, with C_{r-i+1} and D_{s+i-1} as either dummy blades or specific ones.

8.4 Two Clifford algebraic models for Euclidean geometry

In this section we introduce two of the most often used Clifford algebraic models for Euclidean geometry. Besides Euclidean geometry, there are Clifford algebraic models for spherical geometry (Hestenes and Sobczyk 1984), hyperbolic geometry (Li 1997c), conformal geometry (Hestenes 1991), etc. A *Clifford algebraic model* for a geometry refers to a Clifford algebra where each basic geometric entity or constraint in the geometry is represented by an algebraic entity or operator, and geometric relations among geometric entities can be obtained by means of algebraic computations.

The Clifford model $\mathcal{G}_n$. This is the most often used model for n-dimensional Euclidean geometry $\boldsymbol{E}^n$ (Hestenes 1987). Let O be a point in $\boldsymbol{E}^n$. By setting O to be the zero vector, we can represent $\boldsymbol{E}^n$ with $\boldsymbol{R}^n$: a point A is represented by the vector from O to A, denoted by $\mathbf{A}$. Henceforth points are always represented by vectors, and are all boldfaced. A direction is the difference of two point vectors, and is represented by a vector in $\mathcal{G}_n$. The Clifford algebra $\mathcal{G}_n$, which is generated by $\boldsymbol{R}^n$, serves as a model for the n-dimensional Euclidean geometry. In this model:

- A straight line $\mathbf{AB}$ is represented by a pair: its direction $\mathbf{A}-\mathbf{B}$, and a point on the line. Alternatively it can be represented parametrically by $\mathbf{A}+\lambda(\mathbf{B}-\mathbf{A})$, where $\lambda \in \boldsymbol{R}$.

- The distance between points $\mathbf{A}$ and $\mathbf{B}$ is $|\mathbf{A}-\mathbf{B}|$.

- An i-dimensional plane passing through point $\mathbf{B}$ and parallel to the i-dimensional vector subspace spanned by vectors $\mathbf{A}_1, \ldots, \mathbf{A}_i$ is represented by a pair $(\mathbf{A}_1 \wedge \cdots \wedge \mathbf{A}_i, \mathbf{B})$. Alternatively it can be represented parametrically by $\mathbf{B}+\lambda_1\mathbf{A}_1+\cdots+\lambda_i\mathbf{A}_i$. As special cases, an i-dimensional space x_i, where x_i is an i-blade, can be represented by $(x_i, 0)$; a point $\mathbf{A}$ can be taken as a 0-dimensional plane, and be represented by the pair $(1, \mathbf{A})$.

- For an i-dimensional plane $(x_i, \mathbf{A})$ and a j-dimensional plane $(y_j, \mathbf{B})$, where $0 \leq i \leq j \leq n-1$,

  - they represent the same plane if and only if $i=j$, x_i and y_j are equal up to a non-zero scalar factor, and $(\mathbf{A}-\mathbf{B}) \wedge x_i = 0$;

- they are parallel if and only if the space x_i is a subspace of the space y_j and $(\mathbf{A} - \mathbf{B}) \wedge y_j = 0$ (when they are parallel, the distance between them is $|(\mathbf{A} - \mathbf{B}) \wedge y_j|/|y_j|$);
- they are perpendicular if and only if $x_i \cdot y_j = 0$.

In the Clifford model, the difference between the representations of a point and a direction is that the former depends on the choice of the origin $\mathbf{O}$.

The Grassmann model $\mathcal{G}_{n+1}$. This model can be used in n-dimensional Euclidean, affine, and projective geometries (Hestenes and Ziegler 1991). In the Clifford model, the observer $\mathbf{O}$ is inside $\boldsymbol{E}^n$. What about moving it outside $\boldsymbol{E}^n$? Then we need to embed $\boldsymbol{E}^n$ into $\boldsymbol{E}^{n+1}$ and take $\boldsymbol{E}^n$ as a hyperplane of $\boldsymbol{E}^{n+1}$. Every point in $\boldsymbol{E}^n$ is on a unique straight line passing through $\mathbf{O}$ and the point; conversely, every straight line passing through $\mathbf{O}$ intersects with $\boldsymbol{E}^n$ at a unique point, if the intersection occurs. Every straight line passing through $\mathbf{O}$ and parallel with $\boldsymbol{E}^n$ represents a direction in $\boldsymbol{E}^n$, which is the direction of the line; conversely, every direction in $\boldsymbol{E}^n$ can be represented uniquely by the line passing through $\mathbf{O}$ and along the direction. Therefore, a point A can be represented uniquely by the vector from $\mathbf{O}$ to A, denoted by $\mathbf{A}$, or by any vector along the vector $\mathbf{A}$ up to a non-zero scalar factor; a direction can be represented by a vector along the direction, which is unique up to a non-zero scalar factor. The Clifford algebra $\mathcal{G}_{n+1}$, which is generated by $\boldsymbol{R}^{n+1}$, serves as a model for the n-dimensional Euclidean, affine, and projective geometries.

Let the distance from the origin $\mathbf{O}$ to $\boldsymbol{E}^n$ be 1, the vector from $\mathbf{O}$ to the foot drawn from $\mathbf{O}$ to $\boldsymbol{E}^n$ be $\mathbf{e}_0$. Let $\partial : \mathcal{G}_{n+1} \longrightarrow \mathcal{G}_{n+1}$ be defined by

$$\partial = \mathbf{e}_0 \cdot .$$

Then $\partial \circ \partial = (\mathbf{e}_0 \wedge \mathbf{e}_0)\cdot = 0$. This linear operator has the following properties.

(1) A vector $\mathbf{A}$ in $\boldsymbol{R}^{n+1}$ represents a direction in $\boldsymbol{E}^n$ if $\partial(\mathbf{A}) = 0$, or a point in $\boldsymbol{E}^n$ if $\partial(\mathbf{A}) = 1$. When $\partial(\mathbf{A}) \neq 0$, $\mathbf{A}$ represents the point $\mathbf{A}/\partial(\mathbf{A})$.

(2) For any multivector x in $\mathcal{G}_{n+1}$, $\partial(x)$ is in $\mathcal{G}_n$; on the other hand, x is in $\mathcal{G}_n$, the Clifford algebra generated by $\boldsymbol{R}^n$, if and only if $\partial(x) = 0$.

(3) For a multivector x in $\mathcal{G}_{n+1}$, if there exists a point $\mathbf{A}$ in $\boldsymbol{E}^n$ satisfying $\mathbf{A} \wedge x = 0$, then

$$x = \mathbf{A} \wedge \partial(x).$$

(4) For $(n+1)$-blades x_{n+1} and y_{n+1}, $x_{n+1}/y_{n+1} = \partial(x_{n+1})/\partial(y_{n+1})$. In particular, $x_{n+1} = y_{n+1}$ if and only if $\partial(x_{n+1}) = \partial(y_{n+1})$.

As to representing geometric objects and relations, in the Grassmann model:

- A line **AB** is represented by $\mathbf{A} \wedge \mathbf{B}$.
- An i-dimensional affine plane generated by points $\mathbf{A}_1, \ldots, \mathbf{A}_{i+1}$ is represented by $x_{i+1} = \mathbf{A}_1 \wedge \cdots \wedge \mathbf{A}_{i+1}$. The scalar $|x_{i+1}|$ is the volume of the i-dimensional simplex generated by these points. When $i = n$, $x_{i+1}^{\sim}$ is the signed volume of the simplex.
- For an i-dimensional plane x_{i+1} and an j-dimensional plane y_{j+1}, where $i \leq j \leq n-1$,
  - they represent the same plane if and only if $i = j$, x_{i+1} and y_{j+1} are equal up to a non-zero scalar factor;
  - they are parallel if and only if the space $\partial(x_{i+1})$ is a vector subspace of the space y_{j+1};
  - they are perpendicular if and only if $\partial(x_{i+1}) \cdot \partial(y_{j+1}) = 0$.

In the Grassmann model, the inner product of two vectors has geometric meaning in $\boldsymbol{E}^n$ if and only if they belong to $\mathcal{G}_n$.

8.5 Vector equation solving

One typical feature of Clifford algebra in equation solving is its ability to represent the solutions of a set of (multi)vector equations in a way that is geometrically invariant. Because of this, when used in theorem proving, Clifford algebra can significantly enhance the geometric computations and interpretability of the proofs. In this section we present some general formulas for vector equation solving.

The following is a fundamental formula proposed by Hestenes and Sobczyk (1984).

FUNDAMENTAL TYPE. $uxv - y = 0$, where the order of expressions is $u, v, y \prec x$, u, v, x, y are all multivectors, x is the leading basic variable, and u, v, y do not involve x.

Solution:

$$x = u^{-1}yv^{-1}.$$

Non-degeneracy conditions: u, v are invertible.

Mourrain (1992) proposed a formula for vector equation solving in Grassmann–Cayley algebra: let $\mathbf{X}_1 \prec \cdots \prec \mathbf{X}_m$ be points and $Z_j^i = Z_j^i(\mathbf{X}_1, \ldots, \mathbf{X}_{i-1})$ be blades.

TYPE M. $\mathbf{X}_i \wedge Z_j^i = 0$, for $j = 1, \ldots, k_i$, where i is fixed.

Solution:

$$\mathbf{X}_i \simeq \gamma(Z_1^i \vee \cdots \vee Z_{k_i}^i, \mathbf{X}_i),$$

where "$\simeq$" means that the two sides of the symbol are equal up to a non-zero scalar factor. The *Cramer operator* γ, which is linear with respect to its first argument, is defined as

$$\gamma(\mathbf{Y}_1 \wedge \cdots \wedge \mathbf{Y}_s, \mathbf{Y}_0) := \sum_{i=1}^{m} (-1)^{i+1} [\mathbf{Y}_0 \wedge \mathbf{Y}_1 \wedge \cdots \check{\mathbf{Y}}_i \wedge \cdots \mathbf{Y}_m] \mathbf{Y}_i$$

for the vectors $\mathbf{Y}_i$.

Non-degeneracy conditions: $Z_1^i \vee \cdots \vee Z_j^i \neq 0$ for $j = 2, \ldots, k_i$.

The following are formulas proposed by Li and Cheng (1997) and completed by Li (2000).

Formulas for the Clifford model $\mathcal{G}_n$. Let $\mathbf{X}$ and the $\mathbf{A}_i$ be vectors, the c_j be scalars, the B_k be 2-vectors, the C_l be 3-vectors, the Γ_m be homogeneous multi-vectors, the K_p be $(n-1)$-vectors, and the J_q be n-vectors. Here i, j, k, l, m, p, q are indices. Also let $\mathbf{A}_i, c_j, B_k, C_l, \Gamma_m, K_p, J_q$ be $\prec \mathbf{X}$ and do not involve $\mathbf{X}$.

TYPE C1. $\mathbf{X} \cdot \mathbf{A}_1 - c_1 = 0$, $\mathbf{X} \wedge \mathbf{A}_2 - B_2 = 0$.

Solution:

$$\mathbf{X} = \frac{c_1 \mathbf{A}_2 - \mathbf{A}_1 \cdot B_2}{\mathbf{A}_1 \cdot \mathbf{A}_2}.$$

Non-degeneracy condition: $\mathbf{A}_1 \cdot \mathbf{A}_2 \neq 0$.

TYPE C2. $\mathbf{X} \cdot \mathbf{A}_1 - c_1 = 0, \ldots, \mathbf{X} \cdot \mathbf{A}_n - c_n = 0$.

Solution:

$$\mathbf{X} = \frac{\sum_{i=1}^{n} c_i (\mathbf{A}_1 \wedge \cdots \wedge \check{\mathbf{A}}_i \wedge \cdots \wedge \mathbf{A}_n)^{\sim}}{(\mathbf{A}_1 \wedge \cdots \wedge \mathbf{A}_n)^{\sim}}.$$

Non-degeneracy condition: $\mathbf{A}_1 \wedge \cdots \wedge \mathbf{A}_n \neq 0$.

TYPE C3. $\mathbf{X}\wedge\mathbf{A}_1 - B_1 = 0$, $\mathbf{X}\wedge\mathbf{A}_2 - B_2 = 0$.

Formula 1. When $n \geq 3$, let $B = \mathbf{A}_1\wedge\mathbf{A}_2$. Solution:

$$\begin{cases} \mathbf{X} = \dfrac{(B_1\wedge\mathbf{A}_2)\cdot B - B_1\cdot B\ \mathbf{A}_2 + B_2\cdot B\ \mathbf{A}_1}{B\cdot B}, \\ B_1\wedge\mathbf{A}_2 + B_2\wedge\mathbf{A}_1 = 0. \end{cases}$$

Non-degeneracy condition: $B \neq 0$.

Formula 2. When $n = 3$, let $\mathbf{Y}$ be a vector that satisfies $\mathbf{Y}\wedge\mathbf{A}_1\wedge\mathbf{A}_2 \neq 0$. Solution:

$$\begin{cases} \mathbf{X} = \dfrac{(B_2\wedge\mathbf{Y})^{\sim}\mathbf{A}_1 - (\mathbf{A}_2\wedge\mathbf{Y})^{\sim}\cdot B_1}{(\mathbf{A}_1\wedge\mathbf{A}_2\wedge\mathbf{Y})^{\sim}}, \\ B_1\wedge\mathbf{A}_2 + B_2\wedge\mathbf{A}_1 = 0. \end{cases}$$

Non-degeneracy condition: $\mathbf{A}_1\wedge\mathbf{A}_2\wedge\mathbf{Y} \neq 0$.

TYPE C4. $\mathbf{X}\cdot\mathbf{A}_1 - c_1 = 0, \ldots, \mathbf{X}\cdot\mathbf{A}_{n-1} - c_{n-1} = 0$.

Parametric solution (with parameter λ): let $K = \mathbf{A}_1\wedge\cdots\wedge\mathbf{A}_{n-1}$; then

$$\mathbf{X} = \lambda K^{\sim} + \sum_{i=1}^{n-1}\frac{c_i}{K\cdot K}(\mathbf{A}_1\wedge\cdots\wedge\check{\mathbf{A}}_i\wedge\cdots\wedge\mathbf{A}_{n-1})\cdot K.$$

Non-degeneracy condition: $K \neq 0$.

TYPE C5. $\mathbf{X}\wedge\mathbf{A}_1\wedge\cdots\wedge\mathbf{A}_r - \Gamma = 0$, where $r < n$ and Γ is an $(r+1)$-vector.

Parametric solution (with parameters λ_i): let $\Gamma_1 = \mathbf{A}_1\wedge\cdots\wedge\mathbf{A}_r$; then

$$\mathbf{X} = \sum_{i=1}^{r}\lambda_i\mathbf{A}_i + \frac{\Gamma\cdot\Gamma_1}{\Gamma_1\cdot\Gamma_1}.$$

Non-degeneracy condition: $\Gamma_1 \neq 0$.

Formulas for the Grassmann model $\mathcal{G}_{n+1}$. The leading variable $\mathbf{X}$ below is always assumed to be a point vector.

TYPE G1. $\mathbf{X}\wedge K - J = 0$, $\mathbf{X}\wedge B - C = 0$.

Solution:

$$\mathbf{X} = \frac{K\vee\partial(C) - K\vee B + J^{\sim}\partial(B)}{(\partial(B)\wedge K)^{\sim}}.$$

Non-degeneracy condition: $\partial(B)\wedge K \neq 0$.

TYPE G2. $\mathbf{X} \wedge K_1 - J_1 = 0, \ldots, \mathbf{X} \wedge K_{n-1} - J_{n-1} = 0$.

Solution: let $\mathbf{A} = K_1 \vee \cdots \vee K_{n-1}$; then

$$\mathbf{X} = \frac{\mathbf{A} + \sum_{i=1}^{n-1} (-1)^i J_i^\sim \partial(K_1 \vee \cdots \vee \check{K}_i \vee \cdots \vee K_{n-1})}{\partial(\mathbf{A})}.$$

Non-degeneracy condition: $\partial(\mathbf{A}) \neq 0$.

TYPE G3. $\mathbf{X} \wedge B_1 - C_1 = 0, \mathbf{X} \wedge B_2 - C_2 = 0$.

Formula 1. When $n \geq 4$ and B_1 is a blade, let $B_1 = \mathbf{A}_1 \wedge \mathbf{A}_2$. Let $B = \partial(B_1) \wedge \partial(B_2)$. Solution:

$$\begin{cases} \mathbf{X} = (-\partial(C_2) \cdot B\ \partial(B_1) + \partial(C_1) \cdot B\ \partial(B_2) \\ \qquad + \partial(\mathbf{A}_1 \wedge B_2) \cdot B\ \mathbf{A}_2 - \partial(\mathbf{A}_2 \wedge B_2) \cdot B\ \mathbf{A}_1 \\ \qquad - (\partial(C_1) \wedge \partial(B_2)) \cdot B)/B \cdot B, \\ \partial(C_2) \wedge \partial(B_1) + \partial(C_1) \wedge \partial(B_2) = \partial(B_1 \wedge B_2). \end{cases}$$

Non-degeneracy condition: $B \neq 0$.

Formula 2. When $n = 4$, let $\mathbf{Y}$ be a vector satisfying $\partial(B_1) \wedge B_2 \wedge \mathbf{Y} \neq 0$. Solution:

$$\begin{cases} \mathbf{X} = \dfrac{(B_2 \wedge \mathbf{Z}) \vee (\partial(C_1) - B_1) + (C_2 \wedge \mathbf{Z})^\sim\ \partial(B_1)}{(\partial(B_1) \wedge B_2 \wedge \mathbf{Z})^\sim}, \\ C_1 \wedge \partial(B_2) + C_2 \wedge \partial(B_1) = B_1 \wedge B_2. \end{cases}$$

Non-degeneracy condition: $\partial(B_1) \wedge B_2 \wedge Y \neq 0$.

TYPE G4. $\mathbf{X} \wedge \mathbf{A}_0 \wedge \cdots \wedge \mathbf{A}_r - \Gamma = 0$, where $r < n-1$ and Γ is an $(r+2)$-vector.

Parametric solution (with parameters λ_i): let $\Gamma_1 = \partial(\mathbf{A}_0 \wedge \cdots \wedge \mathbf{A}_r) \neq 0$ and $\partial(\mathbf{A}_0) \neq 0$; then

$$\mathbf{X} = \frac{\mathbf{A}_0 + \sum_{i=1}^{r} \lambda_i \partial(\mathbf{A}_0 \wedge \mathbf{A}_i)}{\partial(\mathbf{A}_0)} - \frac{\partial(\Gamma) \cdot \Gamma_1}{\Gamma_1 \cdot \Gamma_1}.$$

Non-degeneracy conditions: $\Gamma_1, \partial(\mathbf{A}_0) \neq 0$.

Examples of applying these formulas to AGTP can be found in the next section. From the above list of formulas, one sees that they are all on linear (multi)vector equations. In application, when the hypothesis of a geometric problem is expressed by a set of linear (multi)vector equations, the previous formulas can lead to solutions that are geometric invariant. What if there are non-linear equations?

When there are both linear and non-linear equations for a (multi)vector variable, if the subset of linear equations cannot provide a solution for the variable, the parametric equation solving technique (Li 1994; Li and Cheng 1997; in projective geometry, Cramer's rule as used by Mourrain 1992) can help provide a parametric solution from the subset of linear equations. These parameters serve as "geometric coordinates" for the unknown variable because of their geometric properties. The set of (multi)vector equations in the unknown variable becomes a set of equations of the parameters. Then Wu's method or the Gröbner basis method can be applied to these equations for further algebraic manipulations.

If there is no linear equation at all, then some algebraic manipulations, e.g., pseudo-divisions, can be made prior to vector equation solving to obtain linear equations. Parametric representations of non-linear entities, e.g., conics and quadrics, can also be resorted for "geometric coordinates," see Li and Cheng (1997) for illustrations.

8.6 Grassmann–Cayley algebra and Wu's method

Mourrain (1992) proposed a method integrating Grassmann–Cayley algebra manipulations with Wu's method for proving certain kinds of theorems in projective geometry. The method is to generate a set of substitution rules from the construction of a projective configuration of points using Grassmann–Cayley algebra. For certain kinds of projective configurations, proving the generic correctness of an equality in the Grassmann–Cayley algebra associated with the projective configuration is reduced to checking whether the equality is true after the set of substitution rules is applied to it.

Mourrain's Algorithm B: (Generation of rules from the construction of a projective configuration of points)

Input: a list of projective points $\mathbf{X}_1, ..., \mathbf{X}_m$ in the order of the constructions; the set H of constructions of points in Grassmann algebra formalism, i.e.,

$$\begin{cases} \mathbf{X}_i \wedge Z_i^1 = 0, \\ \quad \vdots \\ \mathbf{X}_i \wedge Z_i^{j_i} = 0, \end{cases}$$

where $i \in \{1, \ldots, m\}$, and for $1 \le k \le j_i$, Z_i^k is an extensor depending on points in $\{\mathbf{X}_1, \ldots, \mathbf{X}_{i-1}\}$ only.

Output: R, the set of substitution rules.

M1. Set $R \leftarrow \varnothing$ and $i \leftarrow 1$.

M2. If there is no construction on $\mathbf{X}_i$, then set $i \leftarrow i+1$ and go back to M2.

Set $W \leftarrow \{Z_i^1, \ldots, Z_i^{j_i}\}$. Apply the rules in R to W and denote the result by W'. If $W' = \{0\}$, then set $i \leftarrow i+1$ and go back to M2.

M3. Assume that $W' = \{Z_i^{1'}, \ldots, Z_i^{k'_i}\}$, and use a modification of Mourrain's algorithm A to compute $W'' = Z_i^{1'} \triangledown \ldots \triangledown Z_i^{k'_i}$: in M1 of Mourrain's algorithm A, apply the rules in R to C after every evaluation of C and then replace C with the result from the substitution.

W'' is non-zero as long as the construction of $\mathbf{X}_i$ is correct.

M4. Set $R \leftarrow R \cup \{\mathbf{X}_i \rightarrow \gamma(W'', \mathbf{X}_i)\}$. ■

Example 8.1: (Mourrain 1992)

Let $\mathbf{X}, \mathbf{Y}, \mathbf{Z}$ be points on a plane, $\mathbf{A}, \mathbf{A}'$ be points on line $\mathbf{XY}$, $\mathbf{B}, \mathbf{B}'$ be points on line $\mathbf{YZ}$, and $\mathbf{C}, \mathbf{C}'$ be points on line $\mathbf{XZ}$. Let $\mathbf{P} = \mathbf{AB} \cap \mathbf{A'B'}$, $\mathbf{Q} = \mathbf{AB} \cap \mathbf{YZ}$, $\mathbf{U} = \mathbf{AC} \cap \mathbf{A'C'}$, $\mathbf{V} = \mathbf{BC} \cap \mathbf{B'C'}$, and $\mathbf{R} = \mathbf{YU} \cap \mathbf{ZV}$. Then $\mathbf{P}, \mathbf{Q}, \mathbf{R}$ are collinear.

The list of variables: $\mathbf{X}, \mathbf{Y}, \mathbf{Z}, \mathbf{A}, \mathbf{A}', \mathbf{B}, \mathbf{B}', \mathbf{C}, \mathbf{C}', \mathbf{U}, \mathbf{V}, \mathbf{P}, \mathbf{Q}, \mathbf{R}$.

The constructions are

$$\begin{cases} \mathbf{A} \wedge \mathbf{X} \wedge \mathbf{Y} = 0, & \mathbf{U} \wedge \mathbf{A} \wedge \mathbf{C} = 0, & \mathbf{A}' \wedge \mathbf{X} \wedge \mathbf{Y} = 0, \\ \mathbf{U} \wedge \mathbf{A}' \wedge \mathbf{C}' = 0, & \mathbf{B} \wedge \mathbf{X} \wedge \mathbf{Z} = 0, & \mathbf{V} \wedge \mathbf{B} \wedge \mathbf{C} = 0, \\ \mathbf{B}' \wedge \mathbf{X} \wedge \mathbf{Z} = 0, & \mathbf{V} \wedge \mathbf{B}' \wedge \mathbf{C}' = 0, & \mathbf{C} \wedge \mathbf{Y} \wedge \mathbf{Z} = 0, \\ \mathbf{C}' \wedge \mathbf{Y} \wedge \mathbf{Z} = 0, & \mathbf{P} \wedge \mathbf{A} \wedge \mathbf{B} = 0, & \mathbf{P} \wedge \mathbf{A}' \wedge \mathbf{B}' = 0, \\ \mathbf{Q} \wedge \mathbf{A} \wedge \mathbf{Z} = 0, & \mathbf{Q} \wedge \mathbf{B}' \wedge \mathbf{Y} = 0, & \mathbf{R} \wedge \mathbf{U} \wedge \mathbf{Z} = 0, \\ \mathbf{R} \wedge \mathbf{V} \wedge \mathbf{Y} = 0. & & \end{cases}$$

The conclusion is $\mathbf{P} \wedge \mathbf{Q} \wedge \mathbf{R} = 0$.

By Mourrain's algorithm B, the following rules are generated:

$$\begin{cases} \mathbf{R} & \rightarrow & -\mathbf{V}[\mathbf{U} \wedge \mathbf{Z} \wedge \mathbf{Y}] + \mathbf{Y}[\mathbf{V} \wedge \mathbf{U} \wedge \mathbf{Z}], \\ \mathbf{Q} & \rightarrow & -\mathbf{B}'[\mathbf{A} \wedge \mathbf{Z} \wedge \mathbf{Y}] + \mathbf{Y}[\mathbf{B}' \wedge \mathbf{A} \wedge \mathbf{Z}], \\ \mathbf{P} & \rightarrow & \mathbf{B}'[\mathbf{B} \wedge \mathbf{A}' \wedge \mathbf{A}] + \mathbf{A}'[\mathbf{B}' \wedge \mathbf{B} \wedge \mathbf{A}], \\ \mathbf{V} & \rightarrow & \mathbf{C}'[\mathbf{C} \wedge \mathbf{B}' \wedge \mathbf{B}] + \mathbf{B}'[\mathbf{C}' \wedge \mathbf{C} \wedge \mathbf{B}], \\ \mathbf{U} & \rightarrow & \mathbf{C}'[\mathbf{C} \wedge \mathbf{A}' \wedge \mathbf{A}] + \mathbf{A}'[\mathbf{C}' \wedge \mathbf{C} \wedge \mathbf{A}], \\ \mathbf{C}' & \rightarrow & \mathbf{Z}[\mathbf{C}' \wedge \mathbf{Y}] - \mathbf{Y}[\mathbf{C}' \wedge \mathbf{Z}], \\ \mathbf{C} & \rightarrow & \mathbf{Z}[\mathbf{C} \wedge \mathbf{Y}] - \mathbf{Y}[\mathbf{C} \wedge \mathbf{Z}], \\ \mathbf{B}' & \rightarrow & \mathbf{Z}[\mathbf{B}' \wedge \mathbf{X}] - \mathbf{X}[\mathbf{B}' \wedge \mathbf{Z}], \\ \mathbf{B} & \rightarrow & \mathbf{Z}[\mathbf{B} \wedge \mathbf{X}] - \mathbf{X}[\mathbf{B} \wedge \mathbf{Z}], \\ \mathbf{A}' & \rightarrow & \mathbf{Y}[\mathbf{A}' \wedge \mathbf{X}] - \mathbf{X}[\mathbf{A}' \wedge \mathbf{Y}], \\ \mathbf{A} & \rightarrow & \mathbf{Y}[\mathbf{A} \wedge \mathbf{X}] - \mathbf{X}[\mathbf{A} \wedge \mathbf{Y}]. \end{cases}$$

The conclusion is proved by direct substitution of the above rules.

8.7 Clifford algebra and Wu's method

In 1994, Li and Cheng proposed a method that combines Clifford algebraic representation with Wu's method. This method is complete for AGTP in Euclidean, non-Euclidean and differential geometries, see Li (1997b), Li and Shi (1997), Li and Cheng (1998a).

Li–Cheng's Algorithm: (Theorem proving with Clifford algebra)

The hypothesis and conclusion are assumed to be in Clifford algebraic equality form. The conclusion is assumed to be an equality.

L1. [Computing a triangular sequence.] Triangulate the hypothesis using substitutions, pseudo-divisions and vector equation solving. The parametric equation solving (that is, solving in which the solution contains new parameters) is not allowed in this step. The result is called a *triangular sequence*.

L2. [Proving by the triangular sequence.] Carry out substitutions and pseudo-divisions on the conclusion with the triangular sequence. If the conclusion is reduced to $0 = 0$, the algorithm terminates.

L3. [Computing a parametric triangular sequence.] Triangulate the triangular sequence with parametric equation solving, substitutions, pseudo-divisions and vector equation solving. The result is called a *parametric triangular sequence.*

L4. [Proving by the parametric triangular sequence.] Do substitutions and pseudo-divisions to the conclusion with the parametric triangular sequence. If the conclusion is reduced to $0 = 0$, the algorithm terminates.

L5. [Computing a characteristic set.] Select a coordinate system and translate all expressions in the parametric triangular sequence into polynomials. Use Wu's method to compute a characteristic set.

L6. [Proving by the characteristic set.] Same as in Wu's method (Chap. 5). ■

Example 8.2: (Li 2000; Boy de la Tour et al. 1999)

Let **ABCD** be a tetrahedron. There is a plane which intersects the lines **AB**, **AC**, **DC**, **DB** at points $\mathbf{M}, \mathbf{N}, \mathbf{E}, \mathbf{F}$ respectively. If **MNEF** is always a parallelogram when the plane moves, then the center **O** of the parallelogram is always on a fixed straight line.

In the Grassmann model $\mathcal{G}_4$ for the space, the hypothesis consists of

$$\begin{cases} \mathbf{M} - \mathbf{N} = \mathbf{F} - \mathbf{E}, & (\mathbf{MNEF} \text{ is a parallelogram}) \\ \mathbf{A} \wedge \mathbf{B} \wedge \mathbf{M} = 0, & (\mathbf{A}, \mathbf{B}, \mathbf{M} \text{ are collinear}) \\ \mathbf{A} \wedge \mathbf{C} \wedge \mathbf{N} = 0, & (\mathbf{A}, \mathbf{C}, \mathbf{N} \text{ are collinear}) \\ \mathbf{C} \wedge \mathbf{D} \wedge \mathbf{E} = 0, & (\mathbf{C}, \mathbf{D}, \mathbf{E} \text{ are collinear}) \\ \mathbf{B} \wedge \mathbf{D} \wedge \mathbf{F} = 0, & (\mathbf{B}, \mathbf{D}, \mathbf{F} \text{ are collinear}) \\ \mathbf{O} = \dfrac{\mathbf{M} + \mathbf{E}}{2}, & (\mathbf{O} \text{ is the midpoint of } \mathbf{ME}) \\ \mathbf{A} \wedge \mathbf{B} \wedge \mathbf{C} \wedge \mathbf{D} \neq 0. & (\mathbf{A}, \mathbf{B}, \mathbf{C}, \mathbf{D} \text{ are not coplanar}) \end{cases}$$

Under the order $\mathbf{A} \prec \mathbf{B} \prec \mathbf{C} \prec \mathbf{D} \prec \mathbf{M} \prec \mathbf{N} \prec \mathbf{E} \prec \mathbf{F} \prec \mathbf{O}$, the parametric triangular sequence is

$$\begin{cases} 2\mathbf{O} - \mathbf{A} - \mathbf{D} + \lambda\mathbf{A} + \lambda\mathbf{D} - \lambda\mathbf{B} - \lambda\mathbf{C}, \\ \mathbf{F} - \mathbf{D} + \lambda\mathbf{D} - \lambda\mathbf{B}, \\ \mathbf{E} - \mathbf{D} + \lambda\mathbf{D} - \lambda\mathbf{C}, \\ \mathbf{N} - \mathbf{A} + \lambda\mathbf{A} - \lambda\mathbf{C}, \\ \mathbf{M} - \mathbf{A} + \lambda\mathbf{A} - \lambda\mathbf{B}. \end{cases}$$

The non-degeneracy conditions are

$$\mathbf{A} \wedge \mathbf{B} \wedge \mathbf{C} \wedge \mathbf{D}, \qquad \mathbf{B} - \mathbf{A} \neq 0,$$

which are all guaranteed by the original hypothesis.

The conclusion of the theorem cannot be algebraized. However, it is obvious from the first term in the parametric triangular sequence. The theorem is proved actually by finding the locus equation of the point **O**. If the conclusion is stated more precisely as "the center **O** is on the line connecting the middle points of **AD** and **BC**," then the theorem may be proved directly (see Boy de la Tour et al. 1999).

8.8 Proving geometric theorems of constructive type

The algorithm of Li and Cheng provides a general principle for AGTP, and is complete in the sense that Wu's coordinate-based method is used. As the solution formulas are for linear (multi)vector equations, the algorithm can produce coordinate-free proofs mainly for geometric theorems whose Clifford algebraic representations do not involve non-linear (multi)vector equations. When it fails in finding a coordinate-free proof, the algorithm has to use coordinates and Wu's method for the involved polynomial computation and reduction. In this case, though the triangulation with Clifford algebraic calculus can often simplify the hypothesis expressions for the subsequent use of Wu's method, the conclusion cannot be proved by the reduction using the (parametric) triangular sequence. Therefore, it would be better to know *a priori* for which kinds of theorems proofs can be achieved without introducing coordinates (or without using Wu's method).

Based on the above observation, Wang (1997) restricted himself to a class of geometric theorems that can be stated constructively by introducing new geometric entities step by step, and that each newly introduced dependent geometric entity χ_i can be explicitly represented in terms of the previously constructed geometric entities $\boldsymbol{\mu}$ and $\chi_1, \ldots, \chi_{i-1}$

$$\chi_i = f_i(\boldsymbol{\mu}, \chi_1, \ldots, \chi_{i-1}), \quad 1 \leq i \leq r, \tag{8.2}$$

using Clifford algebraic operators. Here $\boldsymbol{\mu} = (\mu_1, \ldots, \mu_d)$ are free *geometric entities*, and $\chi_1, \ldots, \chi_r$ are constrained by the geometric hypotheses (called *dependents*). Examples of geometric entities are points, areas of triangles and other geometric invariants. A set of popular geometric constructions and the corresponding representations are provided in Wang (1997) and Fèvre and Wang (1998a). In fact, such representations can be easily obtained from the general solution formulas given in Sect. 8.5.

The advantage of using constructions and explicit representations lies in the fact that the steps of (multi)vector equation solving and triangulation are no longer needed during the proof. The proof process for any theorem within the class is a simple substitution of the expressions of the dependents into a Clifford algebraic expression

$$g = g(\boldsymbol{\mu}, \chi_1, \ldots, \chi_r) = 0$$

for the conclusion of the theorem. The theorem will be proved, under the non-degeneracy conditions that the denominator d_i of f_i in (8.2) does not vanish for $1 \leq i \leq r$, if the substituted expression g^* is reduced to 0.

In case one is allowed to used coordinates for the geometric entities as in Li and Cheng's algorithm, the theorem is true if and only if g^* automatically evaluates to 0. However, using coordinates is not of interest in the present case where geometric proofs are sought. So the question becomes how to simplify to 0 a Clifford algebraic expression that is identically 0 without using coordinates. An answer to this question was given by Wang (1997) using the techniques of term rewriting. The following simple algorithm can be implemented to prove geometric theorems from the class under discussion.

Wang's Algorithm: (Proving theorems of constructive type)

W1. Let $f_1^* = f_1$.

$$f_i^* = f_i^*(\boldsymbol{\mu}) = f_i|_{\chi_1 = f_1^*, \ldots, \chi_{i-1} = f_{i-1}^*} = f_i(\boldsymbol{\mu}, f_1^*, \ldots, f_{i-1}^*).$$

W2. Substitute all $\chi_i = f_i^*$ into g:

$$g^* = g^*(\boldsymbol{\mu}) = g|_{\chi_1 = f_1^*, \ldots, \chi_r = f_r^*} = g(\boldsymbol{\mu}, f_1^*, \ldots, f_r^*).$$

Let h be the numerator of g^*.

W3. Apply a term-rewriting system with suitably chosen rewrite rules to h. If h is rewritten to 0, then the theorem is true under the non-degeneracy conditions $d_i \neq 0$ for $1 \leq i \leq r$. If h does not evaluate to 0 when (pseudo-) coordinates are used, then the theorem is false under the conditions $d_i \neq 0$.

The proposed use of a rewriting system in step W3 has been further investigated in Fèvre and Wang (1998a,b, 1999) and Boy de la Tour et al. (1999). This will be discussed briefly in Sect. 8.9.1.

8.9 Clifford algebraic simplification and normalization

Suppose that one has obtained a triangular sequence from the hypothesis of a theorem and one wants to verify the conclusion from it. After substitutions and pseudo-divisions of the conclusion by the triangular sequence, one may obtain an expression g of independent variables. It is now a general question how to verify whether such a Clifford algebraic expression g is identically equal to zero. It is trivial to do so in coordinate representation, but not so in Clifford algebra. This is because g may involve various operators other than scalar product and summation; various computational rules in some order are needed to evaluate it to zero. Even in the case of coordinate representation, the evaluation of large Clifford algebraic expressions may also be quite time-consuming. In order to maintain geometrically invariant representation and computation, it is highly desirable to prove such identities with formal reasoning in the language of Clifford algebra. When applying Clifford algebra to geometric reasoning, Fèvre and Wang (1998a,b, 1999), Boy de la Tour et al. (1999), Yang (1998), and Yang et al. (1998, 1999) have proposed techniques to simplify and normalize Clifford algebraic expressions.

8.9.1 Term rewriting

One approach to achieve formal proving for Clifford algebraic identities is to use techniques and systems of term rewriting as suggested in step W3 of Wang's algorithm. Rewriting techniques and their applications form a sub-area of automated reasoning and have been well developed. The study of the rewriting approach for AGTP with Clifford algebra started in Wang (1997), where the following four groups of computational rules are proposed:

α. Trivial rules; β. Elementary rules;
γ. Advanced rules; δ. Coordinate rules.

Here are two examples of such rules:

- $(a^\sim)^\sim \to -a$ and $(a \wedge b)^\sim \to a \cdot b^\sim$ for $n = 2$ and any vectors a, b.
- $a \cdot (b \wedge C) \to a \cdot b C - b \wedge (a \cdot C)$ for any vectors a, b and multivector C.

To verify whether a Clifford algebraic expression of independent variables is identical to 0, Wang (1997) proposed to use the rules in α, β and γ successively. The rules in δ are applied when the first three groups of rules fail to evaluate the expression to 0. This simple rewriting process was implemented in Maple and used to prove a set of geometric theorems in Wang (1997).

Soon after that, Fèvre and Wang (1998a,b) refined the rewriting rules and studied the rewriting approach in more detail. In view of the rather poor performance of prevailing computer algebra systems such as Maple and Mathematica for rewriting, they chose to implement functions for Clifford term rewriting in Objective Caml, while keeping Clifford algebraic computation in Maple; an interface has been built between the two systems.

The current version of the (conditional) rewriting system for 3-dimensional space as described in Boy de la Tour et al. (1999) consists of four rewriting steps reviewed below.

Basic simplification. This is done by first

- rewriting trivial "zero terms" such as $A \wedge 0$ and $0^{\sim}$ to 0, and then
- applying the distributivity rules to sums, to which grades cannot be attributed.

By using the simplification rules, an expression is rewritten either to 0 or to a sum of terms whose subterms are non-zero and of grades between 0 and 3.

Reduction to normal forms. This step uses five groups of rules: distributivity rules, linearity rules, duality rules, ordering rules and reduction rules. By the system R consisting of these as well as the basic simplification rules, any Clifford algebraic expression can be rewritten to an irreducible term, called a *R-normal form*. Such R-normal forms have a simple grammar:

$$\begin{aligned}
L_0 &::= \tilde{\boldsymbol{K}} | \mathcal{V} \cdot \mathcal{V} | \mathcal{V} \cdot (\mathcal{V} \cdot \mathcal{V}^{\sim}) | L_0 + L_0 | L_0 L_0, \\
L_1 &::= \mathcal{V} | \mathcal{V} \cdot \mathcal{V}^{\sim} | L_1 + L_1 | L_0 L_1, \\
L_2 &::= \mathcal{V}^{\sim} | \mathcal{V} \wedge \mathcal{V} | L_2 + L_2 | L_0 L_2, \\
L_3 &::= I_3 | L_3 + L_3 | L_0 L_3,
\end{aligned}$$

where L_i is the language of terms of grade i, $\tilde{\boldsymbol{K}}$ an extension of the ground field and $\mathcal{V}$ a finite set of 3-dimensional vectors.

Further simplification. A R-normal form does not necessarily appear as 0 even if it is identically equal to 0 in the Clifford algebra. This is because the system R is not complete for rewriting terms to 0, due to the fact that R is not confluent. Thus, Boy de la Tour et al. (1999) proposed to perform standard algebraic computation using an external computer algebra system and to add geometric and ordering rules to R. The obtained system (with an innermost strategy for applying the rules) may provide simple normal forms for Clifford terms (though it is still not complete for proving equalities). The idea and difficulty of applying standard techniques to complete the rewriting system by adding rules for solving critical pairs were also discussed.

Pseudo-coordinates. The last resort to make the rewriting system complete for proving equalities is to use so-called *pseudo-coordinates* introduced by these authors. The main idea is take two non-zero and non-collinear vectors a, b and represent any other vector c in terms of a and b. With some simple reasoning, the following *pseudo-coordinate expression* for c may be established:

$$c = \alpha a + \beta b + \gamma (a \cdot b^{\sim})$$

with

$$\alpha = \frac{(c \cdot a)(b \cdot b) - (c \cdot b)(a \cdot b)}{\kappa},$$

$$\beta = \frac{(c \cdot b)(a \cdot a) - (c \cdot a)(a \cdot b)}{\kappa},$$

$$\gamma = \frac{c \cdot (a \cdot b^{\sim})}{\kappa};$$

$$\kappa = (a \cdot a)(b \cdot b) - (a \cdot b)^2.$$

The scalars α, β, γ are considered as the *pseudo-coordinates* of c relative to a and b. After all vectors other than a and b are replaced by their pseudo-coordinate expressions, a Clifford equality can always be proved by the rewriting system.

The use of pseudo-coordinates is equivalent to the introduction of the usual coordinates for vectors. However, pseudo-coordinates remain as expressions of the Clifford algebra in question.

8.9.2 Normal forms

Yang (1998) and Yang et al. (1998, 1999) developed another technique to normalize a Clifford algebraic expression. The rules used by them (in the 2-dimensional case) are listed as follows.

LAW 1. The inner product and the outer product are multilinear.

LAW 2. For any scalar λ and multivector A,

$$\lambda \wedge A = A \wedge \lambda = \lambda A, \qquad \lambda \cdot A = A \cdot \lambda = 0.$$

LAW 3. For any r-vector A_r and s-vector B_s,

$$A_r \cdot B_s = (-1)^{rs - \min(r,s)} B_s \cdot A_r, \qquad A_r \wedge B_s = (-1)^{rs} B_s \wedge A_r.$$

LAW 4. For any vectors a, b,

$$\begin{aligned} a \wedge (b \cdot I_2) &= a \cdot b\, I_2, \\ a \cdot (b \cdot I_2) &= (a \wedge b) \cdot I_2, \\ (a \cdot I_2) \cdot I_2 &= -a, \\ (a \wedge b) \cdot I_2\, I_2 &= -a \wedge b, \\ I_2 \cdot I_2 &= -1. \end{aligned}$$

LAW 5. For any vectors a, b, c, d,

$$\begin{aligned} a \cdot (b \wedge c) &= a \cdot b\, c - a \cdot c\, b, \\ (a \wedge b) \cdot (c \wedge d) &= b \cdot c\, a \cdot d - b \cdot d\, a \cdot c. \end{aligned}$$

LAW 6. For any vectors a, b, c, d,

$$\begin{aligned} (a \wedge b) \cdot I_2\; (c \wedge d) \cdot I_2 &= b \cdot d\, a \cdot c - b \cdot c\, a \cdot d, \\ (a \wedge b) \cdot I_2\; c \cdot I_2 &= b \cdot c\, a - a \cdot c\, b, \\ (a \wedge b) \cdot I_2\; c &= a \cdot c\, b \cdot I_2 - b \cdot c\, a \cdot I_2, \\ (a \wedge b) \cdot I_2\; c \wedge d &= b \cdot c\, a \cdot d\, I_2 - b \cdot d\, a \cdot c\, I_2. \end{aligned}$$

LAW 7. Let $\{e_1, e_2\}$ be an orthonormal basis of the plane; then for any vector v,

$$v = v_1 e_1 + v_2 e_2, \qquad I_2 = e_1 \wedge e_2,$$

where v_1, v_2 are the coordinates of v with respect to the basis.

Given a Clifford algebraic expression, Laws 1, 2, 3 are automatically applied. Then several functions are called to apply Laws 4, 5, 6. Now the expression is a sum of terms of the following forms:

$$\lambda \prod_{i=1}^{j} (a_i \cdot b_i); \qquad \lambda \prod_{i=1}^{j} (a_i \cdot b_i)((c \wedge d) \cdot I_2); \qquad \lambda \prod_{i=1}^{j} (a_i \cdot b_i) I_2;$$

$$\lambda \prod_{i=1}^{j} (a_i \cdot b_i)(c \wedge d); \qquad \lambda \prod_{i=1}^{j} (a_i \cdot b_i) c; \qquad \lambda \prod_{i=1}^{j} (a_i \cdot b_i)(c \cdot I_2).$$

Here λ is a scalar and a_i, b_j, c, d are vectors. The above forms are called *normal forms.* For many problems in practice, to verify whether an expression is identical to 0, it is enough to transform it into its normal form. Unfortunately, the six normal forms are not independent. To obtain a set of independent normal forms, one needs to apply Law 7 to introduce coordinates.

8.10 Gröbner bases and other related work

The problem of computing the Gröbner basis of a set of polynomials in bracket algebra (or Grassmann–Cayley algebra) has been extensively studied by Sturmfels (1993a) and some powerful algorithms have been proposed. The techniques have been used to prove theorems in projective geometry by Crapo and Richter-Gebert (1995).

For a Grassmann algebra, let $v_1, \ldots, v_m$ be vectors and $\boldsymbol{K}$ a basic field. Construct formally the polynomial ring $P = \boldsymbol{K}[v_1, \ldots, v_m]$ and take the commutation system

$$\mathcal{Q} = \{v_i \wedge v_j + v_j \wedge v_i | \ 1 \leq i < j \leq m\}.$$

It is easy to see that $\mathcal{I}(\mathcal{Q})$, the two-sided ideal in P generated by $\mathcal{Q}$, does not contain any non-zero commutative polynomial; therefore,

$$R = \boldsymbol{K}[v_1, \ldots, v_m]/\mathcal{I}(\mathcal{Q})$$

is a polynomial ring of solvable type. Let $\mathcal{I}$ be the two-sided ideal in R generated by

$$S = \{v_1 \wedge v_1, \ldots, v_m \wedge v_m\},$$

a two-sided Gröbner basis in R. Then, the ideal membership problem in the quotient algebra $R/\mathcal{I}$ can be treated as follows: Given the generating set F of a two-sided ideal $\mathcal{J}$ and a polynomial f in $R/\mathcal{I}$, construct a Gröbner basis G of $\mathcal{I} + \mathcal{J}$ from $S \cup F$ and compute a normal form h of f modulo G. Then $f \in \mathcal{J}$ if and only if $h = 0$. See Kandri-Rody and Weispfenning (1990) and Wang (1997) for more details.

Wang (1997, 1998b) investigated the application of the above method of non-commutative Gröbner bases in Grassmann algebra to AGTP. Using a Gröbner basis implementation in the MAS system, he was able to prove several geometric theorems and to show how the Gröbner basis technique may be applied to problems in computer vision.

Other related work include the early vector approaches of Chou et al. (1993b) and Stifter (1993), the recent study of area in Grassmann geometry by Fearnley-Sander and Stokes (1997), and the axiomatic setup of the so-called *Euclid ring*. The Euclid ring models the behavior of geometric objects and has Clifford algebra as one of its substructures in the plane (Fearnley-Sander 1999). It provides a way for formal representation and calculus in plane geometry.

Along the general line of producing geometrically interpretable proofs, we should mention the extensive work done by Chou et al. (1994a). A review of several point elimination methods proposed by them may be found in Chap. 7. The problem of finding geometric interpretations for algebraic expressions has been considered by Balbiani et al. (1994) and Fèvre (1997).

Chapter 9

PROVING BY EXAMPLES

Xiaorong Hou

In this chapter, we review the methods developed for proving plane geometric theorems by examples. The main purpose of this survey is to provide researchers with an overview of the technique. Some new, improved algorithms are discussed.

9.1 Introduction

Let us draw a triangle and the bisectors of its three angles on a sheet of paper. One can see that the three bisectors are concurrent if the construction has enough precision. Trying more examples, one can come to the conclusion that the bisectors of the three angles of any triangle are concurrent. Can experiments like this also be used in mathematics, as they are used in physics? In other words, can a mathematical theorem be proved by verifying numerical (or approximate) examples? The word "prove" is to be used in a deterministic sense, of course, not in a probabilistic sense (see Schwartz 1980).

In elementary plane geometry, diagrams or numerical examples are often used as counterexamples to disprove conjectures and used as intuitive models to find a rigorous proof of a theorem. On the other hand, experimental methods and inductive reasoning seem unacceptable in mathematics.

However, there is a class of mathematical theorems which can be proved by checking one or several special examples. In this case, concrete examples satisfying certain conditions imply general conclusions, and approximate computation may lead to exact results.

We know that there are many methods for deciding whether an algebraic equality in one variable is identical. For example, given a_i and b_k, to see whether the equality

$$(x + a_1)(x + a_2)(x + a_3)(x + a_4) = b_0x^4 + b_1x^3 + b_2x^2 + b_3x + b_4 \qquad (9.1)$$

holds, we can use any one of the following methods.

MATHEMATICS MECHANIZATION AND APPLICATIONS
ISBN 0-12-734760-7

(1) Formal calculation: expand the left-hand side of the equality, merge similar terms, and then compare the coefficients of the two sides of the equality.

(2) Multi-example numerical verification: take five distinct values $x_1, \ldots, x_5$ of x, and substitute each of them for x in (9.1), respectively. If (9.1) holds for each $x = x_i$ $(i = 1, \ldots, 5)$, then (9.1) always holds; for (9.1) is an algebraic equation with degree not greater than 4, which has 4 roots at most.

(3) Single-example numerical verification I: take a large enough positive number N and x_0 such that $x_0 > N$. If (9.1) holds for $x = x_0$, then (9.1) always holds. Otherwise, by transporting and packing terms we may infer the equation

$$c_0x^4 + c_1x^3 + c_2x^2 + c_3x + c_4 = 0 \tag{9.2}$$

from (9.1). Assume that $c_0 \neq 0$. If x_0 were the root of (9.2), we should have

$$|x_0| = |(c_1x_0^3 + c_2x_0^2 + c_3x_0 + c_4)/(c_0x_0^3)| \leq (|c_1| + \cdots + |c_4|)/|c_0|$$

when $|x_0| \geq 1$. Now take $N = (|c_1| + \cdots + |c_4|)/|c_0|$; then $|x_0| > N$ implies that x_0 is not a root of (9.2).

(4) Single-example numerical verification II: assume that (9.1) is of integral coefficients. If (9.1) holds for $x = p^{1/5}$, where p is a prime, then (9.1) must be an identity because $p^{1/5}$ does not satisfy any equation with integral coefficients and of degree less than 5.

These trivial results may be regarded as the basis of a new idea, *proving-by-examples*.

According to the idea of Wu (1978, 1984a,b) for mechanical theorem proving, we can reduce the proving of geometric theorems to that of algebraic identities (see Sect. 9.2, Theorem 9.1, Theorem 9.6 and Chap. 5).

Along the line of single-example numerical verification I, Hong (1986a,b,c) established a method for proving a class of plane geometric theorems by computing one example for each theorem, and discussed what a small number should be regarded to be absolutely equal to 0. Later, Wang (1988) clarified the irreducibility issue for Hong's method and analyzed the computational complexity of the method for a smaller class of geometric theorems, which are called theorems of *Hilbert intersection point type* by Wu (1982). He also showed that the inclusion relation of two varieties may be determined by only a finite number of points. Further discussion and implementation of Hong's method are pending.

From the viewpoint of multi-example numerical verification, Zhang et al. (1988, 1990) proposed the *parallel numerical method* (PNM) for proving a class of plane geometric theorems by computing a certain number of examples for each theorem. For the so-called *class L* (see Sect. 9.5) PNM involves integer operations only, so the results obtained are error-free. A prover based on PNM for geometric statements of class L was developed by Yang, Zhang, and Li (1992).

Based on single-example numerical verification II, Hou (1995; also Yang et al. 1996) introduced a *single-example experimental algorithm* (SEA) with a computer implementation to prove constructive geometric statements (with ruler and compass) by computing approximately one example for each theorem. The software works also for geometric statements involving reducibility.

Other approaches related to proving-by-examples have also been proposed for geometry theorem proving: for instance, Schwartz (1980) proposed the technique of probabilistic verification, of which a computer implementation for a class of plane geometry statements was made by Carrá Ferro et al. (1997); Collins (1975) introduced a semi-numerical algorithm, i.e., the algorithm of cylindrical algebraic decomposition; Yang et al. (1999) devised a semi-numerical algorithm to prove and discover geometric inequalities.

The method of proving-by-examples has received less attention than the work devoted to algebraic and logic methods. In spite of the great success of algebraic methods, research on proving-by-examples also remains attractive, because

- the idea provides a new methodology and has not been developed sufficiently,
- using counterexamples is one of the most efficient techniques for disproving conjectures,
- numerical computation uses less memory than symbolic computation, and
- as a new methodology, proving-by-examples has potential applications in other fields (see Yang et al. 1996).

However, a comprehensive treatment of proving-by-examples is far beyond what can be done in this chapter. We will describe mainly the basic principles of proving-by-examples and two algorithms, i.e., PNM and SEA mentioned above, with computer implementations.

9.2 The problem

A geometric statement without involving order relation may lead to several configurations involving order relations. Let us consider the following example.

On the three sides of a triangle ABC, three equilateral triangles BCA_1, CAB_1, and ABC_1 are drawn. Prove that the lines AA_1, BB_1, and CC_1 are concurrent.

In fact, the statement leads to 8 configurations as some of the equilateral triangles may be inward and the others outward to triangle ABC. The statement is true only for two configurations, that is, three equilateral triangles all inward or all outward. In this case, the statement is said to be *partially true.*

One problem with the above statement is that it does not mention certain *non-degeneracy* (ndg) conditions that are implicitly assumed. Generally speaking, a geometric theorem is true only under some non-explicitly given ndg conditions (see Wu 1978, 1984a,b, and Chap. 5). In this case, the statement (or theorem) is said to be *generically true.*

For a geometric statement the problem we are concerned with in this chapter is how to determine the configurations on which the statement is generically true. To see how to prove by examples, we need to describe the problem from an algebraic viewpoint.

Suppose that we know how to choose coordinate systems for transforming geometric statements into algebraic expressions. Let us consider only those statements whose hypotheses can be expressed as polynomial equations of the form

$$H_1(y_1, \ldots, y_n) = 0, \ldots, H_r(y_1, \ldots, y_n) = 0,$$

and whose conclusions can be expressed, without loss of generality, as a single polynomial equation of the form

$$C(y_1, \ldots, y_n) = 0,$$

where the variables $y_1, \ldots, y_n$ are geometric entities like the coordinates of points and all the polynomials have their coefficients in a fixed field $\boldsymbol{K}$ associated with the geometry.

By Wu–Ritt's well-ordering principle and changing notations for the variables appropriately, the set $PS = \{H_1, \ldots, H_n\}$ of polynomials may be transformed into a certain "triangular form," or more precisely an *ascending set*

$$AS: \quad \begin{cases} A_1(\boldsymbol{u}, x_1), \\ A_2(\boldsymbol{u}, x_1, x_2), \\ \quad \cdots\cdots \\ A_s(\boldsymbol{u}, x_1, x_2, \ldots, x_s), \end{cases}$$

where $\boldsymbol{u} = (u_1, \ldots, u_d)$ are regarded as parameters and x_j is called the *leading variable* of A_j for $j = 1, \ldots, s$. Correspondingly, $C(y_1, \ldots, y_n)$ can be rewritten as $G(\boldsymbol{u}, x_1, \ldots, x_s)$.

Recall the notations ini and Zero, which have been defined in Chap. 1. Let $J = \text{ini}(A_1) \cdots \text{ini}(A_s)$. We have

$$\text{Zero}(AS/J) \subseteq \text{Zero}(PS) \subseteq \text{Zero}(AS)$$

according to Wu (1984a, 1986a).

Put $\mu_j = \deg(A_j, x_j)$ for $j = 1, \ldots, s$, and $\mu = \mu_1 \cdots \mu_s$. Since the system of equations

$$A_1(\boldsymbol{u}, x_1) = 0, \ldots, A_s(\boldsymbol{u}, x_1, \ldots, x_s) = 0 \tag{9.3}$$

is in triangular form, theoretically speaking all the indeterminates $x_1, \ldots, x_s$ as algebraic functions of $\boldsymbol{u}$ can be solved one by one, successively. Let $\bar{\boldsymbol{K}}$ be an algebraic extension of $\boldsymbol{K}$ and algebraically closed. If a geometric statement is well-stated, (9.3) has μ solutions $q_1, \ldots, q_\mu$ in $\bar{\boldsymbol{K}}$, each of which corresponds to a configuration of the geometric statement. In this case, we call AS a *normal ascending set* (see Chap. 2 for a formal definition of the equivalent concept *regular set*). Throughout this chapter, we assume that all ascending sets, which express the hypotheses of well-stated geometric statements, are normal ascending sets, and denote the set of the solutions (branches) of (9.3) by $\mathcal{S}$, i.e.,

$$\mathcal{S} = \{q_1, \ldots, q_\mu\}.$$

The number of the branches (solutions) on which G vanishes is called the *dependency number* of G on AS. If this number is 0, the geometric statement is false. If this number is μ, the statement is generically true. Otherwise, the statement is partially true.

Thus, the problem under consideration is reduced to determining the dependency number of G on AS, and, furthermore, to determining the branches on which G vanishes.

9.3 The principles

This section consists of three subsections, each of which contains several theorems. These theorems, taken together, compose the basis of proving-by-examples. Section 9.3.1 shows how geometric theorem proving can be transformed into algebraic identity proving. Section 9.3.2 presents three methods to prove multivariate algebraic identities by example numerical verification and explains how to determine the dependency number by the three methods. In Sect. 9.3.3, we give some gap theorems, which state what, in the corresponding proving-by-examples, a small number should be regarded to be absolutely equal to 0.

9.3.1 The elimination algorithm

Denote by $\mathrm{res}(F, G, x)$ the resultant of any two polynomials F and G with respect to an indeterminate x, which is a determinant in terms of the coefficients of $F(x)$ and $G(x)$ to verify whether the two polynomials have a common solution.

Given an ascending set $AS = \{A_1, \ldots, A_s\}$ and another polynomial G, put

$$\begin{aligned} R_{s-1} &= \mathrm{res}(G, A_s, x_s), \\ R_{s-2} &= \mathrm{res}(R_{s-1}, A_{s-1}, x_{s-1}), \\ R_{s-3} &= \mathrm{res}(R_{s-2}, A_{s-2}, x_{s-2}), \\ &\cdots\cdots \\ R_1 &= \mathrm{res}(R_2, A_2, x_2), \\ R_0 &= \mathrm{res}(R_1, A_1, x_1). \end{aligned}$$

We call R_0 the *resultant* of G with respect to AS (see Yang, Zhang, and Hou 1992) and denote it simply by $\mathrm{res}(G, AS)$.

Let $\mathrm{ldeg}(P, \lambda)$ denote the lowest degrees of any polynomial P with respect to variable λ. We have the following criterion (see Yang, Zhang, and Hou 1992).

Theorem 9.1:

Given a normal ascending set $AS = \{A_1, \ldots, A_s\}$, a polynomial G and a new variable λ, put

$$k = \mathrm{ldeg}(\mathrm{res}(G + \lambda, AS), \lambda).$$

Then, the dependency number of G on AS is equal to k.

Theorem 9.1 may be used for automated theorem proving in geometry. More precisely, we have the following.

Corollary 9.2:

Assume that a geometric statement is well-stated, and let $A_1 = 0, \ldots, A_s = 0$ and $G = 0$ be the hypothesis and conclusion of the statement, respectively. Then, the statement is true for exact k of the μ configurations led by the statement if and only if

$$k = \mathrm{ldeg}(\mathrm{res}(G + \lambda, \{A_1, \ldots, A_s\}), \lambda).$$

Corollary 9.3:

Given a normal ascending set $\{A_1, \ldots, A_s\}$ and a polynomial G, none of the solutions of $A_1 = 0, \ldots, A_s = 0$ in $\bar{\boldsymbol{K}}$ satisfies $G = 0$ if and only if

$$\mathrm{res}(G, \{A_1, \ldots, A_s\}) \neq 0.$$

Lemma 9.4:

Let $F(x)$ be in $\boldsymbol{K}[x]$ with degree m, roots $x^{(1)}, x^{(2)}, \ldots, x^{(m)}$ and leading coefficient I, and $G(x) \in \boldsymbol{K}[x]$ with $\deg(G, x) = n$. Then

$$\mathrm{res}(F, G, x) = I^n \prod_{i=1}^{m} G(x^{(i)}) \in \boldsymbol{K}.$$

This well-known fact can be found in standard algebra textbooks.

Lemma 9.5: (Yang, Zhang, and Hou 1992)

Let the notations be as above, and set $\nu_s = \deg(G, x_s)$,

$$\nu_j = \deg(\mathrm{res}(G, \{A_{j+1}, A_{j+2}, \ldots, A_s\}), x_j)$$

for $j = 1, \ldots, s-1$, and

$$P_0(\boldsymbol{u}) = [\mathrm{ini}(A_1)]^{\nu_1} \prod_{x \in \mathcal{S}} [\mathrm{ini}(A_j)(\boldsymbol{u}, x)]^{\nu_j} \in \boldsymbol{K}[\boldsymbol{u}].$$

Then

$$\mathrm{res}(G, \{A_1, \ldots, A_s\}) = P_0(\boldsymbol{u}) \prod_{x \in \mathcal{S}} G(\boldsymbol{u}, x) \in \boldsymbol{K}[\boldsymbol{u}]. \tag{9.4}$$

It is not difficult to prove Lemma 9.5 by repeated use of Lemma 9.4.

Theorem 9.6:

Let the notations be as above, and let $AS = \{A_1, \ldots, A_s\}$ *be a normal ascending set,* G *a polynomial and* λ *a new variable,* $\sigma_k(\boldsymbol{u})$ *the* k*th elementary symmetric function of* $G(\boldsymbol{u}, q_1), \ldots, G(\boldsymbol{u}, q_\mu)$*, for* $r = 1, \ldots, \mu$*, i.e.*

$$\sigma_k(\boldsymbol{u}) = \sum_{1 \le i_1 < \cdots < i_k \le \mu} G(\boldsymbol{u}, q_{i_1}) \cdots G(\boldsymbol{u}, q_{i_k}),$$

and

$$\mathrm{res}(G + \lambda, AS) = P_0(\boldsymbol{u})\lambda^\mu + P_1(\boldsymbol{u})\lambda^{\mu-1} + \cdots + P_\mu(\boldsymbol{u}).$$

Then

$$P_k(\boldsymbol{u}) = P_0(\boldsymbol{u})\sigma_k(\boldsymbol{u}) \in \boldsymbol{K}[\boldsymbol{u}].$$

Proof:

The key is to prove Theorem 9.1, from which the other results follow.

When in (9.4) G is replaced by $G + \lambda$, the right-hand side becomes a polynomial $P(\lambda)$ with roots

$$-G(\boldsymbol{u}, q_i), \quad i = 1, \ldots, \mu.$$

The lowest degree of a polynomial is nothing but the number of the vanishing roots, so that among the μ roots, exactly k of them vanish. This completes the proof.

Definition 9.1: (Chou 1988; Wu 1984a,b)

An ascending set $\{A_1, \ldots, A_r\}$ *is said to be irreducible if each* A_i *is irreducible in the polynomial ring*

$$\boldsymbol{K}(\boldsymbol{u})[x_1, \ldots, x_i]/\mathrm{Ideal}(\{A_1, \ldots, A_{i-1}\}),$$

where Ideal$(\{A_1, \ldots, A_{i-1}\})$ *denotes the* ideal *generated by the polynomials* $A_1, \ldots, A_{i-1}$ *in* $\boldsymbol{K}(\boldsymbol{u})$.

Using the decomposition algorithm of Wu (1984a; Chap. 1), for any normal ascending set AS, we can get

$$\mathrm{Zero}(AS) = \mathrm{Zero}(AS_1/J_1) \cup \cdots \cup \mathrm{Zero}(AS_r/J_r),$$

where all the AS_i are irreducible ascending sets and each J_i is the product of the initials of the elements in AS_i.

Theorem 9.7: (Yang, Zhang, and Hou 1992)

Let the notations be as above and G a polynomial. Then for each i, G vanishes on AS_i if and only if $\mathrm{res}(G, AS) = 0$. *Moreover,*

$$\mathrm{res}(G, AS) = \mathrm{res}(G, AS_1) \cdots \mathrm{res}(G, AS_r).$$

See Sect. 2.4 for other relevant results about the resultant of a polynomial with respect to an ascending (or triangular) set.

9.3.2 Determining the dependency number by examples

To determine the dependency number, we have given a simple symbolic algorithm by Theorem 9.1 in Sect. 9.3.1. This section shows how to get the number by examples.

For simplicity, from now on we assume that every polynomial to be mentioned is of integer coefficients.

Definition 9.2:

Let $S_1, \ldots, S_d$ be d finite subsets of $\mathbb{Z}$ and t_i denote the cardinal number of S_i, for $i = 1, \ldots, d$. We call the Cartesian product of the above d subsets,

$$S = S_1 \times \cdots \times S_d,$$

*a d-*dimensional lattice array *on $\mathbb{Z}$, with* size $(t_1, \ldots, t_d)$.

Lemma 9.8:

Let $\Phi(u_1, \ldots, u_d)$ be a polynomial over $\mathbb{Z}$. Assume that the degree of Φ in u_i is not greater than n_i, for $i = 1, \ldots, d$. If there exists a d-dimensional lattice array S on $\mathbb{Z}$, with size $(n_1 + 1, \ldots, n_d + 1)$, such that Φ vanishes over S, i.e.

$$\Phi(\bar{u}_1, \ldots, \bar{u}_d) = 0$$

for any $(\bar{u}_1, \ldots, \bar{u}_d) \in S$, then Φ is identical to zero.

It is easy to prove this lemma by induction on d.

Lemma 9.9: (Zhou 1990)

Let $p_1, \ldots, p_d$ be d primes, pair-wise relatively prime, $n_1, \ldots, n_d$ natural numbers, and $\boldsymbol{Q}$ the field of rational numbers. Then

$$[\boldsymbol{Q}(p_1^{\frac{1}{n_1}}, \ldots, p_d^{\frac{1}{n_d}}) : \boldsymbol{Q}] = n_1 \cdots n_d.$$

Corollary 9.10:

Let $\Phi(u_1, \ldots, u_d)$ be a polynomial over $\mathbb{Z}$. Assume that the degree of Φ in u_i is not greater than n_i, for $i = 1, \ldots, d$. If

$$\Phi(p_1^{\frac{1}{n_1+1}}, \ldots, p_d^{\frac{1}{n_d+1}}) = 0,$$

then Φ is identical to zero.

We define the *weight* of a polynomial to be the sum of the absolute values of all the coefficients of its monomials.

Lemma 9.11: (Hong 1986b)

Let $\Phi(u_1, \ldots, u_d)$ be a polynomial over $\mathbb{Z}$. Assume that the total degree and weight of Φ are not greater than n and w, respectively. If

$$\Phi(nw, (nw)^{1+n}, \ldots, (nw)^{1+n+\cdots+n^{d-1}}) = 0,$$

then Φ is identical to zero.

Put $R_s = G$,

$$R_{k-1} = \text{res}(R_k, A_k, x_k), \quad k = 1, \ldots, s,$$

and $\text{res}(G, \{A_1, \ldots, A_s\}) = R_0$. The upper bounds of degrees of $\text{res}(G, \{A_1, \ldots, A_s\})$ in the u_i can be given by recursion inequalities as follows.

Lemma 9.12:

The following holds for $k = 1, \ldots, s$:

$$\deg(R_{k-1}, x_j) \leq \deg(A_k, x_k)\deg(R_k, x_j) + \deg(R_k, x_k)\deg(A_k, x_j),$$

$$j = 1, \ldots, k-1.$$

$$\deg(R_{k-1}, u_i) \leq \deg(A_k, x_k)\deg(R_k, u_i) + \deg(R_k, x_k)\deg(A_k, u_i),$$

$$i = 1, \ldots, d.$$

This lemma can be easily deduced from the definition of resultants.

By a *normal parameter point* $\boldsymbol{u}_0$ of *AS*, we mean that it satisfies

$$P_0(\boldsymbol{u}_0) \neq 0.$$

Theorem 9.13:

Let $AS = \{A_1, \ldots, A_s\}$ *be a normal ascending set and* G *a polynomial, and write*

$$P(\boldsymbol{u}, \lambda) = \mathrm{res}(G + \lambda, AS),$$

which depends on $\boldsymbol{u}$ *and* λ *only. Assume that*

$$\deg(P(\boldsymbol{u}, \lambda), u_i) \le n_i, \quad i = 1, \ldots, d,$$

and S *is a lattice array on* $\mathbb{Z}$ *with size* $(n_1 + 1, \ldots, n_d + 1)$ *which consists of normal parameter points of* AS*. Denote by* $\kappa(\boldsymbol{u}_0)$ *the dependency number of* $G(\boldsymbol{u}_0)$ *on the ascending set*

$$\{A_1(\boldsymbol{u}_0, x_1), \ldots, A_s(\boldsymbol{u}_0, x_1, \ldots, x_s)\}$$

for every normal parameter point $\boldsymbol{u}_0$ *of* AS*, and let*

$$\kappa = \min_{\boldsymbol{u}_0 \in S} \kappa(\boldsymbol{u}_0).$$

Then, κ *is equal to the dependency number of* $G(\boldsymbol{u}, x_1, \ldots, x_s)$ *on* AS*, i.e.*

$$\kappa = \mathrm{ldeg}(P(\boldsymbol{u}, \lambda), \lambda).$$

Proof:

By Lemma 9.5, we have

$$P(\boldsymbol{u}, \lambda) = P_0(\boldsymbol{u}) \prod_{x \in \mathcal{S}} (G(\boldsymbol{u}, x) + \lambda) = P_0(\boldsymbol{u})\lambda^{\mu} \;+\; P_1(\boldsymbol{u})\lambda^{\mu-1} + \cdots + P_{\mu}(\boldsymbol{u}).$$

Then, by the relation between roots and coefficients, $P_i(\boldsymbol{u}_0) = 0$ for $\boldsymbol{u}_0 \in S$ and $i > \kappa(\boldsymbol{u}_0)$. Making use of Lemma 4.1.2, we have $P_i(\boldsymbol{u}) \equiv 0$ for $i > \kappa$. Consequently,

$$\mathrm{ldeg}(P(\boldsymbol{u}, \lambda), \lambda) \le \kappa.$$

On the other hand, there is at least one $\boldsymbol{u}_0$ such that $P_{\mu-\kappa}(\boldsymbol{u}_0) \ne 0$, i.e. $P_{\mu-\kappa}(\boldsymbol{u})$ is a non-zero polynomial. This means that

$$\mathrm{ldeg}(P(\boldsymbol{u}, \lambda), \lambda) \ge \kappa.$$

Thus $\kappa = \mathrm{ldeg}(P(\boldsymbol{u}, \lambda), \lambda)$, and the proof is complete.

Similarly, we have the following.

Theorem 9.14:

Let $AS = \{A_1, \ldots, A_s\}$ be a normal ascending set and G a polynomial, and write

$$P(\boldsymbol{u}, \lambda) = \text{res}(G + \lambda, AS).$$

Assume that

$$\deg(P(\boldsymbol{u}, \lambda), u_i) \le n_i, \quad i = 1, \ldots, d,$$

and let

$$\boldsymbol{v} = (p_1^{\frac{1}{n_1+1}}, \ldots, p_d^{\frac{1}{n_d+1}}).$$

Denote by κ the dependency number of $G(\boldsymbol{v})$ on the ascending set

$$\{A_1(\boldsymbol{v}, x_1), \ldots, A_s(\boldsymbol{v}, x_1, \ldots, x_s)\}.$$

Then

(1) *v is a normal parameter point;*

(2) *κ is equal to the dependency number of $G(\boldsymbol{u}, x_1, \ldots, x_s)$ on AS, i.e.*

$$\kappa = \text{ldeg}(P(\boldsymbol{u}, \lambda), \lambda).$$

9.3.3 Determining the dependency number by experiments

A natural question that may arise is concerned within the computer representation of numbers. For simplicity we denote $G(\boldsymbol{u}_0, q_i(\boldsymbol{u}_0))$ by $G_i(\boldsymbol{u}_0)$ for $i = 1, \ldots, \mu$. If, via floating point computation, $G_i(\boldsymbol{u}_0)$ is 0 or close to 0 for some i, is $G_i(\boldsymbol{u}_0)$ equal to zero exactly or a number with small absolute value?

Applying the following three gap theorems, we can deal with such problems involving floating point computation.

Theorem 9.15:

Assume that $A_1, \ldots, A_s$ and G are of integer coefficients, and S is a lattice array on $\mathbb{Z}$, which consists of normal parameter points of the ascending set $\{A_1, \ldots, A_s\}$. Let $\sigma_r(\boldsymbol{u}_0)$ be the rth elementary symmetric function of

$$G_1(\boldsymbol{u}_0), \ldots, G_\mu(\boldsymbol{u}_0)$$

for $r = 1, \ldots, \mu$ and $\boldsymbol{u}_0 \in S$. Assume that $|P_0(\boldsymbol{u}_0)| \le D$, where P_0 is as in Theorem 9.6. Estimate $\sigma_r(\boldsymbol{u}_0)$ in the order of $r = \mu, \mu - 1, \ldots$ accurately such that the error is less than $\frac{1}{2D}$. If for a natural number k

$$|\sigma_r(\boldsymbol{u}_0)| < \frac{1}{2D} \text{ for } r > k \quad \text{and} \quad |\sigma_k(\boldsymbol{u}_0)| \ge \frac{1}{2D}$$

hold, then $\mu - k = \kappa(\boldsymbol{u}_0)$.

Proof:

By definition, $P_r(\boldsymbol{u}_0) = \sigma_r(\boldsymbol{u}_0)P_0(\boldsymbol{u}_0)$ for $r = 1, \ldots, \mu$. Since

$$|\sigma_k(\boldsymbol{u}_0)| \geq \frac{1}{2D}$$

and the error of calculation is less than $\frac{1}{2D}$, it is clear that $|\sigma_k(\boldsymbol{u}_0)| > 0$, so that $P_k(\boldsymbol{u}_0) \neq 0$. On the other hand, for $r > k$, if $\sigma_r(\boldsymbol{u}_0) \neq 0$ then $P_r(\boldsymbol{u}_0) \neq 0$ and $|P_r(\boldsymbol{u}_0)| \geq 1$ because $\boldsymbol{u}_0$ is an integral normal parameter point of $\{A_1, \ldots, A_s\}$ and P_r has integral coefficients. Consequently,

$$|\sigma_r(\boldsymbol{u}_0)| \geq \frac{1}{D}$$

because $|P_0(\boldsymbol{u}_0)| \leq D$. Therefore, the approximate value of $|\sigma_r(\boldsymbol{u}_0)|$ must be greater than $\frac{1}{2D}$ because the error of calculation is less than $\frac{1}{2D}$. Thus, $\sigma_r(\boldsymbol{u}_0) = 0$ holds for $r > k$, and so does $P_r(\boldsymbol{u}_0) = 0$. Therefore, $\kappa(\boldsymbol{u}_0) = \mu - k$.

Let $Q_i = u_i^{n_i+1} - p_i \in \mathbb{Z}[\boldsymbol{u}]$, $i = 1, \ldots, d$; U be the set of solutions of the system of equations $Q_1 = 0, \ldots, Q_d = 0$ for the variables $u_1, \ldots, u_d$;

$$\boldsymbol{v} = (p_1^{\frac{1}{n_1+1}}, \ldots, p_d^{\frac{1}{n_d+1}});$$

and

$$\mathrm{res}(G + \lambda, \{A_1, \ldots, A_s\}) = P_0(\boldsymbol{u})\lambda^\mu + P_1(\boldsymbol{u})\lambda^{\mu-1} + \cdots + P_\mu(\boldsymbol{u}).$$

Note that the p_i are primes, so the $P_i(\boldsymbol{u})$ are polynomials ($i = 1, \ldots, d$).

Obviously, $\{Q_1, \ldots, Q_d\}$ is a normal ascending set in the variables $u_1, \ldots, u_d$.

Lemma 9.16:

Let $A_i = \prod_{\boldsymbol{u} \in U, \boldsymbol{u} \neq \boldsymbol{v}} P_i(\boldsymbol{u})$, $i = 0, 1, \ldots, \mu$; *then*

$$R_i = P_i(\boldsymbol{v})A_i = \mathrm{res}(P_i(\boldsymbol{u}), \{Q_1, \ldots, Q_d\}) \in \mathbb{Z}.$$

It is not difficult to prove Lemma 9.16 by using Lemma 9.5.

Theorem 9.17:

Let the notations be as above and assume that $|A_i| \leq D$. *Estimate* $P_r(\boldsymbol{v})$ *in the order of* $r = \mu, \mu - 1, \ldots$ *accurately such that the error is less than* $\frac{1}{2D}$. *If for a natural number* k

$$|P_r(\boldsymbol{v})| < \frac{1}{2D} \quad \text{for } r > k \quad \text{and} \quad |P_k(\boldsymbol{v})| \geq \frac{1}{2D}$$

hold, then $\mu - k = \kappa$.

Proof:

Since $|P_k(\boldsymbol{v})| \geq \frac{1}{2D}$ and the error of calculation is less than $\frac{1}{2D}$, it is clear that $|P_k(\boldsymbol{v})| > 0$. On the other hand, for $r > k$, if $P_r(\boldsymbol{v}) \neq 0$ then according to Lemma 9.9 $A_r \neq 0$ and $|P_r(\boldsymbol{v})A_r| = |R_r| \geq 1$ because R_r is integer. Consequently,

$$|P_r(\boldsymbol{v})| \geq \frac{1}{D}$$

because $|A_r| \leq D$. Therefore, the approximate value of $|P_r(\boldsymbol{v})|$ must be greater than $\frac{1}{2D}$ because the error of calculation is less than $\frac{1}{2D}$. Thus, $P_r(\boldsymbol{v}) = 0$ holds for $r > k$. It follows that $\kappa = \kappa(\boldsymbol{v}) = \mu - k$.

The above two theorems give two algorithms based on floating point computation to determine the dependency number.

By the definition of $P_0(\boldsymbol{u}_0)$,

$$P_0(\boldsymbol{u}) = [\mathrm{ini}(A_1)]^{\nu_1} \prod_{x \in \mathcal{S}} [\mathrm{ini}(A_j)(\boldsymbol{u}, x)]^{\nu_j},$$

we can find a upper bound of $P_0(\boldsymbol{u}_0)$ by estimating $\mathrm{ini}(A_j)$ and ν_j for $j = 1, \ldots, s$.

According to Theorem 9.7, we could give a gap theorem to decide whether a polynomial vanishes on a certain branch of a normal ascending set or not. But it is difficult to determine the irreducible decomposition of an ascending set by using only numerical algorithm, so we give the following theorem to decide that a polynomial does not vanish on a certain branch of a normal ascending set.

Theorem 9.18:

Let $G(\boldsymbol{u}_1, \ldots, u_d)$ be a continuous function of $u_1, \ldots, u_d$. For given complex numbers $z_1, \ldots, z_d$, if

$$|\mathrm{evalf}(G(z_1, \ldots, z_d))| > |\mathrm{error}(G(z_1, \ldots, z_d))|,$$

where $\mathrm{evalf}(G(z_1, \ldots, z_d))$ *means to evaluate to* $G(z_1, \ldots, z_d)$ *using floating-point arithmetic and* $\mathrm{error}(G(z_1, \ldots, z_d))$ *means to estimate* $\mathrm{evalf}(G(z_1, \ldots, z_d))$ *an error bound, then*

$$G(z_1, \ldots, z_d) \neq 0.$$

9.4 Algorithms

This section presents two numerical algorithms for computing the dependency number of a polynomial on a normal ascending chain. The first one is based upon Theorem 9.17 and Theorem 9.18, and the other is based upon Theorem 9.15. They are similar in form, but we describe them in different styles. Furthermore, the first algorithm S may decide the configurations led by a statement, on which the statement are generically true.

For a geometric statement in the two classes defined in Sect. 9.5, we may easily transform it into an algebraic statement, of which the hypothesis corresponds to a normal ascending set. Therefore for such a geometric statement, each of the two algorithms below embodies a method of proving by examples.

Algorithm S: (SEA)

Given a polynomials G and a normal ascending set $\mathit{AS} = \{A_1, \ldots, A_s\}$, this algorithm computes C, the set of the branches, on which G vanishes.

S1. For $i = 1, \ldots, d$ choose n_i such that $n_i \geq \deg(P(\boldsymbol{u}, \lambda), u_i)$.

S2. Set $N \leftarrow 0$ and $\mu \leftarrow \deg(A_1, x_1) \cdots \deg(A_s, x_s)$.

For $i = \mu, \ldots, 0$ do:

S2$_0$. Set

$$\begin{aligned}&\text{Digits} \leftarrow 10,\\ &\boldsymbol{v} \leftarrow (p_1^{1/n_1}, \ldots, p_s^{1/n_s}),\\ &D \leftarrow |\text{evalf}(A_i)| + |\text{error}(A_i)|,\\ &a \leftarrow |\text{evalf}(P_i(\boldsymbol{v}))|,\\ &e \leftarrow |\text{error}(P_i(\boldsymbol{v}))|.\end{aligned}$$

If $a > e$ then go to step S3. If $a + e \geq \frac{1}{D}$ then set

$$\text{Digits} \leftarrow \text{Digits} + 10$$

and repeat step S2$_0$ for this i else set $N \leftarrow N + 1$.

S3. If $N = 0$, then set $C \leftarrow \varnothing$ and stop. Let C be the set of branches of AS at $\boldsymbol{u} = \boldsymbol{v}$. If $N = \mu$, then stop.

While $|C| > N$ do:

For $c \in C$ do: if $|\text{evalf}(G(\boldsymbol{v}))| > \text{error}(G(\boldsymbol{v}))$ then set $C \leftarrow C \setminus \{c\}$.

Set Digits $\leftarrow$ Digits + 10.

In the above algorithm, evalf($G(\boldsymbol{v})$) means to evaluate to $G(\boldsymbol{v})$ using floating-point arithmetic, error($G(\boldsymbol{v})$) means to estimate evalf($G(\boldsymbol{v})$) an error bound, Digits means the digits of floating-point arithmetic, and $|C|$ is the number of elements of C. See Sect. 9.3.3 for the definitions of the other notations.

Algorithm P: (PNM)

Given a polynomials G and a normal ascending set $AS = \{A_1, \ldots, A_s\}$, this algorithm computes the dependency number N of G on AS.

P1. Estimate the upper bounds for degrees of res(G, AS), i.e. $\deg(P(\boldsymbol{u}, T), u_i)$ for $i = 1, \ldots, d$.

P2. Take n_i such that $\deg(P(\boldsymbol{u}, T), u_i) \leq n_i$ for $i = 1, \ldots, d$ and let S be a lattice array consists of normal parameter points, with size $(n_1+1, \ldots, n_d+1)$.

P3. For every $\boldsymbol{u}_0 \in S$, do the following: Substitute $\boldsymbol{u}_0$ for $\boldsymbol{u}$ in the system of equations $A_1 = 0, \ldots, A_s = 0$, which gives

$$\begin{cases} A_1(\boldsymbol{u}_0, x_1) = 0, \\ A_2(\boldsymbol{u}_0, x_1, x_2) = 0, \\ \quad \cdots\cdots \\ A_s(\boldsymbol{u}_0, x_1, x_2, \ldots, x_s) = 0. \end{cases} \tag{9.5}$$

Let the solutions of (9.5) for $x_1, \ldots, x_s$ be

$$x^{(i_1\, i_2\, \cdots\, i_s)}(\boldsymbol{u}_0) = (x_1^{(i_1)}(\boldsymbol{u}_0), x_2^{(i_1\, i_2)}(\boldsymbol{u}_0), \ldots, x_s^{(i_1\, i_2\, \cdots\, i_s)}(\boldsymbol{u}_0)),$$

where i_j ranges from 1 to μ_j, for $j = 1, \ldots, s$. Then, substitute the solutions for $(x_1, \ldots, x_s)$ in $G(\boldsymbol{u}_0, x_1, \ldots, x_s)$ to determine the number $\kappa(\boldsymbol{u}_0)$ of the solutions which make $G = 0$.

P4. $N \leftarrow \min\limits_{\boldsymbol{u}_0 \in S} \kappa(\boldsymbol{u}_0)$.

Remark 9.1:

Of course we need not get the detailed expression for $P(\boldsymbol{u}, T)$. Lemma 9.12 gives a simple algorithm for this step, without resultant computations. It is very easy to write a program for this and, in many cases, the computation is so simple that it can be done by hand.

Remark 9.2:

The elements of S can be chosen arbitrarily; only the the size is important.

Remark 9.3:

Since the $\boldsymbol{u}_0$ are constants, system (9.5) may be easily solved by computer.

Remark 9.4:

By Theorem 9.15, the dependency number κ of $G(\boldsymbol{u}, x_1, \ldots, x_s)$ on AS is obtained immediately.

Algorithm P has been implemented by Yang, Zhang, and Li (1992) and is efficient in practice. However, it has a defect — how to take S so that all its elements are normal parameter points. This problem has not yet been fixed theoretically.

9.5 Geometric statements

Without restriction on the type of statements, a lot of numerical computing, sometimes with high complexity, may be needed for our purpose of proving by examples. However, if one restricts oneself to the statements within some special classes, the corresponding algorithms may be of lower complexity, so that they can be used to prove such statements efficiently.

Constructive geometric statements with ruler and compass. Algorithm S given in Sect. 9.4 has been implemented for constructive geometric statements with ruler and compasses (class C) whose definition is as follows.

A *construction* is one of the following six ways of introducing points.

C_1. Take an arbitrary point on a plane.

C_2. Take an arbitrary point on a line.

C_3. Take an arbitrary point on a circle.

C_4. Take the intersection of two lines.

C_5. Take the intersection of a line and a circle.

C_6. Take the intersection of two circles.

In addition, suppose that for any two points, the line passing through them and the circles centered at one point and passing through the other point exist.

A *constructive series* is represented by a list

$$C = (c_1, \ldots, c_k),$$

where $c_1, \dots, c_k$ are k constructions such that the point(s) introduced by each c_i must be different from the points introduced by c_j, $j = 1, \dots, i-1$, and the other points occurring in c_i must have been introduced before.

A *conclusion* is represented by $P_1 = P_2$, where P_1 and P_2 are two points, which can be introduced by two different constructive series, respectively. A *constructive statement* consists of a constructive series and a conclusion. C is also called the *hypothesis* of the statement.

In general, either of C_5 and C_6 introduces two points, which are indistinguishable in the geometry without involving order relation, and therefore a statement may lead to several configurations involving order relations.

A prover that implements Algorithm S translates geometric statements of class C into algebraic ones automatically.

Geometric statements of class L. Algorithm P described in Sect. 9.4 has been implemented for geometric statements of *class L* whose definition is as follows.

A *construction* in the geometry of class L is one of C_1, C_2, C_4 above or one of the following four ways of introducing points.

C_3'. Take an arbitrary point on a single circle specially designated.

C_5'. Take one intersection point of a line and a circle, while the other intersection point has been taken.

C_6'. Take one intersection point of two circles, while the other intersection point has been taken.

C_7'. Take a dividing point of a segment, whose endpoints A and B have been taken, with rational ratio (i.e., $PA : PB = m : n$, where m and n are integers).

For the definitions of other concepts such as *constructive series*, *conclusion*, and *constructive statement*, see class C.

In geometry of class L, every construction introduces only one point, and therefore a statement leads to one configuration. In this case, every statement can be verified simply by integer operations, so the results obtained are error-free.

A prover implementing Algorithm P of Yang, Zhang and Li (1992) also translates geometric statements of class L into algebraic ones automatically. Class L is narrower than the class of constructive geometric statements with ruler and compass. In practice, many well-known and non-trivial theorems can be proven by this prover, however. Amongst the 512 theorems proven by the prover of Chou (1988), most of them can be stated as theorems in class L. See Yang et al. (1992) for more details. Wang (1988) considered another smaller class of linearly constructive geometric statements and gave some bound for using the method of single-example numerical verification without implementation.

9.6 Experiments

We have implemented Algorithm S in C on PC, and with our software *SEA*, a number of non-trivial theorems including the Simson theorem, Butterfly theorem, Feuerbach theorem and Steiner theorem have been proved.

Example 9.1: (Simson theorem)

Let D, A, B, C be four points on a circle and E, F, G the perpendicular feet of D on lines BC, CA, AB, respectively. Show that the points E, F, G are collinear.

Input:

circularpoint(D)	
circularpoint(A)	
circularpoint(B)	
circularpoint(C)	(Take four points D, A, B, C on a circle)
foot(D, B, C, E)	(Take the perpendicular foot E of D on line BC)
foot(D, C, A, F)	(Take the perpendicular foot F of D on line CA)
foot(D, A, B, G)	(Take the perpendicular foot G of D on line AB)
collinear(E, F, G)	(Show that E, F, G are collinear)

Output: "The theorem is generically true."

The running time is almost 0 second on a PII/333.

Example 9.2: (Butterfly theorem)

Let A and D be two points on a circle O, and M the midpoint of AD. Draw two secant lines of circle O, passing through M and intersecting the circle at B, E and C, F, respectively. Assume that lines BF and CE intersect AD at P and Q. Show that M is the midpoint of the segment PQ.

Input:

circle(O)	(Draw a circle with center O)
circularpoint(A)	(Take a point A on circle O)
circularpoint(D)	(Take a point D on circle O)
midpoint(A, M, D)	(Take the midpoint M of AD)
circularpoint(B)	(Take a point B on circle O)
circularpoint(C)	(Take a point C on circle O)
secant(O, B, E, M)	(Draw two secant lines of O, passing through M and
secant(O, C, F, M)	intersecting the circle at B, E and C, F, respectively)
interll(B, F, A, D, P)	(Take the intersection P of lines BF and AD)
interll(C, E, A, D, Q)	(Take the intersection Q of lines CE and AD)
midpoint(P, M, Q)	(Show that M is the midpoint of segment PQ)

Output: "The theorem is generically true."

The running time is also almost 0 second on the same PII/333.

Example 9.3: (Pascal theorem)

Let A, B, C, D, E, F be six points on a circle, $P = AB \cap DE$, $Q = BC \cap EF$, and $R = CD \cap FA$, where $\cap$ denotes the intersection. Show that the points P, Q, R are collinear.

Input:

circularpoint(A)	
circularpoint(B)	
circularpoint(C)	
circularpoint(D)	
circularpoint(E)	
circularpoint(F)	(Take six points A, B, C, D, E, F on a circle)
interll(A, B, D, E, P)	(Take the intersection P of lines AB and DE)
interll(B, C, E, F, Q)	(Take the intersection Q of lines BC and EF)
interll(C, D, F, A, R)	(Take the intersection R of lines CD and FA)
collinear(P, Q, R)	(Show that P, Q, R are collinear)

Output: "The theorem is generically true."

The running time is 1 second on the PII/333.

Example 9.4: (Nine-point circle)

Let D, E, F be the midpoints of the segments BC, CA, AB, respectively, and L the perpendicular foot of A on BC. Show that the four points D, E, F, L are concyclic.

Input:

point3(A, B, C)	(Take three points on a plane)
midpoint(B, D, C)	(Take the midpoint D of BC)
midpoint(C, E, A)	(Take the midpoint E of CA)
midpoint(A, F, B)	(Take the midpoint F of AB)
foot(A, B, C, L)	(Take the perpendicular foot L of A on line BC)
concyclic(D, E, F, L)	(Show that D, E, F, L are concyclic)

Output: "The theorem is generically true."

The running time is 2 seconds on the PII/333.

Example 9.5:

On the three sides of a triangle ABC, three equilateral triangles BCA_1, CAB_1, and ABC_1 are drawn. Show that the lines AA_1, BB_1, and CC_1 are concurrent.

Input:

point3(A, B, C)	(Take three points on a plane)
intercc(A, B, B, A, C_1)	(Take the intersection C_1 of circles AB and BA)
intercc(B, C, C, B, A_1)	(Take the intersection A_1 of circles BC and CB)
intercc(C, A, A, C, B_1)	(Take the intersection C_1 of circles CA and AC)
interll(A, A_1, B, B_1, D)	(Take the intersection D of lines AA_1 and BB_1)
collinear(C, D, C_1)	(Show that C, D, C_1 are collinear)

where circle AB means the circle centered at A and passing through B.

Output: "The theorem is generically true on 2 of the 8 branches."

The running time is 2 seconds on the PII/333.

Acknowledgments. The work was supported in part by National "973 Project" of China, by the Special Funds of Academia Sinica, and by CNRS and Academia Sinica under a cooperation project between LEIBNIZ and CICA. The author thanks Dongming Wang for his helpful suggestions on an earlier version of this chapter.

Chapter 10

SEARCH METHODS REVISITED

Xiao-Shan Gao

In this chapter, we review the work of automated geometry theorem proving based on heuristic search methods and report some recent advances in this direction, especially the *geometry deductive database* which shows surprising power in terms of proving and discovering difficult geometric theorems. We also discuss several key issues in developing an efficient geometry theorem prover with heuristic search methods.

10.1 Introduction

There are mainly two approaches to automated geometry theorem proving (AGTP): the synthetic approach based on *search and heuristics* initiated by H. Gelernter (1959) in the late 1950s and the approach based on *algebraic computation* initiated by A. Tarski (1951) in the late 1930s, reshaped and popularized in the 1980s by W.-t. Wu (1978, 1984a,b). Most of the recent work on AGTP is inspired by Wu's method and focuses on using algebraic methods. Generally speaking, the algebraic approaches are very powerful decision procedures. Does this make the search methods obsolete? The answer is negative. Despite its "weakness," the methods based on search and heuristics are still worth improving, because this may lead to techniques useful to automated reasoning in the general case. Even for AGTP, improving the synthetic methods has many positive aspects:

(1) Proofs produced by synthetic methods are generally easier to understand than proofs given by algebraic computations.

(2) Using search methods, it is possible to generate multiple and shortest proofs for a given geometric theorem.

(3) Using predicates only (no algebraic computation) makes the reaching of fixpoints possible (see Sect. 10.3.2). As a result, new theorems may be discovered.

MATHEMATICS MECHANIZATION AND APPLICATIONS
ISBN 0-12-734760-7

(4) Although algebraic methods can prove a much greater number of theorems, there still exist theorems (Example 10.3) which can be proved by the synthetic approaches elegantly, but have not been proved with the algebraic approaches because of the excessively large computer memory needed.

In this chapter, we review some of the main work and techniques for the synthetic approach of AGTP.

Using the search method to prove geometric theorems goes back to as early as the late 1950s in the work by Gelernter et al. (1960), which has been considered classic in the field of artificial intelligence (AI). While several basic concepts of AGTP are introduced, the prover reported in this work seems unable to prove a large number of moderately difficult theorems. It is noted by Feigenbaum and Feldman (1963) that the "fascination with mechanical theorem proving for most of the researchers working in this area lies less with the end (the production of theorems, perhaps new and important) than with the means (a thorough understanding of the organization of information processing activity in mathematical discovery)." In our opinion, both the ends and the means are important, and the lesser success of Gelernter's work in proving difficult theorems has its reasons. We will discuss some of the main issues involved in Sect. 10.5.

Gelernter's work led to many subsequent investigations on AGTP using similar approaches (Evans 1968; Gilmore 1970; Anderson 1981; Coelho and Pereira 1986; Koedinger and Anderson 1990). Although many improvements were made, they are all within the framework presented by Gelernter. The theorem prover of Nevins (1975) is a notable exception, in which a combination of forward chaining (see next section for definition) and backward chaining is used with emphasis on the forward chaining. As a consequence of using forward chaining, numerical diagrams are not used, because each derived fact in the forward chaining is valid.

In Chou et al. (1995c), a prover is implemented based on the theory of deductive databases, which escapes Gelernter's framework completely and is the first synthetic prover capable of proving a large number of difficult geometric theorems. This prover can be used to find the *fixpoint* for a geometric configuration, i.e., the system can find all the properties of the configuration that can be deduced using a fixed set of geometric rules.

In Chou et al. (1994a), the *area method* for automated generation of human-readable proofs of geometric theorems is introduced. This method is deterministic in that the method eliminates points in a rigid way. In Chou et al. (1996b,c), a relaxed search strategy is used to introduce some kind of non-determinism into the area method. Using this strategy has two positive aspects: (1) it allows the program to generate multiple and the shortest proofs for the same theorem; (2) the area method can be extended to non-constructive geometric statements. Most of the previous methods of AGTP generally produce one proof. In the case of the area method, the proofs generated are short and readable, so one can talk about generating multiple and shortest proofs for geometric theorems; this is important for application of the method to *geometry education.*

In the rest of this chapter, we shall discuss Gelernter's work in Sect. 10.2, the deductive database approach in Sect. 10.3, the method for generating multiple and shortest proofs in Sect. 10.4, and various technical issues in synthetic methods in Sect. 10.5.

10.2 Gelernter's geometry theorem proving machine

Gelernter's geometry theorem proving machine (Gelernter 1959) uses a *backward chaining search.* Let G_0 be the conclusion to be established by the proof. It will be called the problem goal. If G_i is a formal statement with the property that G_{i-1} may be immediately inferred from G_i, then G_i is said to be a sub-goal of order i for the problem. The problem solving graph (Fig. 10.1) has as nodes the G_i, with each G_i joined to at least one G_{i-1} by a direct link. The problem is solved when any G_i can be immediately inferred from the premises of the problem and axioms. If, as is generally the case in geometry, a given sub-goal is a conjunction of statements, the graph splits at that point, and each parallel sub-goal must be separately established.

Without any improvements, this kind of brute force search has been shown to be too much time-consuming for so simple a logic as propositional calculus (Newell et al. 1963). It is a fortiori out of the question for any of the more interesting logics. A remaining alternative is to have the machine rely upon heuristic methods.

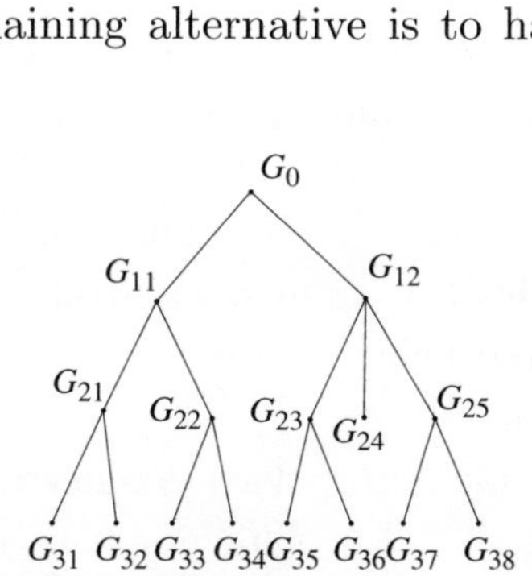

Figure 10.1

The main heuristic of Gelernter is the use of numerical diagrams as semantic models. There are two benefits the prover derives from a numerical diagram.

(1) The diagram is used as a filter to reject goals not consistent with its numerical representation as early as possible. This will reduce the search space drastically, because without this heuristic, most of the deduction paths generated will end up as false results.

(2) The diagram is used to determine order relations among points, or points and lines. These relations are necessary for the prover to find a proof.

The first benefit is very important to the backward search and is not useful in the forward chaining (see Sect. 10.3). The second benefit is related to the problem of producing diagram-independent proofs, of which a discussion can be found in Sect. 10.5.2.

With the backward chaining search and the main heuristic determined, now the search for a proof actually becomes the search for a valid sub-goal in the problem solving tree. There are two methods of doing this search: the *breadth-first search* and the *depth-first search.*

- In the breadth-first search, one will search the sub-goals in order $1, 2, \ldots$ respectively. In Fig. 10.1, the search order is: $G_0, G_{11}, G_{12}, G_{21}, \ldots$.

- In the depth-first search, one searches the left-most branch of the problem solving tree until one cannot go further. If the last sub-goal is not established to be valid, this sub-goal is deleted and one goes back to its parent sub-goal. Now the search continues with this sub-goal as the starting point and with the same style. The step of returning to the parent of a sub-goal is called *backtracking*. In Fig. 10.1, the search order is $G_0, G_{11}, G_{21}, G_{31}, G_{32}, \ldots$.

The breadth-first search is *complete* in the sense that if a valid sub-goal does exist in the tree then it will be found by the breadth-first search. On the other hand, if the search tree is infinite then the depth-first search may end up in an infinite branch without a valid sub-goal and thus cannot find the valid sub-goal in other branches. In Gelernter's proving machine, a breadth-first search is used. Wang (1991c) suggested using algebraic methods to reject false sub-goals without further investigation.

In his proving machine, Gelernter used geometric axioms about congruent triangles as the basic geometric rules. This might be one of the reasons that his proving machine did not succeed in proving many difficult geometric theorems.

Without adding techniques about auxiliary points, the rules about triangle congruence can be used to prove only a very limited number of high-school level theorems involving straight lines. Most of the basic results in geometry such as the centroid theorem, the orthocenter theorem, and Simson's theorem are beyond the scope of these rules. In Sect. 10.5, we will discuss the concerns with selecting a set of "good" geometric rules.

Gelernter's proving machine was used to prove more than fifty theorems from the high school textbooks. These theorems are quite simple compared with the theorems proved by the current theorem provers based on algebraic methods (Chap. 5) or the deductive database method (Sect. 10.3). The following is one of the most difficult theorems proved by the machine.

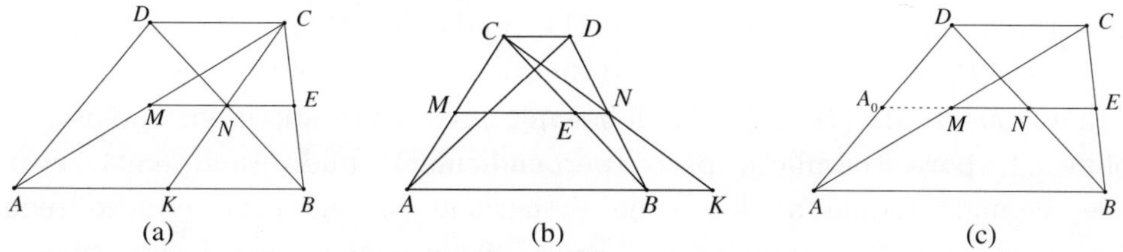

Figure 10.2

Example 10.1:

$ABCD$ is a trapezoid such that $AB \parallel CD$. M and N are the midpoints of AC and BD. Let E be the intersection of MN and BC. Show that E is the midpoint of BC (Fig. 10.2a).

The proving machine is unable to find a proof for the theorem at the present form. Then a new point $K = CN \cap AB$ is added by the user. Now, the proving machine is able to find a proof based on the key fact that $\triangle NBK$ is congruent to $\triangle NDC$. Also, many facts, such as $\angle ABD = \angle CDB$, are derived from the numerical diagram without a rigorous proof.

Now it is widely known that traditional synthetic proofs of geometric theorems may be not rigorous due to the missing of *non-degeneracy* or *subsidiary conditions* (Wu 1984a,b; Chap. 5). The above trapezoid theorem becomes false or meaningless when $ABCD$ is a parallelogram. This unexpected (not really) degenerated case was discovered automatically by an algebraic prover (see Wang 1996c).

10.3 A deductive database approach

Most of the subsequent work of AGTP based on heuristics uses a similar approach to that of Gelernter's: both use a backward chaining search and axioms about congruent triangles. The theorem prover by Nevins (1975) uses a combination of forward chaining and backward chaining. He also used rules about congruent triangles. In Chou et al. (1995c), a geometry theorem prover based on the theory of deductive database is implemented which uses forward chaining to achieve fixpoints and a set of geometric rules about full-angles.

10.3.1 Geometric rules

Although not specified explicitly, Gelernter used a special kind of deduction rules in his proving machine: Horn clauses. A rule is called a *definite Horn clause* if it has the following form:

$$Q(x) :\text{-}\quad P_1(x), \ldots, P_k(x) \qquad \text{meaning } \forall x[P_1(x) \wedge \cdots \wedge P_k(x) \Rightarrow Q(x)]$$

where the x are points occurring in the predicates $P_1, \ldots, P_k$, and Q. Also $P_1(x) \wedge \cdots \wedge P_k(x)$ and $Q(x)$ are called the *body* and *head* of the rule respectively.

In Chou et al. (1995c), the following predicates are used: points, coll (collinear), para (parallel), perp (perpendicular), midp (midpoint), cyclic, circle, eqangle (equal angle), cong (congruent of segment), eqratio (equal ratio), simtri (similar triangle), and contri (congruent triangle). The rules are highly complicated from the viewpoint of a deductive database (Bancilhon and Ramakrishnan 1986): all the predicates are mutually recursive and most of them are not linear. The central concept is eqangle. Here the angle is not the ordinary angle but the full-angle. Intuitively, a *full-angle* $\angle[u, v]$ is the angle from line u to line v. Two full-angles $\angle[l, m]$ and $\angle[u, v]$ are equal if there exists a rotation K such that $K(l) \parallel u$ and $K(m) \parallel v$. If A, B and C, D are distinct points on l and m respectively, then $\angle[l, m]$ is denoted by $\angle[AB, CD]$.

The introduction of full-angles simplifies the predicate of the angle congruence. For instance, we have the following rule about parallel lines.

R1. $AB \parallel CD$ if and only if $\angle[AB, PQ] = \angle[CD, PQ]$ (Fig. 10.3).

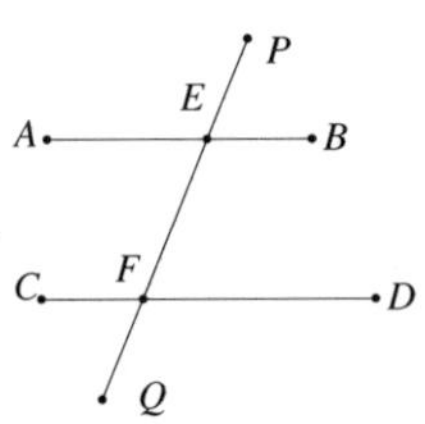

Figure 10.3

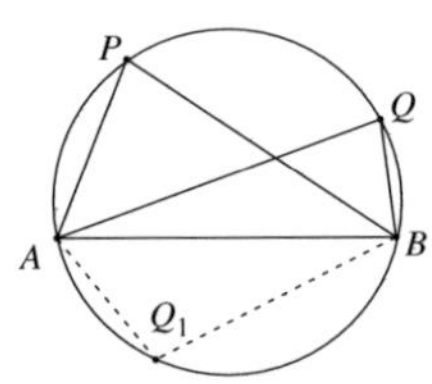

Figure 10.4

Using ordinary angles, we need to specify the relations among eight angles and we need to use order relations (inequalities) to distinguish the cases. For instance, we have: "if points B, D are on the same side of line PQ and points P, C are on the different sides of line AB (the order relations), then $AB \parallel CD \Leftrightarrow \angle PEB = \angle PFD$." This rule is very difficult to use and may lead to branchings during the deduction. The following two rules also show why full-angle is crucial to this approach.

R2. $\angle[PA, PB] = \angle[QA, QB]$:- cyclic(A, B, P, Q).

R3. cyclic(A, B, P, Q) :- $\angle[PA, PB] = \angle[QA, QB]$, $\neg$ coll(P, Q, A, B).

In rule R2, using the ordinary angle, we need two conditions (Fig. 10.4), $\angle APB = \angle AQB$ or $\angle APB + \angle AQ_1B = 180°$, and to distinguish these two cases we need to know that "points P and Q are on the same or different sides of line AB." Using full-angles, the two cases can be treated uniformly. For the treatment of the negation in rule R3, see Chou et al. (1995c).

10.3.2 Forward chaining and fixpoints

Let D_0 be the hypotheses of a geometric statement and R the geometric rule (or axiom) set. Basically speaking, the *forward chaining search* works as follows:

$$\boxed{D_0} \overset{R}{\subset} \quad \boxed{D_1} \overset{R}{\subset} \quad \cdots \quad \overset{R}{\subset} \quad \boxed{D_k} \quad \text{(Fixpoint)}.$$

For each rule r in R, we apply r to D_0 to obtain new facts. Let D_1 be the union of D_0 and the set of new facts obtained in this way. Repeat the above process for D_1 to obtain D_2, etc. If at certain step $D_k = D_{k+1}$, then we say that a *fixpoint* for D_0 and R is reached, i.e., D_k is a fixpoint of reasoning rules R:

$$R(D_k) = D_k.$$

In the case of datalog (Gallaire 1984) (i.e., when we assume that the rules are Horn clauses without function symbols) the fixpoint can always be reached if the rule set R (i.e., the intensional database) and the initial fact set D_0 (i.e., the extensional database) are finite.

Since D_1 contains D_0 as a subset, the derivation of D_2 from D_1 clearly repeats all the previous deductions used to derive D_1 from D_0. The *semi-naive evaluation* is proposed to solve this problem (Bancilhon and Ramakrishnan 1986). Note that in the forward chaining searches, the main loop of process is to search the rule set R. We call such search strategies *rule-based search strategies*. In Chou et al. (1995c), a *data-based search strategy* is presented, in which a new-fact-list is kept and for each fact d in the list we find and apply all the rules whose bodies contain the predicate of d.

Data-Based Search Algorithm:

S1. Set the hypotheses of the statement to be the initial new-fact-list and the initial database. While the new-fact-list is not empty do step S2.

S2. Let d be the first new fact in the list. Delete it from the list, add it to the database, and do step S3.

S3. Let r be a rule whose body contains a predicate P_0 of the fact d. To apply the rule r, we need to instantiate other predicates in r. Since predicate P_0 will be instantiated as the new fact d, other predicates in r need to be instantiated for all the facts in the database. For all the predicate forms of fact d (notice that a fact could have many predicate forms) and for all the facts of the other predicates in r, do step S4.

S4. Apply rule r to obtain a fact d'. If d' is in the database, do nothing. Otherwise, add it to the end of the new-fact-list.

Since the hypothesis set of a geometric statement is finite and we use a finite rule set without function symbols, a fixpoint will always be reached.

10.3.3 Structured database

In the traditional way of representing a relational database, each n-ary predicate is associated with an n-dimensional array. Since most geometric predicates satisfy some special properties, building databases according to this way will lead to very large databases. In Chou et al. (1995c), the concept of *structured database* is introduced to reduce the size of the database. The following are three such principles.

Use canonical form for predicates. One geometric property can be represented as many predicate forms. For instance, the predicate coll satisfies the following rules:

$$\text{coll}(A, B, C) \text{ :- } \text{coll}(A, C, B), \qquad \text{coll}(A, B, C) \text{ :- } \text{coll}(B, A, C).$$

Thus from $\text{coll}(A, B, C)$, one can obtain five "new" facts: $\text{coll}(A, C, B)$, $\text{coll}(B, A, C)$, $\text{coll}(B, C, A)$, $\text{coll}(C, A, B)$, $\text{coll}(C, B, A)$. We represent predicates as canonical forms by assigning an order to the points in the geometric statement. With such an order, predicates can be represented uniquely. Using canonical forms, the above rules are not needed explicitly in the deduction steps. This will reduce the number of rules used in the deduction process.

Use equivalent classes to represent predicates. For instance, the fact that points $A_1, A_2, \ldots, A_n$ are on the same line can be represented by a sequence of points. In predicate form, we need $n(n-1)(n-2)$ different forms like $\text{coll}(A_i, A_j, A_k)$ to represent this fact.

Use representative elements for equivalent classes. For instance, the fact that the line containing points $A_1, \ldots, A_n$ is parallel to the line containing points $B_1, \ldots, B_k$ can be represented by $l_1 \parallel l_2$ if l_1 and l_2 are used to represent the two lines. If using predicates $\text{para}(A_i, A_j, B_k, B_l)$ to represent this fact, we need $2n(n-1)k(k-1)$ predicates.

For 160 geometric problems solved by the prover, the average size of the databases is 221 if the above structure is used. Using the predicate form, the average size would be 242 117; i.e., one thousand times larger. For many configurations such as Example 10.4, the fixpoint cannot be reached within reasonable time if the above structure is not used.

10.3.4 Experimental results and examples

Table 10.1 contains the timing and database size statistics for 160 geometric theorems proved by the program. The timing is collected on a Sparc-20 workstation.

Time (seconds)		Structured DB		Relational DB	
Time	Theorems	Size	Theorems	Size	Theorems
≤ 0.1	30%	≤ 50	16%	$\leq 10\,000$	11%
≤ 1	69%	≤ 100	42%	$\leq 50\,000$	43%
≤ 10	94%	≤ 200	66%	$\leq 100\,000$	59%
≤ 60	98%	≤ 500	91%	$\leq 1\,000\,000$	95%
≤ 650	100%	≤ 4021	100%	$\leq 5\,041\,102$	100%
8.37 (average)		221 (average)		242 117 (average)	

Table 10.1

The 160 geometric theorems proved by the prover are mainly chosen from about 600 theorems in Chou (1988) and Chou et al. (1994a). We can find fixpoints for each of the 600 theorems, but only about 160 of them can be proved in this way. Some well-known theorems such as Pappus' theorem and Pascal's theorem are beyond the scope of the program, though the fixpoints can be reached for both of them.

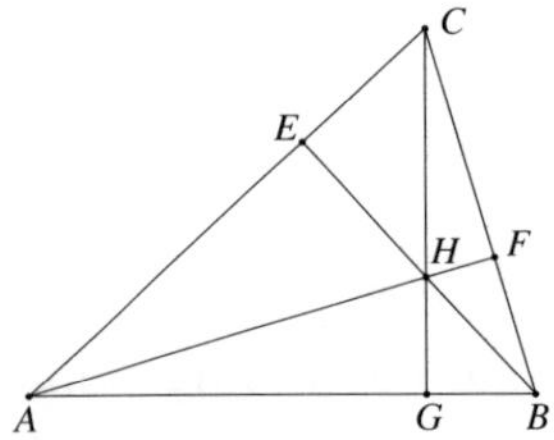

Figure 10.5

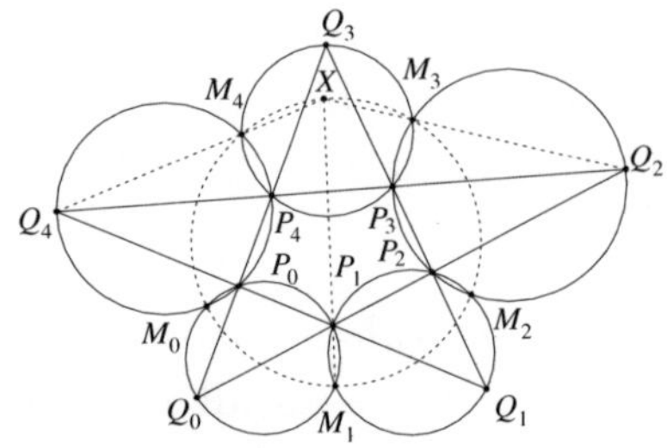

Figure 10.6

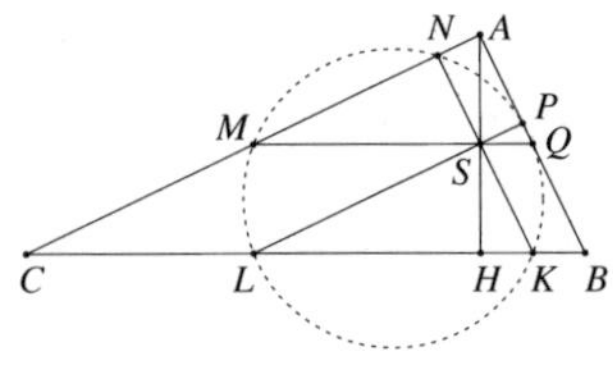

Figure 10.7

Example 10.2: (Orthocenter theorem)

Show that the three altitudes of a triangle are concurrent (Fig. 10.5).

The hypotheses (extensional database) are: points(A, B, C), coll(E, A, C), perp(B, E, A, C), coll(F, B, C), perp(A, F, B, C), coll(H, A, F), coll(H, B, E), coll(G, A, B), coll(G, C, H).

Reaching the fixpoint costs the program 0.75 second. The size of the fixpoint is 146 if the structured database is used. In predicate form, the size of the fixpoint would be 56 940. The fixpoint contains two of the most often encountered properties of this configuration: perp(C, G, A, B) (the conclusion) and $\angle[GF, GC] = \angle[GC, GE]$. Another amazing fact is that this simple configuration contains 105 non-trivial ratios!

The following is the proof for the fact perp(C, G, A, B), which is *automatically* generated by the prover. In the proof, "(hyp)" means that the corresponding fact is from the hypotheses.

The machine proof:

1. perp$[CG, AB]$:- (hyp)perp$[BC, AF]$, (hyp)coll$[GCH]$, (2)$\angle[BC, AF] = \angle[AB, CH]$.
2. $\angle[BC, AF] = \angle[AB, CH]$:- (3)$\angle[BC, AB] = \angle[AF, CH]$.
3. $\angle[BC, AB] = \angle[AF, CH]$:- (4)$\angle[BC, AB] = \angle[FE, AC]$, (5)$\angle[AF, CH] = \angle[FE, AC]$.
4. $\angle[BC, AB] = \angle[FE, AC]$:- (hyp)coll$[CBF]$, (hyp)coll$[CEA]$, (6)$\angle[BF, BA] = \angle[EF, EA]$.
5. $\angle[AF, CH] = \angle[FE, AC]$:- (hyp)coll$[AHF]$, (hyp)coll$[AEC]$, (7)$\angle[HF, HC] = \angle[EF, EC]$.
6. $\angle[BF, BA] = \angle[EF, EA]$:- (8)cyclic$[AFBE]$.
7. $\angle[HF, HC] = \angle[EF, EC]$:- (9)cyclic$[CFEH]$.

8. cyclic$[AFBE]$:- (hyp)perp$[FB, FA]$, (hyp)perp$[EB, EA]$.
9. cyclic$[CFEH]$:- (hyp)perp$[FH, FC]$, (hyp)perp$[EH, EC]$.

Example 10.3: (Five-circle theorem)

$P_0P_1P_2P_3P_4$ is a pentagon. $Q_i = P_{i-1}P_i \cap P_{i+1}P_{i+2}$, $M_i = \text{circle}(Q_{i-1}P_{i-1}P_i) \cap \text{circle}(Q_iP_iP_{i+1})$ (the subscripts are understood to be mod 5). Show that points M_0, M_1, M_2, M_3, M_4 are cyclic (Fig. 10.6).

The fixpoint is reached in 3.89 seconds and contains 541 facts (220 680 in predicate form). Besides the fact that M_0, M_1, M_2, M_3, and M_4 are cyclic, the program finds the following new result: the ten groups of lines below

$$P_{i+1}M_{i+1}, Q_{i-1}M_{i-1}, Q_{i+2}M_{i-2}; \qquad P_{i-1}M_{i-2}, P_iM_{i+1}, Q_{i-1}M_{i+2}, \quad i = 0, \ldots, 4,$$

are concurrent and their ten intersection points are on the circle determined by M_0, M_1, M_2, M_3, and M_4, i.e., this circle contains 15 points. The three dotted lines in Fig. 10.6 represent one group of concurrent lines.

Example 10.4:

In the right triangle ABC, $\angle A = 90°$; $AH \perp BC$ and H is on line BC; S is the midpoint of AH; K and N are points on BC and AC such that $KN \parallel AB$; P and L are points on AB and BC such that $PL \parallel AC$; Q and M are points on AB and AC such that $QM \parallel BC$. Show that the six points P, Q, K, L, M, N are on the same circle (Fig. 10.7).

It takes the program 461.06 seconds to reach the fixpoint which contains 1326 facts (5 041 102 in predicate form). Without using the structured database, the fixpoint is too big to be reached within reasonable time.

10.4 Multiple and shortest proof generation

Generating multiple proofs is an inherent property of search methods. But the precondition is that the proving method itself should be very fast to generate a large number of proofs in a short period of time. Another precondition is that the proofs generated should be readable; otherwise there is no reason to generate multiple proofs. The area method (Chap. 7) and the deductive database method are used to generate multiple proofs. In this section, we will introduce briefly how to use the area method to generate multiple proofs.

10.4.1 Basic lemmas about signed areas and ratio of segments

The signed area S_{ABC} in Euclidean geometry is used as a basic (undefined) geometric quantity described by the following properties.

L1. $S_{ABC} = S_{CAB} = S_{BCA} = -S_{BAC} = -S_{CBA} = -S_{ACB}$.

L2. Points A, B, and C are collinear iff $S_{ABC} = 0$.

L3. $PQ \parallel AB$ iff $S_{PAB} = S_{QAB}$.

L4. For points A, B, C, and D, we have $S_{ABC} = S_{ABD} + S_{ADC} + S_{DBC}$.

L5. If points A, B, C, and D are four collinear points such that $A \neq B$, and P is any point not on line AB, then

$$\frac{\overline{CD}}{\overline{AB}} = \frac{S_{PCD}}{S_{PAB}}.$$

Here we introduce another geometric quantity $\frac{\overline{CD}}{\overline{AB}}$, the ratio of two directed segments on the same line, which satisfies L7 below.

The *signed area of a quadrilateral* $ABCD$ is defined to be

$$S_{ABCD} = S_{ABC} + S_{ACD}.$$

By lemmas L4 and L1, we have:

Q1. $S_{ABCD} = S_{ABC} + S_{ACD} = S_{ABD} - S_{CBD}$.

Q2. $S_{ABCD} = S_{BCDA} = S_{CDAB} = S_{DABC} = -S_{ADCB} = -S_{DCBA} = -S_{CBAD} = -S_{BADC}$.

L6 (Co-side theorem). *Let M be the intersection of two lines AB and PQ and $Q \neq M$. Then*

$$\frac{\overline{PM}}{\overline{QM}} = \frac{S_{PAB}}{S_{QAB}}, \qquad \frac{\overline{PM}}{\overline{PQ}} = \frac{S_{PAB}}{S_{PAQB}}, \qquad \frac{\overline{QM}}{\overline{PQ}} = \frac{S_{QAB}}{S_{PAQB}}.$$

L7. *For four distinct collinear points P, Q, A, and B, $\frac{\overline{PQ}}{\overline{AB}}$ is a real number which satisfies*

(1) $$\frac{\overline{PQ}}{\overline{AB}} = -\frac{\overline{QP}}{\overline{AB}} = \frac{\overline{QP}}{\overline{BA}} = -\frac{\overline{PQ}}{\overline{BA}};$$

(2) $$\frac{\overline{PQ}}{\overline{AB}} \cdot \frac{\overline{AB}}{\overline{PQ}} = 1;$$

(3) $$\frac{\overline{PQ}}{\overline{AB}} = 0 \quad \textit{iff} \quad P = Q; \quad \textit{and}$$

(4) $$\frac{\overline{AP}}{\overline{AB}} + \frac{\overline{PB}}{\overline{AB}} = 1.$$

For the relation of these lemmas to other axiom systems of geometries, refer to Chou et al. (1994a).

In the prover, each geometric lemma is represented as a *rule* or a *clause*. For instance, lemma L3 can be written as the following Prolog rule.

el_rule([area,P,A,B],[area,Q,A,B]) :- para(A,B,P,Q),P $\neq$ Q,

where [area,P,A,B] is S_{PAB} and the predicate para(A,B,P,Q) means that $AB \parallel PQ$. This lemma means that if $AB \parallel PQ$, [area,P,A,B] will be replaced by [area,Q,A,B].

10.4.2 The prover

The prover is implemented in Prolog. The geometric lemmas can be put as inference rules in Prolog, and the (backward) logic engine of Prolog is used to do as much inference as possible. At the highest level, the prover uses a typical *backward chaining* search strategy.

Step 1. The prover first transforms the conclusion predicate into an equation of geometric quantities $\alpha = \beta$.

Step 2. The prover keeps replacing geometric quantities in α and β with expressions in new geometric quantities by using the rules introduced in the preceding subsection until no new replacement can be made. After applying a rule to α or β to obtain α' and β', the common factors of α' and β' are removed.

Step 3. Let the final equation be $\alpha'' = \beta''$. If α'' is literally equal to β'' then the theorem is true. Otherwise, it is not known whether the statement is true or not.

Example 10.5: (Centroid theorem)

The three medians of a triangle meet at a point, and each median is trisected by this point.

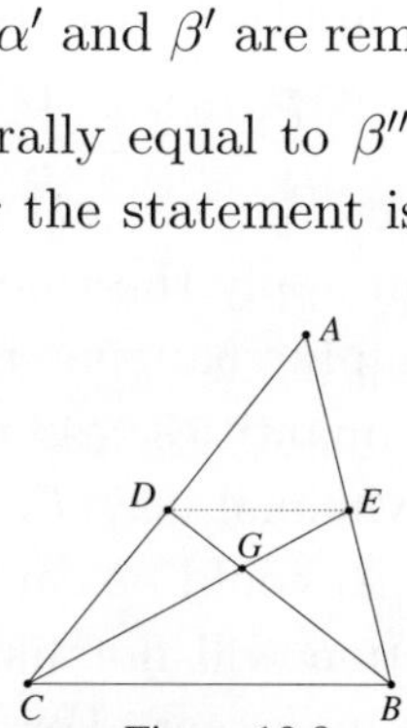

Figure 10.8

As in Fig. 10.8, we need to prove $-2 \cdot \frac{\overline{GD}}{\overline{GB}} = 1$. Non-degeneracy conditions are that A, B, C are not collinear.

The machine proof:

$$(-2) \cdot \frac{\overline{GD}}{\overline{GB}}$$
$(\frac{\overline{GD}}{\overline{GB}} = \frac{S_{EDC}}{-S_{ECB}}$, because collinear($B,D,G$), collinear($C,E,G$). (L6))
$$= \frac{2 \cdot S_{EDC}}{S_{ECB}}$$
$(S_{EDC} = \frac{1}{-2}(S_{ECA})$, because midpoint($D,C,A$). (L5))
$(S_{ECB} = \frac{1}{2}(S_{CBA})$, because midpoint($E,B,A$). (L5))
$$= \frac{2 \cdot S_{ECA}}{(-1) \cdot S_{CBA}}$$
$(S_{ECA} = \frac{1}{-2}(S_{CBA})$, because midpoint($E,B,A$). (L5))
$$= \frac{S_{CBA}}{S_{CBA}} = 1.$$
(Simplification: removing the factor: S_{CBA}.)

By introducing an order among the points, the area and ratio can be represented canonically. We can thus introduce an order or a rank among the geometric quantities.

Control Strategy 1. A rule can be used only if it reduces an invariant to an expression in invariants with lower ranks.

Control Strategy 1 has a two-fold purpose. First, it guarantees the termination. Second, it makes the production of short proofs possible. Without this strategy, the prover may do endless searches before reaching the shortest proof. This can be seen from Example 10.6.

10.4.3 Multiple proof generation

There are two ways to generate different proofs. First, for an algebraic quantity

$$\alpha = \frac{P_1 \cdot P_2 \cdots P_n}{R_1 \cdot R_2 \cdots R_m}$$

we can apply the rules to the polynomials P_i and R_i in different orders. Even by simply changing the order of applying the same set of rules to the algebraic quantities, essentially different proofs may be generated, because when a polynomial, say P_1, becomes Q_1 with R_1 unchanged, it might happen that $Q_1 = R_1$ could be cancelled. If we change R_1 to S_1 at the same time, the above reduction will not happen. The second way of generating different proofs is more important. For each geometric invariant, we may apply different rules to it to obtain different results. The following strategies to control the order of elimination are used.

Control Strategy 2. To apply a set of rules to an algebraic expression

$$\alpha = \frac{P_1 \cdot P_2 \cdots P_n}{R_1 \cdot R_2 \cdots R_m}$$

means either to apply one rule to one of $P_1, \ldots, P_n, R_1, \ldots, R_m$, or to apply one rule to one of the polynomials in the numerator and to apply another rule to one of the polynomials in the denominator.

Control Strategy 3. If two rules are used to eliminate the same point from both the numerator and the denominator of an algebraic expression or used to change the types of the geometric quantities, they will not be applied separately to the same algebraic expression, since the separate use of these rules will not lead to cancelling of geometric quantities.

Strategies 1, 2 and 3 are quite effective to control the number and the length of the proofs.

Example 10.6: (Continued from Example 10.5)

If all the three control strategies 1, 2 and 3 are used, the prover gives 70 proofs with proof lengths ranging from 4 to 7. If we relax the control by dropping control strategy 3, then the prover gives 212 proofs with lengths ranging from 4 to 8. If we further relax the control by dropping the control strategy 1, then the prover gives 24 360 proofs before it exhausts the memory of the computer. Many more proofs are expected, because in the 24 360 proofs there are only two proofs with length less than or equal to 8.

10.4.4 Shortest proof generation

Using the *breadth-first search*, the first available proof is naturally a shortest proof. Moreover, using the breadth-first search, for some difficult geometric problems, the prover may exhaust the computer space before it obtains the first (shortest) proof. This problem is solved by using the following technique: if the shortest proof is needed, the prover will remember the length of a proof when it is generated and prohibit any backtracking that can generate proofs with lengths longer than or equal to the recorded length.

Another approach is to use the *depth-first iterative deepening search* (Korf 1985; Stickel 1985). In this approach, one first tries to find a proof with depth 1, then depth 2, and so on by repeated use of depth-first search. Experiences show that both of the above approaches to find shortest proofs are quite successful (Chou et al. 1996b,c).

Example 10.7: (Continued from Example 10.6)

The following proof for the centroid theorem is the shortest proof found by the prover. This proof cannot be generated by the previous area method.

The machine proof:

$$(-2)\cdot\frac{\overline{GD}}{\overline{GB}}$$

$$\left(\frac{\overline{GD}}{\overline{GB}} = \frac{S_{EDC}}{-S_{ECB}},\ \text{because collinear}(B,D,G),\ \text{collinear}(C,E,G).\ \text{(L6)}\right)$$

$$= \frac{2\cdot S_{EDC}}{S_{ECB}}$$

$$\left(S_{EDC} = \frac{1}{-2}(S_{ECA}),\ \text{because midpoint}(D,C,A).\ \text{(L5)}\right)$$

$$\left(S_{ECB} = -S_{ECA},\ \text{because midpoint}(E,B,A).\ \text{(L5)}\right)$$

$$= \frac{S_{ECA}}{S_{ECA}}$$

$$(\text{Simplification: removing the factor: } S_{ECA}.\ )$$

$$= 1$$

10.5 Issues concerning search methods

Now it is time to discuss several key issues in developing an efficient geometry theorem prover with the synthetic approaches.

10.5.1 Skolemization and adding of auxiliary points

Since most synthetic geometric reasoning systems use Horn clauses as rules, the systems have no ability to generate auxiliary points. Thus it is important that one can prove at least a large portion of the geometric theorems with the chosen geometric rules. The rules used by the deductive database approach are quite good in this respect. Most of the 160 theorems proved by these rules without adding auxiliary points cannot be proved with the rules about congruent triangles if no auxiliary points are added. On the other hand, with functions of adding auxiliary points the prover will clearly be able to prove more theorems.

In the theory of logic, constructing new points corresponds to the skolemization of the existential quantifiers. The use of this idea in AGTP is discussed in Robinson (1983) and Reiter (1977), but seems not to have been implemented.

In Chou et al. (1995c), about forty rules are used to add auxiliary points. Two strategies are proposed to control the adding of new points to achieve effectiveness. The first strategy is to separate the process of adding new points from the process of reaching the fixpoints. The program works as follows. For a geometric theorem, it first finds a fixpoint without adding auxiliary points. If the conclusion is already in the database then the program terminates. Otherwise, the program will try to construct an auxiliary point and find a new fixpoint. The program will repeat the above process until either the conclusion is in the new database or there exist no new auxiliary points.

The second strategy is to use two heuristics to control the constructing of too many auxiliary points.

(1) After an auxiliary point is added and a new fixpoint is reached, we will check whether new properties about the original diagram are found. If they are, we will keep this auxiliary point; otherwise, the auxiliary point will be deleted.

(2) No recursive auxiliary points are allowed; i.e., to construct an auxiliary point, we can only use points occurring in the original statement. The second condition guarantees that the program will terminate.

For Example 10.1, the program by Chou et al. (1995c) automatically adds an auxiliary point: A_0 which is the midpoint of AD. With this auxiliary point, the proof of the conclusion seems easier:

The machine proof:

1. midp$[E, BC]$:- (2)para$[CD, EN]$, (hyp)midp$[N, BD]$.
2. para$[CD, EN]$:- (3)coll$[ENMA_0]$, (4)para$[CD, MA_0]$.
3. coll$[ENMA_0]$:- (hyp)line$[MNE]$, (5)line$[MNA_0]$.
4. para$[CD, MA_0]$:- (hyp)midp$[M, AC]$, (hyp)midp$[A_0, DA]$.
5. line$[MNA_0]$:- (6)para$[A_0M, A_0N]$.
6. para$[A_0M, A_0N]$:- (7)para$[A_0M, AB]$, (8)para$[A_0N, AB]$.
7. para$[A_0M, AB]$:- (4)para$[CD, MA_0]$, (hyp)para$[AB, CD]$.
8. para$[A_0N, AB]$:- (hyp)midp$[N, BD]$, (hyp)midp$[A_0, DA]$.

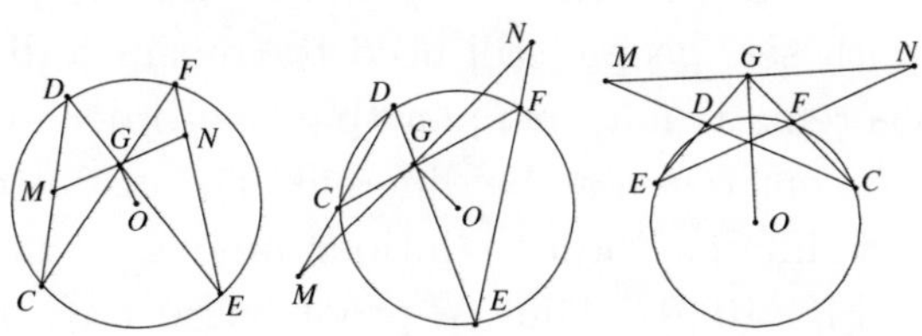

Figure 10.9

Example 10.8: (Butterfly theorem)

C, D, E, and F are on the same circle with center O. G is the intersection of CF and DE. The line passing through G and perpendicular to OG meets CD and EF at M and N respectively. Show that G is the midpoint of NM (Fig. 10.9).

The conclusion is not in the first fixpoint. The program automatically adds an auxiliary point A_0 which is the intersection of the line passing through C and parallel to MN and the circle O. With the point A_0, it takes the program 0.4 second on a Sparc-20 workstation to reach the fixpoint which contains the conclusion.

10.5.2 Producing diagram-independent proofs

The validity of most theorems in elementary geometry involving only equalities is independent of the relative order positions of the points involved. Such geometric theorems belong to the so-called *unordered geometry.* This idea originated from Wu's algebraic method of automated reasoning (Wu 1984a,b). In unordered geometry, the proofs of these theorems can be very simple. However, the ordinary proofs of these theorems involve the order relation; hence they are not only complicated but also not strict. The method based on congruent triangles is not for unordered geometry. In the work using this method, the order relations needed are either derived from a numerical diagram or given in the input. On the other hand, the rules used in the deductive database are for unordered geometry.

For instance, Gelernter's proof depends on the fact $\angle ABD = \angle BDC$, which is true in Fig. 10.2a. But in a different figure of the same theorem (Fig. 10.2b), this fact is no longer valid. Such facts are not isolated cases in AGTP. On the contrary, they exist in almost all geometric theorems. Nevins (1975) claimed that he had got rid of this drawback by adding the ordering relations to the hypotheses of the statement. Nevins' approach makes the situation much clearer, but still does not solve the problem completely. First, to prepare for the order relation in the hypotheses, people still need to consult a diagram. Second, for some geometric theorems it may happen that the order of points in *different diagrams of the same theorem* may be different. For instance, there are at least three diagrams with different order relations among points and lines for the Butterfly theorem (Fig. 10.9). Thus, a proof based on a fixed order relation among points and lines is valid only for some special cases of the statement. On the other hand, using a set of rules for unordered geometry, a uniform proof of this theorem can be given which is valid for all figures.

10.5.3 Avoiding redundant deductions

A *deduction is redundant* if it generates a fact that is already in the database. The redundant deduction is a major hurdle for speeding up the search, and eliminating redundancies is proposed as a basic research problem by Wos (1988). It is proved that, in general cases, the problem of eliminating all redundancies is undecidable (Helm 1990). So the best we can do is to design strategies to reduce the redundancies.

There are three kinds of redundant deductions. First, repeated application of the same rule to the same fact will generate the same result. This kind of redundancy can be solved by the semi-naive search or the data-based search.

Second, some *redundant deductions* are logically guaranteed. For example, if a rule r in a rule set R is a logical consequence of $R \setminus \{r\}$ then each fact deduced using rule r will also be deduced by other rules in R. For a given rule set, the problem of obtaining a minimal and logically equivalent rule set is undecidable (Sagiv 1988). Many useful partial methods are given in Buntine (1988), Helm (1990) and Sagiv (1988).

Third, two logically irrelevant deductions may give the same fact if the input facts satisfy certain conditions. We call this kind of redundancy the *conditional redundancy*. There seems no general method to control conditional redundancies.

In Chou et al. (1995c), many heuristics are used to control redundant deductions.

Chapter 11

FIRST-ORDER POLYNOMIAL BASED THEOREM PROVING

Jinzhao Wu

This chapter focuses on first-order polynomial based theorem proving. We present the notion of well-behaved inference rules and two concrete such inference rules termed superposition I and superposition II. The SOS-strategy and the L-strategy for well-behaved inference rules are developed, and both of them are proved to be complete. We introduce the LM-strategy, the confluent LM-strategy, the MM-strategy, as well as the IM-strategy for superpositions I and II, and show that the last four are complete whereas the first is complete for clausal case. Finally, two open problems related to the Gröbner basis method and the odd-strategy are discussed.

11.1 Introduction

The Boolean ring or first-order polynomial based theorem proving began with the work of Hsiang (1982, 1985). Hsiang extended the idea of using Boolean polynomials to represent propositional formulas to the case of first-order predicate calculus. Based on the completion procedure of Knuth and Bendix (1970), the N-strategy was proposed. Later, by imitating the framework of Buchberger's algorithm to compute the Gröbner bases of polynomial ideals (Buchberger 1985), Kapur and Narendran (1985) developed another approach which is also referred to as the Gröbner basis method.

One obvious advantage of using Boolean polynomials is that every propositional formula has a unique representation, and sometimes it is easily generalized to some non-classical logic systems (Chazarain et al. 1991; Wu and Tan 1994). Stimulated by them, some approaches and results have been reported (Bachmair and Dershowitz 1987; Dietrich 1986; Kapur and Zhang 1989; Wu and Liu 1998; Zhang 1988, 1994a).

In this chapter we review the work by J.-Z. Wu and Z.-J. Liu on theorem proving based on first-order polynomials, including the notion of well-behaved

MATHEMATICS MECHANIZATION AND APPLICATIONS
ISBN 0-12-734760-7

inference rules and the strategies proposed in Wu and Liu (1998), the strategies developed in Wu (1996), as well as two open problems discussed in Wu (1996) and Wu and Liu (1996b).

It is well known that the completeness of inference rules is one of the most important theoretical aspects in theorem proving. In general, after inference rules are presented, their completeness should be considered immediately. This is usually boring and sometimes difficult. Based on this consideration, the concept of well-behaved inference rules is proposed. The category of such inference rules is pretty rich. For example, roughly speaking, most known inference rules such as resolution (Robinson 1965), generalized resolution and the NC-resolution (Liu and Jiang 1987; Murray 1982; Wang and Liu 1982), as well as the remainder method (Wu and Liu 1996a) are well-behaved.

We also present two concrete well-behaved inference rules named superposition I and superposition II.

As a matter of fact, superposition I comes from generalized resolution and the NC-resolution. The odd-strategy of Zhang (1994a) is indeed a further restriction on it. Superposition II comes from an idea in the method of Wu (1984a,b; Chap. 1). Wu's method actually provides us with some techniques of using single overlaps and parallel overlaps. It is worthwhile to apply its frame to first-order theorem proving. Originally we noted that the procedure of resolution can be so considered, in fact, as to eliminate a variable. So the idea comes to employing the similar remainder computing in Wu's method as inference rule. It is not hard to see that the Boolean polynomial with respect to $P \vee Q$ is just the remainder of those with respect to $P \vee R$ and $Q \vee \neg R$. This is quite an encouraging phenomenon. Moreover, the Boolean polynomials with respect to P and Q are two remainders of zero and that with respect to $P \wedge Q$, where we utilize zero to denote the Boolean polynomial corresponding to the idempotency rule B_2.

Superpositions I and II are non-clausal, and employ single overlaps and parallel overlaps instead of multiple overlaps and sequential overlaps.

As is known, the N-strategy and the Gröbner basis method use multiple overlaps and sequential overlaps, and the former is in fact a kind of clausal theorem proving. However, multiple overlaps generate more inferences which are not possible for single overlaps, the searching space is thus increased. Also, multiple overlaps are much more expensive to compute than single overlaps because the operation mentioned is associative and commutative.

Zhang proved that only single overlaps are needed in the N-strategy. Generally people are not interested in whether parallel overlaps are better than sequential overlaps. They only investigate whether sequential overlaps can be replaced by parallel overlaps or vice versa. For various overlaps, the reader is referred to Muller and Socher-Ambrosius (1988a,b) and Zhang (1994a).

As the inference rules described are non-clausal, we need not convert the given formulas into clauses. For instance, in this case the association law and the formula of form $\phi_1 \leftrightarrow \phi_2$ are much simpler than their two respective clausal forms.

On the other hand, developing strategies is a crucial work in theorem proving. Indeed, one challenge is to give strategies which are not only efficient but also complete. We know that the basic ideas of many famous strategies for resolution, such as the linear, set-of-support, semantic and lock resolutions (Boyer 1971; Loveland 1969; Slagle 1967; Wos et al. 1965), do not depend upon resolution itself. For example, up to overlaps, we can view the N-strategy (as well as the odd-strategy) and the Gröbner basis method as the strategies utilized in the semantic (or more strictly, negative) resolution and the lock resolution, respectively. The difficulty lies in how to establish inference rules in the underlying version. Obviously each of these ideas has its own advantages, and also these advantages do not rely on the expressing style of knowledge. Here we do not concern ourselves with the detailed improvements on the strategies. We are only interested in extending the basic ideas to first-order polynomial based theorem proving.

The SOS-strategy, L-strategy, LM-strategy, confluent LM-strategy, IM-strategy and MM-strategy are presented according to this consideration. Some of them are also designed for emphasizing application issues. Although it seems to have lost its flavor from the viewpoint that it is not effective in practice, theorem proving in first-order logic has vast applications in artificial intelligence, logic programming, deductive databases, program verification and synthesis, specification analysis, and so on. An interested reader may find clues in Chang and Lee (1973), Lloyd (1984), Murray (1982), Robinson (1965), and Slagle (1974).

We show that well-behaved inference rules are complete for both the L-strategy and the SOS-strategy. Superpositions I and II are complete for the confluent LM-strategy, MM-strategy and IM-strategy. The LM-strategy for them is complete in clausal case.

Finally, two problems are discussed. One is about the completeness of odd superposition II posed by Zhang, and the other is on the overlaps and sufficiency of unifications on maximal atoms in the Gröbner basis method. Indeed, the IM-strategy and MM-strategy are motivated by these two problems.

This chapter is arranged as follows. In Sect. 11.2, we describe some basic notions and results in first-order polynomial based theorem proving. In Sect. 11.3, an overview of the previous work is given, followed by a review on our current work in this field. We present the definition of well-behaved inference rules, and show that resolution is in fact well-behaved. Then, two concrete inference rules are given, and strategies are considered in Sect. 11.4. We show that well-behaved inference rules are complete for both the SOS-strategy and the L-strategy. We also introduce four other strategies, the LM-strategy, confluent LM-strategy, MM-strategy and IM-strategy, for the two concrete inference rules; two related problems are discussed. Section 11.5 is an overview of non-clausal theorem proving. We have the conclusion in Sect. 11.6.

11.2 Basic knowledge

For those concepts that are used but not defined in this chapter, the reader may consult Chang and Lee (1973), Kapur and Narendran (1985), and Hsiang (1985). In this section, we describe some fundamental concepts and results on which our following discussions are directly based.

11.2.1 First-order polynomials

Let Z_2 be the residue class ring modulo 2. We define a *first-order polynomial* as a polynomial over Z_2 with atoms as indeterminates. A first-order polynomial is ground if the atoms in it are ground. We still call the variables in atoms variables. For a first-order polynomial c, we denote by $\Phi(c)$ the one obtained by removing the power of each atom in c.

For two first-order polynomials c_1 and c_2, we have the following simple facts:

$$\Phi(c_1 + c_2) = \Phi(c_1) + \Phi(c_2), \Phi(c_1 c_2) = \Phi(\Phi(c_1)\ \Phi(c_2)),$$
$$\Phi(c_1 + c_2 + c_1 c_2) \neq 0 \text{ if } \Phi(c_1) \neq 0.$$

For arbitrary skolemized formulas F_j $(j = 1, 2)$, let

$$P(F_j) = F_j, \text{ if } F_j \text{ is an atom;}$$
$$P(\neg F_j) = \Phi(1 + P(F_j));$$
$$P(F_1 \wedge F_2) = \Phi(P(F_1)P(F_2));$$
$$P(F_1 \vee F_2) = \Phi(P(F_1) + P(F_2) + P(F_1)P(F_2));$$
$$P(F_1 \rightarrow F_2) = \Phi(1 + P(F_1) + P(F_1)P(F_2));$$
$$P(F_1 \leftrightarrow F_2) = \Phi(1 + P(F_1) + P(F_2)).$$

Note that $P(1)$ and $P(0)$ are defined as 1 and 0, respectively.

By the above definition, we can thus convert a skolemized formula into a first-order polynomial. Let S be a given skolemized formula set, and suppose

$$PS = \{P(F) + 1 | \; F \in S\}.$$

Then S is unsatisfiable if and only if PS is unsatisfiable (Kapur and Narendran 1985; Wu and Liu 1996a). Therefore, it is sufficient for us to consider whether a first-order polynomial set is unsatisfiable. Like the Herbrand theorem in resolution (Chang and Lee 1973), we have the following.

Theorem 11.1:

PS is unsatisfiable if and only if there is $j \in \{0, 1, \ldots\}$, such that $PS|H_j$ is unsatisfiable.

Here H_j are the j-level constants of S. $PS|H_j$ represents the set which is constructed from PS by replacing each variable in the atoms in each member of PS by all the elements of H_j.

11.2.2 Notations

Throughout this chapter, all the atoms are not 0 and 1 unless there is a specific statement. We always suppose that PS is a first-order polynomial set, and all the first-order polynomial sets mentioned contain a zero. For a first-order polynomial c, A_c denotes the set of all atoms occurring in $\Phi(c)$, $A_c\theta = \{m\theta | \; m \in A_c\}$, where θ is a substitution. Moreover, a, b, c, d, e, r always represent first-order polynomials, m, n denote atoms, and $\theta, \sigma, \lambda, \alpha$ stand for substitutions. Sometimes a non-zero first-order polynomial, in which the powers of all the atoms are one, is directly written in the form $am + b$, where $m \notin A_b$.

11.3 Inference rules

An *inference rule* is defined as an algorithm, when applied to two given first-order polynomials, which yields finite first-order polynomials.

11.3.1 Historic development

Since this chapter concentrates only on our own work, now we only present a brief overview of the other primary work on first-order polynomial based theorem proving, to make the chapter clearer and to enable the reader to handle the whole situation.

(1) *N-strategy.* If F is a clause, we call $P(F)+1$ a *clausal first-order polynomial*. Therefore c is clausal if and only if c is either 0 or 1 or of the form $(m_1 + i_1)\cdots(m_k+i_k)$, where each $i_j \in \{0,1\}$. A product of atoms is called a *monomial*.

The method developed by Hsiang and termed N-strategy uses clausal first-order polynomials, in which the inference rule is

$$\frac{\begin{array}{l} a_1, \\ a_2t + b_2 \end{array}}{\Phi((vb_2)\sigma)}$$

where a_1, a_2 and t are monomials, $a_2t + b_2$ is clausal, σ is a most general BN-unifier such that $(ua_1)\sigma = (vt)\sigma$, and u, v are the extra variables obtained in the BN-unification (Hsiang 1985).

The N-strategy is similar to the negative resolution. It also has some features of unit resolution. A more important difference between the N-strategy and other resolution methods is that the N-strategy is a reduction method while resolution strategies are not. Together with a factoring computation the N-strategy is complete (Hsiang 1985). Furthermore, Zhang (1994a) proved that it is still complete with reduction(when only N-rules and P-rules are used). Improvements of the N-strategy can be found in Hsiang (1987) and Muller and Socher-Ambrosius (1988a,b).

(2) *Gröbner basis method.* In this method proposed by Kapur and Narendran, new first-order polynomials are generated from any two first-order polynomials by unifying their maximal monomials under a term ordering. The main inference rule is

$$\frac{\begin{array}{l} a_{i_1} + \cdots + a_{i_k} + b_1, \\ a_{j_1} + \cdots + a_{j_l} + b_2 \end{array}}{\Phi(c_1b_1\sigma + c_2\sum_{t=1,t\neq v}^{l} a_{j_t}\sigma + c_1\sum_{s=1,s\neq u}^{k} a_{i_s}\sigma + c_2b_2\sigma)}$$

where $a_{i_1}, \cdots, a_{i_k}$ $(a_{j_1}, \cdots, a_{j_l})$ are all maximal monomials in $a_{i_1} + \cdots + a_{i_k} + b_1$ (respectively $a_{j_1} + \cdots + a_{j_l} + b_2$), u, v meet the conditions that $a_{i_u} = d_1e_1$ and $a_{j_v} = d_2e_2$ such that e_1 and e_2 can be unified by a most general unifier (*mgu* for short) σ, and $c_1 = d_2\sigma, c_2 = d_1\sigma$.

In addition, one has to consider the actions of the idempotency rule B_2 and self-inverse rule B_3 (Kapur and Narendran 1985).

The above inference rule is in fact similar to resolution but is more powerful and allows a lot more flexibility. The major difference between the Gröbner basis method and N-strategy is that the former is based on the Gröbner basis computation whereas the latter is based on the extensions of the Knuth–Bendix completion procedure for handling associative and commutative operators.

(3) *Odd-strategy.* This method proposed by Zhang (1994a) may be seen as an extension of Hsiang's N-strategy to non-clausal case, in which the inference rule is

$$\frac{\begin{array}{l} a_1 m_1 + b_1, \\ a_2 m_2 + b_2 \end{array}}{\Phi((b_2(a_1 + b_1))\sigma)}$$

where $a_1 m_1 + b_1$ is odd (namely the number of its monomials is odd), σ is an mgu of m_1 and m_2, and $m_1\sigma$ is maximal in $(a_1 + b_1)\sigma$ under a given partial order of atoms.

One advantage of the odd-strategy is that it uses single overlaps. However, both the N-strategy and the Gröbner basis method use multiple overlaps. Also, together with odd-factoring, the odd-strategy is complete.

Before closing this subsection, we remark that for more informations on the implementations, and on what kind of problems that they are more effective, in addition to the three papers mentioned above, the reader may consult Bachmair and Dershowitz (1987), Kapur and Zhang (1989), and Zhang (1988).

11.3.2 Well-behaved inference rules

Hereafter we mention the work of Wu and Liu, beginning with the notion of well-behaved inference rules.

We use $c(m \leftarrow i)$ to denote the first-order polynomial derived from c by replacing m with i, where $i \in \{0, 1\}$. Clearly,

$$\Phi(c(m \leftarrow i)) = \Phi(c)(m \leftarrow i).$$

We call $\Phi(c\theta)$ an *instance* of c. If two or more atoms in A_c have an mgu σ, $\Phi(c\sigma)$ is called a *factor* of c.

From now on we assume that the inference rule IR under discussion is sound. By $IR(c_1, c_2)$ we denote the set of the results obtained by applying IR to c_1 and c_2. IR is said to be *well-behaved* if it meets the following four conditions.

Unit condition. If c is ground and $m \in A_c$, then $c(m \leftarrow i) \in IR(m + i, c)$.

Stable condition. For ground c_j $(j = 1, 2)$, and any $r' \in IR(c_1(m \leftarrow i), c_2(m \leftarrow i))$, there exists $r \in IR(c_1, c_2)$, such that $r(m \leftarrow i) = r'$.

Split condition. Let c_j $(j = 1, 2)$ be ground, $A_{c_j} \neq \varnothing$ and $A_{c_1} \cap A_{c_2} = \varnothing$. Then there exists d_1 such that $A_{d_1} \subset A_{c_1}$, and $d_1 c_2 + 1 \in IR(c_1 c_2 + 1, e)$, where $e \in \{c_1 c_2 + 1, 0\}$.

Lifting condition. Suppose that c'_j $(j = 1, 2)$ are ground instances of c_j respectively, and $r' \in IR(c'_1, c'_2)$. Then there exist d_j, such that d_j are factors of c_j and c'_j are instances of d_j, and there exists $r \in IR(d_1, d_2)$, r' is an instance of r.

One main reason to introduce the unit condition is that we have at least to guarantee that the empty clause should be derived from m and $\neg m$. It is a little similar to the one-literal rule in the method of Davis and Putnam (1960). The stable condition, in fact, makes sure that one can obtain a deduction from a related case in which there are fewer atoms. The split condition ensures that we are able to derive both of the irrelevant formulas from their conjunction and some known information. Just like the lifting lemma in resolution, the lifting condition ensures that we may lift the propositional case to the predicate case.

Now we show that the resolution of first-order polynomial form is well-behaved.

Clearly we only need to deal with clausal first-order polynomials in this case. The inference rule of resolution is computing binary resolvents of two clauses, which can be expressed in the following manner (in the viewpoint of deleting tautologies):

For two clausal

$$c_1 = (m + i) \prod_{k=1}^{s} (m_k + i_k) \quad \text{and} \quad c_2 = (n + j) \prod_{k=1}^{t} (n_k + j_k),$$

where each $i, j, i_k, j_k \in \{0, 1\}$, let σ be an mgu of m and n. Then, if $i + j = 1$,

$$r = \Phi(\prod_{k=1}^{s} (m_k\sigma + i_k) \prod_{k=1}^{t} (n_k\sigma + j_k))$$

is defined as a *binary resolvent* of c_1 and c_2 upon $n\sigma$.

Obviously, r is still clausal, and $m = n$ if c_1, c_2 are ground. If $s = 0$, then

$$r = \Phi(\prod_{k=1}^{t} (n_k + j_k)),$$

$r = c_2(m \leftarrow i)$, and hence the unit condition holds.

Suppose now that there exist p and q, $1 \le p \le s$, $1 \le q \le t$, such that $m_p = n_q$. Then, if $c_1(m_p \leftarrow l)$ and $c_2(n_q \leftarrow l)$ $(l = 0, 1)$ are not zero, the binary resolvent of them upon n is

$$r' = \Phi((i_p + l)(j_q + l) \prod_{k \ne p} (m_k + i_k) \prod_{k \ne q} (n_k + j_k)) = r(m_p \leftarrow l).$$

Thus, the stable condition holds.

We see that there are no c_1 and c_2, which are clausal, such that $A_{c_1} \neq \varnothing$, $A_{c_2} \neq \varnothing$, $A_{c_1} \cap A_{c_2} = \varnothing$, and $c_1c_2 + 1$ is clausal. The split condition therefore holds naturally.

Now suppose that c_1 and c_2 have no common variables, $c_1' = a_1'(m'+1), c_2' = a_2'm'$ are ground, $c_1' = \Phi(c_1\theta), c_2' = \Phi(c_2\theta)$, and $r' = \Phi(a_1'a_2') \in IR(c_1', c_2')$. Let $d_1 = a_1(m_1+1), d_2 = a_2m_2$ be factors of c_1, c_2 respectively, such that

$$\Phi(d_1\theta) = c_1', \Phi(d_2\theta) = c_2', m_1\theta = m_2\theta = m',$$

and for any $n \in A_{a_1} \cup A_{a_2}$, $n\theta \neq m'$. Then,

$$a_1' = \Phi(a_1\theta), a_2' = \Phi(a_2\theta).$$

Let σ be an mgu of m_1 and m_2, and $\theta = \sigma \circ \alpha$. Then, $m' = m_1\sigma \circ \alpha$. Therefore, $m_1\sigma \notin A_{a_1\sigma} \cup A_{a_2\sigma}$. Further, let $r = \Phi((a_1a_2)\sigma)$. Then

$$r \in IR(d_1, d_2) \quad \text{and} \quad \Phi(r\alpha) = \Phi(a_1'a_2') = r'.$$

The lifting condition holds.

In the sequel a sequence $c_1, \ldots, c_k$ is called a *deduction* of c_k from *PS* if for any $1 \leq i \leq k$, either $c_i \in PS$, or $c_i \in IR(c_s, c_t)$, where $s, t < i$. If $c_k = 1$, we call this sequence a *refutation* from *PS*.

Theorem 11.2:

Suppose that PS is unsatisfiable. If IR is well-behaved, then there exists a refutation from PS.

This theorem indicates that well-behaved inference rules are complete. In fact, there are stronger results, which will be seen in Sect. 11.4.1. The reader is referred to Wu and Liu (1998) for a proof.

11.3.3 Two examples

Now we describe two concrete inference rules that are well-behaved. They are non-clausal and use single and parallel overlaps rather than multiple and sequential overlaps.

We call the following inference rule *superposition I*:

$$\frac{\begin{array}{l} a_1m_1 + b_1, \\ a_2m_2 + b_2,\ \sigma \text{ is an mgu of } m_1 \text{ and } m_2 \end{array}}{\Phi((b_1(a_2+b_2))\sigma), \Phi((b_2(a_1+b_1))\sigma)}$$

The idea of superposition I comes, in fact, from generalized resolution and the NC-resolution. Let F_1 and F_2 be generalized clauses, and R_1 and R_2 be the two generalized resolvents of F_1 and F_2 upon $m_j\sigma$. If $m_j \in A_{P(F_j)}$ $(j = 1, 2)$, then $P(R_1)+1$ and $P(R_2)+1$ are two results obtained by applying superposition I to $P(F_1) + 1$ and $P(F_2) + 1$.

It is worth converting these two types of resolutions into first-order polynomial forms. In doing so, the reductions defined in the NC-resolution can, for example, be applied naturally, and we will see later that the similar semantic and lock strategies still keep complete. However, the NC-resolution is incomplete for the lock and semantic strategies (Liu and Sun 1994). Also, the inference rule of odd-strategy is indeed superposition I with some restrictions.

Superposition I is sound since

$$\Phi((b_1(a_2 + b_2))\sigma) = \Phi(((a_2 + b_2)(a_1m_1 + b_1) + a_1m_1(a_2m_2 + b_2))\sigma),$$
$$\Phi((b_2(a_1 + b_1))\sigma) = \Phi(((a_1 + b_1)(a_2m_2 + b_2) + a_2m_2(a_1m_1 + b_1))\sigma).$$

The following inference rule is called *superposition II*:

$$\frac{\begin{array}{l} a_1m_1 + b_1, \\ a_2m_2 + b_2,\ \sigma \text{ is an mgu of } m_1 \text{ and } m_2 \end{array}}{\Phi((a_1b_2 + a_2b_1)\sigma)}, \qquad \frac{\begin{array}{l} am + b, \\ 0 \end{array}}{\Phi(b(a + b))}$$

In fact, superposition II is constructed from the idea of pseudo-remainder computation of two polynomials in Wu's method. Using $a_1m + b_1$ to divide $a_2m + b_2$, one obtains the pseudo-remainder $a_1b_2 + a_2b_1$. Using $am + b$ to divide $m^2 + m$, one gets the pseudo-remainder $b(a + b)$ (Wu 1984a,b). We see that the zero here represents the idempotency rule B_2. Actually all the proof systems developed by us employed superposition II, which was called the *remainder method* in Wu and Liu (1996a).

As is known, one advantage of the Gröbner basis method is that a partial monomial order is introduced. We thus do not have to consider every monomial in $a_1m_1 + b_1$ for possible unification with every monomial in $a_2m_2 + b_2$. We will see later that a partial atom ordering can also be introduced and we can only utilize the unifications on maximal atoms. Besides, there are other two major differences between superposition II and the critical pair computation used in the Gröbner basis method. As we have pointed out, one is that single overlaps and parallel overlaps are employed in superposition II, another is that the use of the idempotency rule B_2 is quite different. The reader may easily find these differences by comparing superposition II with the inference rule employed in the Gröbner basis method.

Superposition II is sound since

$$\begin{aligned}\Phi((a_1b_2 + a_2b_1)\sigma) &= \Phi((a_1(a_2m_2 + b_2))\sigma + (a_2(a_1m_1 + b_1))\sigma),\\ \Phi(b(a+b)) &= \Phi(a(am+b)) + \Phi(am+b).\end{aligned}$$

Theorem 11.3:

Both superposition I and superposition II are well-behaved.

For a proof of this theorem, the reader is referred to Wu and Liu (1998).

11.4 Strategies

In this section, we discuss two strategies for well-behaved inference rules and four strategies for superpositions I and II. We are only interested in the theoretical results here, i.e., the completeness of various strategies.

11.4.1 SOS-strategy and L-strategy

We have shown that well-behaved inference rules are complete. As a matter of fact, we even have much stronger results.

The set-of-support strategy (SOS-strategy) for resolution was introduced by Wos et al. (1965). It is designed to restrict the application of inference rules. Generally it is considered as the most powerful restriction strategy available. The use of this strategy enables programs to prove theorems in far less computer memory and time than would be required if it were not used. This advance has resulted in a lot of impressive successes though there are still some problems to be investigated further. In the 33 basic research problems proposed by Wos, the first and the second ones are related directly to the SOS-strategy, and the second is to see whether there exists a restriction strategy that is more powerful. These problems indicate that the SOS-strategy is basic and important (Wos 1988).

The linear strategy (L-strategy for short) for resolution was developed in 1970 by Loveland and Luckham, separately. Its two special appeals are that it is natural and is of very simple structure. Above all, various tree searching techniques and heuristic methods can be applied with it easily and conveniently (Loveland 1978). Its basic idea is frequently used in practical applications such as logic programming and deductive databases (Das 1992; Lloyd 1984).

However, these advantages do not depend on resolution itself. We extend these two resolution strategies to first-order polynomial based theorem proving, and show that well-behaved inference rules are complete for both of them.

Suppose that $T \subseteq PS$. A sequence $c_0, \ldots, c_k$ is called an *SOS-deduction* of c_k from PS with support T, if for $0 \leq s \leq k$, either c_s is a factor of an element of $PS \cup \{c_t|\ t < s\}$, or $c_s \in IR(c_p, c_q)$, where $p, q < s$ and c_p or c_q is not a factor of some element of $PS \setminus T$. If $c_k = 1$, we say that this sequence is an *SOS-refutation* from PS with support T.

Assume that c_0 is given. A sequence $c_0, \ldots, c_k$ is called an *L-deduction* of c_k from PS with top c_0, if for $0 < s \leq k$, either c_s is a factor of c_{s-1}, or $c_s \in IR(c_{s-1}, e)$, where e is a factor of an element of $PS \cup \{c_t|\ t < s\}$. Similarly, this sequence is called an *L-refutation* from PS with top c_0 if $c_k = 1$.

Now we consider an example to illustrate these two definitions. Suppose that IR is superposition II, and

$$PS = \left\{ \begin{array}{l} c_0 = P(b)P(i(b)) + P(b) + 1, \\ c_1 = P(x)P(y)M(x, i(y), z)P(z) + P(x)P(y)M(x, i(y), z), \\ c_2 = M(e, x, x) + 1, \\ c_3 = M(x, i(x), e) + 1 \end{array} \right\}.$$

Let $T = \{c_0, c_2, c_3\}$. Obviously

$$\begin{array}{l} P(e)P(i(y)) + P(e) \in IR(c_1, c_2), \\ P(i(b)) \in IR(c_0, 0), \\ P(e) \in IR(P(e)P(i(y)) + P(e), P(i(b))), \\ P(x)P(y)M(x, i(y), e) \in IR(P(e), c_1), \\ P(x) \in IR(P(x)P(y)M(x, i(y), e), c_3), \\ 1 \in IR(P(x), c_0). \end{array}$$

Thus the sequence

$$c_1, c_2, P(e)P(i(y)) + P(e), c_0, 0, P(i(b)), P(e), P(x)P(y)M(x, i(y), e), \\ c_3, P(x), 1$$

is an SOS-refutation from PS with support T. Also we know

$$\begin{array}{l} P(i(b)) \in IR(c_0, 0), \\ P(x)P(y)M(x, i(y), i(b)) \in IR(P(i(b)), c_1), \\ P(e)P(b) \in IR(P(x)P(y)M(x, i(y), i(b)), c_2), \\ P(e) \in IR(P(e)P(b), c_0), \\ P(x)P(y)M(x, i(y), z) \in IR(P(e), c_1), \\ P(x) \in IR(P(x)P(y)M(x, i(y), z), c_3), \\ 1 \in IR(P(x), c_0). \end{array}$$

So the sequence

$$c_0, P(i(b)), P(x)P(y)M(x, i(y), i(b)), P(e)P(b), P(e), P(x)P(y)M(x, i(y), z), \\ P(x), 1$$

is an L-refutation from *PS* with top c_0.

Theorem 11.4:

Suppose that PS is unsatisfiable, and IR is well-behaved.

(1) *If* $PS \setminus T$ *is satisfiable, then there exists an SOS-refutation from PS with support* T.

(2) *If* $PS \setminus \{c_0\}$ *is satisfiable, then there exists an L-refutation from PS with top* c_0.

Proof:

We may use mathematical induction on the number of the elements of $\cup_{c \in PS|H_j} A_c$, where $PS|H_j$ is the same as that of Theorem 11.1. Indeed, the unit, stable, and split conditions ensure that the theorem holds for the ground case. Then, by using the lifting condition, we get the result as required. For more details, see Wu and Liu (1998).

Theorems 11.2 and 11.4 show that well-behaved inference rules are not only complete, but also complete for the SOS-strategy and L-strategy. On the other hand, these facts also indicate that it is not asking too much to require that the inference rules are well-behaved.

11.4.2 LM-strategy, confluent LM-strategy, MM-strategy and IM-strategy

In the previous section we proposed two well-behaved inference rules, superposition I and superposition II. In this section, we consider some other strategies for them by introducing a partial order between atoms.

In fact, the MM-strategy is motivated by the considerations of using not only single overlaps, but also unifications on maximal atoms, rather than using multiple overlaps and unifications on maximal monomials as in the Gröbner basis method.

We have shown how Wu's method applies to propositional logic (Wu and Tan 1994). The IM-strategy is the initial work to extend the whole framework of Wu's method to first-order theorem proving. In addition, it is also a further restriction of the SOS-strategy.

One reason why we design the LM-strategy and confluent LM-strategy lies in application issues. For instance, using first-order polynomials to represent knowledge may increase the power of logic programming. Let us utilize normal program $S = \{\neg p \to p\}$ as an example. The completion of P is inconsistent, we therefore cannot obtain p from S using the SLDNF-resolution. However, we may derive p from $PS = \{p + 1\}$. On the other hand, roughly speaking most procedural semantics, for example, various SLD-resolutions, employ the L-strategy together with an atom ordering. So we want to introduce an atom ordering into the L-strategy.

Now let us go into the real topics. Hereafter, IR stands for superposition I or superposition II. First of all, we need to further classify the elements of $IR(c_1, c_2)$.

Suppose that $c_j = a_j m_j + b_j$ $(j = 1, 2)$, and σ is an mgu of m_1 and m_2. If IR is superposition I, we call $\Phi((b_1(a_2+b_2))\sigma)$ and $\Phi((b_2(a_1+b_1))\sigma)$ two *superposants* of c_1 and c_2 upon $m_1\sigma$. If IR is superposition II, we call $\Phi((a_1b_2 + a_2b_1)\sigma)$ a *superposant* of c_1 and c_2 upon $m_1\sigma$. And in the latter case $\Phi(b_1(a_1 + b_1))$ is called a *superposant* of c_1 and zero upon m_1.

We then have the following facts.

(1) Let c be ground, $n \in A_c$, and $n \notin A_{c(m \leftarrow i)}$. Then there exists $b \in \{c, 0\}$, such that for a superposant r of c and b upon n, $A_r \subset A_c$, $n \notin A_r$, and $c(m \leftarrow i) = r(m \leftarrow i)$.

(2) Suppose that $c' = c\theta$ is ground. For $n \in A_c$, if $n\theta \notin A_{c'}$, and for an arbitrary $m \in A_c - \{n\}$, $m\theta \neq n\theta$, then there exist $b \in \{c, 0\}$ and a superposant r of c and b upon n, such that the number of the elements of A_r is less than that of A_c, and c' is an instance of r.

In what follows, by $>_p$ we denote a partial ordering (a transitive and irreflexive binary relation) on atoms, $m >_p 1 >_p 0$ holds for any m, and $>_p$ is stable, namely, for arbitrary m, n and θ, $m >_p n$ implies that $m\theta >_p n\theta$. We say that m is *maximal* (*minimal*) in an atom set A, if for any $n \in A$, $n \not>_p m$ ($m \not>_p n$).

Now let I be a Herbrand interpretation of PS. For $m_1, m_2 \in A_c$, assume that σ is an mgu of m_1 and m_2. If $m_1\sigma$ is maximal in $A_c\sigma$, we call $\Phi(c\sigma)$ an *M-factor* of c. Moreover, if c is not satisfied by I, $\Phi(c\sigma)$ is said to be an *IM-factor* of c.

Suppose that r is a superposant of c_1 and a factor of c_2 upon $m_1\sigma$. If $m_1\sigma$ is maximal in $A_{c_1}\sigma$, we call r an *M-superposant* of c_1 and c_2 (upon $m_1\sigma$). Also, if c_1 is not satisfied by I, r is called an *IM-superposant* of c_1 and c_2 (upon $m_1\sigma$). For a superposant r of c_1 and c_2 upon $m_1\sigma$, if $m_1\sigma$ is maximal in both $A_{c_1}\sigma$ and $A_{c_2}\sigma$, we say that r is an *MM-superposant* of c_1 and c_2 (upon $m_1\sigma$).

A sequence $c_1, \ldots, c_k$ is called an *M-deduction* of c_k from $\{c_1\}$, if for $1 < t \leq k$, c_t is either an M-factor of c_{t-1}, or an M-superposant of c_{t-1} and e, where $e = c_{t-1}$ if IR is superposition I, and $e = 0$ if IR is superposition II.

A sequence $c_0, \ldots, c_k$ is defined as a *basic confluent LM-deduction* of c_k from *PS* with top c_0, if c_s is an M-factor of c_{s-1} or an M-superposant of c_{s-1} and b_{s-1}, where $b_{s-1} \in PS$ $(s = 1, \ldots, k)$.

We also have the following recursive definition.

(1) A basic confluent LM-deduction of c_1 from *PS* with top c_0 is a confluent LM-deduction of c_1 from *PS* with top c_0.

(2) Suppose that D_p $(p = 1, 2)$ are confluent LM-deductions of c_p from *PS* with top c_0, and c_{12} is an MM-superposant of c_1 and c_2; then the sequence D_1, D_2, c_{12} is a confluent LM-deduction of c_{12} from *PS* with top c_0.

(3) Suppose that D_p is a confluent LM-deduction of c_p from *PS* with top c_0, and D_q is a confluent LM-deduction of c_q from *PS* with top c_p. Then the sequence D_p, D_q is a confluent LM-deduction of c_q from *PS* with top c_0.

A sequence $c_1, \ldots, c_k$ is called an *MM-deduction* of c_k from *PS*, if for any $t = 1, \ldots, k$, either $c_t \in PS$, or c_t is an M-factor of c_s $(s < t)$, or c_t is an MM-superposant of c_p and c_q $(p, q < t)$.

A sequence $c_1, \ldots, c_k$ is called an *IM-deduction* of c_k from *PS*, if for any $t = 1, \ldots, k$, either $c_t \in PS$, or c_t is an IM-factor of c_s $(s < t)$, or c_t is an IM-superposant of c_p and c_q $(p, q < t)$.

Similarly, if $c_k = 1$ in the three definitions above, we call the sequences a *confluent LM-refutation* from *PS* with top c_0, an *MM-refutation* from *PS* and an *IM-refutation* from *PS*, respectively.

To illustrate these definitions, we consider the example in Sect. 11.4.1 again. Here we suppose $M(x, y, z) >_p P(w)$. Then,

$M(b, i(b), z)(P(z) + 1)$ is an M-superposant of c_0 and c_1,
$P(e) + 1$ is an M-superposant of $M(b, i(b), z)(P(z) + 1)$ and c_3.

Thus the sequence

$$c_0, M(b, i(b), z)(P(z) + 1), P(e) + 1$$

is a basic confluent LM-deduction of $P(e) + 1$ from *PS* with top c_0. Also

$P(i(b))$ is an M-superposant of c_0 and zero,
$P(x)P(y)M(x, i(y), i(b))$ is an M-superposant of $P(i(b))$ and c_1,
$P(e)P(b)$ is an M-superposant of $P(x)P(y)M(x, i(y), i(b))$ and c_2.

Therefore the sequence

$$c_0, P(i(b)), P(x)P(y)M(x, i(y), i(b)), P(e)P(b)$$

is a basic confluent LM-deduction of $P(e)P(b)$ from PS with top c_0. In addition,

$P(b)$ is an MM-superposant of $P(e)+1$ and $P(e)P(b)$,
1 is an M-superposant of $P(b)$ and c_0.

So the sequence

$$P(b), 1$$

is a basic confluent LM-deduction of 1 from PS with top $P(b)$.

The sequence

$$c_0,\ M(b,i(b),z)(P(z)+1),\ P(e)+1,\ c_0,\ P(i(b)),\\ P(x)P(y)M(x,i(y),i(b)), P(e)P(b),\ P(b),\ P(b),\ 1$$

is thus a confluent LM-refutation from PS with top c_0. We also know that

$P(e)P(y)(P(i(y))+1)$ is an MM-superposant of c_1 and c_2,
$P(i(b))$ is an MM-superposant of c_0 and zero,
$P(e)P(b)$ is an MM-superposant of $P(e)P(y)(P(i(y))+1)$ and $P(i(b))$,
$P(x)(P(e)+1)$ is an MM-superposant of c_1 and c_3,
$P(b)P(x)$ is an MM-superposant of $P(e)P(b)$ and $P(x)(P(e)+1)$,
$P(b)$ is an M-factor of $P(b)P(x)$,
1 is an MM-superposant of $P(b)$ and c_0.

Therefore the sequence

$$c_1, c_2,\ P(e)P(y)(P(i(y))+1),\ c_0,\ 0,\ P(i(b)),\ P(e)P(b),\\ c_3,\ P(x)(P(e)+1),\ P(b)P(x),\ P(b),\ 1$$

is an MM-refutation from PS.

Now let I be the positive Herbrand interpretation of PS. Then

$P(i(b))$ is an IM-superposant of c_0 and zero,
$P(x)P(y)M(x,i(y),i(b))$ is an IM-superposant of $P(i(b))$ and c_1,
$P(e)P(b)$ is an IM-superposant of $P(x)P(y)M(x,i(y),i(b))$ and c_2,
$P(e)$ is an IM-superposant of $P(e)P(b)$ and c_0,
$P(x)P(y)M(x,i(y),e\ )$ is an IM-superposant of $P(e)$ and c_1,
$P(x)$ is an IM-superposant of $P(x)P(y)M(x,i(y),e)$ and c_3,
1 is an IM-superposant of $P(x)$ and c_0.

Thus the sequence

$$c_0,\ 0,\ P(i(b)),\ c_1,\ P(x)P(y)M(x,i(y),i(b)),\ c_2,\ P(e)P(b),\\ P(e),\ P(x)P(y)M(x,\ i(y),\ e),\ c_3,\ P(x),\ 1$$

is an IM-refutation from PS.

Here we note that an M-deduction of c_k from $\{c_1\}$ is in fact a (basic) confluent LM-deduction of c_k from $\{c_1\}$ with top c_1. It is also an MM-deduction of c_k from $\{c_1\}$. And an M-deduction of $c_k(m \leftarrow i)$ from $\{c_1(m \leftarrow i)\}$ can be improved to be an IM-deduction of c_k from $\{c_1\}$ if $c_1(m \leftarrow i)$ is not satisfied by I and $i \in I$.

According to the above facts and from the unit, stable, split, and lifting conditions, we have the following two theorems.

Theorem 11.5:

Suppose that PS is unsatisfiable. Let IR be superposition I or superposition II.

(1) *If PS $\setminus \{c_0\}$ is satisfiable, then there exists a confluent LM-refutation from PS with top c_0.*

(2) *There exists an MM-refutation from PS.*

(3) *There exists an IM-refutation from PS.*

This theorem shows that superpositions I and II are complete for the confluent LM-strategy, MM-strategy and IM-strategy.

If we require $b_{s-1} \in PS \cup \{c_t |\ 0 \leq t \leq s-1\}$ in the definition of a basic confluent LM-deduction of c_k from *PS* with top c_0, then the sequence is called an *LM-deduction* of c_k from *PS* with top c_0. We can define an *LM-refutation* from *PS* with top c_0 similarly.

Theorem 11.6:

Suppose that PS is clausal and unsatisfiable. Let IR be superposition I or superposition II. If PS $\setminus \{c_0\}$ is satisfiable, then there exists an LM-refutation from PS with top c_0.

This theorem demonstrates that the LM-strategy is complete for clausal theorem proving. However, we do not know whether or not it is complete in the general case.

For the proofs of Theorem 11.5 (1) and Theorem 11.6, the reader is referred to Wu (1996). The proofs of Theorem 11.5 (2) and (3) are similar, for which one may consult Wu (1996) and Wu and Liu (1998). We also remark that the proof of Theorem 11.5 (1) is quite different from those of Theorem 11.5 (2) and (3).

We see that the advantage of the LM-strategy over the L-strategy is achieved by introducing a partial ordering $>_p$. Some superposant computation may thus be avoided. Although the confluent LM-strategy has the above advantage and it is complete, the number of first-order polynomials we have to consider sometimes may increase rapidly. For the case where there are more atoms and fewer first-order polynomials, the confluent LM-strategy may be better. Conversely, when there are more first-order polynomials and fewer atoms, the L-strategy may be better.

11.4.3 Two problems

In this section, we briefly discuss two problems related to the Gröbner basis method and odd-strategy.

(1) *On the Gröbner basis method.* The Gröbner basis method developed by Kapur and Narendran uses multiple overlaps and sequential overlaps. There are still some questions about it:

Can parallel overlaps be used instead of sequential overlaps? Are multiple overlaps needed? And are only the unifications on maximal atoms sufficient?

Dershowitz et al. (1991), Kapur and Narendran (1985), and Zhang (1994a) contain more details. These questions are also posed as Problem #36 in the RTA list of open problems (see the web site www.lri.fr/~rtaloop/).

As we have seen, in the MM-strategy for superposition II what we use are single overlaps and parallel overlaps, and we only need the unifications on maximal atoms since the inference rule is computing the MM-superposants and M-factors. Furthermore, we see that there are two main differences between the MM-strategy and the Gröbner basis method. One is that parallel overlaps are used in the MM-strategy, and the other is that the use of the idempotency rule B_2 is quite different. We have the following.

Theorem 11.7: (Wu and Liu 1996a)

For clausal theorem proving, parallel overlaps can be used instead of sequential overlaps, and only single overlaps and the unifications on maximal atoms are needed if parallel overlaps are used in the Gröbner basis method.

For non-clausal theorem proving, if we convert the use of the idempotency rule B_2 in the Gröbner basis method to what we did in the MM-strategy, we see that the conclusions above still hold.

(2) *The completeness of odd-superposition II.* Parenthetically, we discuss here whether or not odd-superposition II together with odd-factoring is complete. This is posed as an open question by Zhang (1994a) though it does not seem so difficult.

For clausal theorem proving, this proof system is clearly complete since odd-superposition II coincides with odd-superposition in this case. Now let us think of the IM-strategy for superposition II we presented before. In the definition of the IM-superposants, if we are not allowed to use arbitrary factors of c_2 as well as the IM-superposants of c_1 and zero, and choose I to be the positive Herbrand interpretation of *PS* (c_1 is thus odd), then the IM-strategy for superposition II is just odd-superposition II. So, if the answer to the question were still positive in non-clausal case, it would be unnecessary for us to define the IM-superposants of a first-order polynomial and zero. However, the following example shows that this is not true even for propositional logic.

Example 11.1:

$PS = \{c_1 = pq + 1, c_2 = pr + r + 1\}, r >_p q >_p p.$

PS is unsatisfiable, and in *PS* only c_2 is odd. If we follow the above requirements, then c_1 and c_2 cannot conduct the computation of superposition II, and the unique result applying superposition II to c_2 and itself is zero. So we are unable to derive 1 from *PS*.

We therefore have

Theorem 11.8:

Odd-superposition II with odd-factoring is incomplete.

Indeed, odd-superposition II cannot offer all the necessary information derived from the idempotency rule B_2. So we have to figure out such information by defining the IM-superposants of first-order polynomials and zero. Also, we have to use the factors of c_2 when we compute the IM-superposants of c_1 and c_2 in the IM-strategy. We do have a question: Is the IM-strategy still complete, if we do not compute the factors of c_2? For superposition I, the answer is positive (Zhang 1994a).

For more information about this problem and the proof system using odd-superposition II, the reader is referred to Wu and Liu (1996b).

11.4.4 On simplification rules

How to introduce simplification rules is a critical topic in theorem proving. So far it has not been considered carefully. We only remark here that the completeness of the MM-strategy and IM-strategy for superpositions I and II can also be shown using the technique of shrinking semantic trees (Chang and Lee

1973; Hsiang 1985). Thus, we can prove that superposition I and superposition II are still complete if P-rules and N-rules are used as simplification rules, or the reduction defined in Zhang (1994a) is applied. Also, the reader can consult Bachmair and Ganzinger (1997) on introducing simplification rules for the similar superpositions I and II.

11.4.5 A remark

B. Ma from Peking University has implemented the L-strategy for superposition II using Mathematica on a computer without considering too much data structures and efficiency, and made some experiments. Table 11.1 is a statistic list for several classical theorems chosen from Chang and Lee (1973) and Hsiang (1985), in which, "PS" stands for the number of the input first-order polynomials, "New Polys" the number of the new first-order polynomials generated in the whole proof procedure, and "Proof" the number of the first-order polynomials which constitute a proof of the theorem. Certainly this proof contains all the input first-order polynomials used in the procedure.

Examples	PS	New Polys	Proof
Ex1	4	6	5
Ex2	9	26	14
Ex3	5	12	4
Ex4	6	42	11
Ex5	4	38	7
Ex6	4	57	7
Ex7	8	14	4
Ex8	8	20	7
Ex9	7	15	6
Ex10	8	44	10
Ex11	9	49	11

Table 11.1

Ex1: If a symmetric and transitive relation R has the property $\forall x \exists y R(x, y)$ then R is reflexive.
Ex2: A verification condition from Hoare's FIND program.
Ex3: There is a right identity in an associative algebraic system with left and right solutions.
Ex4: Associative systems with identity e where the square of each element is e are commutative.
Ex5: In a group left identity is right identity.
Ex6: In a group with left identity and left inverse every element has a right inverse.
Ex7: S ($\neq \emptyset$) is a subset of a group such that $x, y \in S$ implies $xy^{-1} \in S$. Then the identity $e \in S$.
Ex8: S ($\neq \emptyset$) is a subset of a group such that $x, y \in S$ implies $xy^{-1} \in S$. Then $x^{-1} \in S$ for $x \in S$.
Ex9: If a is prime and $a = b^2/c^2$ then a divides b.
Ex10: There exist infinitely many primes.
Ex11: Any number greater than 1 has a prime divisor.

We remark that more first-order polynomials might be produced since we did not introduce any simplification rules. As a matter of fact, our aim here is just testing whether the L-strategy is practically available rather than comparing the efficiency.

11.5 On non-clausal theorem proving

An advantage of non-clausal theorem proving is that one does not need to convert the input formulas into clauses before they are transformed into first-order polynomials. In addition to various non-clausal theorem proof systems based on Boolean ring or first-order polynomials, there are mainly two that are complete. They are generalized resolution and the NC-resolution. All these proof systems are different from each other. We do not discuss their individual advantages and defects here. Hereafter, we only say something simple and direct about the relationship between them.

Generalized resolution and the NC-resolution are similar except for the fact that reduction and polarity are employed in the NC-resolution. Unfortunately, the NC-resolution is incomplete for some strategies.

It is not hard to see that both superposition I and superposition II are generalizations of resolution, and superposition I is indeed the generalized resolution or NC-resolution in first-order polynomial version.

Zhang's odd-strategy may be considered as a restriction strategy for superposition I, in which if c_1 is odd then $\Phi((b_1(a_2+b_2))\sigma)$ can be discarded.

Like generalized resolution, we know that it is useful to define superposition I. We can thus consider various different strategies. This is quite significant since different strategies have different advantages. For example, we can easily use various tree searching techniques if the linear strategy is applied. However, if we want to guarantee the completeness of the proof systems, generally speaking, such a superposant has to consist of two first-order polynomials. This is also like the NC-resolution and generalized resolution. Therefore, in the proof procedure, more first-order polynomials may be produced.

As to the relationship between generalized resolution or the NC-resolution and superposition II, we only point out the following simple facts.

Let ψ, ψ_1 and ψ_2 be generalized clauses, and ϕ_1 and ϕ_2 be the generalized resolvent of ψ_1 and ψ_2 upon $m_j\sigma$, ϕ the generalized resolvent of ψ and itself upon m. If m appears in $p\psi$ and m_j appears in $p\psi_j$ $(j=1,2)$, then for the superposant r of $p\psi_1+1$ and $p\psi_2+1$ upon $m_j\sigma$, $r=p\phi_1+p\phi_2$. For the superposant r of $p\psi+1$ and zero upon m, we have $r=p\phi+1$.

Basically, using superposition II we do not have to produce two first-order polynomials or generalized resolvents each time, However, we cannot neglect the results of idempotency rule B_2 in this case.

11.6 Conclusion

In this chapter, we are concerned with theorem proving based on first-order polynomials. We defined well-behaved inference rules by describing four conditions. It was shown that well-behaved inference rules are not only complete but also complete for both the SOS-strategy and the L-strategy. We also presented two such concrete inference rules: superposition I and superposition II. The example in Sect. 11.3.2 demonstrates that resolution is actually well-behaved. Further, we introduced the confluent LM-strategy, LM-strategy, MM-strategy and IM-strategy for superpositions I and II, and discussed their completeness.

Before closing the chapter, we propose some topics that are worthy of being investigated further.

(1) In Bachmair and Ganzinger (1997), a rather general notion of redundancy for inferences was developed. One may investigate the behaviors of well-behaved inference rules combined with this powerful notion.

(2) We know that one advantage of the N-strategy and the Gröbner basis method is that they are able to be generalized to built-in equational theories in a natural way. This aspect may be considered for the remainder method.

(3) As we have pointed out, superposition II is similar to computing pseudo-remainders of two polynomials in Wu's method. One may investigate whether the entire framework of Wu's method — (a) computing basic chains, (b) computing characteristic sets, and (c) decomposing zero sets — can be applied to first-order theorem proving to obtain a complete method.

(4) It is interesting to study logic programming using first-order polynomials. First one may focus on operational semantics. For definite programs, it seems that the resolution in polynomial form similar to the SLD-resolution is still available. The next thing is how to cope with negative information. For more general programs, we might also talk about these issues since we have the L-strategy, etc. as the foundation.

Finally, there are two other technical questions: Do there exist other inference rules which are well-behaved? Is there an inference rule that uses single overlaps and is not well-behaved?

Part III

Algebraic Computation

Chapter 12

GREATEST COMMON RIGHT DIVISORS, LEAST COMMON LEFT MULTIPLES, AND SUBRESULTANTS OF ORE POLYNOMIALS

Ziming Li

Ore polynomials establish a general mathematical setting for manipulating linear differential and difference polynomials. In this chapter, we summarize some recent progress in computing greatest common right divisors (gcrd's) and least common left multiples (lclm's) of Ore polynomials. We describe a subresultant theory for Ore polynomials, and outline fraction-free and modular algorithms for computing gcrd's and lclm's.

12.1 Introduction

Polynomial gcd-computation is one of the most fundamental building blocks in computer algebra systems. The lcm of two polynomials A and B can easily be obtained from their gcd through the well-known formula:

$$\operatorname{lcm}(A, B) = \frac{AB}{\gcd(A, B)}. \tag{12.1}$$

The study of efficient gcd-computation dates back to the early 1960s when computer algebra was born. The first attempt was to implement the Euclidean algorithm for computing the gcd of two univariate polynomials over the integers. It was immediately observed that this algorithm was only of classroom use, because integral coefficients of intermediate polynomials grow exponentially (see Knuth 1997, pp. 432–433). The Euclidean algorithm for univariate polynomials over the rational numbers was also slow, because it required too many integral gcd-calculations. To solve this bottle-neck problem, Collins (1966, 1967), Brown and Traub (1971) developed a subresultant theory, which shows that any polynomial produced by a Euclidean algorithm is linearly dependent on certain determinants. These determinants are called *subresultants* and can be expanded recursively. The subresultant theory provides us with

MATHEMATICS MECHANIZATION AND APPLICATIONS
ISBN 0-12-734760-7

a fraction-free Euclidean algorithm (called the *subresultant algorithm*), which has a polynomial complexity (see Knuth 1997, pp. 432–433). Besides gcd-computation, the subresultant theory is applicable to many other problems such as real root isolation (Collins and Loos 1982), computation in algebraic extensions (Rubald 1973), cylindrical algebraic decomposition (Collins 1975), geometric coding theory (Shen 1992) and polynomial system solving (Manocha and Canny 1992; Wang 1998a). Attempts to extend the subresultant theory to multivariate polynomials were also made for different purposes by Gonzàlez-Vega (1991) and Mandache (1994).

Soon after the birth of the commutative subresultant theory, Brown (1971) and Collins developed a modular algorithm for computing gcd's (see, also, Knuth 1997, pp. 453–455). The idea is to map arithmetic over $\mathbb{Z}$ (the ring of integers), which is costly, to arithmetic over $\mathbb{Z}_p$ for a number of primes p via modular homomorphisms, and combine modular images by the Chinese remainder algorithm (see any book on computer algebra). This idea has been developed further and applied to many other problems such as: computing gcd's of multivariate polynomials over the integers, gcd's of polynomials over algebraic number fields (Encarnación 1995) and multivariate resultants (Collins 1971).

To manipulate linear homogeneous differential and difference equations, we regard these equations as non-commutative polynomials. A general algebraic setting for linear operators from analysis is called *Ore polynomial rings* (Ore 1933). Gcrd's and lclm's are basic objects in the computation of Ore polynomials. For instance, the gcrd and lclm of two linear differential operators correspond to the intersection and sum of their solution spaces, respectively.

Gcrd's and lclm's can be computed by the non-commutative Euclidean algorithm and extended Euclidean algorithm, respectively (Ore 1933; Bronstein and Petkovšek 1994, 1996). But lclm's cannot be computed as easily as in the usual commutative case, because (12.1) does not hold for non-commutative polynomials. As in the commutative case, naive implementations of these two algorithms are very inefficient. This is not a surprise because integral univariate polynomials are linear differential (difference) operators with constant coefficients. To obtain better implementations, we extend the subresultant algorithm and modular techniques developed for efficient gcd-computation to Ore polynomials. It turns out that the commutative subresultant theory can be fully generalized to Ore polynomials, and, moreover, subresultants are essentially used in applying modular techniques to computing gcrd's. Recall that a modular gcd-algorithm may be designed without using subresultants.

This chapter discusses various algorithms for computing gcrd's and lclm's of Ore polynomials, and describe the key role played by subresultants in efficient gcrd- and lclm-computation. We present basic ideas and descriptions of these algorithms, and compare them by empirical timings. Some proofs are merely sketched. The reader is referred to the articles cited for details.

The organization of this chapter is as follows. Sect. 12.2 reviews Ore polynomial rings and classical algorithms for computing gcrd's and lclm's. Sect. 12.3 describes a subresultant theory for Ore polynomials and fraction-free methods for computing gcrd's and lclm's. Sect. 12.4 outlines a modular method for computing gcrd's. Sect. 12.5 contains a few remarks on possible extensions of the present methods.

12.2 Basic operations in an Ore polynomial ring

The notion of Ore polynomial rings was introduced by Ore (1933). His goal was to provide a general mathematical setting to study decompositions of various linear operators from analysis. The use of Ore polynomials in computer algebra is pointed out by Bronstein and Petkovšek (1994, 1996), because an algorithm working for Ore polynomials generates a family of algorithms working for linear operators of various kinds such as: linear differential, shift (difference), and q-difference operators. For example, Chyzak and Salvy (1998) use Ore algebras to formulate non-commutative elimination and apply this elimination machinery to prove multivariate identities. Bronstein and Petkovšek also express Ore polynomials as pseudo-linear operators acting on some vector spaces. Hence, we benefit from Ore polynomials not only as a labor-saving device but also as an interplay between algebraic operations and geometric meanings.

The goal of this section is to describe basic notions and operations of Ore polynomial rings. Sect. 12.2.1 presents basic definitions and Sect. 12.2.2 presents classical algorithms for computing gcrd's and lclm's.

12.2.1 Ore polynomial rings

Let $\mathcal{R}$ be a commutative domain and σ an injective endomorphism of $\mathcal{R}$, which is called *conjugate operator*. An endomorphism δ of the additive group $\mathcal{R}$ is called a *pseudo-derivation* with respect to σ if

$$\delta(ab) = \sigma(a)\delta(b) + \delta(a)b, \quad \text{for all } a, b \in \mathcal{R}.$$

Regard $\mathcal{R}[X]$ as a left module over $\mathcal{R}$ and define the (non-commutative) multiplication by the commutation rule

$$Xa = \sigma(a)X + \delta(a), \quad \text{for all } a \in \mathcal{R}. \tag{12.2}$$

The triple $(\mathcal{R}[X], \sigma, \delta)$ is called an *Ore polynomial ring*. For brevity we denote this triple by $\mathcal{R}[X]$.

Example 12.1:

Let a be in $\mathcal{R}$. Then (12.2) yields

$$X(aX) = (Xa)X = (\sigma(a)X + \delta(a))\, X = \sigma(a)X^2 + \delta(a)X.$$

For $A, B \in \mathcal{R}[X]$, the product of A and B is denoted by AB and the degree of AB is equal to the sum of the degrees of A and B. The conjugate operator σ and pseudo-derivation δ can be uniquely extended to $\mathcal{F}$, the quotient field of $\mathcal{R}$, by letting

$$\sigma\left(\frac{a}{b}\right) = \frac{\sigma(a)}{\sigma(b)} \quad \text{and} \quad \delta\left(\frac{a}{b}\right) = \frac{b\delta(a) - a\delta(b)}{\sigma(b)b},$$

for $a, b \in \mathcal{R}$ with $b \neq 0$. It follows that $(\mathcal{F}[X], \sigma, \delta)$ is also an Ore polynomial ring. A detailed verification of this statement can be found in Li Z. (1996, Chap. 2).

Some terminology will be used later. For $n \in \mathbb{N}$ (the set of natural numbers) and $a \in \mathcal{R}$, the *nth σ-factorial* of a is defined to be the product $\prod_{i=0}^{n-1} \sigma^i(a)$ and denoted by $a^{[n]}$. In addition, $a^{[0]}$ is 1. For a non-zero element A of $\mathcal{F}[X]$, the leading coefficient of A is denoted by $\mathrm{lc}(A)$. It follows from (12.2) that

$$\mathrm{lc}(X^k A) = \sigma^k(\mathrm{lc}(A)) \quad \text{for } k \in \mathbb{N}. \tag{12.3}$$

Moreover, A is *monic* if $\mathrm{lc}(A) = 1$; A is *fraction-free* if $A \in \mathcal{R}[X]$; and A is *primitive* if R is a unique factorization domain, $A \in \mathcal{R}[X]$ and the coefficients of A are relatively prime.

If non-zero $A, B \in \mathcal{F}[X]$ with $\deg A = m \geq \deg B = n \geq 0$, the (right) division of A and B is performed as follows: let

$$Q_0 = \frac{\mathrm{lc}(A)}{\sigma^{m-n}(\mathrm{lc}(B))} X^{m-n}.$$

The leading monomial of $Q_0 B$ is $\mathrm{lc}(A)X^m$ by (12.3), so we can recursively divide $A - Q_0 B$ by B on the right, obtaining $Q_1, R \in \mathcal{F}[X]$ such that

$$A - Q_0 B = Q_1 B + R \quad \text{and} \quad \deg R < n.$$

Thus, we obtain the formula

$$A = QB + R \tag{12.4}$$

where $Q = Q_0 + Q_1$. We call Q and R the *right quotient* and *right remainder* of A and B, and denote them by $\text{quo}(A, B)$ and $\text{rem}(A, B)$, respectively.

If A and B belong to $\mathcal{R}[X]$ and one wants to perform division in $\mathcal{R}[X]$, one can define the right pseudo-division in a similar way, which leads to the pseudo-division formula:

$$\text{lc}(B)^{[m-n+1]} A = QB + R \quad \text{and} \quad \deg R < n, \tag{12.5}$$

where $Q, R \in \mathcal{R}[X]$. We call Q and R the *right pseudo-quotient* and *right pseudo-remainder* of A and B, and denote them by $\text{pquo}(A, B)$ and $\text{prem}(A, B)$, respectively. Left division and pseudo-division of two Ore polynomials exist when σ is an automorphism of $\mathcal{R}$. They can be considered as right division and pseudo-division via adjoint construction. Hence, we will focus on the right (pseudo) division and omit the word "right" when no confusion occurs.

Since formulas (12.4) and (12.5) imply that

$$\text{pquo}(A, B) = \text{lc}(B)^{[m-n+1]}\text{quo}(A, B)$$

and

$$\text{prem}(A, B) = \text{lc}(B)^{[m-n+1]}\text{rem}(A, B),$$

there is little theoretical difference between the division and pseudo-division. However, these two operations are quite different from a computational point of view, because the pseudo-division only requires arithmetic within the coefficient domain $\mathcal{R}$ when A and B belong to $\mathcal{R}[X]$. Recall that arithmetic in $\mathcal{R}$ is usually cheaper than that in $\mathcal{F}$.

Finally, we present two examples of Ore polynomial rings. The reader is referred to Chyzak and Salvy (1998) for more examples.

Example 12.2:

If D is a derivation operator on $\mathcal{R}$, then D is a pseudo-derivation with respect to the identity mapping $\mathbf{1}$ because $D(ab) = aD(b) + D(a)b$, for $a, b \in \mathcal{R}$. Hence, $(\mathcal{R}[X], \mathbf{1}, D)$ is an Ore ring with the multiplication given by $Xa = aX + D(a)$, for $a \in \mathcal{R}$. This ring is isomorphic to the ring of linear homogeneous differential polynomials in one differential indeterminate over $\mathcal{R}$.

Example 12.3:

Let σ be an injective endomorphism of $\mathcal{R}$ and δ be the null mapping $\mathbf{0}$. Then $(\mathcal{R}[X], \sigma, \mathbf{0})$ is an Ore ring with the multiplication given by $Xa = \sigma(a)X$, for $a \in \mathcal{R}$. This ring is isomorphic to the ring of linear homogeneous difference polynomials in one difference indeterminate (with respect to σ) over $\mathcal{R}$.

12.2.2 Classical algorithms for computing gcrd's and lclm's

For $A, B \in \mathcal{F}[X]$, B is a right factor of A if there exists $L \in \mathcal{F}[X]$ such that $A = LB$; B is a left multiple of A if there exists $L \in \mathcal{F}[X]$ such that $B = LA$. A common right factor of A and B, with the highest degree, is called a *gcrd* of A and B. A common left multiple of A and B, with the lowest degree, is called an *lclm* of A and B. Two gcrd's (lclm's) of A and B are unique up to a non-zero element in $\mathcal{F}$. A gcrd and an lclm of A and B are denoted by $\text{gcrd}(A, B)$ and $\text{lclm}(A, B)$, respectively.

Ore (1933) shows that, for non-zero $A, B \in \mathcal{F}[X]$, we have

$$\deg A + \deg B = \deg \text{lclm}(A, B) + \deg \text{gcrd}(A, B). \tag{12.6}$$

If A and B are linear differential operators with respective solutions spaces $\mathcal{V}(A)$ and $\mathcal{V}(B)$ over the constant field C, where $C = \{a \in \mathcal{F} \mid \delta(a) = 0\}$, this equality specializes to the well-known fact

$$\dim_C \mathcal{V}(A) + \dim_C \mathcal{V}(B) = \dim_C \mathcal{V}(\text{gcrd}(A, B)) + \dim_C \mathcal{V}(\text{lclm}(A, B)),$$

because

$$\mathcal{V}(\text{gcrd}(A, B)) = \mathcal{V}(A) \cap \mathcal{V}(B) \quad \text{and} \quad \mathcal{V}(\text{lclm}(A, B)) = \mathcal{V}(A) + \mathcal{V}(B).$$

The (extended) Euclidean algorithm for computing gcrd's (lclm's) reads exactly the same as its commutative counterpart.

Algorithm E: (Greatest common right divisor of two Ore polynomials)

Given non-zero A and B in $\mathcal{F}[X]$, this algorithm calculates a gcrd of A and B.

E1. [Division.] Compute the (right) remainder R of A and B.

E2. [Check the termination.] If $R = 0$ then the algorithm terminates with B as the desired result. Otherwise, replace A by B, and B by R. Go back to E1. ■

Numbering intermediate remainders in Algorithm E properly, we obtain a sequence of non-zero polynomials

$$A, B, A_3, A_4, \ldots, A_n$$

in which $A_i = \text{rem}(A_{i-2}, A_{i-1})$, for $i = 3, 4, \ldots, n$, (setting $A_1 = A$ and $A_2 = B$), A_n is a gcrd of A and B. A sequence of non-zero polynomials (in $\mathcal{F}[X]$):

$$A, B, B_3, B_4, \ldots, B_n$$

is called a *polynomial remainder sequence* (abbreviated as *prs*) of A and B if B_i and A_i are linearly dependent over $\mathcal{F}$, for $i = 3, 4, \ldots, n$.

There are many ways to normalize remainders appearing in Algorithm E. For example, if one changes the division (rem) to the pseudo-division (prem) in Algorithm E, one gets a variant of the Euclidean algorithm which we call the *pseudo-Euclidean algorithm*.

The extended Euclidean algorithm yields a gcrd and an lclm of two Ore polynomials altogether.

Algorithm X: (Extended Euclidean algorithm)
Given non-zero A and B in $\mathcal{F}[X]$, this algorithm determines G, U and V in $\mathcal{F}[X]$ such that G is a gcrd of A and B and that both UA and VB are lclm's of A and B. The computation makes use of auxiliary vectors (R_1, R_2, R_3) (S_1, S_2, S_3) and (T_1, T_2, T_3); all vectors are manipulated in such a way that the relations

$$R_1A + R_2B = R_3, \qquad S_1A + S_2B = S_3, \qquad T_1A + T_2B = T_3$$

hold throughout the calculations.

X1. [Initialize.] Set $(R_1, R_2, R_3) \leftarrow (1, 0, A)$, $(S_1, S_2, S_3) \leftarrow (0, 1, B)$.

X2. [Is $S_3 = 0$?] If $S_3 = 0$, set $G \leftarrow R_3$, $U \leftarrow S_1$, $V \leftarrow S_2$ and exit.

X3. [Divide, subtract.] Set $Q \leftarrow \text{quo}(R_3, S_3)$, and then set

$$(T_1, T_2, T_3) \leftarrow (R_1, R_2, R_3) - Q(S_1, S_2, S_3),$$
$$(R_1, R_2, R_3) \leftarrow (S_1, S_2, S_3), \qquad (S_1, S_2, S_3) \leftarrow (T_1, T_2, T_3).$$

Return to X2. ■

It is clear that G is a gcrd of A and B. When $S_3 = 0$, we have $S_1A + S_2B = 0$, so that both UA and VB are left common multiples of A and B. An easy induction on degrees proves $\deg U = \deg B - \deg G$ and $\deg V = \deg A - \deg G$. It then follows from (12.6) that both UA and VB are lclm's of A and B. We may omit the second components R_2, S_2 and T_2 in Algorithm X. The resulting algorithm is called the *half-extended Euclidean algorithm*, which produces $G = \text{gcrd}(A, B)$ and $UA = \text{lclm}(A, B)$.

Another method for computing lclm's is as follows: let non-zero A and B be in $\mathcal{F}[X]$ with $\deg A = m$ and $\deg B = n$. Regard the sequence

$$X^n A, X^{n-1} A \ldots, A, X^m B, X^{m-1} B, \ldots, B \tag{12.7}$$

as $(m+n+2)$ vectors in an $(m+n+1)$-dimensional vector space over $\mathcal{F}$. Then there exists a non-trivial linear relation among these vectors. Set up the linear homogeneous system

$$\lambda_n X^n A + \lambda_{n-1} X^{n-1} A + \ldots + \lambda_0 A + \mu_0 B + \ldots + \mu_{m-1} X^{m-1} B + \mu_m X^m B = 0 \tag{12.8}$$

which consists of $(m+n+2)$ unknowns λ_i $(0 \le i \le n)$ and μ_j $(0 \le j \le m)$ and $(m+n+1)$ equations. Find the largest integer r such that

$$(\underbrace{0, \ldots, 0}_{r}, \lambda^*_{n-r}, \lambda^*_{n-r-1}, \ldots, \lambda^*_0, \mu^*_0, \ldots, \mu^*_{m-r-1}, \mu^*_{m-r}, \underbrace{0, \ldots, 0}_{r})$$

is a non-trivial solution of (12.8). Then

$$\lambda^*_{n-r} X^{n-r} A + \cdots + \lambda^*_0 A \quad \text{and} \quad \mu^*_0 B + \cdots + \mu^*_{m-r} X^{m-r} B$$

are lclm's of A and B.

Example 12.4:

Let the identity mapping and differential operator $\frac{d}{dt}$ be the conjugate operator and pseudo-derivation on $\mathbb{Z}[t]$, respectively. Let X be an indeterminate over $\mathbb{Z}[t]$. Then $\mathbb{Z}[t][X]$ is $\mathbb{Z}[t]$-isomorphic to the ring of linear differential operators over $\mathbb{Z}[t]$. If

$$A = X^4 \quad \text{and} \quad B = (t^2 - t)X^3 - 3tX + 6,$$

then $(tX - 2)$ is a gcrd of A and B. Observe that

$$t^2 A = (tX^3 - X^2)(tX - 2) \quad \text{and} \quad B = ((t-1)X^2 - 3)(tX - 2).$$

Hence the product of two primitive polynomials is *not* necessarily primitive, and there does *not* exist F in $\mathbb{Z}[t][X]$ such that $A = F(tX - 2)$.

Example 12.5:
Let σ be the endomorphism of $\mathbb{Z}[t]$ over $\mathbb{Z}$ that sends t to $(t+1)$. Let σ and the null mapping be the conjugate operator and pseudo-derivation on $\mathbb{Z}[t]$, respectively. Then $\mathbb{Z}[t][X]$ is $\mathbb{Z}[t]$-isomorphic to the ring of linear shift operators over $\mathbb{Z}[t]$. If

$$A = t(t+1)X^2 - 2t(t+2)X + (t+1)(t+2)$$

and

$$B = (t-1)X^2 - (3t-2)X + 2t,$$

then $tX - (t+1)$ is the primitive gcrd of A and B w.r.t. X. Notice that the gcd of $\mathrm{lc}(A)$ and $\mathrm{lc}(B)$ is *not* a multiple of the leading coefficient of $\mathrm{gcrd}(A, B)$.

These two examples warn us to be careful about Ore polynomials, although the (half-extended) Euclidean algorithm for Ore polynomials is a straightforward generalization of its commutative counterpart.

Let $\mathcal{R}$ be the ring of commutative univariate polynomials in t over the integers. The pseudo-Euclidean algorithm computes a prs whose coefficients are in $\mathcal{R}$. Unfortunately, both degree (in t) and the integral length of the coefficients in this prs grow exponentially. The proof of this assertion can be found, for instance, in Geddes et al. (1992) for the commutative case. The assertion remains true for Ore polynomials, because commutative polynomials are a special type of Ore polynomials.

If the input polynomials have a trivial gcrd, then the method for computing lclm's by solving linear equations has a very good performance, thanks to fraction-free Gaussian elimination (Bareiss 1968). But if the input polynomials have non-trivial gcrd's, we have to apply the method several times to find the "minimal" linear relation among the members of (12.7).

12.3 Subresultants of Ore polynomials

In this section we present a subresultant theory for Ore polynomials (Li Z. 1996, 1998). The subresultant theorem (Theorem 12.1) is the major result. This theorem has two important consequences: Theorem 12.2 and Theorem 12.3; the former is a fraction-free Euclidean algorithm with polynomial complexity, the latter presents determinant formulas for gcrd's and lclm's.

This section is organized as follows. In Sect. 12.3.1 we discuss the motivation of subresultants. The subresultant theory is summarized in Sect. 12.3.2. We show how to use the subresultant algorithm to compute gcrd's and lclm's in Sect. 12.3.3.

12.3.1 Motivation

Our goal is to design algorithms for computing gcrds and lclms of two Ore polynomials in $\mathcal{R}[X]$. These algorithms operate in $\mathcal{R}[X]$ and have polynomial complexities in terms of coefficients.

Let A_1 and A_2 be $\mathcal{R}[X]$ with $\deg A_1 \geq \deg A_2 \geq 2$. If we compute

$$A_3 = \mathrm{prem}(A_1, A_2) \quad \text{and} \quad A_4 = \mathrm{prem}(A_2, A_3),$$

then

$$A_4 = \mathrm{lc}(A_2)^{[d]} B$$

for some integer $d > 0$. If one does not remove the factor $\mathrm{lc}(A_2)^{[d]}$, it will swell exponentially as the pseudo-division is performed repeatedly, so it is preferable to use B instead of A_4 to compute the next remainder. One observes further that $\mathrm{prem}(A_3, B)$ also has a non-trivial content. The subresultant theory shows how certain σ-factorials of leading coefficients enter systematically all following remainders and the subresultant algorithm removes all these σ-factorials to obtain a variant of the Euclidean algorithm with a polynomial complexity. The idea of commutative subresultants is to express the members of a prs as determinantal polynomials in the coefficients of A_1 and A_2. We show that the same essentially holds for Ore polynomials in the next section.

The notion of subresultants has appeared in the literature since the 19th century (Weber 1898). According to Loos (1982), the commutative subresultant theorem was first stated by Habicht (1948). Not being aware of Habicht's work, Collins, Brown and Traub rediscovered the subresultant theorem and, moreover, discovered the fundamental theorem of prs's (Brown and Traub 1971; Collins 1966, 1967). Their motivation was to obtain efficient algorithms for computing commutative prs's. Recall that Habicht's motivation was to generalize the notion of resultants by asking for the conditions that two polynomials have a gcd of a specified degree. Chardin (1991) defined subresultants of two linear differential operators and studied how a subresultant chain specializes to others. He also presented a differential subresultant algorithm and claimed that the correctness of his algorithm was a word-by-word copy of its commutative counterpart. The author (Li 1996, 1998) generalized the notion of subresultants to (inhomogeneous) linear differential and difference, and Ore polynomials, and presented a subresultant theorem with a detailed proof. The subresultant algorithm for Ore polynomials is an easy consequence of the subresultant theorem.

12.3.2 Subresultant theory

In this section we define subresultants of two Ore polynomials. First, let us recall the definition of determinant polynomials (Loos 1982; Mishra 1993).

Definition 12.1:

Let M be an $r \times c$ matrix over $\mathcal{R}$. Assume that $r \leq c$. The determinant polynomial *of M is*

$$| M |= \sum_{i=0}^{c-r} \det(M_i) X^i,$$

where M_i is the $r \times r$ matrix whose first $(r-1)$ columns are the first $(r-1)$ columns of M and whose last column is the $(c-i)$th column of M, for $i = 0$, $\ldots$, $c-r$.

We will encounter determinants whose last columns contain polynomials in $\mathcal{R}[X] \setminus \mathcal{R}$. When expanding such a determinant, we place the product of entries in $\mathcal{R}$ on the left-hand side of the entry in $\mathcal{R}[X] \setminus \mathcal{R}$. With this convention we can express determinant polynomials by determinants. Let the matrix M in Definition 12.1 be

$$\begin{pmatrix} m_{11} & \cdots & m_{1,r-1} & m_{1r} & \cdots & m_{1c} \\ m_{21} & \cdots & m_{2,r-1} & m_{2r} & \cdots & m_{2c} \\ \cdot & \cdots & \cdot & \cdot & \cdots & \cdot \\ m_{r1} & \cdots & m_{r,r-1} & m_{r,r+1} & \cdots & m_{rc} \end{pmatrix}$$

and, for $i = 1, \ldots, r$, let

$$H_i = m_{i1} X^{c-1} + \cdots + m_{ir} X^{c-r} + \cdots + m_{ic}.$$

Then

$$| M |= \det \begin{pmatrix} m_{11} & \cdots & m_{1,r-1} & H_1 \\ m_{21} & \cdots & m_{2,r-1} & H_2 \\ \cdot & \cdots & \cdot & \cdot \\ m_{r1} & \cdots & m_{r,r-1} & H_r \end{pmatrix}.$$

Thus, a determinant polynomial is alternative and multi-linear for its rows. Let $\mathcal{A} : A_1, A_2, \ldots, A_r$ be a sequence of polynomials in $\mathcal{R}[X]$ and let $\max_{1 \leq i \leq r} (\deg A_i)$

be d. The matrix associated with $\mathcal{A}$, denoted by $\mathrm{mat}(\mathcal{A})$, is the $r \times (d+1)$ matrix whose entry in the ith row and jth column is the coefficient of X^{d+1-j} in A_i, where $1 \le i \le r$ and $1 \le j \le d+1$. If $r \le d+1$, then the determinant polynomial of $\mathcal{A}$ is defined to be $|\mathrm{mat}(\mathcal{A})|$, which is denoted by $\mid A_1, A_2, \ldots, A_r \mid$ or $\mid \mathcal{A} \mid$.

For convenience, we let A and B be in $\mathcal{R}[X]$ with $\deg A = m$, $\deg B = n$, and $m \ge n > 0$ in the rest of Sect. 12.3.

Definition 12.2:

The nth subresultant of A and B is B. For $j = n-1, n-2, \ldots, 0$, the jth subresultant of A and B, $\mathrm{sres}_j(A, B)$, is

$$\mid \underbrace{X^{n-j-1}A, \ldots, XA, A,}_{n-j} \underbrace{X^{m-j-1}B, \ldots, XB, B}_{m-j} \mid,$$

The sequence $\mathcal{S}(A, B)$:

$$A, B, \mathrm{sres}_{n-1}(A, B), \ldots, \mathrm{sres}_0(A, B),$$

is called the subresultant sequence *of A and B.*

Example 12.6:

Let $A = a_2X^2 + a_1X + a_0$ and $B = b_2X^2 + b_1X + b_0$. Then

$$\mathrm{sres}_1(A, B) = \mid A, B \mid = \begin{vmatrix} a_2 & A \\ b_2 & B \end{vmatrix},$$

and

$$\begin{aligned} \mathrm{sres}_0(A, B) &= \mid XA, A, XB, B \mid \\ &= \begin{vmatrix} \sigma a_2 & \delta a_2 + \sigma a_1 & \delta a_1 + \sigma a_0 & XA \\ 0 & a_2 & a_1 & A \\ \sigma b_2 & \delta b_2 + \sigma b_1 & \delta b_1 + \sigma b_0 & XB \\ 0 & b_2 & b_1 & B \end{vmatrix}. \end{aligned}$$

Note that the jth subresultant has degree no greater than j, and that all the subresultants of A and B are contained in the left ideal generated by A and B.

From now on, we denote by S_j the jth subresultant of A and B, for $0 \le j \le n$.

Definition 12.3:

The jth subresultant S_j is said to be regular *if $\deg S_j = j$ and otherwise S_j is* defective. *In particular, the nth subresultant S_n is always regular.*

The relations among the subresultants of A and B are described in the following theorem. A detailed proof of this theorem can be found in Li Z. (1996, 1998).

Theorem 12.1: (Subresultant theorem)

Let $\alpha_i = \text{lc}(S_i)$, *for* $i = n,\ n-1,\ \ldots,\ 0$; $\beta_n = \sigma\text{lc}(S_n)^{[m-n]}$, *and* $\beta_i = \sigma\text{lc}(S_i)$, *for* $i = n-1,\ \ldots,\ 0$. *If* S_{j+1} *is regular, for some* j *with* $n-1 \geq j \geq 0$, *and* $\deg S_j = r$, *then*

(1) *If* $S_j = 0$, *then*

$$S_i = 0, \quad (j-1 \geq i \geq 0). \tag{12.9}$$

(2) *If* $S_j \neq 0$, *then*

$$S_i = 0, \quad (j-1 \geq i \geq r+1), \tag{12.10}$$

$$\beta_{j+1}^{[j-r]} S_r = \beta_j^{[j-r]} S_j, \tag{12.11}$$

and

$$\alpha_{j+1}\beta_{j+1}^{[j-r+1]} S_{r-1} = (-1)^{j-r}\text{prem}(S_{j+1}, S_j). \tag{12.12}$$

The subresultant theorem essentially says the following. If S_{j+1} is regular and S_j is of degree r, then S_j and S_r are linearly dependent over $\mathcal{R}$; such a linear relation, as described in (12.11), can readily be obtained from previous subresultants. To compute S_{r-1}, we first compute $\text{prem}(S_{j+1}, S_j)$, and then remove the extraneous factor $\alpha_{j+1}\beta_{j+1}^{[j-r+1]}$ from the remainder, as indicated by (12.12). A pictorial description of the subresultant theorem is given in Fig. 12.1.

Now, we outline the idea for proving Theorem 12.1. Let R be the pseudo-remainder of A and B. The proof is done by induction on the subsequences of regular subresultants of $\mathcal{S}(A, B)$. As in the commutative case, the key technique is to substitute $X^{n-k-i}R$ for $X^{n-k-i}A$ in

$$S_i = | X^{n-1-i}A, \ldots, X^{n-k-i}A, A, \ldots, X^{m-i-1}B, \ldots, B |$$

where $1 \leq k \leq n-i$, so that

$$S_i = r\ \text{sres}_j(B, R) \tag{12.13}$$

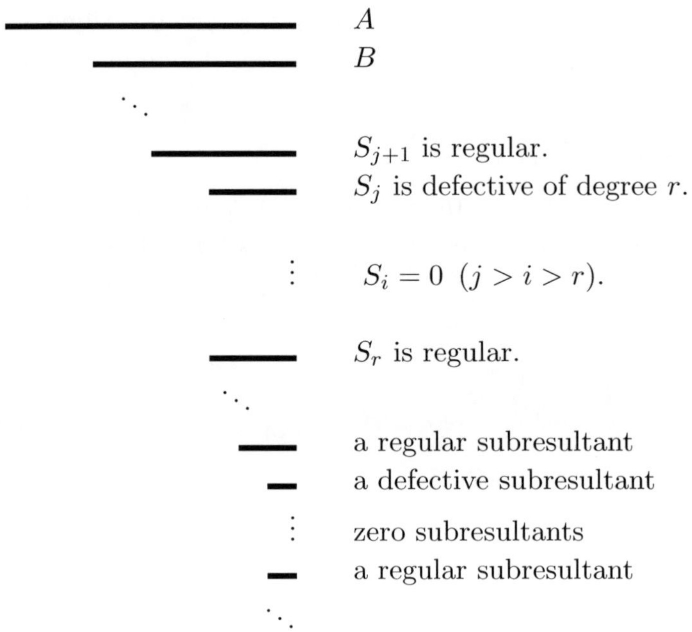

Figure 12.1 The gap structure of a subresultant chain

for some $r \in \mathcal{R}$ and some j with $0 \leq j < i$. Such substitutions are possible for Ore polynomials, because

$$\mathrm{prem}(X^{n-k-i}A, X^{n-k-i}B) - X^{n-k-i}R$$

is an $\mathcal{R}$-linear combination of $X^{n-k-1-i}A, \ldots, A$ and $X^{m-i-1}B, \ldots, B$ (see Lemma 4.2 of Li and Nemes 1997). Using these substitutions, we move from one regular subresultant of A and B to a regular subresultant of B and R with lower order. Due to the presence of σ-factorials, r in (12.13) is much more involved than in the commutative case. This complication makes the proof of Theorem 12.1 quite tedious.

There are two prs's contained in $\mathcal{S}(A, B)$, as shown in following definition.

Definition 12.4:

The subresultant sequence of A and B of the first kind consists of the following polynomials: A, B, and S_j if S_{j+1} is regular and S_j is non-zero. The subresultant sequence of A and B of the second kind consists of A, B, and other regular subresultants of $\mathcal{S}(A, B)$. We denote the subresultant sequences of A and B of the first and second kinds by $\mathcal{S}_1(A, B)$ and $\mathcal{S}_2(A, B)$, respectively.

As illustrated in Fig. 12.1, if S_j is defective of degree r, then S_j is in $\mathcal{S}_1(A, B)$ and S_r is in $\mathcal{S}_2(A, B)$, and if two successive subresultants in $\mathcal{S}(A, B)$ are regular then the second one must in both $\mathcal{S}_1(A, B)$ and $\mathcal{S}_2(A, B)$. In particular, $\mathcal{S}_1(A, B)$ and $\mathcal{S}_2(A, B)$ coincide if all the subresultants of A and B are regular. The members of these two sequences are linearly dependent in order. We prefer to compute $\mathcal{S}_1(A, B)$, because the coefficients of its members are determinants with smaller order than those of corresponding members in $\mathcal{S}_2(A, B)$. The sequence $\mathcal{S}_1(A, B)$ is computed by the subresultant algorithm given below.

Theorem 12.2: (Subresultant algorithm)

(1) *If S_{n-1} is non-zero, then the third member of $\mathcal{S}_1(A, B)$ is*

$$S_{n-1} = (-1)^{m-n+1}\text{prem}(A, B).$$

(2) *If S_k is the fourth member of $\mathcal{S}_1(A, B)$, and let l be $(n - \deg S_{n-1})$, then*

$$S_k = \text{prem}(B, S_{n-1})/e, \tag{12.14}$$

where $e = (-1)^{l+1}\sigma(\text{lc}(B))^{[m-n][l]}\text{lc}(B)$.

(3) *If S_i, S_j and S_k are three consecutive members in $\mathcal{S}_1(A, B)$, where $n > i > j > k$, and let l be $(\deg S_i - \deg S_j)$, then*

$$S_k = \text{prem}(S_i, S_j)/e, \tag{12.15}$$

where $e = (-1)^{l+1}\sigma(\text{lc}(S_{j+1}))^{[l]}\text{lc}(S_i)$.

Note that divisions occurring in the subresultant algorithm are *exact*, because each S_k belongs to $\mathcal{R}[X]$. This theorem also implies that the last member of $\mathcal{S}_1(A, B)$ is a gcrd of A and B.

Theorem 12.2 is a consequence of Theorem 12.1. Indeed, the first assertion is immediate from the definition of S_{n-1}. The second is (12.12) when $j = n - 1$. The last is proved as follows. By the gap-structure, S_i and S_{j+1} are $\mathcal{R}$-linearly dependent. Hence, S_k is $\mathcal{R}$-linearly dependent on $\text{prem}(S_{j+1}, S_j)$. The coefficients of this linear relation can be obtained from (12.11). Equation (12.15) then follows from (12.12).

The subresultant algorithm achieves a two-fold goal: first, it computes a prs in a fraction-free fashion as the pseudo-Euclidean algorithm does; second, it controls the growth of coefficients of intermediate polynomials within polynomial size, because all the coefficients of members in $\mathcal{S}_1(A, B)$ are determinants whose orders are less than $\deg A + \deg B$ and whose entries are coefficients of $X^{n-1}A$, $X^{n-2}A$, …, A, $X^{m-1}B$, $X^{m-2}B$, …, B.

12.3.3 Computing gcrd's and lclm's by subresultants

First, we extend the resultant of two differential operators in Berkovich and Tsirulik (1986) and Chardin (1991).

Definition 12.5:

The subresultant S_0 *is called the* Sylvester resultant *of* A *and* B *and denoted by* $\mathrm{res}(A, B)$. *The matrix associated with the sequence* $X^{n-1}A$, $X^{n-2}A$, $\ldots$, A, $X^{m-1}B$, $X^{m-2}B$, $\ldots$, B *is called the* Sylvester matrix *of* A *and* B.

The next theorem is important for efficient gcrd- and lclm-calculations.

Theorem 12.3:

Let $d = \deg \mathrm{gcrd}(A, B)$. *Then the following hold.*

(1) S_d *is a gcrd of* A *and* B.

(2) *If* r *is the rank of the Sylvester matrix of* A *and* B, *then* $r + d = m + n$. *In particular,* $d = 0$ *if and only if* $\mathrm{res}(A, B) \neq 0$.

(3) UA *is an lclm of* A *and* B, *where* U *is the determinant of order* $(m + n - 2d + 2)$ *whose first* $(m + n - 2d + 1)$ *columns are the first* $(m + n - 2d + 1)$ *columns of*

$$\mathrm{mat}(X^{n-d}A, \ldots, XA, A, X^{m-d}B, \ldots, XB, B)$$

and whose last column is the transpose of the vector $(X^{n-d}, \ldots X, 1, \underbrace{0, 0, \ldots, 0}_{m-d+1})$.

Proof:

The last subresultant in $\mathcal{S}_2(A, B)$ is a gcrd of A and B by Theorem 12.2 and the gap-structure. This subresultant is regular and of degree d, so it must be S_d.

Since S_d is non-zero, $X^{n-d-1}A$, $\ldots$, A, $X^{m-d-1}B$, $\ldots$, B are linearly independent as vectors in an $(m + n - d)$-dimensional space over $\mathcal{F}$. It follows that $X^{n-d-1}A$, $\ldots$, A, $X^{m-1}B$, $\ldots$, $X^{m-d}B$, $X^{m-d-1}B$, $\ldots$, B are linearly independent in an $(n+m)$-dimensional space over $\mathcal{F}$. Hence, $r \geq m+n-d$. On the other hand, there are F and G in $\mathcal{F}[X]$ such that $A = FS_d$ and $B = GS_d$. Thus, each vector X^iA or X^jB, for $0 \leq i \leq n-1$ or $0 \leq j \leq m-1$, is an $\mathcal{F}$-linear combination of $X^{m+n-d-1}S_d, \ldots, S_d$. So, $r \leq m + n - d$.

To prove the last assertion, we first show that UA is right divisible by B. Let M be the matrix whose first $(m+n-2d+1)$ columns are the same as those of U, and its last column is the transpose of the vector

$$(X^{n-d}A, \ldots, XA, A, X^{m-d}B, \ldots, XB, B).$$

Then $|M|$ is zero because $|M| = S_{d-1}$ when $d > 0$, and the last column of M can be reduced to zero by the previous columns when $d = 0$. Expanding $|M|$ according to its last column yields $UA + VB = 0$, for some $V \in \mathcal{R}[X]$, that is, UA is right divisible by B. By (12.6), it suffices to prove that the degree of U is equal to $(n-d)$, that is, the cofactor of X^{n-d} in the determinant U is non-zero. From the definition of U one sees that this cofactor is the product of $\pm\sigma^{m-d}(\mathrm{lc}(B))$ and the coefficient of X^d in S_d. Since S_d is of degree d, the cofactor is non-zero.

Example 12.7:

Let A and B be the same as those in Example 12.4. Since the degree of $\mathrm{gcrd}(A, B) = 1$, Theorem 12.3 implies that a gcrd of A and B is

$$\begin{vmatrix} 1 & 0 & 0 & 0 & 0 & 0 \\ 0 & 1 & 0 & 0 & 0 & 0 \\ t^2-t & 4t-2 & -3t+2 & 0 & 0 & 0 \\ 0 & t^2-t & 2t-1 & -3t & 3 & 0 \\ 0 & 0 & t^2-t & 0 & -3t & 6 \end{vmatrix}.$$

and that an lclm of A and B is

$$\begin{vmatrix} 1 & 0 & 0 & 0 & 0 & 0 & X^2 \\ 0 & 1 & 0 & 0 & 0 & 0 & X \\ 0 & 0 & 1 & 0 & 0 & 0 & 1 \\ t^2-t & 8t-4 & -3t+12 & -6 & 0 & 0 & 0 \\ 0 & t^2-t & 4t-2 & -3t+2 & 0 & 0 & 0 \\ 0 & 0 & t^2-t & 2t-1 & -3t & 3 & 0 \\ 0 & 0 & 0 & t^2-t & 0 & -3t & 0 \end{vmatrix} A.$$

The determinant formula for gcrd's is a generalization of the corresponding formula for gcd's in Perron (1951). The determinant formula for lclm's is possibly new. With these two formulas, one can estimate the coefficient and degree bounds for gcrd's and lclm's.

If $\mathcal{R}$ is a unique factorization domain, then

$$\sigma^{m-d}(\mathrm{lc}(B))\sigma^{n-d}(\mathrm{lc}(A))\mathrm{lc}(S_d)$$

is a multiple of the leading coefficient of the primitive lclm of A and B, because UA is an lclm of A and B and $\mathrm{lc}(U)$ is $\sigma^{m-d}(\mathrm{lc}(B))\mathrm{lc}(S_d)$.

Now, we present an algorithm for computing lclm's using the subresultant algorithm. For a later convenience, we modify our notation. Let $A_1 = A$, $A_2 = B$ and $\mathcal{S}_1(A, B)$ be the sequence:

$$A_1, A_2, A_3, \ldots, A_k.$$

Let $U_1 = 1$ and $U_2 = 0$. If A_i is the jth subresultant of A and B, then we let U_i be the $(m+n-2j) \times (m+n-2j)$ determinant whose first $(m+n-2j-1)$ column is the same as those in

$$\mathrm{mat}(X^{n-j-1}A, \ldots, A, X^{m-j-1}B, \ldots, B).$$

and whose last row is the transpose of

$$(X^{n-j-1}, \ldots X, 1, \underbrace{0, 0, \ldots, 0}_{m-j}).$$

Note that the index j decreases as the index i increases. The sequence

$$U_1, U_2, U_3, \ldots, U_k$$

is called the *first co-sequence* associated with $\mathcal{S}_1(A, B)$, because

$$U_i A_1 \equiv A_i \bmod A_2 \quad \text{for } i = 1, 2, \ldots, k,$$

i.e., $U_i A_1 - A_i$ is right-divisible by A_2. Another property of the first co-sequence is that $\deg U_i < (n - \deg A_i)$, for $i = 3, \ldots, k$, because $\deg A_i \leq j$ if A_i is the jth subresultant of A and B. By Theorem 12.2 we have, for $i = 3, 4, \ldots, k$,

$$l_i A_{i-2} = Q_i A_{i-1} + e_i A_i,$$

where

$$l_i = \mathrm{lc}(A_{i-1})^{[\deg A_{i-2} - \deg A_{i-1} + 1]},$$

e_3 is $(-1)^{m-n+1}$ and e_i is the extraneous factor e removed in (12.14) or (12.15) when we use the subresultant algorithm to compute A_i. The next theorem provides a fraction-free version of the half-extended Euclidean algorithm.

Theorem 12.4:

With the notation just introduced, we have

$$U_i = (l_{i-1} U_{i-2} - Q_i U_{i-1})/e_i$$

for $i = 3, \ldots, k$. *Furthermore*

$$L = (l_k U_{k-1} - Q_k U_k) A_1$$

is an lclm of A_1 *and* A_2.

The proof of this theorem is done by a routine induction on i. It makes essential use of (12.6) and Theorem 12.2. A detailed proof may be found in Li Z. (1996, 1998).

We present an experimental comparison among three algorithms for computing lclm's of A and B. The first algorithm (labeled L) is based on linear algebra, which has two time-consuming costs:

- compute the solution space $\mathcal{V}$ of the linear homogeneous system whose matrix is the transpose of

$$M = \text{mat}(X^n A, \ldots, A, X^m B, \ldots, B)$$

- adjust the parameters in $\mathcal{V}$ to find a linear relation (over $\mathcal{R}$) among the polynomials $X^{n-r}A, \ldots, A, X^{m-r}B, \ldots, B$ with r as large as possible.

The second algorithm (labeled X) is the extended Euclidean algorithm given in Sect. 12.2.2. The implementation of Algorithm L we used is the function `LCLM` in the Maple package `DEtools`, which computes an lclm of two linear differential operators. The implementation of Algorithm X we used is the function `Annihilators` in the Maple package `Ore_algebra`, which computes an lclm of two Ore polynomials. The implementations based on Theorem 12.4 (labeled FF) are the functions `dlclm` and `slclm` in the package `FracFree`, which compute an lclm of two linear differential and shift operators, respectively. The package `FracFree` can be downloaded at

`http://www.gmd.de/SCAI/alg/cade/software.html`.

We generated three random polynomials in $\mathbb{Z}[t][X]$, A, B and C, with respective degrees d_A, d_B, and d_C in X, regarded A, B and C as linear differential operators, and computed the primitive lclm of AC and BC. Our experiment was carried out in Maple V (Release 5) on a Sun-workstation (ultra 2). The random polynomials in the experiment were generated by the Maple function `randpoly`. When $d_C = 0$, we set $C = 1$. All the timings are Maple CPU time and given in seconds.

Figure 12.2 shows that if $d_C = 0$, Algorithm L is slightly faster than Algorithm FF. This is because both algorithms triangularize the Sylvester matrix

d_A	d_B	d_G	L	FF
6	5	0	24.4	30.2
6	5	1	33.2	17.5
6	5	2	35.9	4.6
6	5	3	25.1	1.3

Figure 12.2 L vs S for sparse differential operators

d_A	d_B	d_G	X	FF
6	5	0	219.0	22.5
6	5	1	327.9	58,3
6	5	2	59.2	14.3
6	5	3	0.9	1.2

Figure 12.3 X vs S for sparse differential operators

d_A	d_B	d_G	X	FF
5	5	0	300.0	28.0
5	5	1	71.3	19.7
5	5	2	18.8	6.0
5	5	3	1.4	1.1

Figure 12.4 X vs S for sparse shift operators

but LCLM uses an implementation of fraction-free Gaussian elimination that takes advantage of the sparsity of the matrix. However, the higher d_C is, the slower Algorithm L is.

Figure 12.3 shows Algorithm FF is much more efficient than Algorithm X unless d_C is very close d_B. It is expected that the timings will be more significant if A and B are of higher degrees.

The timings in Fig. 12.4 are similar to those in Fig. 12.3. We set $d_A = 5$ in the shift case as the function `Annihilators` returned the message "object too large" for some A and B with $d_A = 6$.

12.4 A modular algorithm for computing gcrd's

In this section, we outline a modular algorithm for computing gcrd's over the domain of commutative univariate integral polynomials (Li and Nemes 1997). This algorithm leads to a modular algorithm for computing lclm's, because an lclm of two Ore polynomials can be written as a determinant by Theorem 12.3 if the degree of a gcrd is known. A determinant with polynomial entries can be expanded by modular techniques.

We will extend the techniques used in modular gcd-algorithms as much as we can (Brown 1971; Geddes et al. 1992). Two new problems that cannot be tackled by the classical techniques, are that

- evaluation mappings are *not* Ore ring homomorphisms;
- the normalization of leading coefficients is different from that in modular gcd-algorithms.

The first problem will be solved by Theorem 12.3; the second one by rational number reconstruction (RNR) and rational function reconstruction (RFR).

In this section, p is a prime, and $\mathbb{Z}_p$ is the finite field with p elements. For an indeterminate t, $\mathbb{Z}[t]$ and $\mathbb{Z}_p[t]$ are the rings of usual commutative polynomials in t over $\mathbb{Z}$ and $\mathbb{Z}_p$, respectively. Let X be a new indeterminate. We regard $\mathbb{Z}[t][X]$ (resp. $\mathbb{Z}_p[t][X]$) as a left $\mathbb{Z}[t]$-module (resp. $\mathbb{Z}_p[t]$-module). For non-zero F in $\mathbb{Z}[t][X]$ or $\mathbb{Z}_p[t][X]$, the leading coefficient of F in X is denoted by $\mathrm{lc}(F)$, the leading coefficient of $\mathrm{lc}(F)$ in t is called the *head coefficient* of F and denoted by $\mathrm{hc}(F)$, the degree of F in X is denoted by $\deg F$, and the degree of F in t by $\deg_t F$.

From now on, we assume that the triple $(\mathbb{Z}[t][X], \sigma, \delta)$ is an Ore polynomial ring over $\mathbb{Z}[t]$. For brevity we denote this triple by $\mathbb{Z}[t][X]$. If $A, B \in \mathbb{Z}[t][X]$, then the *normalized* gcrd of A and B is the gcrd of A and B that is in $\mathbb{Z}[t][X]$ and primitive with respect to X, and has positive head coefficient. If A and B are in the Ore polynomial ring $\mathbb{Z}_p[t][X]$, then the *normalized* gcrd of A and B is the gcrd of A and B that is in $\mathbb{Z}_p[t][X]$ and primitive with respect to X, and has head coefficient 1. The normalized gcrd of A and B is denoted by $\mathrm{gcrd}(A, B)$.

Our goal is to compute the normalized gcrd of A and B in $\mathbb{Z}[t][X]$. To this end, we will regard A and B as elements in $\mathbb{Z}_p[t][X]$ for a number of primes. The use of RNR and RFR will, sometimes, force us to operate in $\boldsymbol{Q}[t][X]$ and $\mathbb{Z}_p(t)[X]$, respectively. If $A, B \in \mathbb{Z}_p[t][X]$, we will first compute the monic gcrd of A and B, which is in $\mathbb{Z}_p(t)[X]$, and, then, convert it into $\mathrm{gcrd}(A, B)$. If

$A, B \in \mathbb{Z}[t][X]$, we will first compute the primitive gcrd with head coefficient is 1 in $\boldsymbol{Q}[t][X]$, and convert it to $\text{gcrd}(A, B)$.

A modular mapping $\phi_p : \mathbb{Z}[t][X] \longrightarrow \mathbb{Z}_p[t][X]$ is a left-module homomorphism (over $\mathbb{Z}[t]$) defined for p by $\phi_p(A) = A \bmod p$ for $A \in \mathbb{Z}[t][X]$. An evaluation mapping $\psi_k : \mathbb{Z}_p[t][X] \longrightarrow \mathbb{Z}_p[X]$ is a left-module homomorphism (over $\mathbb{Z}_p[t]$) defined for $k \in \mathbb{Z}_p$ by $\psi_k(A(t, X)) = A(k, X)$ for $A \in \mathbb{Z}_p[t][X]$.

The following lemma implies that ϕ_p usually induces a ring structure in $\mathbb{Z}_p[t][X]$. The proof of this lemma is straightforward (Li and Nemes 1997).

Lemma 12.5:

If p is not a divisor of $\text{hc}(\sigma(t))$, then the triple $(\mathbb{Z}_p[t][X], \sigma_p, \delta_p)$ is an Ore polynomial ring and ϕ_p is a ring homomorphism, where σ_p and δ_p are defined by $\sigma_p(\phi_p(f)) = \phi_p(\sigma(f))$ and $\delta_p(\phi_p(f)) = \phi_p(\delta(f))$ for $f \in \mathbb{Z}[t]$.

From now on, we assume that p does not divide $\text{hc}(\sigma(t))$. Hence, $\mathbb{Z}_p[t][X]$ is an Ore polynomial ring and ϕ_p is a ring homomorphism. What about evaluation mappings? The next example may convince the reader that most evaluation mappings *cannot* be regarded as ring homomorphisms no matter how we define multiplication on $\mathbb{Z}_p[X]$.

Example 12.8:

Let $\mathbb{Z}_p[t][X]$ be the ring of differential operators over $\mathbb{Z}_p[t]$ and k an element of $\mathbb{Z}_p$. Then the multiplicative rule implies

$$\psi_k(Xt) = \psi_k(tX + 1) = kX + 1.$$

On the other hand, since $k = \underbrace{1 + 1 + \cdots + 1}_{k \text{ times}}$, we have

$$\psi_k(X)\psi_k(t) = Xk = kX.$$

Hence, ψ_k is not a ring homomorphism.

This example raises the question: If $A, B \in \mathbb{Z}_p[t][X]$ and $k \in \mathbb{Z}$ such that k is not a root of $\text{lc}(A)\text{lc}(B)$, can we compute $\psi_k(\text{gcrd}(A, B))$ *without* computing $\text{gcrd}(A, B)$? An answer is as follows. Assume that d is the degree of $\text{gcrd}(A, B)$. Then $\text{sres}_d(A, B)$ is a gcrd of A and B by Theorem 12.3. It is cheap to form the determinantal formula $\text{sres}_d(A, B)$, which is a polynomial in X whose coefficients are determinants. Apply ψ_k to the entries and, then, expand these evaluated determinants to obtain $\psi_k(\text{gcrd}(A, B))$.

To guess the degree of gcrd(A, B), we form the Sylvester matrix of A and B, say M, and apply ψ_k to each entry of M to obtain a matrix M_k. The rank of M_k is less than or equal to the rank of M. It follows from Theorem 12.3 that

$$d_k = m + n - \text{rank}(M_k) \geq m + n - \text{rank}(M) = d.$$

Hence, we are able to compute d_k efficiently such that $d_k \geq d$. Moreover, if $d_k > d$, then k is a root of $\text{lc}(\text{sres}_d(A, B))$, which is a non-zero minor of the resultant of A and B. So there are only few k's such that $d_k > d$.

Example 12.9:

Let $A = X^4$ and $B = (t^2 - t)X^3 - 3tX + 6$ be two linear differential operator in $\mathbb{Z}_{11}[t][X]$, $d = \deg \text{gcrd}(A, B)$, and let M be the Sylvester matrix of A and B, that is,

$$\begin{pmatrix} 1 & 0 & 0 & 0 & 0 & 0 & 0 \\ 0 & 1 & 0 & 0 & 0 & 0 & 0 \\ 0 & 0 & 1 & 0 & 0 & 0 & 0 \\ t^2 - t & 6t - 3 & -3t + 6 & -3 & 0 & 0 & 0 \\ 0 & t^2 - t & 4t - 2 & -3t + 2 & 0 & 0 & 0 \\ 0 & 0 & t^2 - t & 2t - 1 & -3t & 3 & 0 \\ 0 & 0 & 0 & t^2 - t & 0 & -3t & 6 \end{pmatrix}.$$

For $k = 2$, M is mapped by ψ_2 to

$$\begin{pmatrix} 1 & 0 & 0 & 0 & 0 & 0 & 0 \\ 0 & 1 & 0 & 0 & 0 & 0 & 0 \\ 0 & 0 & 1 & 0 & 0 & 0 & 0 \\ 2 & 9 & 0 & 8 & 0 & 0 & 0 \\ 0 & 2 & 6 & 7 & 0 & 0 & 0 \\ 0 & 0 & 2 & 3 & 5 & 3 & 0 \\ 0 & 0 & 0 & 2 & 0 & 5 & 6 \end{pmatrix},$$

whose rank is 6, and hence the guessed degree for d is

$$\deg(A) + \deg(B) - 6 = 1,$$

which is greater than or equal to d. Apply ψ_2 to the matrix associated with $\text{sres}_1(A, B)$ to get

$$\begin{pmatrix} 1 & 0 & 0 & 0 & 0 & 0 \\ 0 & 1 & 0 & 0 & 0 & 0 \\ 2 & 6 & 7 & 0 & 0 & 0 \\ 0 & 2 & 3 & 5 & 3 & 0 \\ 0 & 0 & 2 & 0 & 5 & 6 \end{pmatrix}.$$

The determinant polynomial of this matrix is $g = 10X + 1$. One can easily check $g/\mathrm{lc}(g)$ is the monic image of $\mathrm{gcrd}(A, B)$ under ψ_2. In the same way, we can compute images of the monic gcrd of A and B under ψ_k, for $k = 3, 4, 5, 6, 7$. We find that the rank of $\psi_k(M)$ is 6, for $k = 2, \ldots, 7$. Thus, the guessed degree for d is kept as 1. Moreover, the image of $|\mathrm{sres}_1(A, B)|$ under ϕ_k is of degree 1, for $k = 2, \ldots, 7$. Therefore, without knowing d, we can tell that either these images are correct or the guessed degree is greater than d. For ψ_8, our method produces a wrong image, because the rank of $\psi_8(M)$ is 6 but $\psi_8(|\mathrm{sres}_1(A, B)|)$ is zero. If an image is of degree lower than the guessed one, it is discarded.

For $A, B \in \mathbb{Z}_p[t][X]$ with $\mathrm{gcrd}(A, B) = G$ and $k \in \mathbb{Z}_p$ with $\psi_k(\mathrm{lc}(A)\mathrm{lc}(B)) \neq 0$, the method described above (named Method S) produces

- either a message "FAIL" (because the degree of the image is less than the guessed one), or
- a monic polynomial g in $\mathbb{Z}_p[X]$ such that $\deg g \geq \deg G$, and $g = \psi_k(G/\mathrm{lc}(G))$ if $\deg g = \deg G$.

Similarly to modular gcd-algorithms, we can decide if $\deg g > \deg G$ when one correct image is produced. In addition, $G = 1$ if $g = 1$.

An element $k \in \mathbb{Z}_p$ is *unlucky* for A and B if

- either k is a root of $\sigma_p^i(\mathrm{lc}(B))$ for some i with $0 \leq i \leq m - 1$, or
- Method S fails, or
- Method S produces g with $\deg g > \deg G$.

The first case can certainly be excluded; the second is detected after an execution of Method S; the last can only be detected when Method S has produced a correct image for some evaluation point. The next lemma asserts that unlucky elements are rare.

Lemma 12.6:

If $A, B \in \mathbb{Z}_p[t][X]$, $\deg A = m$, $\deg B = n$, and $m \geq n$, then there are at most

$$\deg_t \left(\prod_{i=0}^{m-1} \sigma_p^i(\mathrm{lc}(B)) \right) + m \deg_t B + n \deg_t A \tag{12.16}$$

unlucky elements for A and B.

Proof:

If k is unlucky for A and B, then k is a root of

$$\left(\prod_{i=0}^{m-1} \sigma_p^i(\mathrm{lc}(B))\right) \mathrm{lc}(\mathrm{sres}_d(A, B)),$$

where $d = \deg \mathrm{gcrd}(A, B)$, and (12.16) is a degree bound for this polynomial.

By Lemma 12.6, Method S is ensured to succeed in producing correct images when p is larger than the bound given in (12.16).

The second problem is to combine evaluation images of C, where C is the monic gcrd of A and B. The typical situation is as follows. We have a pair (g, M) where g is in $\mathbb{Z}_p[t][X]$ and monic, and M is in $\mathbb{Z}_p[t]$. We call M a *modulus* and g the *combined image* with respect to M. We assume that

$$g \equiv C \mod M \tag{12.17}$$

For a new evaluation point $k \in \mathbb{Z}_p$ with $M(k) \neq 0$, Method S yields a monic polynomial $g_1 \in \mathbb{Z}_p[t][X]$ with the assumption that

$$g_1 \equiv C \mod (t - k). \tag{12.18}$$

If $\deg g_1 > \deg g$, then (12.18) fails, so we discard g_1 and apply Method S to a new element in $\mathbb{Z}_p$. If $\deg g_1 < \deg g$, then (12.17) fails, so we discard g, set $g = g_1$ and $M = t - k$, and apply Method S to a new element in $\mathbb{Z}_p$. If $\deg g_1 = \deg g$, then the Chinese remainder algorithm (CRA) computes g' in $\mathbb{Z}_p[t][X]$ such that

$$g' \equiv C \mod M(t - k).$$

Update (g, M) as $(g', M(t - k))$. Our assumption (12.17) holds as long as we have encountered a lucky k.

We choose C as our pre-image, for known multiplicative bounds for the denominators of C are loose. The same choice was made by Encarnación (1995) for computing polynomial gcds over algebraic number fields. The reader is referred to Brown (1971) or text books on computer algebra (Geddes et al. 1992; Winkler 1996) for the choice of pre-images in modular gcd-algorithms over the integers.

The third problem is how to recover C from the combined image g (with respect to M). Notice that g is fraction-free but C is not. The technique we use is RFR which computes a polynomial $\tilde{g}$ in $\mathbb{Z}_p(t)[X]$ having the properties:

- $\tilde{g} \equiv g \mod M$;
- the numerators of the coefficients of $\tilde{g}$ are of degree (in t) less than or equal to $(\deg_t M)/2$;
- the denominators of the coefficients of $\tilde{g}$ are of degree $< (\deg_t M)/2$;
- $\tilde{g}$ with the above-listed three properties is unique if existent.

The polynomial C will be reached by RFR as deg M increases. The algorithm for RFR can be found in the computer algebra system Maple (function `Ratrecon`) and Li Z. (1996).

When two successive RFR's yield the same result we attempt a trial-division of both A and B by the result, using more evaluations if the division is not exact. The idea for computing gcrd's over $\mathbb{Z}_p[t][X]$ is illustrated by the following example.

Example 12.10:

Let A and B be the same as those in Example 12.9. Compute $\text{gcrd}(A, B)$ in $\mathbb{Z}_{11}[t][X]$. Denote by g_k the result produced by Method S for A, B and $k \in \mathbb{Z}_{11}$. We find $g_2 = X + 10$, so that we have a pair

$$(g = X + 10,\ M = t - 2).$$

But RFR for g fails. Applying Method S to $k = 3$ yields $g_3 = X + 3$. Combining g_2 and g_3 modulo $(t-2)(t-3)$ yields

$$(g = X + 4t + 2,\ M = (t-2)(t-3)).$$

RFR yields $\tilde{g} = g$. Combining g and g_4 yields

$$(g = X + 10t^2 + 9t + 7,\ M = (t-2)(t-3)(t-4)),$$

and applying RFR to g yields $\tilde{g} = X + 9/t$. Repeating the same process for g and $g_5 = X + 4$, we find that RFR produces the same result $X + 9/t$. Two trial divisions prove that $X + 9/t$ is the monic gcrd of A and B. Thus, $\text{gcrd}(A, B) = tX + 9$.

Combining Method S, CRA and RFR, we obtain a modular approach to computing $\text{gcrd}(A, B)$ in $\mathbb{Z}_p[t][X]$. This approach fails only when there are not sufficient evaluation points in $\mathbb{Z}_p$. To compute $\text{gcrd}(A, B)$ in $\mathbb{Z}[t][X]$, we are free to choose primes. By Lemma 12.6, this modular approach will never fail if p is "sufficiently" large.

Since modular mappings are ring homomorphisms, we can compute gcrd's in $\mathbb{Z}[t][X]$ by computing gcrd's in $\mathbb{Z}_p[t][X]$ for sufficiently many primes. The combining process consists of CRA and RNR by Wang (1981). See also Wang et al. (1992), and Collins and Encarnación (1995). We combine the normalized gcrd's of $\phi_p(A)$ and $\phi_p(B)$, which is an element in $\mathbb{Z}[t][X]$ whose head coefficient is 1. We then recover rational numbers from numerical coefficients of the combined image by RNR. When two successive RNR's yield the same result, we attempt a trial-division of both A and B by the result, using more primes if the division is not exact. When the division is exact, we obtain a gcrd of A and B which is in $\boldsymbol{Q}[t][X]$. We then convert this gcrd to $\mathrm{gcrd}(A, B)$. Once again, we have to detect wrong modular images (unlucky primes) carefully (Li and Nemes 1997).

Example 12.11:

Let A and B be the same as those in Example 12.9. We compute $\mathrm{gcrd}(A, B)$ in $\mathbb{Z}[t][X]$. Denote by h_p the normalized gcrd of A and B in $\mathbb{Z}_p[t][X]$. h_{11} and h_{13} are $tX + 9$ and $tX + 11$, respectively. Combining them modulo 11×13 yields $h = tX + 141$. Rational number reconstruction returns $\tilde{h} = tX - 2$. Letting $p = 17$ and repeating the process, we find that $\tilde{h}$ stays the same. Trial divisions verify that $\mathrm{gcrd}(A, B) = tX - 2$.

Next, we compare the modular algorithm (labeled M) and the Euclidean algorithm (labeled E) for computing gcrd's. The implementation of Algorithm M was the function `dgcrd_mod` of the package `FracFree` and that of Algorithm E was the function `GCRD` of the Maple package `DEtools`.

We generated three random polynomials in $\mathbb{Z}[t][X]$, A, B and C, with respective degrees d_A, d_B, and d_C in X, regarded A, B and C as linear differential operators, and computed $\mathrm{gcrd}(AC, BC)$. Our experiment was carried out in Maple V (Release 5) on a Sun-workstation (ultra 2). The random polynomials in the experiment were generated by the Maple function `randpoly`. When $d_C = 0$, we set $C = 1$. Some of the timings are summarized in Fig. 12.5. All the timings are Maple CPU time and given in seconds. The timings show that the modular method works more efficiently when the gcrd is of small degree. It seems that the trade-off point is around $d_C = [d_B/2] + 1$. We can use the same trick as that used in Method S to guess d_C.

d_A	d_B	d_C	E	M
10	9	0	227.4	0.5
10	9	1	231.6	5.6
10	9	2	149.9	12.9
10	9	3	127.4	22.8
10	9	4	49.9	20.3
10	9	5	19.7	29.7
10	9	6	8.6	31.7
10	9	7	2.6	30.8
10	9	8	1.0	34.3

Figure 12.5 E vs M for differential operators

12.5 Future study

In this chapter the coefficient domain of Ore polynomials is commutative. It would be interesting to extend notions of gcrd's, lclm's and subresultants to Ore polynomials with non-commutative coefficients. Such extensions could result in efficient algorithms for handling linear partial differential operators and linear differential-difference operators (Ore 1931, 1933).

It seems possible to combine the idea in Encarnación (1995) with the modular method in Sect. 12.4 to design a modular algorithm for computing gcrd's of Ore polynomials over an algebraic extension of $\boldsymbol{Q}(t)$.

Chapter 13

ALGEBRAIC FACTORIZATION AND GCD COMPUTATION

Lihong Zhi

This chapter describes several algorithms for factorization and GCD (greatest common divisor) computation of polynomials over algebraic extension fields. These algorithms are common in using the method of characteristic sets introduced in the first chapter, and have found encouraging applications to geometric theorem proving, irreducible decomposition of algebraic varieties and other problems. Some information about implementations and performance of the presented algorithms is provided.

13.1 Introduction

Factoring polynomials over algebraic extension fields can be traced back to Kronecker (1882). A similar algorithm can also be found in van der Waerden (1949), which was adopted and improved by Trager (1976). Further improvements have been given by Encarnación (1997) and Noro and Yokoyama (1997). Using the Chinese remainder theorem, Hensel lemma, and lattice techniques, several different approaches were given in Wang (1976), Weinberger and Rothschild (1976), Lenstra (1982, 1987), Landau (1985), and Abbott (1989).

The study of algebraic factorization in Wu's research group started in 1984, motivated by the need for it in the method of Wu (1984a,b, 1987a,b) for geometric theorem proving (GTP). Two different methods were proposed in Hu and Wang (1986), Wang (1992d) and Wu (1994c), and applied to GTP and irreducible decomposition of algebraic varieties (see Wang 1992a, 1994). Investigations along this line have been furthered by Zhi (1996), who has been trying to work out an optimized algorithm by incorporating and improving different techniques.

Let $\mathbb{Z}$ denote the ring of integers and $\boldsymbol{Q}$ the field of rational numbers, and let $u_1, u_2, \ldots, u_d$ be a set of transcendental elements, abbreviated as $\boldsymbol{u}$. The

MATHEMATICS MECHANIZATION AND APPLICATIONS
ISBN 0-12-734760-7

transcendental extension field obtained from $\boldsymbol{Q}$ by adjoining $\boldsymbol{u}$ is denoted by $\boldsymbol{K}_0$, i.e., $\boldsymbol{K}_0 = \boldsymbol{Q}(\boldsymbol{u})$. The algebraic elements $\eta_1, \eta_2, \ldots, \eta_r$, abbreviated as $\boldsymbol{\eta}$, are defined by an *irreducible ascending set AS*

$$[A_1(\boldsymbol{u}, y_1), A_2(\boldsymbol{u}, y_1, y_2), \ldots, A_r(\boldsymbol{u}, y_1, y_2, \ldots, y_r)]$$

with $A_i \in \boldsymbol{Q}[\boldsymbol{u}, y_1, \ldots, y_i]$, $\deg(A_i, y_i) = m_i > 0$, and $\deg(A_i, y_j) < \deg(A_j, y_j)$ for each pair $j < i$. Here $\deg(A_i, y_j)$ denotes the *degree* of A_i in y_j as usual. A_i, as a polynomial in $\boldsymbol{K}_{i-1}[y_i]$, is irreducible, where $\boldsymbol{K}_{i-1} = \boldsymbol{K}_{i-2}(\eta_{i-1})$, with A_{i-1} the minimal polynomial of η_{i-1} for each $i \geq 2$. The field $\boldsymbol{K}_r$ is called an *algebraic extension field* of $\boldsymbol{K}_0$ defined by *AS* (or simply by A_1 when $r = 1$). If $d = 0$, and thus $\boldsymbol{K}_0 = \boldsymbol{Q}$, then $\boldsymbol{K}_r$ is called an *algebraic number field*; otherwise it is called an *algebraic function field.* Sometimes, when *AS* is specified as above, we simply write $\boldsymbol{K}_{i-1}(y_i)$ for $\boldsymbol{K}_i$ without explicitly introducing the algebraic element η_i.

The problem amounts to factorizing a polynomial

$$F(\boldsymbol{u}, \boldsymbol{\eta}, x_1, \ldots, x_t) \in \boldsymbol{K}_r[x_1, \ldots, x_t]$$

over $\boldsymbol{K}_r$.

By choosing a main variable x, suppose x_1 without loss of generality, one can write F in the form

$$F = f_0 x^n + f_1 x^{n-1} + \cdots + f_n$$

with $f_i \in \boldsymbol{K}_r[x_2, \ldots, x_t]$, for $i = 0, 1, \ldots, n$. Here $f_0 = \mathrm{lc}(F, x)$ is the leading coefficient of F in x. The *content* of F with respect to x is the greatest common divisor of $f_0, \ldots, f_n$; F is *primitive* if its content is 1. F is said to be *squarefree* if it has no repeated factors. In what follows, F is assumed to be squarefree and primitive with respect to its main variable. In the first two methods to be presented, $x_2, x_3, \ldots, x_t$ are treated as transcendental elements and are absorbed in $\boldsymbol{K}_r$, while in the third method we distinguish the x's from the u's. For non-zero polynomials A, B over $\boldsymbol{K}_r$ with $\deg(A, x) = m \geq \deg(B, x) = n \geq 0$, one defines the pseudo-division by the formula

$$\mathrm{lc}(B, x)^{m-n+1} A = QB + R, \qquad \text{and} \qquad \deg(R, x) < n,$$

where Q, R are polynomials over $\boldsymbol{K}_r$. We call Q and R the *pseudo-quotient* and *pseudo-remainder* of A with respect to B in x, denoted by $\mathrm{pquo}(A, B, x)$ and $\mathrm{prem}(A, B, x)$, respectively. Similarly, one defines the pseudo-remainder of A with respect to the ascending set *AS* as

$$\mathrm{prem}(A, AS) = \mathrm{prem}(\cdots(\mathrm{prem}(A, A_r, y_r), \cdots), A_1, y_1).$$

13.2 Method of undetermined coefficients

Suppose that $F(x)$ can be factorized over $\boldsymbol{K}_r$ as

$$F(x) \equiv f_0 \cdot G(x) \cdot H(x) \mod AS,$$

where

$$\begin{aligned} G(x) &= x^s + g_1 x^{s-1} + \cdots + g_s, \\ H(x) &= x^t + h_1 x^{t-1} + \cdots + h_t \end{aligned} \qquad s+t=n, 1 \le s,t \le n-1,$$

and $\equiv$ means that $\operatorname{prem}(F - f_0 GH, AS) = 0$. The above g_i and h_j can be written as

$$\begin{aligned} g_i &= \sum_{\substack{0\le k_l \le m_l -1\\ 1\le l\le r}} g_{ik_1\cdots k_r} {y_1}^{k_1} \cdots {y_r}^{k_r}, & g_{ik_1\cdots k_r}, h_{jk_1\cdots k_r} \in \boldsymbol{K}_0, \\ h_j &= \sum_{\substack{0\le k_l \le m_l -1\\ 1\le l\le r}} h_{jk_1\cdots k_r} {y_1}^{k_1} \cdots {y_r}^{k_r}, & i=1,\ldots,s, j=1,\ldots,t. \end{aligned} \tag{13.1}$$

Here, the number of $g_{ik_1\cdots k_r}$ and $h_{jk_1\cdots k_r}$ is $(s+t)m_1\cdots m_r = M$. We rename these indeterminate coefficients with a fixed order as $z_1 \prec z_2 \prec \cdots \prec z_M$. Now expand $F - f_0 GH$, compute its pseudo-remainder R with respect to AS, and equate the coefficients of all the power products of R in $y_1, \ldots, y_r, x$ to 0; we obtain a system of M polynomial equations

$$\begin{cases} V_1(\boldsymbol{u}, z_1, \ldots, z_M) = 0 \\ V_2(\boldsymbol{u}, z_1, \ldots, z_M) = 0 \\ \qquad \cdots\cdots \\ V_M(\boldsymbol{u}, z_1, \ldots, z_M) = 0 \end{cases} \tag{13.2}$$

with coefficients in $\boldsymbol{K}_0$. Thus, whether the polynomial F can be factorized over $\boldsymbol{K}_r$ is equivalent to whether the above system of polynomial equations has a solution for $z_1, \ldots, z_M$ in the field $\boldsymbol{K}_0$, which can be determined by using the characteristic set method (see Chap. 1).

Algorithm $\mathbf{F}_A$: (Hu and Wang 1986)

Given an irreducible ascending set $AS = [A_1, \ldots, A_r]$ that defines the field $\boldsymbol{K}_r$ and a polynomial $F \in \boldsymbol{K}_r[x]$ of degree $m > 1$ which is irreducible over $\boldsymbol{K}_0$ and reduced with respect to AS, this algorithm calculates an irreducible factorization of F over $\boldsymbol{K}_r$.

F1. If m is even then set $\bar{m} \leftarrow m/2$ else set $\bar{m} \leftarrow (m-1)/2$.

F2. For $s = 1, \ldots, \bar{m}$ do:

F2.1. Set $t \leftarrow m - s$, and

$$G \leftarrow x^s + g_1 x^{s-1} + \cdots + g_s, \qquad H \leftarrow x^t + h_1 x^{t-1} + \cdots + h_t,$$

where g_i, h_j are defined by (13.1).

F2.2. Expand $R \leftarrow \mathrm{prem}(F - f_0 GH, \mathcal{AS})$, equate the coefficients of all power products of R in $y_1, \ldots, y_r, x$ to 0 and obtain a system (13.2) of polynomial equations.

F2.3. Solve (13.2) for $x_1, \ldots, x_M$ in $\boldsymbol{Q}(\boldsymbol{u})$ by using the characteristic set method. If there is no solution then go back to F2 for the next s. Otherwise, let $x_1 = \bar{x}_1, \ldots, x_M = \bar{x}_M$ be any solution of (13.2), set

$$G \leftarrow G|_{x_1=\bar{x}_1,\ldots,x_M=\bar{x}_M}, \qquad H \leftarrow H|_{x_1=\bar{x}_1,\ldots,x_M=\bar{x}_M},$$

and go to F4.

F3. Return "F is irreducible over $\boldsymbol{K}_r$."

F4. Factorize G and H in $\boldsymbol{K}_r[x]$ and return

$$F \leftarrow f_0 \cdot \mathrm{F}_A(G, \mathcal{AS}) \cdot \mathrm{F}_A(H, \mathcal{AS}).$$

Example 13.1:

Factorize the polynomial $F = x^3 + 3yx^2 - x + 6y$ over $\boldsymbol{K}$, where $\boldsymbol{K} = \boldsymbol{Q}(y)$ is the algebraic extension field defined by $A = y^2 + 2$.

Suppose that F has a factorization of the form

$$F = [x + (x_1 y + x_2)][x^2 + (x_3 y + x_4)x + (x_5 y + x_6)], \quad i = 1, \ldots, 6, \quad (13.3)$$

where the x_i are unknowns to be determined in $\boldsymbol{Q}$. Comparing the coefficients of x on the two sides of (13.3), we obtain

$$\begin{aligned} x_1 y + x_2 + x_3 y + x_4 &= 3y, \\ (x_1 y + x_2)(x_3 y + x_4) + x_5 y + x_6 &= -1, \\ (x_1 y + x_2)(x_5 y + x_6) &= 6y. \end{aligned}$$

Expanding the above equalities, reducing them by A and then comparing the coefficients of y on the two sides of the obtained equalities, we get a set of polynomial equations

$$\begin{aligned} x_1 + x_3 &= 3,\\ x_2 + x_4 &= 0,\\ x_2x_3 + x_1x_4 + x_5 &= 0,\\ x_2x_4 - 2x_1x_3 + x_6 &= -1,\\ x_2x_5 + x_1x_6 &= 6,\\ x_2x_6 - 2x_1x_5 &= 0. \end{aligned} \tag{13.4}$$

Using the characteristic set method, we may find a rational solution:

$$(x_1, x_2, x_3, x_4, x_5, x_6) = (2, 0, 1, 0, 0, 3).$$

Therefore, F can be factorized as

$$F = (x + 2y)(x^2 + xy + 3)$$

over $\boldsymbol{Q}$. Factorizing $x^2 + xy + 3$ by using the same method, we shall find that it is irreducible.

The above method proposed by Hu and Wang (1986) has been improved in Wu (1994c) by introducing only one polynomial H and performing $R = \mathrm{prem}(\mathrm{prem}(F, H, x), \mathit{AS})$. For H to be a factor of F it is necessary and sufficient that $R = 0$.

Algorithm $\mathbf{F}_A^*$: (Wu 1994c)

Given an irreducible ascending set $\mathit{AS} = [A_1, \ldots, A_r]$ that defines the field $\boldsymbol{K}_r$ and a polynomial $F \in \boldsymbol{K}_r[x]$ of degree $m > 1$, irreducible over $\boldsymbol{K}_0$ and reduced with respect to AS, this algorithm calculates an irreducible factorization of F over $\boldsymbol{K}_r$.

F1. If m is even then set $\bar{m} \leftarrow m/2$ else set $\bar{m} \leftarrow (m-1)/2$.

F2. For $t = 1, \ldots, \bar{m}$ do:

F2.1. Set $H \leftarrow x^t + h_1x^{t-1} + \cdots + h_t$, where h_j are defined by (13.1).

F2.2. Expand $R \leftarrow \mathrm{prem}(\mathrm{prem}(F, H, x), \mathit{AS})$, equate the coefficients of all the power products of R in $y_1, \ldots, y_r, x$ to 0, and obtain a system of polynomial equations as in (13.2) with $M = tm_1 \cdots m_r$.

F2.3. Solve the polynomial system for $x_1, \ldots, x_M$ in $\boldsymbol{Q}(\boldsymbol{u})$ using the characteristic set method. If there is no solution then go back to F2 for the next t. Otherwise, let $x_1 = \bar{x}_1, \ldots, x_M = \bar{x}_M$ be any such solution, set

$$H \leftarrow H|_{x_1=\bar{x}_1,\ldots,x_M=\bar{x}_M}, \qquad G \leftarrow \mathrm{pquo}(F, H, x),$$

and go to F4.

F3. Return "F is irreducible over $\boldsymbol{K}_r$."

F4. Factorize G and H in $\boldsymbol{K}_r[x]$ and return

$$F \leftarrow f_0 \cdot \mathrm{F}_A^*(G, AS) \cdot \mathrm{F}_A^*(H, AS).$$

Now we apply this improved method to Example 13.1 again. Suppose that F has a factor $H = x + (x_1 y + x_2)$ of degree 1; then

$$\begin{aligned} R &= \mathrm{prem}(\mathrm{prem}(F, H, x), A, y) \\ &= 6y + x_1 y - 3x_1 x_2^2 y + x_2 + 3x_2^2 y - x_2^3 - 6yx_1^2 + 2yx_1^3 + 6x_1^2 x_2 - 12x_1 x_2. \end{aligned}$$

Let $R = 0$, i.e., let the coefficients of y in R be all zero; we obtain

$$\begin{aligned} 6x_1^2 x_2 - x_2^3 + x_2 - 12x_1 x_2 &= 0, \\ -3x_1 x_2^2 + 6 - 6x_1^2 + 2x_1^3 + 3x_2^2 + x_1 &= 0. \end{aligned} \tag{13.5}$$

Solving this system of equations by the characteristic set method, we find a unique rational solution $(x_1, x_2) = (2, 0)$. Therefore, F can be factorized as

$$F = (x + 2y)(x^2 + xy + 3).$$

From this example one can see that the number of equations is 6 in (13.4) and 2 in (13.5), but the equations in (13.4) are all simpler than those in (13.5).

13.3 Method via transformation and triangularization

The method presented in this section was discovered by Wang (1992d, 1995d) during his implementation of the CharSets package. The basic idea underlying the method is the reduction of polynomial factorization over algebraic extension fields to that over the rational number field via linear transformation and the computation of characteristic sets with respect to a proper variable ordering. The factors over the algebraic extension fields are finally determined via algebraic GCD computation. The following lemma from Wang (2000a) guarantees the correctness of the factoring algorithm described below.

Lemma 13.1:

Let AS and F be as in the preceding section, $c_1, \ldots, c_r$ be r integers,

$$\bar{F} \leftarrow F|_{x = x - c_1 y_1 - \cdots - c_r y_r},$$

and $\overline{CS}$ be a characteristic set of $\overline{AS} = AS \cup \{\bar{F}\}$ over $\boldsymbol{K}_0$ with respect to $x \prec y_1 \prec y_2 \prec \cdots \prec y_r$. Let $\bar{C}$ be the first polynomial in $\overline{CS}$ and

$$C \leftarrow \bar{C}|_{x = x + c_1 y_1 + \cdots + c_r y_r}.$$

If $\overline{CS}$ is irreducible and contains exactly $r+1$ polynomials, then the GCD of F and C is irreducible over $\boldsymbol{K}_r$.

We continue using the above notations and let $\overline{CS} = [\bar{C}_0, \bar{C}_1, \ldots, \bar{C}_r]$ be a characteristic set of $\overline{AS}$. It happens in general that all the polynomials other than $\bar{C}_0$ in $\overline{CS}$ are linear in their leading variables, while $\bar{C}_0$ involves the variables $\boldsymbol{u}$ and x only. If this is the case, $\overline{CS}$ is said to be *quasi-linear*. If $\overline{CS}$ is not quasi-linear, we make a linear transformation by substituting $x - c_1y_1 - \cdots - c_ry_r$ for x, where $c_1, \ldots, c_r$ are randomly chosen integers. The probability of obtaining a quasi-linear characteristic set with such a linear transformation is one (see Wang 1992d).

If $\overline{CS}$ is quasi-linear and $\bar{C}_0$ is irreducible over $\boldsymbol{K}_0$, then the GCD of F and $C_0 = \bar{C}_0|_{x=x+c_1y_1+\cdots+c_ry_r}$ over $\boldsymbol{K}_r$ must be a true irreducible factor of F over $\boldsymbol{K}_r$ according to Lemma 13.1. If $\bar{C}_0$ is reducible over $\boldsymbol{K}_0$, we try to determine possible factors of F over $\boldsymbol{K}_r$ by computing the GCD of F with each $\boldsymbol{K}_0$-factor of C_0 over the algebraic extension field $\boldsymbol{K}_r$. Practically, we can start this determination as soon as $\overline{CS}$ is triangularized, without need to arrive at an exact characteristic set. Note that during the computation one should try to remove some factors (over $\boldsymbol{K}_0$) if possible. Some factors of F over $\boldsymbol{K}_r$ may also be determined from those $\boldsymbol{K}_0$-factors and the initials of the polynomials in $CS = \overline{CS}|_{x=x+c_1y_1+\cdots+c_ry_r}$.

Algorithm $\mathbf{F}_B$: (Wang 1992d, 2000a)

Given an irreducible ascending set $AS = [A_1, \ldots, A_r]$ that defines the field $\boldsymbol{K}_r$ and a polynomial $F \in \boldsymbol{K}_r[x]$ that is reduced with respect to AS and irreducible over $\boldsymbol{Q}$, this algorithm computes an irreducible factorization of F over $\boldsymbol{K}_r$.

F1. Set $AS^* \leftarrow [A_i : \deg(A_i, y_i) > 1, A_i \in AS]$. If AS^* is empty, then the procedure terminates and return F. Otherwise, let $y_{p_1} \prec y_{p_2} \cdots \prec y_{p_s}$ be the leading variables of the polynomials in AS^*. Choose a set of integers $[c_1, \ldots, c_s]$.

F2. Set $\bar{F} \leftarrow F|_{x=x-c_1y_1-\cdots-c_sy_s}$. Compute a characteristic set $\overline{CS}$ of $AS^* \cup \{\bar{F}\}$ with respect to the variable ordering $x \prec y_{p_1} \cdots \prec y_{p_s}$. Let Δ be the set of the irreducible factors (over $\boldsymbol{K}_0$) of the initials of the polynomials in $\overline{CS}$ and Ω the set of the irreducible factors (over $\boldsymbol{K}_0$) of the first polynomial in $\overline{CS}$ which is not included in Δ.

F3. If $\overline{CS}$ is quasi-linear, then go to F4. If Ω is empty, then choose a new set of integers $c_1, \ldots, c_s$ and go to F2. Otherwise, set $\Delta \leftarrow \Delta \cup \Omega$, $\Omega \leftarrow \varnothing$.

F4. Set $G \leftarrow F$, $\Omega \leftarrow \Omega|_{x=x+c_1y_1+\cdots+c_sy_s}$ and $\Delta \leftarrow \Delta|_{x=x+c_1y_1+\cdots+c_sy_s}$. For each $P \in \Omega \cup \Delta$, compute the GCD F_P of G and P over $\boldsymbol{K}_r$ with heuristic normalization, and set $G \leftarrow G/F_P$ over $\boldsymbol{K}_r$. If some true factors of F are found, then apply Algorithm F_B to F_P and obtain an irreducible factorization F_P^* for each $P \in \Delta \cup \{G\}$, then return

$$F^* = \prod_{P\in\Omega} F_P \prod_{P\in\Delta\cup\{G\}} F_P^*.$$

Otherwise, if $\overline{CS}$ is quasi-linear, then return F; otherwise try new $c_1, \ldots, c_s$ and go to F2.

Normalizing a polynomial G by AS amounts to finding a polynomial G^* that differs from G only by a factor in $\boldsymbol{K}_r$, and $\mathrm{lc}(G^*, x) \in \boldsymbol{Q}[\boldsymbol{u}]$. In many cases, G^* is much simpler than G, but the opposite is also true in many other cases. Heuristic use of normalization may improve the efficiency of Algorithm F_B considerably (see Wang 1992d, 2000a). An immediate variation of the above algorithm is to compute not only the characteristic set but also the characteristic series in step F2. The irreducible factors of F are determined from those ascending sets in the series whose irreducibility can easily be verified.

Example 13.2:

Factorize the polynomial

$$\begin{aligned}F = {} & 2x^3 - 2x^2y_2 + 2x^2y_1 + 4x^2y_1y_2 - 2xy_1y_2^2 + 2x^2y_2^2 + y_1x \\ & -6y_1y_2 - xy_1y_2 - 4x^2 - 4xy_2 - 12\end{aligned}$$

over $\boldsymbol{K}$, where $\boldsymbol{K} = \boldsymbol{Q}(y_1, y_2)$ is the algebraic extension field defined by the irreducible ascending set

$$AS = [y_1^2 + 2, 2y_2^3 - y_1y_2 - y_1].$$

We begin by choosing $c_1 = 0, c_2 = 0$; $\bar{F} = F$. A characteristic set of $\{\bar{F}\} \cup AS$ with respect to the ordering $x \prec y_1 \prec y_2$ is quasi-linear and the first polynomial can be factorized over $\boldsymbol{Q}$ into $2G_1G_2$ with

$$\begin{aligned}G_1 &= 2x^6 + 12x^4 - 8x^3 + 13x^2 - 14x + 27, \\ G_2 &= 4x^{12} - 48x^{11} + 292x^{10} - 820x^9 + 1561x^8 - 3490x^7 + 7657x^6 \\ &\quad -14400x^5 + 23778x^4 - 28080x^3 + 3888x^2 + 15552x + 23328.\end{aligned}$$

One may find that the GCD of F and G_1 over $\boldsymbol{K}$ is

$$F_1 = x + y_1 - y_2,$$

and the GCD of F and G_2 over $\boldsymbol{K}$ is

$$F_2 = x^2 - 2x + xy_2^2 + 2xy_1y_2 + 3y_1.$$

Therefore, F is factorized into $2F_1F_2$ over $\boldsymbol{K}$.

13.4 Hybrid method with modular techniques

The methods described previously are both of sufficient generality. However, in the presence of trancendentals they are quite slow. This is mainly because the complexity of computation in $\boldsymbol{Q}(\boldsymbol{u})$ is high. It turns out that in this case modular techniques of using integer substitution and Hensel lifting can be adapted to improve the methods.

Following the standard modular approach to the factorization of a multivariate polynomial over an algebraic function field, we have the following steps:

- Ensure that the polynomial is squarefree.
- Find "lucky" integer substitutions to reduce the multivariate polynomial to a univariate polynomial, and the algebraic function field to an algebraic number field.
- Factorize the univariate polynomial over the algebraic number field.
- Lift the univariate factors as well as the ascending set.
- Check the true factors.

We assume that x_1 is the main variable and the polynomial F is squarefree and primitive with respect to x_1 in the algebraic extension field $\boldsymbol{Q}(\boldsymbol{u}, \boldsymbol{\eta})$. The set of integers $\boldsymbol{b}, \boldsymbol{a}$, with $\boldsymbol{b} = (b_1, \ldots, b_d)$ and $\boldsymbol{a} = (a_2, \ldots, a_t)$, is said to be *lucky* if it satisfies the following two conditions:

(1) $F^{(0)} = F(\boldsymbol{b}, \boldsymbol{\eta}, x_1, \boldsymbol{a})$ remains squarefree and $\deg(F^{(0)}, x_1) = \deg(F, x_1)$.

(2) $\mathcal{AS}^{(0)} = [A_1(\boldsymbol{b}, y_1), \ldots, A_r(\boldsymbol{b}, y_1, \ldots, y_r)]$ is still an irreducible ascending set.

For the first condition, choose $\boldsymbol{a}$ and $\boldsymbol{b}$ such that

$$\operatorname{res}(F, F', x_1)(\boldsymbol{b}, x_1, \boldsymbol{a}) \neq 0.$$

Here, and later, $\operatorname{res}(F, G, x)$ denotes the resultant of the polynomials F and G with respect to the variable x; F' is the partial differential of F with respect to x_1. It is more difficult to choose $\boldsymbol{b}$ such that the ascending set remains irreducible. However, there is a lot of freedom according to the following Hilbert irreducibility theorem (see Abbott 1989) and the primitive element theory.

Proposition 13.2: (Hilbert irreducibility theorem)

Let P be irreducible in $\mathbb{Z}$ and $U(N)$ denote the number of s-tuples $(b_1, \ldots, b_s) \in \mathbb{Z}^s$ such that $|b_i| \leq N$ for $1 \leq i \leq s$. Let $P(b_1, \ldots, b_s, x_1, \ldots, x_t)$ be reducible in $\mathbb{Z}[x_1, \ldots, x_t]$. Then there exist constants a and c (depending on P) such that $U(N) \leq c(2N+1)^{s-a}$ and $0 < a < 1$.

For the Hensel lemma, we refer to Zassenhaus (1969) for details. Now the most important thing is to determine when to stop the lifting and check the true factors. Let us look at two examples.

Example 13.3:

Factorize $F = ux^2 - 2yx - u + 1 \in \boldsymbol{Q}[u, y, x]$ over $\boldsymbol{Q}(u, y)$ with y defined by the polynomial $A = y^2 - u$.

The solution is given by

$$F \equiv u\left(x + \frac{u-y}{u}\right)\left(x - \frac{u+y}{u}\right) \quad \mathrm{mod}\, A.$$

The factors have u appearing in the denominators.

Example 13.4:

Factorize $F = x^2 - u \in \boldsymbol{Q}[u, y, x]$ over $\boldsymbol{Q}(u, y)$ with y defined by $A = u^3y^2 - 1$.

The solution is given by

$$F = (x + u^2y)(x - u^2y) \quad \mathrm{mod}\, A.$$

The factors have powers of u higher than that in the original F.

The above two simple examples show that it is necessary to distinguish the transcendental elements $\boldsymbol{u}$ from the variables $x_1, \ldots, x_t$ because the total degree in $x_1, \ldots, x_t$ is bounded by that of the original F, but this is not true for the total degree in $\boldsymbol{u}$ and, much worse, $\boldsymbol{u}$ can appear in the denominators of the factors. In Abbott (1989) a possible upper bound for the total degree in $\boldsymbol{u}$ was given, but unfortunately the bound is often too large and his proof, based on Trager (1976), is not complete. We adopt an optimal bound; after arriving at that bound, we multiply every factor by a common polynomial and check whether it is a true factor of F over the field $\boldsymbol{K}_r$. This common polynomial can be obtained according to the following discussion taken from Abbott (1989).

Any element in $\boldsymbol{K}_r$ can be represented by the basis

$$\{{\eta_1}^{e_1} \cdots {\eta_r}^{e_r} : \quad 0 \leq e_i < m_i \text{ for all } i\}.$$

The *defect* of this basis for $\boldsymbol{K}_r$ is the largest denominator appearing in the representation of those algebraic functions whose monic minimal polynomials lie in $\mathbb{Z}[\boldsymbol{u}]$. The *discriminant* of the basis for $\boldsymbol{K}_r$ is

$$N_2N_3\cdots(\mathrm{dis}(A_1))N_1N_3\cdots(\mathrm{dis}(A_2))N_1N_2\cdots(\mathrm{dis}(A_3))\cdots,$$

where N_i is the norm map, i.e., the product of the images under the different embeddings from $\boldsymbol{K}_i$ to $\boldsymbol{K}_{i-1}$, and $\mathrm{dis}(A_i) = \mathrm{res}(A_i, A_i{}', y_i)$.

Proposition 13.3: (Abbott 1989)

The square of the defect of the basis for $\boldsymbol{K}_r$ divides its discriminant.

The following algorithms can be considered as variants of the method of P. S. Wang (1976); Algorithm S improves the method of Trager (1976) by using non-squarefree norms of polynomials.

Algorithm $\mathbf{F}_C$: (Zhi 1996; Wang and Zhi 1998)

Given an irreducible ascending set $\mathbb{AS} = [A_1, \ldots, A_r]$ that defines the algebraic function field $\boldsymbol{K}_r$ and a squarefree polynomial $F \in \boldsymbol{K}_r[x_1, \ldots, x_t]$, this algorithm computes an irreducible factorization of F over $\boldsymbol{K}_r$.

F1. Choose lucky integers $\boldsymbol{b} \leftarrow (b_1, \ldots, b_d)$ and $\boldsymbol{a} \leftarrow (a_2, \ldots, a_t)$. Set

$$\begin{aligned} F^{(0)} &\leftarrow F(\boldsymbol{b}, \boldsymbol{\eta}, x_1, \boldsymbol{a}),\\ \mathbb{AS}^{(0)} &\leftarrow [A_1(\boldsymbol{b}, y_1), \ldots, A_r(\boldsymbol{b}, y_1, \ldots, y_r)]. \end{aligned}$$

F2. Use Algorithm U to factorize $F^{(0)}(\boldsymbol{\eta}, x_1)$ over $\boldsymbol{Q}(\boldsymbol{\eta})$ defined by $\mathbb{AS}^{(0)}$:

$$F^{(0)} \equiv G_1^{(0)}(\boldsymbol{\eta}, x_1)\cdots G_m^{(0)}(\boldsymbol{\eta}, x_1) \mod (U, \mathbb{AS}^{(0)}),$$

where $U = (u_1 - b_1, \ldots, u_d - b_d, x_2 - a_2, \ldots, x_t - a_t)$.

F3. Apply Hensel lifting to the factors $G_i^{(0)}$ and $\mathbb{AS}^{(0)}$ such that

$$\begin{aligned} &F \equiv G_1^{(\delta)}(\boldsymbol{u}, \boldsymbol{\eta}, x_1, \ldots, x_t)\cdots G_m^{(\delta)}(\boldsymbol{u}, \boldsymbol{\eta}, x_1, \ldots, x_t) \mod (U^{\delta+1}, \mathbb{AS}^{(\delta)}),\\ &\mathbb{AS} \equiv \mathbb{AS}^{(\delta)} \mod (U^{\delta+1}). \end{aligned}$$

F4. When $\delta > \deg(F, \boldsymbol{u}) + \sum_{i=2}^{t}\deg(F, x_i) + \sum_{i=1}^{r}\deg(A_i, \boldsymbol{u})$ (the degree in $\boldsymbol{u}$ means the *total degree*), try the true factor test to obtain

$$F \leftarrow G_1(\boldsymbol{u}, \boldsymbol{\eta}, x_1, \ldots, x_t)\cdots G_s(\boldsymbol{u}, \boldsymbol{\eta}, x_1, \ldots, x_t).$$

Algorithm U:

Given an irreducible ascending set $\mathit{AS} = [A_1, \ldots, A_r]$ that defines $\boldsymbol{Q}(\boldsymbol{\eta})$ and a squarefree polynomial $F \in \boldsymbol{Q}(\boldsymbol{\eta})[x]$, this algorithm computes an irreducible factorization of F over $\boldsymbol{Q}(\boldsymbol{\eta})$.

U1. Select a set of integers $\boldsymbol{c} \leftarrow (c_1, \ldots, c_r)$ such that a characteristic set CS of $\mathit{AS} \cup \{w - c_1 y_1 - \cdots - c_r y_r\}$ under the variable ordering $w \prec y_1 \prec \cdots \prec y_r$ is irreducible and quasi-linear.

U2. Normalize $\mathit{CS} \leftarrow [C_0(w), y_1 - C_1(w), \ldots, y_r - C_r(w)]$.

U3. Set $F^*(w, x) \leftarrow F(C_1(w), \ldots, C_r(w), x)$ and apply Algorithm S to $F^*(\xi, x)$ over $\boldsymbol{Q}(\xi)$:

$$F^* \leftarrow F_1(\xi, x) \cdots F_s(\xi, x),$$

where ξ has minimal polynomial $C_0(w)$.

U4. Substitute $\xi = \sum_{i=1}^{r} c_i \eta_i$ for ξ in each F_i.

U5. Return $F \leftarrow F_1(\boldsymbol{\eta}, x) \cdots F_s(\boldsymbol{\eta}, x)$.

Algorithm S:

Given a monic minimal polynomial $m(y)$ of α and a squarefree polynomial $F \in \boldsymbol{Q}(\alpha)[x]$, this algorithm computes an irreducible factorization of F over $\boldsymbol{Q}(\alpha)$.

S1. Choose a positive integer s and set

$$\begin{aligned} G(\alpha, x) &\leftarrow F(\alpha, x - s\alpha), \\ R(x) &\leftarrow \operatorname{res}(G(y, x), m(y), y). \end{aligned}$$

S2. If $R(x)$ is squarefree, then go to S3. Otherwise, compute over $\boldsymbol{Q}$ an irreducible factorization

$$R(x)/\gcd(R(x), R'(x)) \leftarrow F_1(x) \cdots F_k(x).$$

If $k = 1$ then go to S1 else set

$$G_i \leftarrow \gcd(F(x), F_i(x + s\alpha))$$

and apply Algorithm S to $F/(G_1 \cdots G_k)$ and $G_i's$:

$$\begin{aligned} G_i &\leftarrow G_{i1}(\alpha, x) \cdots G_{im_i}(\alpha, x), \quad 1 \le i \le k, \\ F/(G_1 \cdots G_k) &\leftarrow G_{01}(\alpha, x) \cdots G_{0m_0}(\alpha, x); \end{aligned}$$

then return

$$F \leftarrow \prod_{\substack{1 \le j \le m_i \\ 0 \le i \le k}} G_{ij}(\alpha, x)$$

and the algorithm terminates.

S3. Factorize R over $\boldsymbol{Q}$:

$$R(x) \leftarrow H_1(x) \cdots H_l(x).$$

If $l = 1$, then return F and the algorithm terminates.

S4. For $i = 1, \ldots, l$ do:

$$\begin{aligned} H_i(\alpha, x) &\leftarrow \gcd(H_i(x), G(\alpha, x)), \\ G(\alpha, x) &\leftarrow G(\alpha, x)/H_i(\alpha, x), \\ H_i(\alpha, x) &\leftarrow H_i(\alpha, x + s\alpha) \end{aligned}$$

over $\boldsymbol{Q}(\alpha)$.

S5. Return $F \leftarrow H_1(\alpha, x) \cdots H_l(\alpha, x)$.

Example 13.5:

Factorize $F = x^2 - y + 1$ over $\boldsymbol{K} = \boldsymbol{Q}(y, a)$ defined by $AS = [a^2 - (y-1)^3]$.

Pick the substitution value 0 for y; then F and AS are mapped to

$$F^{(0)} = x^2 + 1 \quad \text{and} \quad AS^{(0)} = [a^2 + 1]$$

respectively. The ascending set $AS^{(0)}$ is still irreducible. Applying Algorithm U (or S) to $F^{(0)}$, one gets

$$F \equiv (x - a)(x + a) \mod (y, AS^{(0)}).$$

Hensel lifting $AS^{(0)}$ and the two factors of $F^{(0)}$ proceeds as follows:

$$\begin{aligned} F &\equiv (x-a-ay)(x+a+ay) \mod (y^2, AS),\\ AS &\equiv [a^2-3y+1] \mod (y^2),\\ F &\equiv (x-a-ay-ay^2)(x+a+ay+ay^2) \mod (y^3, AS),\\ AS &\equiv [a^2+3y^2-3y+1] \mod (y^3),\\ F &\equiv (x-a-ay-ay^2-ay^3)(x+a+ay+ay^2+ay^3) \mod (y^4, AS),\\ AS &\equiv [a^2-y^3+3y^2-3y+1] \mod (y^4). \end{aligned}$$

Now begin the true factor test. The discriminant of $\boldsymbol{K}$ is

$$\mathrm{dis}(a^2-(y-1)^3) = -4(y-1)^3.$$

Let D be the greatest factor whose square divides the discriminant; clearly $D = y-1$. Take one of the above two factors, e.g.,

$$F_1 = (x-a-ay-ay^2-ay^3).$$

Then, we have

$$F_1^* = DF_1 = (y-1)F_1 \equiv x(y-1)+a \mod (y^4, AS).$$

A simple test shows that $F_1^*/D = x+a/(y-1)$ can divide x^2-y+1. Therefore, we obtain the factorization

$$F = \left(x - \frac{a}{y-1}\right)\left(x + \frac{a}{y-1}\right)$$

over $\boldsymbol{K}$.

13.5 GCD computation over algebraic fields

Using the same notations, we consider the problem of finding the GCD of two polynomials F and G over the algebraic extension field $\boldsymbol{K}_r = \boldsymbol{Q}(\boldsymbol{u}, \boldsymbol{\eta})$. Corresponding to the above three algorithms of algebraic factorization, there are three methods for determining the GCDs of multivariate polynomials over algebraic extension fields. For the modular method of computing GCDs over an algebraic number field, we refer to Langemyr and McCallum (1989).

13.5.1 Method A

For the undetermined coefficients method, Wu (1994c) supposed that

$$\begin{aligned} F(x) &= f_0x^n + f_1x^{n-1} + \cdots + f_n,\\ G(x) &= g_0x^m + g_1x^{m-1} + \cdots + g_m \end{aligned}$$

with $f_i, g_j \in \boldsymbol{K}_r$ already known. Let

$$C_{n-e}(x) = c_0 x^{n-e} + c_1 x^{n-e-1} + \cdots + c_{n-e},$$
$$D_{m-e}(x) = d_0 x^{m-e} + d_1 x^{m-e-1} + \cdots + d_{m-e}$$

be two polynomials satisfying $\mathrm{prem}(D_{m-e}F - C_{n-e}G, \mathcal{AS}) = 0$ and

$$c_i = \sum_{\substack{0 \le k_l \le m_l - 1 \\ 1 \le l \le r}} c_{ik_1 \cdots k_r} {y_1}^{k_1} \cdots {y_r}^{k_r},$$

$$d_j = \sum_{\substack{0 \le k_l \le m_l - 1 \\ 1 \le l \le r}} d_{jk_1 \cdots k_r} {y_1}^{k_1} \cdots {y_r}^{k_r}.$$

Proceeding as in Sect. 13.2, we get a linear system of polynomial equations:

$$\begin{cases} W_1(u, z_1, \ldots, z_{N_1}) = 0, \\ W_2(u, z_1, \ldots, z_{N_1}) = 0, \\ \cdots\cdots \\ W_N(u, z_1, \ldots, z_{N_1}) = 0, \end{cases}$$

where $N = (m+n-e+1)m_1 m_2 \cdots m_r$ and $N_1 = (m+n-2e+1)m_1 m_2 \cdots m_r$. Owing to the linearity in $c_{ik_1 \cdots k_r}$ and $d_{jk_1 \cdots k_r}$, it is easy to see whether or not the system has solutions for $(c_i, d_i) \in \boldsymbol{K}_r$ for any given e. We start now from $e = \min(m, n)$. If there exists no solution, then we proceed to the case of $e - 1$. Proceeding in the same way further and further, we will ultimately get the GCD of F and G as required.

It is interesting that the above algorithm has also been used for computing the approximate GCD for polynomials whose coefficients have errors (see Corless 1995 and Karmarkar and Lakshman 1996).

Example 13.6:

Find the GCD of

$$F = x^2 - u, \qquad G = x^2 + 2u^2 yx + u$$

over $\boldsymbol{Q}(u, y)$ with y defined by $A = u^3 y^2 - 1$.

Let

$$C_1 = x + (c_1 y + c_2), \quad D_1 = x + (d_1 y + d_2), \quad c_i, d_i \in \boldsymbol{Q}(u)$$

such that

$$R = \mathrm{prem}(D_1 F - C_1 G, A) = 0.$$

Expanding R and equating its coefficients in y and x to zero, we get

$$\begin{aligned} d_1 - c_1 - 2u^2 &= 0, \\ d_2 - c_2 &= 0, \\ -2u^3 c_2 &= 0, \\ -2u^2 - 2c_1 &= 0, \\ d_1 + c_1 &= 0, \\ d_2 + c_2 &= 0. \end{aligned}$$

Computing a characteristic set for the above equation system, one gets a set of solutions $\{c_1 = -u^2, c_2 = 0, d_1 = u^2, d_2 = 0\}$. Hence

$$C_1 = x - u^2 y, \qquad D_1 = x + u^2 y,$$

and the GCD of F and G is given by

$$H = \frac{F}{C_1} = \frac{G}{D_1} = x + u^2 y \in \boldsymbol{Q}[u, y, x].$$

13.5.2 Method B

Generalizing the Euclidean algorithm for computing the GCD of $F(x)$ and $G(x)$ over $\boldsymbol{Q}$ to algebraic extension field $\boldsymbol{K}_r = \boldsymbol{Q}(\boldsymbol{u}, \boldsymbol{\eta})$, we form a polynomial remainder sequence (prs) $f_1, f_2, \ldots, f_{k+1}$ as follows:

$$\begin{aligned} &f_1 = F, \qquad f_2 = G, && \text{suppose } \deg(F, x) \geq \deg(G, x), \\ &f_i = \mathrm{prem}(\mathrm{prem}(f_{i-2}, f_{i-1}, x), \mathit{AS}), && \text{for } 3 \leq i \leq k+1, \\ &f_i \neq 0, && \text{for } 1 \leq i \leq k \text{ and } f_{k+1} = 0. \end{aligned}$$

The above procedure is similar to computing a characteristic set of the polynomial set $\mathit{AS} \cup \{F, G\}$ and the last polynomial in the characteristic set corresponds to the GCD of F and G over $\boldsymbol{K}_r$.

Let us look at Example 13.2 again. During the execution of Algorithm F_B, we need to compute the GCD of F and G_1 over $\boldsymbol{K}$. Computing a characteristic set of $\mathit{AS} \cup \{F, G_1\}$ with respect to the ordering $y_1 \prec y_2 \prec x$, we obtain

$$\begin{aligned} \mathit{CS} = [&y_1^2 + 2, \ -2y_2^3 + y_1 y_2 + y_1, 139696 x y_1 y_2^2 + 144205 x y_1 y_2 \\ &- 87277x + 66440 y_1 x - 180960 y_1 y_2 - 98706 y_1 - 61437 y_2 \\ &- 103091 x y_2 - 121347 y_1 y_2^2 - 176301 y_2^2 + 22858 x y_2^2 + 6816]. \end{aligned}$$

Normalizing the last polynomial in CS, we get

$$\mathit{CS}_1 = [y_1^2 + 2, -2y_2^3 + y_1 y_2 + y_1, x + y_1 - y_2].$$

The last polynomial in CS_1 is the GCD of F and G_1 over $\boldsymbol{K}_r$.

13.6 Implementations and applications

The described algorithms F_A and F_B for polynomial factorization have been implemented by Wang (1992d, 1995d) in his CharSets package in the Maple system; Algorithm F_C was implemented by Zhi (1996), also in Maple. The algorithms for computing GCDs by methods A and B were implemented by Zhi (1996) and Wang (1992d, 1995d), respectively, in Maple. A large set of examples has been chosen to compare the performances between these different methods as well as the Maple built-in functions. See the timings in Wang (1992d) and Zhi (1996, 1997). According to these experimental results, the three methods seem to be efficient in different cases. If the degrees of the polynomial to be factorized and those in the ascending set are less than 4, Algorithm F_A works well; F_B is relatively faster when the transcendental elements $\boldsymbol{u}$ do not appear in the *AS*; Algorithm F_C is quite powerful for factorizing multivariate polynomials over algebraic function fields. Combining these three methods and some criteria for irreducibility test from Wang (1992d), a hybrid factoring algorithm is given by Zhi (1996).

As we have pointed out at the beginning of this chapter, our motivation for studying algebraic factorization comes from geometric theorem proving. Following Wu (1984a,b; Chap. 5), one may express a theorem in elementary (unordered) geometry by means of a set *HS* of polynomials for its hypothesis and, without loss of generality, a single polynomial C for its conclusion. Proving the theorem amounts to deciding whether any zero of *HS* is a zero of C, and if not, which parts of the zeros of *HS* are zeros of C. An elementary version of Wu's method proceeds by computing first a characteristic set *CS* of *HS* and then the pseudo-remainder R of C with respect to *CS*. If $R \equiv 0$, then the theorem is proved to be true under the subsidiary condition $J \neq 0$, where J is the product of the initials of the polynomials in *CS*. A large number of geometric theorems can be proved effectively in this way. However, if R happens to be non-zero, one cannot immediately tell whether the theorem is false or not; in this case, one has to examine the reducibility of *CS* and perform further decompositions. See Wu (1984a,b), Chou and Gao (1990b), Wang (1994, 2000a), Sect. 2.5.4 and Chap. 5. *CS* is often reducible when some geometric ambiguities such as bisection of angles and contact of circles are involved in the theorem (see Wu 1987a). To test the irreducibility of *CS* or to decompose *CS* into irreducible ascending sets, it is necessary to factorize polynomials over successive algebraic extension fields. Wang (1994) presented a set of geometric theorems whose automated proofs may require algebraic factorization. For example, when we prove Poncelet's theorem, the following factorization is necessary (see Wang and Zhi 1998):

$$\begin{aligned}x_4x_3^2R^2 - x_3^4 - (x_2^2 + x_1^2)x_3^2 - x_1^2x_2^2 \\ = (2x_3R - 2x_4^2 + 2x_2x_4 + 2x_1x_4 + x_3^2 - x_1x_2) \\ (2x_3R + 2x_4^2 - 2x_2x_4 - 2x_1x_4 - x_3^2 + x_1x_2)\end{aligned}$$

over the algebraic field $\boldsymbol{K} = \boldsymbol{Q}(x_1, x_2, x_3, x_4)$ defined by

$$\begin{aligned}A = 4x_4^4 - 8(x_2 + x_1)x_4^3 - 4(x_3^2 - x_2^2 - 3x_1x_2 - x_1^2)x_4^2 \\ + 4(x_2 + x_1)(x_3^2 - x_1x_2)x_4 - (x_2 + x_1)^2x_3^2.\end{aligned}$$

Algebraic curves and surfaces are geometric objects defined by zeros of systems of algebraic equations in 2- or 3-dimensional space. In modern geometry engineering, like computer-aided geometric design and geometric modeling, it is desirable to decompose such objects into *simpler* and *smaller* sub-objects. In the language of algebraic geometry, the problem is to decompose arbitrary algebraic curves and surfaces into irreducible components. There are several algorithmic methods based on characteristic sets and Gröbner bases for carrying out such decompositions, see Wang (1992a) for instance. In these methods, algebraic factorization is indispensable. Other applications of algebraic factorization include verification of geometric conditions and implicitizations of curves and surfaces. See Wang and Zhi (1998) for more examples and timings.

Polynomial factorization over algebraic fields is one of the most difficult tasks in computer algebra. The polynomials we can factorize so far are still of low degree. How to improve the existing algorithms and to invent new and more effective ones for factoring polynomials of high degree are challenging questions that remain for future research.

Chapter 14

CONVERSION BETWEEN IMPLICIT AND PARAMETRIC REPRESENTATIONS OF ALGEBRAIC VARIETIES

Xiao-Shan Gao

In this chapter, we review algorithms for conversion between implicit and parametric representations of algebraic varieties with emphasis on the implicitization of rational parametric equations (RPEs). For the implicitization of RPEs, we will consider the following problems.

(1) Find a basis for the implicit prime ideal determined by a set of RPEs.

(2) Find a canonical representation for the image of a set of RPEs.

(3) Decide whether the parameters of a set of RPEs are independent, and if not, re-parametrize the RPEs so that the new RPEs have independent parameters.

(4) Compute the inversion maps of a set of RPEs, and as a consequence, decide whether a set of RPEs is proper. If the RPEs are not proper, find a proper re-parametrization for the given RPEs.

(5) Decide whether a set of RPEs is normal. If it is not normal, find a normal re-parametrization of the given RPEs.

For the parametrization of implicitly given varieties, we will discuss how to construct a hypersurface that is birational to a given general variety and how to find a set of RPEs of a plane curve if it is possible. Several open problems remaining for future research are also mentioned.

14.1 Introduction

There are two forms to represent algebraic curves or surfaces: the *implicit form* and the *parametric form*. For example, the unit circle can be given by an implicit

MATHEMATICS MECHANIZATION AND APPLICATIONS
ISBN 0-12-734760-7

algebraic equation

$$x^2 + y^2 - 1 = 0,$$

or by a pair of rational parametric equations (RPEs)

$$x = \frac{t^2 - 1}{t^2 + 1}, \qquad y = \frac{2t}{t^2 + 1}. \tag{14.1}$$

Generally, *a set of RPEs* is as follows:

$$x_1 = \frac{P_1(t_1, \ldots, t_m)}{Q_1(t_1, \ldots, t_m)}, \quad \ldots, \quad x_n = \frac{P_n(t_1, \ldots, t_m)}{Q_n(t_1, \ldots, t_m)} \tag{14.2}$$

where $t_1, \ldots, t_m$ are indeterminates and $P_1, \ldots, P_n$, $Q_1, \ldots, Q_n$ ($Q_i \neq 0$) are polynomials in $\boldsymbol{K}[t_1, \ldots, t_m]$. Note that $\boldsymbol{K}$ denotes a fixed basic field, e.g., the field $\boldsymbol{Q}$ of rational numbers, of characteristic 0. We assume that not all P_i and Q_i are constants and $\gcd(P_i, Q_i) = 1$. The maximum of the degrees of P_i and Q_j is called the *degree* of (14.2).

It is well known that a set of RPEs can always be converted into a set of implicit equations. This conversion procedure is called *implicitization.* On the other hand, the inverse procedure — algorithms for *parametrization* of implicitly given algebraic varieties — is much more difficult and is still not completely solved. Both implicitization and parametrization are classic topics in algebraic geometry. They have connections with the theory of resolvents, the theory of quantifier elimination, Lüroth's theorem, the computation of the genus, resolution of singularities, etc.

The recent extensive study of the implicitization problem is due to the fact that implicitization of RPEs is often used in *solid modeling.* It is recognized that both implicit and parametric representations for rational curves and surfaces have their advantages in solid modeling: the parametric form is best suited for generating points along a curve, whereas the implicit representation is most convenient for determining whether a given point lies on a specific curve. This motivates the search for a means of converting from one representation to the other.

Sederberg and Arnon were the first to discuss the implicitization problem using various resultant theories (Sederberg et al. 1984; Arnon and Sederberg 1984). Methods of implicitization for *polynomial* parametric equations are presented in Buchberger (1987) and Shannon and Sweedler (1988) using the Gröbner basis method. A method to find the implicit approximation of parametric equations of curves and surfaces is presented in Chuang and Hoffmann

(1989). Implicitization problems for rational parametric surfaces are carefully studied in Hoffmann (1990), Chionh (1990), Kalkbrener (1990a), and Manocha and Canny (1992). Methods of implicitization for general RPEs are presented in Ollivier (1989), Kalkbrener (1990b), Gao and Chou (1991a, 1992a), and Wang (1995a).

Most of the above work is to find the implicit ideal (see Sect. 14.2) of the RPEs, which is the most important but not the only task in the implicitization. A method for computing the image of RPEs is given in Wu (1990a) and Li (1989). In Gao and Chou (1991a, 1992a), algorithms are presented to find a canonical representation for the image of a set of RPEs. Wang (1995a) gives another algorithm for image computation. In Gao and Chou (1991a, 1992a), the problems (3), (4), and (5) listed at the beginning of this chapter are discussed.

Finding parametric representation for implicitly given algebraic varieties is much more difficult than implicitization. Only a special class of algebraic varieties has parametric representations. Except for the case of algebraic curves and surfaces, there is still no general method for deciding whether a given algebraic variety has a parametric representation or not.

In Abhyankar and Bajaj (1988), classic results for finding RPEs are improved to provide an effective method of parametrization for plane curves. In Abhyankar and Bajaj (1989), a parametrization method for a special class of space curves has been provided. In Gao and Chou (1991b), the theory of resolvents from Ritt (1950) has been used to give a parametrization algorithm for any algebraic curves. This algorithm, though complete, needs to solve a system of algebraic equations, which is a complicated task. Effective parametrization methods for plane curves have been proposed in Sendra and Winkler (1991).

It is a well-known result in algebraic geometry that an irreducible variety is birational to a hypersurface (Hartshorne 1977). Furthermore, a constructive proof of this result can be given, e.g., on the basis of the theory of resolvents (Ritt 1950; Gao and Chou 1991b). Previous work on finding a plane curve birational to a space curve usually uses the technique of taking a projection of the space curve into a randomly selected direction and verifying whether it is one-to-one (Abhyankar and Bajaj 1989; Garrity and Warren 1989; Kalkbrener 1990a,b). The resolvent method is similar to the projection method and deals with the general case in a deterministic way.

By the results discussed in the preceding paragraph, we need only to find RPEs for hypersurfaces. As we have mentioned, there are several methods for parametrization of plane curves. Parametrization methods for surfaces were proposed by classical Italian geometers (Castelnuvo 1894) and studied recently in Schicho (1995). Parametrization methods for varieties with higher dimensions are still open (Eisenbud 1993).

Since an algebraic variety generally does not have parametric representations, an alternative way is to find an approximate parametric representation for it at a local point using, e.g., Puisuix expansion (Alonso et al. 1992; Li L. 1996).

This chapter is a summary of the results about implicitization of RPEs and parametrization of algebraic varieties with emphasis on the implicitization. Also, we will mainly focus on the work done by members of Wu's group at MMRC, including Wu (1990a), Li (1989), Gao and Chou (1991a,b, 1992a,b), Chou, Gao, and Li (1994), and Wang (1995a). Other results are mentioned mainly for comparison purposes.

14.2 The implicit ideal of a set of RPEs

The *implicit ideal* of (14.2) is defined as

$$\mathrm{ID}(P,Q)=\left\{F\in \boldsymbol{K}[x_1,\ldots,x_n]\;\middle|\; F\left(\frac{P_1}{Q_1},\ldots,\frac{P_n}{Q_n}\right)\equiv 0\right\}.$$

Zero(ID(P,Q)) is called the *implicit variety* of (14.2); see Chaps. 1 and 2 for the notations Zero, Ideal, etc. Sometimes, we write $\boldsymbol{x}$ for $x_1,\ldots,x_n$ or $(x_1,\ldots,x_n)$. It is clear that ID(P,Q) is a prime ideal whose dimension is equal to the transcendental degree of $\boldsymbol{K}(P_1/Q_1,\ldots,P_n/Q_n)$ over $\boldsymbol{K}$.

14.2.1 A method based on Gröbner bases computation

For a set of RPEs of the form (14.2), let

$$F_i=Q_ix_i-P_i,\quad D_i=Q_iz_i-1,\quad i=1,\ldots,n,$$

where the z_i are new variables. Let

$$\mathrm{PS}(P,Q)=\mathrm{Ideal}(\{F_1,\ldots,F_n,D_1,\ldots,D_n\}),$$

i.e., the ideal generated by F_i and D_i in $\boldsymbol{K}[t_1,\ldots,t_m,\boldsymbol{x},z_1,\ldots,z_n]$.

Theorem 14.1:

The implicit ideal of (14.2) *is* $\mathrm{PS}(P,Q)\cap\boldsymbol{K}[\boldsymbol{x}]$.

For proofs of this theorem, see Ollivier (1989), Kalkbrener (1990b), and Gao and Chou (1992a). With the help of the following lemma and Theorem 14.1, we can compute a basis for the implicit ideal of (14.2) using the Gröbner basis method.

Lemma 14.2: (Buchberger 1985)

Let GB be a Gröbner basis of an ideal $\mathfrak{I} \subset \boldsymbol{K}[\boldsymbol{x}, y_1, \ldots, y_k]$ *in the purely lexicographic order with* $x_1 \prec \cdots \prec x_n \prec y_1 \prec \cdots \prec y_k$. *Then* $GB \cap \boldsymbol{K}[\boldsymbol{x}]$ *is a Gröbner basis of* $\mathfrak{I} \cap \boldsymbol{K}[\boldsymbol{x}]$.

Example 14.1:

Consider the following RPEs:

$$x = \frac{u+v}{u-v}, \qquad y = \frac{2v^2+2u^2}{(u-v)^2}, \qquad z = \frac{2v^3+6u^2v}{(u-v)^3}. \tag{14.3}$$

Let

$$PS = \{(v-u)x + v + u, (v-u)^2 y - 2v^2 - 2u^2,$$
$$(v-u)^3 z + 2v^3 + 6u^2 v, (v-u)z_1 - 1\}.$$

Note that $(u-v)^2 z_2 - 1$ and $(u-v)^3 z_3 - 1$ can be omitted because of the appearance of $(v-u)z_1 - 1$. Under the purely lexicographical order with $x \prec y \prec z \prec u \prec v \prec z_1$, the Gröbner basis of Ideal(PS) is

$$\{y - x^2 - 1, z - x^3 + 1, (x+1)v + (-x+1)u, 2uyz_1 + x + 1, 2vz_1 + x - 1\}.$$

By Theorem 14.1 and Lemma 14.2, a basis of the implicit ideal of (14.3) is $\{y - x^2 - 1, z - x^3 + 1\}$.

If $m = 1$, then the D_i can be deleted from PS(P, Q), and Theorem 14.1 is still true. This is because $\gcd(P_i, Q_i) = 1$ implies that the resultant of P_i and Q_i is 1. Thus, we get the same result as Kalkbrener (1990a) for the implicitization of parametric equations of curves.

14.2.2 A method based on characteristic sets

If it is difficult to compute the Gröbner basis in the above method, we may use a method based on Wu–Ritt's characteristic sets (CS) to obtain a CS for the implicit ideal. Using the same notations introduced above, let $PS = \{F_1, \ldots, F_n\}$ and $DS = \{Q_1, \ldots, Q_n\}$. Since PS is of triangular form, we have

$$\text{Zero}(PS/DS) = \text{Zero}(\text{PD}(PS)/DS).$$

For the definition of PD(PS), see Chap 1. By the CS method, we can find an irreducible ascending set AS under a new variable order $x_1 \prec \cdots \prec x_n \prec t_1 \prec \cdots \prec t_m$ such that

$$\text{Zero}(PS/DS) = \text{Zero}(\text{PD}(AS)/DS).$$

AS has the same dimension m as PS. Hence AS contains n polynomials. By changing the order of the variables properly, we can assume AS to be

$$\begin{aligned}&\{A_1(x_1,\ldots,x_{d+1}),\ldots,A_{n-d}(x_1,\ldots,x_n),\\ &B_1(\boldsymbol{x},t_1,\ldots,t_{s+1}),\ldots,B_{m-s}(\boldsymbol{x},t_1,\ldots,t_m)\}\end{aligned} \tag{14.4}$$

where $d+s=m$. The *parameter set* of AS is $\{x_1,\ldots,x_d,t_1,\ldots,t_s\}$.

Theorem 14.3: (Gao and Chou 1991a)
The implicit ideal of (14.2) *is* $\mathrm{PD}(\{A_1,\ldots,A_{n-d}\})$.

An algorithm to compute a basis of $\mathrm{PD}(\{A_1,\ldots,A_{n-d}\})$ can be found in Chou (1988), Gao and Chou (1993a), and Wang (1992a).

Example 14.2:
For the RPEs (14.3), let

$$\begin{aligned}&PS=\{(v-u)x+v+u,(v-u)^2y-2v^2-2u^2,(v-u)^3z+2v^3+6u^2v\},\\ &DS=\{u-v\}.\end{aligned}$$

Using the CS method under the variable order $x\prec y\prec z\prec u\prec v\prec z_1$, we have $\mathrm{Zero}(PS/DS)=\mathrm{Zero}(\mathrm{PD}(AS))$ where

$$AS=\{y-x^2-1,z-x^3+1,(x+1)v+(-x+1)u\}.$$

By Theorem 14.3, the implicit ideal of (14.3) is $\mathrm{PD}(\{y-x^2-1,z-x^3+1\})$, which is the ideal generated by $y-x^2-1$ and $z-x^3+1$.

14.2.3 Techniques of computation

Both Theorems 14.1 and 14.3 provide complete methods for finding a basis for the implicit prime ideal of a set of RPEs. But computing the CS or the Gröbner basis is a task of very high complexity. For some parametric equations, we may use their special property to develop efficient algorithms. In what follows, we consider such a partial algorithm.

Consider a set of parametric equations for a space surface,

$$x=\frac{P_1(t,s)}{Q_1(t,s)},\qquad y=\frac{P_2(t,s)}{Q_2(t,s)},\qquad z=\frac{P_3(t,s)}{Q_3(t,s)}, \tag{14.5}$$

where P_i and Q_i are polynomials in $\boldsymbol{K}[t,s]$. Let

$$F_1=Q_1x-P_1,\qquad F_2=Q_2y-P_2,\qquad F_3=Q_3z-P_3.$$

The essential step in the implicitization is the triangulation of F_1, F_2, and F_3 by pseudo-remainder sequences. There are two tricks. First, we can use Collins' method or a similar result in the multi-polynomial case (Li 1987) to remove extraneous factors produced in the computation. Second, we may choose a "good" way to do pseudo-divisions. The purpose is to keep the degrees of the polynomials occurring in the computation as low as possible. We shall explain this in the examples given below.

Theorem 14.4:

Suppose that the following triangular from $\{G_1, G_2, G_3\}$ is obtained under the variable order $x \prec y \prec z \prec s \prec t$:

$$G_1(x,y,z), \quad G_2 = I_2(x,y,z)s - U_2(x,y,z), \quad G_3 = I_3(x,y,z)t - U_3(x,y,z).$$

If $I_2(\eta)I_3(\eta) \not\equiv 0$ for $\eta = (\frac{P_1}{Q_1}, \frac{P_2}{Q_2}, \frac{P_3}{Q_3})$, then:

(1) *One irreducible factor of G_1, say F_1 satisfying $F_1(\frac{P_1}{Q_1}, \frac{P_2}{Q_2}, \frac{P_3}{Q_3}) \equiv 0$, is the implicit equation of* (14.5).

(2) *Parametric equations* (14.5) *are proper and a set of inversion maps is $t = \frac{U_2}{I_2}, s = \frac{U_3}{I_3}$.*

The definition for a set of parametric equations to be *proper* will be given formally in Sect. 14.5. Consider the following examples from Hoffmann (1990).

Example 14.3:

Find the implicit ideal of the following RPEs:

$$(\text{quadratic}) \begin{cases} F_1 = z + (-9s + 15)t - 12s - 34, \\ F_2 = y + (s - 8)t - 6s^2 - 7, \\ F_3 = x - 3t^2 + (-s + 5)t - 4s^2 + 2s - 4; \end{cases}$$

$$(\text{cubic}) \begin{cases} F_1 = z - 2t^3 + (5s - 1)t + s^3, \\ F_2 = y + (-s^2 + 3)t - 1, \\ F_3 = x + t^3 - 3st - s^3 - s; \end{cases}$$

$$(\text{bi-cubic}) \begin{cases} F_1 = z + (3s^3 - 15s^2 + 15s)t^3 + (3s^3 + 18s^2 - 27s + 3)t^2 \\ \qquad +(-6s^3 - 9s^2 + 18s - 3)t + 3s^2 - 3s, \\ F_2 = y - t^3 - 3t - 3s^3 + 6s^2 - 3s, \\ F_3 = x - 3t^3 + 6t^2 - 3t - s^3 + 3s^2 - 6s + 1. \end{cases}$$

The computation of the quadratic form is easy. To compute the cubic form, we first form the pseudo-remainders

$$F_4 = \mathrm{prem}(F_1, F_2), \qquad F_5 = \mathrm{prem}(F_1, F_3), \qquad F_6 = \mathrm{prem}(F_2, F_3).$$

Then F_6 and F_4 are polynomials in x, y, z, s such that $\deg(F_6, s) = 5$ and $\deg(F_4, s) = 9$. We compute subresultant remainder sequence of F_4 and F_6 to obtain a triangular form. For the bi-cubic, we first form the pseudo-remainders

$$\begin{array}{lll} F_4 = \mathrm{prem}(F_1, F_2), & F_5 = \mathrm{prem}(F_3, F_2), & F_6 = \mathrm{prem}(F_4, F_5), \\ F_7 = \mathrm{prem}(F_2, F_5), & F_8 = \mathrm{prem}(F_6, F_7), & F_9 = \mathrm{prem}(F_5, F_7). \end{array}$$

Then F_8 and F_9 are polynomials in x, y, z, s such that $\deg(F_8, s) = 9$ and $\deg(F_9, s) = 9$. We compute subresultant remainder sequence of F_4 and F_6 to obtain the triangular form. We can check that all the three examples satisfy the conditions in Theorem 14.4. Therefore, it is easy to obtain the implicit equation and the inversion maps for them.

14.3 The images of a set of RPEs

Let $\boldsymbol{E}$ be an extension field of $\boldsymbol{K}$. The *image* of RPEs (14.2) in $\boldsymbol{E}^n$ is defined as

$$\mathrm{IM}(P, Q) = \left\{ (x_1, \ldots, x_n) \in \boldsymbol{E}^n \;\middle|\; \begin{array}{c} \exists\, (t_1, \ldots, t_m) \in \boldsymbol{E}^m \text{ such that} \\ x_i = \dfrac{P_i(t_1, \ldots, t_m)}{Q_i(t_1, \ldots, t_m)} \text{ for } 1 \le i \le n \end{array} \right\}.$$

We might naturally guess that $\mathrm{Zero}(\mathrm{ID}(P, Q)) = \mathrm{IM}(P, Q)$. But this is not true. For instance, $(1, 0)$ is a point on the unit circle, but it is not in the image of (14.1). To compute the image of the RPEs, we need first to introduce the concept of projection.

Let *PS* and *DS* be polynomial sets in $\boldsymbol{K}[\boldsymbol{x}, y_1, \ldots, y_m]$. The *projection* of $\mathcal{Z} = \mathrm{Zero}(PS/DS)$ with $y_1, \ldots, y_m$ is defined as

$$\mathrm{Proj}_{y_1, \ldots, y_m} \mathcal{Z} = \left\{ (a_1, \ldots, a_n) \in \boldsymbol{E}^n \;\middle|\; \begin{array}{c} \exists\, (b_1, \ldots, b_m) \in \boldsymbol{E}^m \text{ such that} \\ (a_1, \ldots, a_n, b_1, \ldots, b_m) \in \mathcal{Z} \end{array} \right\}.$$

If $\boldsymbol{E}$ is algebraically closed, the method of Tarski (1951) for the real field can be used to compute the projection. In Heintz (1983), Seidenberg (1956b) and Wang (1993b), direct methods for algebraically closed fields are presented. In Wu (1990a), a method based on CS is proposed. It is a known result in logic that if an existential quantifier can be eliminated then there is a quantifier elimination theory for algebraically closed fields (Shoenfield 1967, p. 83). Since the projection algorithm actually gives a method of eliminating existential quantifiers, it provides a complete method of quantifier elimination over the field of complex numbers. For details, see Chou, Gao, and Li (1994).

Theorem 14.5: (Wu 1990a; Sects. 1.3.3 and 2.3.1)

Using the notations defined above, one can find polynomial sets PS_i *and polynomials* G_i, $i = 1, \ldots, k$, *such that*

$$\text{Proj Zero}(PS/DS) = \bigcup_{i=1}^{k} \text{Zero}(PS_i/G_i).$$

For RPEs (14.2), we have

$$\text{IM}(P, Q) = \text{Proj}_{t_1,\ldots,t_m}\text{Zero}(PS/QS) \tag{14.6}$$

where $PS = \{x_i Q_i - P_i(t) |\ i = 1, \ldots, n\}$, and $QS = \{Q_i(t) |\ i = 1, \ldots, n\}$. As a direct consequence of (14.6) and Theorem 14.5, we have the following.

Lemma 14.6:

There is an algorithm to find polynomial sets $PS_1, \ldots, PS_t$ *and polynomials* $D_1, \ldots, D_t$ *such that*

$$\text{IM}(P, Q) = \bigcup_{i=1}^{t} \text{Zero}(PS_i/D_i).$$

The following result describes the relation between the image and the implicit variety of a set of RPEs and a canonical representation for the image (Chou, Gao, and Li 1994).

Theorem 14.7:

Let $\mathcal{V}$ *be the implicit variety of* (14.2) *and* d *the dimension of* $\mathcal{V}$. *Then:*

(1) $\text{IM}(P, Q) \subset \mathcal{V}$, *and*

(2) $\mathcal{V} \setminus \text{IM}(P, Q)$ *is a quasi-variety with dimension less than* d;

(3) *one can find irreducible ascending sets* AS *and* AS_i *such that*

$$\text{IM}(P, Q) = \text{Zero}(\text{PD}(AS)) \setminus \bigcup_{i=1}^{k} \text{Zero}(AS_i/J_i D_i),$$

where J_i *is the initial-product of* AS_i *and* D_i *a polynomial for each* i.

We also have:

(*a*) PD(AS) *is the implicit ideal of* (14.2)*;*

(*b*) Zero(AS_i/J_iD_i) $\subset$ Zero(PD(AS)).

Example 14.4:

In the following RPEs, a part of positive dimension is missed from the image

$$\begin{aligned}PS = \{&(u^2+v^2-1)x - u^2+v^2-1, (u^2+v^2-1)y-2uv,\\ &(u^2+v^2-1)z-2u\}\end{aligned}$$

and $DS = \{u^2+v^2-1\}$. Using a program based on Lemma 14.6, we have

$$\begin{aligned}&\text{Proj Zero}(PS/DS)\\ &\quad = \text{Zero}(z^2-y^2-x^2+1/z(x+1)(y^2+x^2-1)) \cup \text{Zero}(\{z,y,x+1\}).\end{aligned}$$

So the following canonical form for the image can be found:

$$\begin{aligned}&\text{Zero}(z^2-y^2-x^2+1)\\ &\quad \setminus [\text{Zero}(\{z, y^2+x^2-1\}/x+1) \cup \text{Zero}(\{x+1, z^2-y^2\}/y)].\end{aligned}$$

14.4 Independent parameters

The parameters $t_1, \ldots, t_m$ of RPEs (14.2) are said to be *independent* if the implicit ideal of (14.2) is of dimension m, or equivalently the transcendental degree of the field $\boldsymbol{K}(P_1/Q_1, \ldots, P_n/Q_n)$ over $\boldsymbol{K}$ is m.

Lemma 14.8:

Suppose that (14.4) *has been constructed. Then the transcendental degree of* $\boldsymbol{K}' = \boldsymbol{K}(P_1/Q_1, \ldots, P_n/Q_n)$ *over* $\boldsymbol{K}$ *is* $d = m - s > 0$. *Therefore, the parameters are independent iff* $s = 0$.

Theorem 14.9: (Gao and Chou 1991a)

If the parameters of (14.2) *are not independent, then one can find a set of new RPEs*

$$x_1 = \frac{P_1'}{Q_1'}, \ldots, x_n = \frac{P_n'}{Q_n'} \tag{14.7}$$

which has the same implicit variety as (14.2) *but with independent parameters.*

Proof:

Since (14.4) is irreducible, we may assume that the initials I_i of B_i in (14.4) and the initials J_j of A_j in (14.4) are polynomials in $x_1, \ldots, x_d, t_1, \ldots, t_s$. Since Q_i is not in $\mathrm{PD}(F_1, \ldots, F_n) = \mathrm{PD}(AS)$, we can find a non-zero polynomial q_i in the parameters of AS, i.e., $x_1, \ldots, x_d$ and $t_1, \ldots, t_s$, such that

$$q_i \in \mathrm{Ideal}(\{A_1, \ldots, A_{n-d}, B_1, \ldots, B_{m-s}, Q_i\}).$$

Let $M = \prod_{i=1}^{m-s} I_i \cdot \prod_{j=1}^{n} q_j$. Then M is a polynomial of $x_1, \ldots, x_d, t_1, \ldots, t_s$. Let $h_1, \ldots, h_s$ be integers such that when t_i is replaced by h_i, $i = 1, \ldots, s$, M becomes a non-zero polynomial of $x_1, \ldots, x_d$. Let P_i' and Q_i' be the polynomials obtained from P_i and Q_i by replacing t_i by h_i, $i = 1, \ldots, s$. Now we have obtained (14.7).

The computation in the above theorem can be done with the Gröbner basis method (Gao and Chou 1992a).

Example 14.5:

For Example 14.1, we have $d = 1$ and $s = 1$; hence the parameters u and v are not independent. To re-parametrize (14.3), by Theorem 14.9, we have to compute M. Since $\mathrm{prem}(u - v, AS) = 2u$ and $M = 2(x + 1)u$, selecting a value of u, say 1, we get a set of new parametric equations

$$x = \frac{v+1}{1-v}, \qquad y = \frac{2v^2+2}{(1-v)^2}, \qquad z = \frac{2v^3+6v}{(1-v)^3},$$

which has the same implicit ideal as (14.3) and has an independent parameter v.

14.5 Inversion maps and proper parametric equations

The *inversion problem* is given a point $(a_1, \ldots, a_n)$ on the image of (14.2), to find a set of values $(\tau_1, \ldots, \tau_m)$ for the t such that

$$a_i = \frac{P_i(\tau_1, \ldots, \tau_m)}{Q_i(\tau_1, \ldots, \tau_m)}, \quad i = 1, \ldots, n. \tag{14.8}$$

This problem can be reduced to an equation solving problem (Buchberger 1987). In the following, we show that, in certain cases, we can find a closed form solution to the inversion problem.

Inversion maps for (14.2) are functions

$$t_1 = f_1(\boldsymbol{x}), \ldots, t_m = f_m(\boldsymbol{x})$$

such that $x_i \equiv P_i(f_1, \ldots, f_m)/Q_i(f_1, \ldots, f_m)$ are true on the implicit variety $\mathcal{V}$ of (14.2) except for a subset of $\mathcal{V}$ which has a lower dimension than that of $\mathcal{V}$.

The inversion problem is closely related to whether a set of parametric equations is proper. RPEs (14.2) are said to be *proper* if for each $(a_1, \ldots, a_n) \in$ IM(P, Q) there exists only one $(\tau_1, \ldots, \tau_m) \in \boldsymbol{E}^m$ such that (14.8) holds. Let us assume that the parameters $t_1, \ldots, t_m$ of (14.2) are independent, i.e., $s = 0$. Then (14.4) becomes

$$\{A_1(x_1, \ldots, x_{m+1}), \ldots, A_{n-m}(x_1, \ldots, x_n),\ B_1(\boldsymbol{x}, t_1), \ldots, B_m(\boldsymbol{x}, t_1, \ldots, t_m)\}.$$

Theorem 14.10:

Using the above notations, we have:

(a) $B_i(\boldsymbol{x}, t_1, \ldots, t_i) = 0$ *determine* t_i $(i = 1, \ldots, m)$ *as functions of* $\boldsymbol{x}$ *which are a set of inversion maps for* (14.2)*;*

(b) (14.2) *is proper iff* $B_i = I_i t_i - U_i$ *are linear in* t_i *for* $i = 1, \ldots, m$, *and if this is the case, the inversion maps are*

$$t_1 = \frac{U_1}{I_1}, \ldots, t_m = \frac{U_m}{I_m},$$

where the I_i *and* U_i *are polynomials in* $\boldsymbol{K}[\boldsymbol{x}]$.

Theorem 14.11: (Gao and Chou 1991a)

If $m = 1$ *and RPEs* (14.2) *are not proper, one can find a new parameter* $s = f(t_1)/g(t_1)$, *where* f *and* g *are in* $\boldsymbol{K}[t_1]$, *such that the re-parametrization of* (14.2)

$$x_1 = \frac{F_1(s)}{G_1(s)}, \quad \ldots, \quad x_n = \frac{F_n(s)}{G_n(s)} \tag{14.9}$$

in terms of s *is proper.*

Proof:

Let $\boldsymbol{K}' = \boldsymbol{K}(P_1/Q_1, \ldots, P_n/Q_n)$. Since $P_1(t_1) - Q_1(t_1)l = 0$ where

$$l = P_1(t_1)/Q_1(t_1) \in \boldsymbol{K}',$$

t_1 is algebraic over $\boldsymbol{K}'$. Let $f(y) = a_r y^r + \cdots + a_0$ be an irreducible polynomial in $\boldsymbol{K}'[y]$ for which $f(t_1) = 0$. Then at least one of a_i/a_r, say $\eta = a_s/a_r$, is not in $\boldsymbol{K}$. By a proof of Lüroth's theorem (Walker 1950, p. 149), we have

$\boldsymbol{K}' = \boldsymbol{K}(\eta)$. This means that $x_i = P_i/Q_i$ can be expressed as rational functions of η and η also can be expressed as a rational function of $x_i = P_i/Q_i$, i.e., there is a one-to-one correspondence between the values of the $x_i = P_i/Q_i$ and η. Therefore, η is the new parameter we seek. To compute η, by Theorem 14.10, we can find an inversion map $B_1(\boldsymbol{x}, t_1) = 0$ of the curve. Then $B_1'(y) = B_1(P_1/Q_1, \ldots, P_n/Q_n, y) = 0$ is a polynomial in $\boldsymbol{K}'[y]$ with lowest degree in y such that $B_1'(t_1) = 0$, i.e., $B_1'(y)$ can be taken as $f(y)$. So s can be obtained as follows. If B_1 is linear in t_1, nothing needs to be done. Otherwise let

$$B_1 = b_r t_1^r + \cdots + b_0,$$

where the b_i are in $\boldsymbol{K}[\boldsymbol{x}]$. By (14.2), b_i can also be expressed as rational functions $a_i(t_1)$, $i = 1, \ldots, r$. At least one of a_i/a_r, say a_0/a_r, is not an element in $\boldsymbol{K}$. Let $s = a_0/a_r$. Eliminating t_1 from (14.2) and $a_r s - a_0$, we can get (14.9). Note that a_i comes from b_i by substituting x_j with P_j/Q_j, $j = 1, \ldots, n$; thus $s = b_0/b_r$ is an inversion map of (14.9).

Theorem 14.11 provides a new constructive proof for Lüroth's theorem. In other words, we have the following.

Proposition 14.12:

Let $g_1(t), \ldots, g_r(t)$ be elements of $\boldsymbol{K}(t)$. Then one can find a $g(t) \in \boldsymbol{K}(t)$ such that $\boldsymbol{K}(g_1, \ldots, g_r) = \boldsymbol{K}(g)$.

Example 14.6:

Consider the parametric equations for a Bézier curve (Sederberg 1986):

$$x = \frac{P_1}{Q}, \qquad y = \frac{P_2}{Q} \tag{14.10}$$

with

$$\begin{aligned} P_1 &= 8s^6 - 12s^5 + 32s^3 + 24s^2 + 12s, \\ P_2 &= 24s^5 + 54s^4 - 54s^3 - 54s^2 + 30s, \\ Q &= s^6 - 3s^5 + 3s^4 + 3s^2 + 3s + 1. \end{aligned}$$

Let $HS = \{Qx - P_1, Qy - P_2, Qz - 1\}$. Under the variable order $y \prec s \prec z$, the

Gröbner basis of Ideal(HS) in $\boldsymbol{Q}(x)[s, y, z]$ consists of

$$\begin{aligned}
G_1 &= 224y^3 + (-2268x + 7632)y^2 + (-54x^2 - 1512x - 480384)y + 34263x^3 \\
&\quad - 424224x^2 + 1200960x, \\
G_2 &= (15273x^2 + 1098792x - 9767808)s^2 + (7280y^2 + (-27006x - 125592)y \\
&\quad - 174069x^2 + 598788x - 9767808)s - 7280y^2 + (27006x + 125592)y \\
&\quad + 189342x^2 + 500004x, \\
G_3 &= (488736x + 39071232)z + (33488y^2 + (-95718x + 1701432)y \\
&\quad - 712134x^2 + 9970488x - 34187328)s + 27888y^2 \\
&\quad + (-81210x + 1297128)y - 584109x^2 + 8885196x - 39071232.
\end{aligned}$$

To find a set of proper parametric equations for $G_1 = 0$, by Theorem 14.11, we select a new parameter

$$\begin{aligned}
t_1 &= \frac{(7280y^2 + (-27006x - 125592)y - 174069x^2 + 598788x - 9767808)}{(15273x^2 + 1098792x - 9767808)} \\
&= \frac{s^2 + 1}{1 - s}.
\end{aligned}$$

Eliminating s from (14.10) and the above equation, we have

$$x = \frac{8t_1^3 + 12t_1^2 - 36t_1 + 16}{t_1^3 + 3t_1^2 - 3t_1}, \qquad y = \frac{-24t_1^2 + 78t_1 - 54}{t_1^3 + 3t_1^2 - 3t_1},$$

which form a set of proper parametric equations of $G_1 = 0$.

14.6 Normal RPEs

Generally speaking, the image of a set of RPEs is a quasi-algebraic set. In this section, we discuss when the image of a set of parametric equations is an algebraic set.

RPEs (14.2) are called *normal parametric equations* if IM(P, Q) is the implicit variety of (14.2). The following result is a consequence of Theorem 14.7.

Theorem 14.13:

One can decide in a finite number of steps whether RPEs (14.2) *are normal parametric equations.*

The method in Theorem 14.13, though complete, usually needs tedious computation. In what follows, we give some simple criteria for normal parametrization which can be used without any computational cost.

Theorem 14.14: (Gao and Chou 1991b)

Let

$$x_1 = \frac{u_1(t)}{v_1(t)}, \ldots, x_n = \frac{u_n(t)}{v_n(t)}$$

be parametric equations of an algebraic curve. If $\deg(u_i, t) > \deg(v_i, t)$ *for some* i*, then the parametric equations are normal. As a consequence, any set of polynomial parametric equations of a curve is normal.*

If a set of parametric equations (14.2) is not normal, then naturally we may ask whether one can find a set of normal parametric equations which has the same implicit variety as (14.2). This problem is unsolved in general. In case the implicit variety of the parametric equations is a conic, we have a solution to the problem (Gao and Chou 1991b).

Example 14.7:

The image of the following parametric equations

$$x = \frac{t^4 - 4t^2 + 1}{t^4 + 1}, \qquad y = \frac{2\sqrt{2}(-t^3 + t)}{t^4 + 1}$$

is $\text{Zero}(x^2 + y^2 - 1)$, i.e., this is a set of normal parametric equations for the unit circle. It is proved in Gao and Chou (1991b) that there exist no real coefficients normal quadratic parametric equations for the unit circle.

14.7 Parametrization of algebraic varieties

Finding RPEs for algebraic varieties usually involves two steps:

(1) Construct a hypersurface which is birational to the given irreducible variety and birational transformations between the hypersurface and the variety.

(2) Find RPEs for the hypersurface obtained in step (1). A set of RPEs for the algebraic variety can be obtained from the RPEs of the hypersurface and the birational transformations.

There are complete results for the first step. The second step has been solved only for algebraic curves and surfaces.

14.7.1 The theory of resolvents

We will introduce a constructive proof for the following theorem (Hartshorne 1977) with the concept of resolvents (Ritt 1950).

Theorem 14.15:

Any irreducible variety of dimension r is birational to a hypersurface in $\boldsymbol{E}^{r+1}$.

A prime ideal distinct from Ideal($\{1\}$) and Ideal($\{0\}$) is said to be *non-trivial.* In what follows, we assume that *ID* is a non-trivial prime ideal in $\boldsymbol{K}[\boldsymbol{x}]$. We can divide the x_k into two sets, $\boldsymbol{u} = (u_1, \ldots, u_q)$ and $(y_1, \ldots, y_p)$, $p + q = n$, such that no non-zero polynomial of *ID* involves $\boldsymbol{u}$ alone, while, for each $j = 1, \ldots, p$, there is a non-zero polynomial in *ID* involving y_j and $\boldsymbol{u}$ alone. We call $\boldsymbol{u}$ a *parameter set* of *ID*. A CS of *ID*, e.g., the minimal CS in Gao (1989), under the variable order $u_1 \prec \cdots \prec u_q \prec y_1 \prec \cdots \prec y_p$ is of the form

$$AS = \{A_1(\boldsymbol{u}, y_1), A_2(\boldsymbol{u}, y_1, y_2), \ldots, A_p(\boldsymbol{u}, y_1, \ldots, y_p)\},$$

where each A_i is a polynomial involving y_i effectively, and we also have $ID = \mathrm{PD}(AS)$. It is clear that the dimension of $ID = \mathrm{PD}(AS)$ is equal to the number of parameters of *ID*.

Lemma 14.16:

Let the notations be the same as above. Then for a new variable w, there exist polynomials $M_1, \ldots, M_p, G$ of $\boldsymbol{u}$, such that

(1) *two distinct zeros of ID with $\boldsymbol{u}$ taking the same values for which G does not vanish give different values for $Q = M_1 y_1 + \cdots + M_p y_p$, and*

(2) *a CS of the prime ideal $ID_1 = \mathrm{Ideal}(ID \cup \{w - Q\})$ under the variable order $u_1 \prec \cdots \prec u_q \prec w \prec y_1 \prec \cdots \prec y_p$ is of the form*

$$\{A(\boldsymbol{u}, w), A_1(\boldsymbol{u}, w, y_1), \ldots, A_p(\boldsymbol{u}, w, y_p)\},$$

where A is an irreducible polynomial in w and each A_i is linear in y_i.

See Ritt (1950, p. 85) for a proof of this result, or Gao (1991) for general ideals. According to Ritt, we call the equation $A = 0$ a *resolvent* of *ID*.

Theorem 14.17:

Let ID be a prime ideal in $\boldsymbol{Q}[\boldsymbol{u}, y_1, \ldots, y_p]$ where $\boldsymbol{u}$ are the parameters of ID, and let $A(\boldsymbol{u}, w) = 0$ be a resolvent of ID. Then Zero(*ID*) *is birational to the hypersurface* Zero(A).

Proof:

Use the same notations as in Lemma 14.16. We define a morphism

$$MP_1 : \quad \mathrm{Zero}(ID) \to \mathrm{Zero}(A)$$

by setting $MP_1(\boldsymbol{u}, y_1, \ldots, y_p) = (\boldsymbol{u}, M_1 y_1 + \cdots + M_p y_p)$ where the M_i are the same as in Lemma 14.16. By Lemma 14.16 (2), we can assume that $A_i = I_i y_i - U_i$, $i = 1, \ldots, p$, where I_i and U_i are polynomials in $\boldsymbol{u}$ and w. We can further assume that the I_i are free of w. We define another morphism

$$MP_2 : \quad \mathrm{Zero}(A) \to \mathrm{Zero}(ID)$$

by setting $MP_2(\boldsymbol{u}, w) = (\boldsymbol{u}, U_1/I_1, \ldots, U_p/I_p)$. Let $I = \prod_{i=1}^{p} I_i$. Then MP_2 is well defined on $D_1 = \mathrm{Zero}(A) \setminus \mathrm{Zero}(I)$. We can prove that MP_1 and MP_2 are birational transformations between the variety and the hypersurface.

The following algorithm provides a constructive proof for Theorem 14.15.

Algorithm A:

Let PS be a polynomial set in $\boldsymbol{K}[\boldsymbol{x}]$. This algorithm decides whether $\mathcal{V} = \mathrm{Zero}(PS)$ is an irreducible variety, and if so, finds an irreducible polynomial H such that $\mathcal{V}$ is birational to the hypersurface $\mathrm{Zero}(H)$.

A1. By Wu–Ritt's zero decomposition algorithm, we have an irredundant decomposition

$$\mathcal{V} = \mathrm{Zero}(PS) = \bigcup_{i=1}^{m} \mathrm{Zero}(AS_i/J_i).$$

$\mathcal{V}$ is an irreducible variety if there exists one component, say $\mathrm{Zero}(AS_1/J_1)$, such that $\dim(AS_1) > \dim(AS_i)$ and $\mathrm{Zero}(AS_i/J_i) \subset \mathrm{Zero}(\mathrm{PD}(AS_1))$ for $i > 1$. In this case, we have $\mathcal{V} = \mathrm{Zero}(\mathrm{PD}(AS_1))$.

A2. Let $AS_1 = \{A_1, \ldots, A_p\}$. We rename the variables x_k. If A_i is of class m_i, then x_{m_i} is renamed y_i; the variables other than the y_i are renamed $u_1, \ldots, u_q$ and abbreviated as $\boldsymbol{u}$, where $q = n - p$.

A3. Let $\lambda_1, \ldots, \lambda_p, w$ be new indeterminates and let $ID = \mathrm{Ideal}(\mathrm{PD}(AS_1) \cup \{w - Q\})$, where $Q = \lambda_1 y_1 + \cdots + \lambda_p y_p$. Clearly, ID is a prime ideal in $\boldsymbol{K}(\boldsymbol{u}, \boldsymbol{\lambda}, w, y_1, \ldots, y_p)$ with parameters $\boldsymbol{u}$ and $\boldsymbol{\lambda} = (\lambda_1, \ldots, \lambda_p)$. Let

$$\{R(\boldsymbol{u}, \boldsymbol{\lambda}, w), R_1(\boldsymbol{u}, \boldsymbol{\lambda}, w, y_1), \ldots, R_p(\boldsymbol{u}, \boldsymbol{\lambda}, w, y_p)\} \tag{14.11}$$

be a CS of ID. As $\boldsymbol{\lambda}$ are indeterminates, by Lemma 14.16 (1), R_i are linear in y_i.

A4. By Wu–Ritt's zero decomposition algorithm, under the variable order $\boldsymbol{u} \prec \boldsymbol{\lambda} \prec w \prec y_1 \prec \cdots \prec y_p$ we have

$$\mathrm{Zero}(AS_1 \cup \{w - Q'\}) = \bigcup_{i=1}^{t} \mathrm{Zero}(\mathrm{PD}(AS_i')).$$

There exists only one component in the above decomposition, say $\mathrm{Zero}(\mathrm{PD}(AS_1'))$, on which $\boldsymbol{u}$ and $\boldsymbol{\lambda}$ are algebraically independent and AS_1' is a CS of ID. For convenience, we assume that AS_1' is (14.11).

A5. We can assume that for each $1 \leq i \leq p$, the initial I_i of R_i involves $\boldsymbol{u}$ and $\boldsymbol{\lambda}$ alone. Let $D = I \prod_{i=1}^{p} I_i$, where I is the initial of R. Then D is a polynomial in $\boldsymbol{u}$ and $\boldsymbol{\lambda}$.

A6. Let $a_1, \ldots, a_p$ be integers for which D becomes a non-zero polynomial in $\boldsymbol{u}$ when each λ_i is replaced by a_i. Then for $\lambda_i = a_i, i = 1, \ldots, p$, (14.11) becomes

$$\{R', R_1', \ldots, R_p'\} \tag{14.12}$$

where R and R' have the same degree in w, and y_i occurs in R_i' effectively.

A7. We can prove that R' is an irreducible polynomial in w, and (14.12) is a CS of $ID'' = \mathrm{Ideal}(\mathrm{PD}(AS_1) \cup \{w - a_1 y_1 - \cdots - a_p y_p\})$. Hence R' is a resolvent of $\mathrm{PD}(AS_1)$, and $\mathrm{Zero}(R')$ is birational to $\mathrm{Zero}(PS)$. The birational transformations can be obtained as in Theorem 14.17. ■

In practice, we may use a probability approach by randomly selecting p integers $a_1, \ldots, a_p$ as values of λ_i in step A3. The success probability of the selection of integers is one.

14.7.2 Parametrization of algebraic curves

An irreducible algebraic curve $\mathcal{C}$ is said to be *rational* if it has a set of RPEs. By the resolvent theory, to find a set of RPEs for $\mathcal{C}$ we only need do so for a plane curve defined by $f(x, y) = 0$.

Theorem 14.18:

Let $x = u(t)/w(t), y = v(t)/w(t)$ be proper parametric equations for a plane curve $f(x, y) = 0$. Assume that $\gcd(u, v, w) = 1$. Then the degree of f is equal to the degree of the parametric equations.

The following method for parametrization of $f(x,y) = 0$ is based on Theorem 14.18 and Lüroth's theorem (Walker 1950).

Algorithm B:

Let $f(x,y)$ be an irreducible polynomial. This algorithm decides whether $f(x,y) = 0$ is a rational irreducible algebraic curve, and if so, finds a set of RPEs for the curve.

B1. By Lüroth's theorem, a rational curve always has a set of proper parametric equations (Walker 1950). By Theorem 14.18, if $f(x,y) = 0$ is rational then it has a set of parametric equations of degree d. Let

$$x = u(t)/w(t), \qquad y = v(t)/w(t),$$

where

$$u(t) = u_d t^d + \cdots + u_0, \quad v(t) = v_d t^d + \cdots + v_0, \quad w(t) = w_d t^d + \cdots + w_0$$

for indeterminates u_i, v_i and w_i.

B2. Replacing x by $u(t)/w(t)$ and y by $v(t)/w(t)$ in $f(x,y)$ and clearing denominators, we obtain a polynomial Q in t whose coefficients are polynomials in u_i, v_i and w_i. Let the set of coefficients of Q as a polynomial of t be $HS = \{P_1, \ldots, P_h\}$.

B3. The curve $f = 0$ has a set of RPEs iff HS has a set of zeros such that, when the coefficients of $u(t)$, $v(t)$ and $w(t)$ are replaced by the zeros, u/w and v/w are not constants in $\boldsymbol{K}$. By step B4, we can decide whether there exist such zeros of HS.

B4. Let $DS_1 = \{u_i w_j - u_j w_i \mid i,j = 0, \ldots, d\}$ and $DS_2 = \{v_i w_j - v_j w_i \mid i,j = 0, \ldots, d\}$. Then $f = 0$ is rational iff $HD = \text{Zero}(HS) \setminus [\text{Zero}(DS_1) \cup \text{Zero}(DS_2)]$ is not empty; in case HD is not empty, each zero of HD provides a set of parametric equations for $f = 0$. ■

In step B4, we need to solve a system of algebraic equations. There are many methods for doing this. One can use, e.g., the method based on Wu–Ritt's CS (Wu 1987b; Chap. 1). For an implementation of Algorithm B, refer to Chap. 20.

14.8 Conclusions

For a set of RPEs of the form (14.2), the following results have been summarized.

(1) A basis for the implicit ideal of (14.2) can be found.

(2) The image of (14.2) can be computed and represented in a canonical form.

(3) One can decide whether the parameters $t_1, \ldots, t_m$ are independent, and if not, re-parametrize (14.2) such that the parameters of the new parametric equations are independent.

(4) If the parameters of (14.2) are independent, one can construct equations

$$B_1(\boldsymbol{x}, t_1) = 0, \quad \ldots, \quad B_m(\boldsymbol{x}, t_1, \ldots, t_m) = 0$$

such that the solutions for the t_i in terms of the x_i are the inversion maps of (14.2), and (14.2) is proper iff the B_i are linear in t_i, $i = 1, \ldots, m$.

(5) If $m = 1$ and (14.2) is not proper, then (14.2) can be re-parametrized such that the new parametric equations are proper. In general, how to find a set of proper re-parametrizations for (14.2) is still an open question.

(6) One can decide whether a set of parametric equations is normal. A normal re-parametrization for a set of abnormal parametric equations of conics can also be found. The general case for this problem is still open.

(7) One can construct a hypersurface which is birational to a given irreducible variety and birational transformations between the hypersurface and the variety.

(8) A set of RPEs can be found for a plane curve if it exists. The problem of finding a set of RPEs for a hypersurface of dimension higher than two is still open.

Part IV

Implementations and Applications

Chapter 15

CENTERS, FOCI, AND LIMIT CYCLES FOR POLYNOMIAL DIFFERENTIAL SYSTEMS

Zhengyi Lu and Shilong Ma

Research interest in the problem of distinguishing between centers and foci and the construction of limit cycles for polynomial differential systems has been ongoing since the early years of the 20th century, largely owing to their tight connection with Hilbert's 16th problem. With application of effective methods and systems for large symbolic computations, active research on this subject has made impressive progress in the past decade. This chapter reviews some symbolic approaches based on characteristic sets, triangular systems, Gröbner bases, resultants and Sturm-like theorems, and summarizes a number of results achieved by using these approaches for cubic differential systems. With emphases on the symbolic computational aspect, our presentation is restricted mainly to the recent studies by Chinese researchers and some closely related work done in other groups.

15.1 Introduction

Consider plane systems of differential equations of the form

$$\frac{dx}{dt} = X(x, y), \quad \frac{dy}{dt} = Y(x, y), \tag{15.1}$$

where X and Y are polynomials of x and y. Let P_n denote the set of all real polynomials of total degree in x and y not greater than n. In 1923, Dulac "proved" that given a specific integer n and polynomials X and Y in P_n, system (15.1) has a finite number of *limit cycles*. However, Dulac's theorem had remained questionable until the later 1980s when its proof was completed by Ecalle et al. (1987) and Il'yashenko (1991). For each $n \geq 1$, we define the *Hilbert number* $\mathrm{H}(n)$ as follows:

$$\mathrm{H}(n) = \sup_{X,Y \in P_n} \{\text{the number of limit cycles of system (15.1)}\}.$$

MATHEMATICS MECHANIZATION AND APPLICATIONS
ISBN 0-12-734760-7

The second part of Hilbert's sixteenth problem is to seek for the bound of H(n) and the position of limit cycles. It is still unknown, for a fixed n, whether H(n) is finite or not.

Since H(1) = 0, the first non-trivial Hilbert number is H(2). In 1939 Bautin showed that H(2) $\geq$ 3; the result was published later in 1952. Landis and Petrovskii (1957) "proved" that H(2) = 3 and for any $n \geq 2$,

$$\mathrm{H}(n) \leq \begin{cases} \frac{1}{2}(6n^3 - 7n^2 - 11n + 6) & \text{when} \quad n \quad \text{odd}, \\ \frac{1}{2}(6n^3 - 7n^2 + n + 4) & \text{when} \quad n \quad \text{even}. \end{cases}$$

Later a mistake was found in the proof of Landis and Petrovskii. In 1979, two examples were given by Chen and Wang (1979) and Shi (1980), respectively, showing that

$$\mathrm{H}(2) \geq 4.$$

The relative location of the four limit cycles is the same for the two examples: three limit cycles enclosing the critical point $(0, 0)$ and the fourth enclosing another critical point $(0, 1)$. It seems very difficult to get upper bounds even for H(2).

For $n = 3$, Li and Li (1985) and Zoładek (1994) showed that H(3) $\geq$ 11. For an arbitrary n, Otrokov (1954) showed that

$$\mathrm{H}(n) \geq \begin{cases} \frac{1}{2}(n^2 + 5n - 14) & \text{when} \quad n \quad \text{odd}, \\ \frac{1}{2}(n^2 + 5n - 26) & \text{when} \quad n \quad \text{even}. \end{cases}$$

There are several other bounds for some specific n. Shen (1986) proved the following result.

Theorem 15.1:

For $n = 4k - 1$,

$$\mathrm{H}(n) \geq \frac{11}{16}(n+1)^2.$$

Shen's theorem yields H(7) $\geq$ 44 and H(15) $\geq$ 176 for $k = 2$ and 4.

Christopher and Lloyd (1995) gave the following bound.

Theorem 15.2:

For $n = 2^k - 1$,

$$\mathrm{H}(n) \geq \frac{1}{2}(n+1)^2[\log_2(n+1) - 3] + 3n.$$

For the first few values of k, we have $\mathrm{H}(7) \geq 25$ and $\mathrm{H}(15) \geq 185$.

Besides these general bounds for general systems, the existence of limit cycles and the related problem of centers and foci have also been studied for some specific systems such as cubic ones in the literature. In this chapter, some problems and results for concrete (cubic) systems are reviewed and symbolic computation and manipulation for polynomials arising from these problems are emphasized. Computer algebra systems, the methods of characteristic sets (Wu 1984a,b; Chap. 1), Gröbner bases (Buchberger 1985) and triangular systems (Wang 1993b; Chap. 2) implemented therein, and Sturm-like theorems (Ma and Ning 1998) are used to handle these problems.

In Sect. 15.2, Poincaré's method for computing focal values, that are polynomials in the coefficients of the corresponding polynomials X and Y in x and y, and several approaches to deal with these polynomials using the methods of characteristic sets, Gröbner bases, triangular systems, resultants as well as Sturm-like theorems are introduced.

For different classes of cubic systems, six, seven and eight small amplitude limit cycles may be constructed. This is presented together with Zoładek's result for eleven limit cycles in Sect. 15.3.

In Sect. 15.4, a particular class of cubic systems, known as *Kukles' system*, is discussed. The relationship among the independently obtained center conditions is explained, and necessary and sufficient conditions for the system to have two centers are also given.

15.2 Symbolic approaches to the center-focus problem

In this section, we briefly review some symbolic approaches to the problem of distinguishing between centers and foci (or the *center-focus problem* for short) for plane differential systems, that is important to the second part of Hilbert's 16th problem, and summarize the relevant work on this problem in recent years.

Deriving center-focus conditions for polynomial differential systems may be reduced to finding real symbolic solutions of systems of polynomial equations and inequalities. The latter can be formally expressed as a problem in the first-order theory of real closed fields, which can be solved in principle by using the well-known methods of Tarski (1951) and Collins (1975) (see also Rabin 1977). Nevertheless, the involved symbolic computations in the present case are very complex, and the above-mentioned methods, which have exponential complexity (Fischer and Rabin 1974), are not able to solve such systems in practice. Thus, using more effective methods and tools to solve systems of polynomial equations and inequalities was a key issue to the construction of limit cycles and the study of the center-focus problem.

In the late 1980s, Wang applied the methods of characteristic sets (Ritt 1950; Wu 1984a, 1986a) and Gröbner bases (Buchberger 1985) to deal with the involved polynomial systems with practical success. Meanwhile, Lloyd and his collaborators also worked on the same type of problems using computer algebra systems, where resultant computation was used as a major elimination tool. With the help of these computing devices, Wang, Lloyd and their co-workers obtained several new results, and their symbolic approaches have been considered a significant development.

15.2.1 The center-focus problem

Consider differential systems of *center-focus* type. In canonical coordinates, such systems are of the form

$$\begin{cases} \dot{x} = \lambda x + y + P_2(x,y) + \cdots + P_n(x,y), \\ \dot{y} = -x + \lambda y + Q_2(x,y) + \cdots + Q_n(x,y), \end{cases} \tag{15.2}$$

where P_k and Q_k are homogeneous polynomials of degree k ($k = 2, 3, \ldots$), $\dot{x} = dx/dt$ and $\dot{y} = dy/dt$. It is well-known that there is a function $F(x,y)$, called a *Liapunov function*, defined in a neighborhood of the origin:

$$F(x,y) = \sum_{j=2}^{\infty} F_j(x,y),$$

where

$$F_2(x,y) = x^2 + y^2 \quad \text{or} \quad F_2(x,y) = \frac{1}{2}(x^2 + y^2),$$

$$F_j(x,y) = \sum_{s=0}^{j} f_{j-s,s} x^{j-s} y^s, \quad j \geq 3,$$

and the coefficients $f_{j-s,s}$ remain to be determined (Nemyskii and Stepannov 1960; Zhang et al. 1985).

Let us require that dF/dt along the orbits of (15.2) is of the form

$$\frac{dF(x,y)}{dt} = \eta_2(x^2+y^2) + \eta_4(x^2+y^2)^2 + \eta_6(x^2+y^2)^4 + \cdots, \tag{15.3}$$

or

$$\frac{dF(x,y)}{dt} = v_1 y^2 + v_3 y^4 + v_5 y^6 + \cdots, \tag{15.4}$$

By comparing (15.3) or (15.4) with the expression for dF/dt obtained directly from (15.2), i.e.,

$$\begin{aligned}\frac{dF(x,y)}{dt} &= \frac{\partial F(x,y)}{\partial x}\dot{x} + \frac{\partial F(x,y)}{\partial y}\dot{y}\\ &= \sum_{j=2}^{\infty}\frac{\partial F_j}{\partial x}[\lambda x + y + P_2(x,y) + \cdots + P_n(x,y)] +\\ &\quad \sum_{j=2}^{\infty}\frac{\partial F_j}{\partial y}[-x + \lambda y + Q_2(x,y) + \cdots + Q_n(x,y)],\end{aligned}$$

one can establish a system of linear equations with $f_{j-s,s}$ and η_{2i} or v_{2i-1} as unknowns. This linear system can be solved symbolically, so the explicit expressions for η_{2i} or v_{2i-1} may be obtained. For the details, see Wang (1991a). Note that η_{2i} and v_{2i-1} are all polynomials in the coefficients of $P_k(x,y)$ and $Q_k(x,y)$, and they are called *focal values* or *Liapunov constants*. The origin is said to be a *fine focus* if $\lambda = 0$. In the following, we shall only consider fine foci; therefore we have $\eta_2 = \lambda = 0$ or $v_1 = 0$.

The origin is a *fine focus of order k* for system (15.2) if

$$\eta_{2i} = 0 \quad \text{for} \quad 1 \le i \le k, \quad \text{but} \quad \eta_{2k+2} \ne 0,$$

or

$$v_{2i-1} = 0 \quad \text{for} \quad 1 \le i \le k, \quad \text{but} \quad v_{2k+1} \ne 0.$$

A fine focus of infinite order is called a *center* for system (15.2). Establishing explicit conditions for the origin to be a center or a fine focus of finite order is desirable in several situations, of which one is the construction of limit cycles.

15.2.2 Wang's approach

To deal with the center-focus problem, Wang implemented a system called DEMS (Differential Equations Manipulation System) using the methods of characteristic sets and Gröbner bases (see Wang 1989c,d, 1991a). The system contains programs for computing Liapunov functions and Liapunov constants (or focal values). Wang (1991a) discussed the following problems:

(1) decide whether $v_{2i+1} = 0$ is a formal consequence of $v_1 = v_3 = \cdots = v_{2i-1} = 0$;

(2) if $v_{2i+1} = 0$ is not a consequence of $v_1 = v_3 = \cdots = v_{2i-1} = 0$, simplify v_{2i+1} using $v_1 = v_3 = \cdots = v_{2i-1} = 0$;

(3) find an equivalent but simpler condition instead of the condition given by $v_1 = v_3 = \cdots = v_{2i-1} = 0$;

(4) find particular differential systems such that $v_1 = v_3 = \cdots = v_{2i-1} = 0$ but $v_{2i+1} \neq 0$.

He pointed out that to solve problem (1) one can first compute a zero decomposition for $\{v_1, v_3, \ldots, v_{2i-1}\}$ using Wu–Ritt's method and the pseudo-remainder of v_{2i+1} with respect to each characteristic set in the decomposition, and then check whether or not all the remainders are identically equal to 0; to solve problems (2) and (3) one can use the methods of characteristic sets and Gröbner bases to simplify the Liapunov constants (here the simplicity is measured by the number of terms of polynomials); to solve problem (4) one can derive first the Liapunov constants $v_1, \ldots, v_{2i_1+1}$ for some $i_1 < i$ and simplify the right-hand side of the differential system (15.2) using $v_1 = 0, \ldots, v_{2i_1+1} = 0$, and then derive $v_{2i_1+3}, \ldots, v_{2i_2+1}$ for some $i_1 < i_2 < i$ and simplify the systems again, and so on. In Wang's approach, locating real roots of polynomial equations is not involved; what was considered by Wang mainly is about polynomial manipulation and elimination over algebraically closed fields. He remarked that the major difficulty encountered is how to handle large polynomials.

15.2.3 Approaches by Lloyd and others

Lloyd and his co-workers developed a system called FINDETA in REDUCE 3.3 for computing and reducing focal values (Lloyd and Pearson 1990). They compute each η_{2k} modulo the ideal generated by $\eta_2, \ldots, \eta_{2k-2}$, so the resulting η_{2k} has been simplified by $\eta_2 = 0, \eta_4 = 0, \ldots, \eta_{2k-2} = 0$. The simplification is done symbolically with rational substitutions. They call the simplification process the *reduction* of focal values, and the non-trivial polynomials so obtained *Liapunov quantities*, denoted by $L(0), L(1), \ldots$. That is, $L(k)$ is derived from η_{2k+2} subject to $\eta_{2l} = 0$ for $l \leq k$.

In the work of Lloyd's group, resultant calculation and Sturm's theorem were heavily used for variable elimination, locating real roots of polynomial equations and proving inequalities. Christopher (1994) and Pearson et al. (1996) also used a symbolic approach to systematically search for invariant algebraic curves in order to prove the sufficiency of center conditions. Extensive experiments in REDUCE allowed them to obtain a number of interesting results (see Sect. 15.4.1 and 15.4.2).

However, systems of large polynomial equations and inequalities arise in searching for fine foci of higher order. For example, some large polynomials of degree up to 507 and with integer coefficients having up to 835 digits appear in the work reported in Ma and Ning (1997). The conventional symbolic computation methods do not work in practice when dealing with such highly complex cases. Therefore, it is desirable to find practical and efficient methods (not necessarily complete) for solving systems of large polynomial equations and inequalities.

15.2.4 Ma and Ning's approach

By combining Wu–Ritt's method with Budan–Fourier's theorem, Ma and Ning (1997) gave a simple and efficient method for locating real roots and proving inequalities. Based on this method, they implemented a package in Maple V (Char et al. 1991) on a Sun workstation for generating focal values, deriving center-focus conditions and constructing limit cycles. The package has been used to get some new results; for example, new conditions on the existence of eight limit cycles for a class of cubic differential systems were derived by solving a system of large polynomial equations and inequalities. It does not seem possible to solve the system using the usual symbolic computation methods and tools.

For any polynomial $f(x)$ of degree n with real coefficients and a real number a, we denote by $N(a)$ the *number of sign-changes* in the sequence

$$f(a), f'(a), \ldots, f^{(n)}(a),$$

where $f^{(i)}$ is the ith derivative with respect to x. Let b be another real number greater than a. The integer $\Delta N = N(a) - N(b)$ is called the *number of lost sign-changes* in the sequence of successive derivatives of $f(x)$

$$f(x), f'(x), \ldots, f^{(n)}(x)$$

when going from $x = a$ to $x = b$.

Ma and Ning's method is based on the following theorem (see Demidovich and Maron 1981).

Budan–Fourier's Theorem:

Let $f(x) = 0$ be a polynomial equation of degree n with real coefficients, and a and b ($a < b$) be two real numbers with $f(a) \neq 0$ and $f(b) \neq 0$. Then, the number of real roots of the equation $f(x) = 0$ lying between a and b is equal to $\Delta N = N(a) - N(b)$, or less than ΔN by a positive even number. In particular,

(1) *if $\Delta N = 0$, then $f(x) = 0$ has no real root between a and b;*

(2) *if $\Delta N = 1$, then $f(x) = 0$ has exactly one real root between a and b.*

Noting that the number of sign-changes in $f(a), f'(a), \ldots, f^{(n)}(a)$ is the same as that in $f(a), f'(a), \frac{1}{2!}f''(a), \ldots, \frac{1}{n!}f^{(n)}(a)$, one can obtain $N(a)$ by computing the coefficients $\frac{1}{k!}f^{(k)}(a)$ $(0 \leq k \leq n)$ of the expansion of $f(x)$ at a by means of Horner's scheme (Ma and Ning 1998), and similarly for $N(b)$. Using Horner's scheme to compute the coefficients of the expansion of $f(x)$ at a point is very simple and efficient. Therefore, Budan–Fourier's theorem may be used more efficiently than Sturm's (Collins and Loos 1982) for locating real roots though the method based on the former is incomplete in general.

In some of the conventional approaches, inequalities are transformed into equations by introducing new variables. The introduction of more variables may increase the computational complexity sharply and thus make the method not work for large polynomial inequalities. Another was is to pseudodivide the polynomials on the left-hand side of the inequalities and then prove that the obtained remainders are not zero. This method also does not work when dealing with large polynomials because of the complexity of pseudodivision. In Ma and Ning (1997, 1998), the following rules of reasoning are used for proving inequalities by Budan–Fourier's theorem.

Consider two polynomials $f(x)$ and $g(x)$ and a real number σ. We want to prove one of the following: $f(\sigma) \neq 0$, $f(\sigma) > 0$, $f(\sigma) < 0$, $f(\sigma)/g(\sigma) \neq 0$, $f(\sigma)/g(\sigma) > 0$ and $f(\sigma)/g(\sigma) < 0$. For this purpose, one chooses a neighborhood (a, b) of σ.

To prove that $f(\sigma) \neq 0$ or $f(\sigma)/g(\sigma) \neq 0$, one tries to prove that $f(x)$ or both $f(x)$ and $g(x)$ have no real root in (a, b). For the other cases, one first tries to prove that $f(x)$ or both $f(x)$ and $g(x)$ have no real roots in (a, b), and then chooses a rational number x_0 (sample point) in (a, b) and computes the value $f(x_0)$ or $f(x_0)/g(x_0)$ to see whether it has the same sign as that indicated by the corresponding inequality.

A combination of Wu–Ritt's method and Budan–Fourier's theorem with the above heuristics gives a method for solving systems of polynomial equations and inequalities of the form $S = E_q \cup I_{neq}$, where E_q is a set of polynomial equations and I_{neq} a set of polynomial inequalities (Ma and Ning 1998). This method is explained briefly as follows.

S1. Compute a characteristic set $CS = \{C_1, \ldots, C_k\}$ of the set PS of polynomials in E_q using Wu–Ritt's method (Chap. 1).

S2. Locate all the real roots of $C_1 = 0$ using Budan–Fourier's theorem and then find all the real solutions of $C_1 = 0, C_2 = 0, \ldots, C_k = 0$. If $C_1 = 0$ has no real root, then go to step S1 to compute a characteristic set CS of the enlarged polynomial set $PS \cup \{J_i\}$ (if any), where J_i is the non-constant initial of C_i. If no such a characteristic set can be computed, then go to step S4.

S3. View the inequalities in I_{neq} and $J_1 \neq 0, \ldots, J_k \neq 0$ as constraints on the real solutions of $C_1 = 0, \ldots, C_k = 0$, and select those real solutions satisfying these constraints to obtain the real solutions of $E_q \cup I_{neq}$. [In this step, Budan–Fourier's theorem is used to prove or disprove the inequalities in I_{neq}]. If there are indeed some real solutions satisfying the above constraints, then output those solutions and the procedure terminates. Otherwise, go to step S1 to compute another characteristic set (if any). If no other characteristic set can be computed, then go to step S4.

S4. Report that "$E_q \cup I_{neq}$ has no real solution."

Remark 15.1:

(*a*) The ideal generated by the polynomials in E_q may be of positive dimension, and thus C_1 may contain independent variables. In this case, one can first choose some values of the independent variables and then locate the real roots of $C_1 = 0$ in step S2.

(*b*) To prove or disprove the inequalities in some interval (a, b), one computes ΔN on (a, b). If ΔN is 0 or odd, then one can obtain some conclusion; if ΔN is even, then one can try again by bisecting (a, b) or apply Sturm's theorem on (a, b) until the conclusion is reached.

15.3 Small-amplitude limit cycles for cubic systems

Limit cycles bifurcating from a critical point are called *small amplitude limit cycles*. One of the problems of interest in the theory of polynomial differential systems is the number of small-amplitude limit cycles. Starting with a fine focus of higher order, one can introduce perturbations that reverse the stability of the origin and each of which reduces the order of the fine focus by one. A limit cycle may bifurcate out of the origin corresponding to each perturbation. The above method consists of the following stages:

(1) deriving the Liapunov constants (or focal values);

(2) reducing the Liapunov constants (or focal values);

(3) verifying the order of the fine focus;

(4) introducing appropriate perturbations.

If the origin is a fine focus of order k, then at most k small amplitude limit cycles can bifurcate from it. However, there is no guarantee that it is possible to bifurcate this maximum number of limit cycles. Although Żoładek (1993) announced a theoretical result that there exists a cubic system with 11 amplitude limit cycles, his arguments used in the proof turned out to be incorrect. Later he improved his results in another paper (Żoładek 1995). In the following we shall concentrate ourselves on surveying the work for bifurcating small-amplitude limit cycles for cubic systems using symbolic computation and computer algebra systems. For Żoładek's work, one may refer to the two papers mentioned above.

15.3.1 Wang's results

By means of DEMS, Wang discovered a class of cubic systems with a fine focus of order 6 and showed that six limit cycles can be constructed from the origin (Wang 1990). Consider the following system

$$\begin{cases} \dot{x} = y - ax^2 - 2bxy + ay^2 + \alpha x^2 y + \beta y^3, \\ \dot{y} = -x - bx^2 + 2axy + by^2 + \gamma x^3 + \delta xy^2. \end{cases} \tag{15.5}$$

The first two Liapunov constants are $v_3 = v_5 \equiv 0$ and the third and fourth are

$$v_7 = \frac{4}{21} ab(\alpha + \delta)(\delta - \gamma + \beta - \alpha), \quad v_9 = -\frac{2}{567} ab(\alpha + \delta)\omega,$$

where ω is a polynomial consisting of 18 terms in α, β, γ, and δ. Hence, $v_7 = v_9 = 0$ if and only if (i) $ab(\alpha + \delta) = 0$ or (ii) $\delta - \gamma + \beta - \alpha = 0$, $\omega = 0$. For case (ii), by substitution $\alpha = \delta - \gamma + \beta$ one can write ω in the form

$$\omega = 48(\beta + \gamma)(\delta + 3\gamma + 4\beta).$$

Thus, the polynomial set $PS = \{v_7, v_9\}$ may be decomposed into the following five sets

$$PS_1 = \{a\}, \quad PS_2 = \{b\}, \quad PS_3 = \{\alpha + \delta\},$$
$$PS_4 = \{\delta - \gamma + \delta - \alpha, \beta + \gamma\},$$
$$PS_5 = \{\delta - \gamma + \beta - \alpha, \delta + 3\gamma + 4\beta\}$$

such that

$$\mathrm{Zero}(PS) = \sum_{i=1}^{5} \mathrm{Zero}(PS_i).$$

Consider now the case $PS_5 = 0$ and make a simple variable substitution $A = -\gamma$, $B = -\alpha + 2\beta$. Then system (15.5) is reduced to the form

$$\begin{cases} \dot{x} = y - ax^2 - 2bxy + ay^2 + (7A - 6B)x^2y - (A - 2B)y^3, \\ \dot{y} = -x - bx^2 + 2axy + by^2 - Ax^3 + (7A - 8B)xy^2. \end{cases} \tag{15.6}$$

Since the case of $a = b = 0$ has been considered previously, we assume in what follows that a and b are not simultaneously 0. By DEMS one may compute the next two Liapunov constants v_{11} and v_{13}: they consist of 12 and 28 terms, respectively, and may be factorized as follows

$$v_{11} = -\frac{896}{99}ab(A - B)^3(2a^2 + 2b^2 + 11B), \quad v_{13} = -\frac{1792}{19305}ab(A - B)^3\Delta,$$

where Δ is a polynomial in a, b, A and B. Pseudo-dividing Δ by $2a^2+2b^2+11B$, one can find that $v_{13} \neq 0$ if $2a^2 + 2b^2 + 11B = 0$ and $ab(A - B) \neq 0$. With these calculations, the following result is established.

Theorem 15.3: (Wang 1990)

For the cubic system (15.6), *if* $2a^2 + 2b^2 + 11B = 0$ *and* $ab(A - B) \neq 0$, *then the origin is a fine focus of order 6 and one can construct 6 limit cycles from the origin.*

In other words, a particular three-parameter family of cubic systems with a fine focus of order 6 has been discovered. It was shown in Wang (1990) how the six limit cycles can bifurcate from the origin by small perturbation. According to the stability criteria, the following holds for system (15.6).

Theorem 15.4: (Wang 1990)

(1) *If* $a^2+b^2=0$*, then the origin is stable of center type but not asymptotically stable.*

(2) *If* $ab(A-B)(2a^2+2b^2+11B)<0$*, then the origin is unstable; If* $ab(A-B)(2a^2+2b^2+11B)>0$*, then the origin is asymptotically stable.*

(3) *If* $2a^2+2b^2+11B=0$ *and* $ab(A-B)<0$*, then the origin is unstable; If* $2a^2+2b^2+11B=0$ *and* $ab(A-B)>0$*, then the origin is asymptotically stable.*

After having obtained the above results, Wang found that the class of differential systems with a fine focus of order 6 could be extended. He investigated the following more general cubic differential system

$$\begin{cases} \dot{x}=-y-ax^2-2bxy+ay^2+Fx^3+(7A-6B)x^2y+Exy^2-(A-2B)y^3, \\ \dot{y}=-x-bx^2+2axy+by^2-Ax^3+Dx^2y+(7A-8B)xy^2+Cy^3. \end{cases} \tag{15.7}$$

With calculations by DEMS, Wang arrived at the following theorem.

Theorem 15.5: (Wang 1990)

Let $F=-C$*; then the origin is a fine focus of order 6 for system* (15.7) *if and only if*

$$5C+3D=2a^2+2b^2+11B=0,$$
$$-(a^2-b^2)C[C^2-27(A-B)^2]+18ab(A-B)[C^2-3(A-B)^2]\neq 0.$$

15.3.2 James and Lloyd's results

James and Lloyd discovered a class of cubic systems with 8 small amplitude limit cycles. They wrote the general cubic system in a form that is contained and explained in Lloyd et al. (1988) (the form of the coefficient of y^2 in the $\dot{x}$ equation was found to simplify the computations):

$$\begin{cases} \dot{x}=\lambda x+y+a_1x^2+(a_2+2b_1)xy+(a_3-a_1)y^2+a_4x^3+ \\ \qquad a_5x^2y+(a_6-3b_7)xy^2+a_7y^3, \\ \dot{y}=-x+\lambda y+b_1x^2+(b_2-2a_1)xy-b_1y^2+b_4x^3+ \\ \qquad (b_5-a_6-3a_4)x^2y+(b_6-a_5)xy^2+b_7y^3. \end{cases}$$

By taking $b_2 = 4a_1, a_2 = -4b_1$ and $a_4 = a_6 = b_7 = 0$, the above system is reduced to the form

$$\begin{cases} \dot{x} = \lambda x + y + a_1 x^2 - 2b_1 xy + (a_3 - a_1)y^2 + a_5 x^2 y + a_7 y^3, \\ \dot{y} = -x + \lambda y + b_1 x^2 + 2a_1 xy - b_1 y^2 + b_4 x^3 + b_5 x^2 y + (b_6 - a_5)xy^2. \end{cases} \tag{15.8}$$

To simplify the complicated computations, new coefficients a_8, b_8 and a_9 are introduced in James and Lloyd (1991) by letting

$$a_9 = a_7 - a_5 + \frac{1}{2}b_6, \quad a_8 = b_1^2 - a_1^2, \quad b_8 = b_1^2 + a_1^2.$$

Thus, one should have $b_8 > 0$ and $b_8 > |a_8|$ (or $b_8^2 - a_8^2 > 0$) when $a_1 \neq 0$ or $b_1 \neq 0$; so the complex values of a_1 and b_1 are ruled out. To ensure the origin to be a fine focus of order 8, one must have

$$L(0) = 0,\ L(1) = 0, \ldots, L(7) = 0 \quad \text{and} \quad L(8) \neq 0, \tag{15.9}$$

where $L(k)$ are Liapunov quantities for $k = 0, 1, \ldots$. By taking rational substitutions from $\eta_2 = 0, \cdots, \eta_{2k} = 0$ to reduce $\eta_{2k+2} (k = 1, 2, \ldots)$ and then to obtain $L(k)$, James and Lloyd established the following results.

Theorem 15.6: (James and Lloyd 1991)

Suppose that $a_7 \neq 0$ and $a_9 = \sigma_* a$, where σ_* is the unique real root of

$$\sum_{k=0}^{10} \alpha_k \sigma^k = 0 \quad \textit{in the interval} \quad \left(\frac{666}{97}, \frac{103}{15}\right).$$

Moreover, suppose that the following conditions are satisfied:

$$b_5 = -4a_3 b_1, \quad a_3 = 0, \quad b_4 = -a_7, \quad a_8 = \frac{b_6(13a_9 - 60a_7)}{20c_4},$$

$$b_8 = -\frac{\psi(a_9, a_7)}{48c_4 c_6}, \quad b_6^2 = \frac{25\chi(a_9, a_7)}{36c_6^2 c_7},$$

where ψ and χ are two polynomials. Then the origin is a fine focus of order 8 for system (15.8).

Theorem 15.7: (James and Lloyd 1991)

There are systems of the form (15.8) *with eight small-amplitude limit cycles. Such systems can be obtained by starting with a system in which the conditions in Theorem 15.6 are satisfied, and perturbing a_9, b_6, b_8, a_8, b_4, a_3, b_5 and λ in that order.*

James and Lloyd also gave the following example to show that there are seven small-amplitude limit cycles bifurcating from the origin of the system.

Theorem 15.8: (James and Lloyd 1991)

There are systems of the form

$$\begin{cases} \dot{x} &= \lambda x + y + a_3 y^2 + a_4 x^3 + a_5 x^2 y - 3b_7 x y^2 + a_7 y^3, \\ \dot{y} &= -x + \lambda y + (b_4 - a_7)x^3 + (b_5 - 3a_4)x^2 y + (b_6 - a_5)x y^2 + b_7 y^3. \end{cases}$$

with $3a_5 = -(10a_3^2 + 11a_7)$ *in which there are seven small-amplitude limit cycles.*

15.3.3 Ma and Ning's results

For system (15.8), computing $L(k)$ modulo the ideal generated by $L(0), \ldots,$ $L(k-1)$ for $k \leq 3$, one can obtain $L(0)$, $L(1)$, $L(2)$ and $L(3)$, where

$$L(3) = -a_1 b_1 (a_7 + b_4)(2a_9 + 7b_4 - 9a_7).$$

Let $\lambda = 0$, $b_5 = -4a_3 b_1$ and $a_3 = 0$; then $L(0) = 0$, $L(1) = 0$ and $L(2) = 0$. To make $L(3) = 0$, one may study the two cases $a_7 + b_4 = 0$ and $2a_9 + 7b_4 - 9a_7 = 0$. The former was considered by James and Lloyd (1991), and so was the latter by Ma and Ning (1997). The second case leads to much complicated computations and cannot be approached by conventional symbolic computation techniques. Ma and Ning introduced another new parameter $a_{10} = 2a_9 - 9a_7$. Let $b_4 = -\frac{1}{7}a_{10}$; then $2a_9 + 7b_4 - 9a_7 = 0$. Thus (15.9) is reduced to

$$L(4) = 0,\ L(5) = 0,\ L(6) = 0,\ L(7) = 0,\ L(8) \neq 0,$$

which satisfy $b_8 > 0$ and $b_8^2 - a_8^2 > 0$. Note that $a_{10} = 0$ corresponds to the case of center, so one can let $a_{10} \neq 0$. To reduce the number of parameters, let

$$A_7 = a_7/a_{10},\ A_8 = a_8/a_{10},\ B_6 = b_6/a_{10},\ B_8 = b_8/a_{10}.$$

Now the following can be computed modulo the ideal generated by $L(0), \ldots, L(3)$:

$$\begin{aligned} L(4) &= a_1 b_1 a_{10}^3 P_1(A_8, B_8, B_6, A_7), \\ L(5) &= a_1 b_1 a_{10}^4 P_2(A_8, B_8, B_6, A_7), \\ L(6) &= -a_1 b_1 a_{10}^5 P_3(A_8, B_8, B_6, A_7), \\ L(7) &= -a_1 b_1 a_{10}^6 P_4(A_8, B_8, B_6, A_7), \\ L(8) &= a_1 b_1 a_{10}^7 P_5(A_8, B_8, B_6, A_7), \end{aligned}$$

which satisfy $B_8^2 - A_8^2 = (b_8^2 - a_8^2)/a_{10}^2 > 0$; the sign of a_{10} is the same as that of B_8 (because $a_{10}B_8 = b_8 > 0$). The problem is reduced to solve the following system

$$\begin{cases} P_1(A_8, B_8, B_6, A_7) = 0, \\ P_2(A_8, B_8, B_6, A_7) = 0, \\ P_3(A_8, B_8, B_6, A_7) = 0, \\ P_4(A_8, B_8, B_6, A_7) = 0, \\ P_5(A_8, B_8, B_6, A_7) \neq 0, \\ B_8^2 - A_8^2 > 0, \end{cases} \tag{15.10}$$

where P_1, P_2, P_3, P_4 and P_5 consist of 12, 34, 74, 139 and 237 terms, respectively.

Let $A_8 \succ B_8 \succ B_6 \succ A_7$. A characteristic set $\{C_1, C_2, C_3, C_4\}$ of $\{P_1, P_2, P_3, P_4\}$ can be computed using Wu–Ritt's method:

$$\begin{cases} C_1(A_7) = \sum_{i=0}^{74} \alpha_i A_7^i, \\ C_2(B_6, A_7) = \theta_1(A_7)B_6 + \theta_2(A_7), \\ C_3(B_8, B_6, A_7) = \omega_1(B_6, A_7)B_8 + \omega_2(B_6, A_7), \\ C_4(A_8, B_8, B_6, A_7) = \psi_1(B_8, B_6, A_7)A_8 + \psi_2(B_8, B_6, A_7); \end{cases}$$

see the Appendix in Ma and Ning (1997) for $\theta_1(A_7)$, $\theta_2(A_7)$, $\omega_1(B_6, A_7)$ and $\omega_2(B_6, A_7)$.

Obviously, the initials of the C_i are

$$J_1 = \alpha_{74}, \quad J_2(A_7) = \theta_1(A_7),$$
$$J_3(B_6, A_7) = \omega_1(B_6, A_7), \quad J_4(B_8, B_6, A_7) = \psi_1(B_8, B_6, A_7).$$

Now check whether or not the system of equations $C_1 = 0, \ldots, C_4 = 0$ and inequations $J_1 \neq 0, \ldots, J_4 \neq 0$ has real solutions. Using the method described in Sect. 15.2.4, one can find that $C_1(A_7) = 0$ has 16 real roots σ_i: $-1 < \sigma_1 < \cdots < \sigma_{16} < 1$. Let σ denote any one of these 16 real roots. The following can be proved by using Budan–Fourier's theorem.

(1) $\theta_1(\sigma) \neq 0$, i.e., $J_2(\sigma) \neq 0$. Let $B_6^* = -\theta_2(\sigma)/\theta_1(\sigma)$; then $C_2(B_6^*, \sigma) = 0$.

(2) $\omega_1(B_6^*, \sigma) \neq 0$, i.e., $J_3(B_6^*, \sigma) \neq 0$. Let $B_8^* = -\omega_2(B_6^*, \sigma)/\omega_1(B_6^*, \sigma)$; then $C_3(B_8^*, B_6^*, \sigma) = 0$.

(3) $\psi_1(B_8^*, B_6^*, \sigma) \neq 0$, i.e., $J_4(B_8^*, B_6^*, \sigma) \neq 0$.
Let $A_8^* = -\psi_2(B_8^*, B_6^*, \sigma)/\psi_1(B_8^*, B_6^*, \sigma)$; then $C_4(A_8^*, B_8^*, B_6^*, \sigma) = 0$.

Obviously, $J_1 = \alpha_{74} \neq 0$. Therefore, one obtains 16 real solutions of $C_1 = 0, C_2 = 0, C_3 = 0, C_4 = 0, J_1 \neq 0, J_2 \neq 0, J_3 \neq 0, J_4 \neq 0$. From Wu–Ritt's well-ordering principle, one knows that they are also real solutions of $\{P_1, P_2, P_3, P_4\}$. Now select those satisfying the constraints $B_2^2 - A_8^2 > 0$ and $P_5(A_8, B_8, B_6, A_7) \neq 0$ from the 16 real solutions. Note that, from (2) one can obtain the following by back-substituting:

$$B_8 = (7A_7 + 1)\sum_{i=0}^{282} c_i A_7^i / [4(\sum_{i=0}^{182} d_i A_7^i)\theta_1(A_7)],$$

$$A_8 = (7A_7 - 1)\sum_{i=0}^{282} f_i A_7^i / [16(\sum_{i=0}^{182} d_i A_7^i)\theta_1(A_7)].$$

Using the method explained in Sect. 15.2.4, one can verify that $B_8^2 - A_8^2|_{A_7=\sigma} > 0$ for

$$\sigma \in \Omega = \{\sigma_1, \sigma_2, \sigma_4, \sigma_5, \sigma_{10}, \sigma_{11}, \sigma_{13}, \ldots, \sigma_{16}\}.$$

Now consider $P_5(A_8, B_8, B_6, A_7)$. First pseudodivide P_5 by $\{C_2,\ C_3,\ C_4\}$ and then prove that the pseudo-remainder is not zero at σ using Budan–Fourier's theorem. We have

$$[J_4(\sigma)]^5[J_3(\sigma)]^5[J_2(\sigma)]^{15} P_5|_{A_7=\sigma} =$$

$$-91750400 \times 466560000(7\sigma + 1)^{12}(7\sigma - 1)^3(\sum_{i=0}^{22} \gamma_i \sigma^i)(\sum_{i=0}^{507} \beta_i \sigma^i).$$

Similarly, one can verify that

$$\sum_{i=0}^{507} \beta_i \sigma^i \neq 0 \quad \text{for} \quad \sigma \in \Omega.$$

One can also prove

$$\sum_{i=0}^{22} \gamma_i \sigma^i \neq 0$$

in the same way. Therefore $P_5(A_8^*, B_8^*, B_6^*, \sigma) \neq 0$. Thus, 10 real solutions of system (15.10) have been obtained.

From the above discussions the following is obtained (note that $a_{10} = 2a_9 - 9a_7$):

$$\begin{cases} \lambda = 0, \; b_5 = -4a_3 b_1, \; a_3 = 0, \\ b_4 = -\frac{1}{7}a_{10} = -\frac{1}{7}(2a_9 - 9a_7), \; a_8/(2a_9 - 9a_7) = A_8^*, \\ b_8/(2a_9 - 9a_7) = B_8^*, \; b_6/(2a_9 - 9a_7) = B_6^*, \\ a_7/(2a_9 - 9a_7) = A_7 = \sigma. \end{cases} \tag{15.11}$$

Theorem 15.9: (Ma and Ning 1997)

The origin is a fine focus of order 8 for system (15.8), *if* $a_1 \neq 0$, $b_1 \neq 0$, $a_9 \neq 0$, *and one of the following three sets of conditions holds:*

(*i*) $a_9 < \frac{9}{2}a_7, a_7 > 0$ *and* (15.11) *with* $\sigma \in \{\sigma_2, \sigma_4, \sigma_{10}, \sigma_{14}\}$.

(*ii*) $a_9 > \frac{9}{2}a_7, a_7 < 0$ *and* (15.11) *with* $\sigma \in \{\sigma_1, \sigma_5, \sigma_{11}, \sigma_{13}\}$.

(*iii*) $a_9 > \frac{9}{2}a_7, a_7 > 0$ *and* (15.11) *with* $\sigma \in \{\sigma_{15}, \sigma_{16}\}$.

Perturbing A_7, B_6, B_8, A_8, b_4, a_3, b_5 and λ, respectively, and using Budan–Fourier's theorem, one can prove that eight limit cycles may bifurcate from the origin for system (15.8). The details can be found in Ma and Ning (1997).

Theorem 15.10: (Ma and Ning 1997)

For $a_1 \neq 0$, $b_1 \neq 0$ *and* $a_9 \neq 0$, *and*

(*i*) $\sigma \in \{\sigma_2, \sigma_4, \sigma_{10}, \sigma_{14}\}$, $a_9 < \frac{9}{2}a_7$ *and* $a_7 > 0$, *or*

(*ii*) $\sigma \in \{\sigma_1, \sigma_5, \sigma_{11}, \sigma_{13}\}$, $a_9 > \frac{9}{2}a_7$ *and* $a_7 < 0$, *or*

(*iii*) $\sigma \in \{\sigma_{15}, \sigma_{16}\}$, $a_9 > \frac{9}{2}a_7$ *and* $a_7 > 0$,

the parameters of system (15.8) *can be perturbed for eight limit cycles to bifurcate from the origin.*

When verifying the conditions on the existence of eight limit cycles derived by James and Lloyd (1991) by solving a system of polynomial equations and inequalities in a different ordering, Ning et al. (1994) found another real solution to the system, which leads to the following improvement on the conditions in James and Lloyd (1991). Let

$$\begin{cases} \lambda = 0,\ b_5 = -4a_3b_1,\ a_3 = 0,\ b_4 = -a_7, & a_7 = \dfrac{a_9(13b_6 - 20a_8)}{20(3b_6 - 4a_8)}, \\ a_9 = \dfrac{24b_6b_8(3b_6 - 4a_8)(15b_6 - 28a_8)}{D_8}, & b_8^2 = \dfrac{5D_8^2 D_{10}}{64D_{11}}, \end{cases} \tag{15.12}$$

$$\sum_{k=0}^{10} \alpha_k b_6^k a_8^{10-k} = 0, \tag{15.13}$$

where

$$\begin{aligned} D_8 &= 3213b_6^3 - 6720a_8^3 - 12156b_6^2a_8 + 15920b_6a_8^2, \\ D_{10} &= 115b_6^2 - 120b_6a_8 - 144a_8^2, \\ D_{11} &= 1957968450b_6^6 - 1539283680b_6^5a_8 + 5123792400a_8^2b_6^4 - 9370286592a_8^3b_6^3 \\ &\quad +10052919040a_8^4b_6^2 - 6038323200a_8^5b_6 + 1580544000a_8^6. \end{aligned}$$

Theorem 15.11: (Ning et al. 1994)

Let $a_7 \neq 0$ and $a_9 = \sigma_1 a_7$ or $a_9 = \sigma_4 a_7$, where $287/125 < \sigma_1 < 284/125$ and $666/97 < \sigma_4 < 103/15$. If the conditions (15.12) are satisfied, then the origin is a fine focus of order eight for system (15.8).

Theorem 15.12: (Ning et al. 1994)

For system (15.8), if

(i) $a_7 > 0$,

(ii) $a_9 = \sigma_1 a_7$ *or* $a_9 = \sigma_4 a_7$, *and*

(iii) (15.12) *and* (15.13) *are satisfied,*

then eight limit cycles can bifurcate from the origin by perturbing a_9, b_6, b_8, a_8, b_4, a_3, b_5 and λ successively.

15.4 Kukles' system

Now come to the following Kukles system of center-focus type

$$\begin{cases} \dot{x} = y, \\ \dot{y} = -x + Q_2 + Q_3, \end{cases} \tag{15.14}$$

and its corresponding system

$$\begin{cases} \dot{x} = y, \\ \dot{y} = -x + \lambda y + Q_2 + Q_3, \end{cases} \tag{15.15}$$

where

$$Q_2 = a_{20}x^2 + a_{11}xy + a_{02}y^2, \quad Q_3 = a_{30}x^3 + a_{21}x^2y + a_{12}xy^2 + a_{03}y^3. \tag{15.16}$$

This particular class of systems, studied first by I. Kukles 55 years ago, has been investigated by several researchers in the past decade.

15.4.1 One-center conditions

The conditions for $E_0 = (0,0)$ of (15.14) to be a center were given by Kukles (1944). He "proved" that the origin is a center of (15.14) if and only if one of the following four sets of conditions holds

$$\begin{cases} \alpha = a_{30}^2 a_{11} + a_{21}\lambda = 0, \\ \beta = (3a_{03}\lambda + \lambda^2 + a_{12}a_{11}^2)a_{21} - 3a_{03}\lambda^2 - a_{12}a_{11}^2\lambda = 0, \\ \gamma = \lambda + a_{20}a_{11} + a_{21} = 0, \\ \theta = 9a_{12}a_{11}^2 + 2a_{11}^4 + 9\lambda^2 + 27a_{03}\lambda = 0; \end{cases} \tag{15.17}$$

$$a_{03} = \alpha = \beta = \gamma = 0; \tag{15.18}$$

$$a_{03} = a_{11} = a_{21} = 0; \tag{15.19}$$

$$a_{03} = a_{02} = a_{20} = a_{21} = 0, \tag{15.20}$$

where $\lambda = a_{11}a_{02} + 3a_{03}$. Cherkas (1978) found the incompleteness of Kukles conditions and gave the following set of conditions instead of (15.17):

$$\begin{cases} \gamma = 0, \\ \theta_1 = 9(6a_{20}a_{03} + a_{20}a_{11}a_{02} - a_{21}a_{02} - a_{11}a_{12} - 2a_{30}a_{11}) - 2a_{11}^3 = 0, \\ \theta_2 = 3(6a_{30}a_{03} - 3a_{20}^2a_{03} + a_{30}a_{11}a_{02} + a_{20}a_{02}a_{21} \\ \qquad +a_{20}a_{12}a_{11} - a_{21}a_{12} - a_{30}a_{21}) - 2a_{11}^2a_{21} = 0, \\ \theta_3 = 3(a_{30}a_{21}a_{02} - 6a_{20}a_{30}a_{03} + a_{30}a_{11}a_{12} + a_{20}a_{21}a_{12}) - 2a_{11}a_{21}^2 = 0, \\ \theta_4 = 9(a_{30}a_{21}a_{12} - 3a_{30}^2a_{03}) - 2a_{21}^3 = 0. \end{cases} \tag{15.21}$$

Recent research on Kukles' system started in 1990 when Jin and Wang rediscovered the incompleteness of Kukles' conditions with the following example

$$\begin{cases} a_{20} \neq 0, & a_{11} = 0, \quad a_{02} = -2a_{20}, \quad a_{30} = -\frac{a_{20}^2}{3}, \\ a_{21}^2 = \dfrac{a_{20}^4}{2}, & a_{12} = 0, \quad a_{03} = -\frac{a_{21}}{3}. \end{cases} \tag{15.22}$$

Other conditions which are not covered by Kukles' were also found. Christopher (1994) and Lloyd and Pearson (1992) derived the following conditions

$$\begin{cases} \kappa_1 = 81a_{20}^3 a_{02} - 2(18a_{11}^2 r - 4a_{11}^4 - 27a_{11}^2 a_{20}^2 - 81a_{20}^4) = 0, \\ \kappa_2 = 9\eta a_{30} + 36a_{11}^2 r + 8a_{11}^4 + 90a_{11}^2 a_{20}^2 + 243a_{20}^4 = 0, \\ \kappa_3 = \eta a_{21} - a_{20}a_{11}(27r - 2a_{11}^2 - 9a_{20}^2) = 0, \\ \kappa_4 = 81a_{20}^2 \eta a_{12} + 2a_{11}^2(144a_{11}^2 r - 567a_{20}^4 - 270a_{11}^2 a_{20}^2 \\ \qquad +243a_{20}^2 r - 32a_{11}^4) = 0, \\ \kappa_5 = 3\eta a_{03} + a_{11}(a_{02}\eta + 27a_{20}r + 14a_{20}a_{11}^2 + 72a_{20}^3) = 0, \end{cases} \tag{15.23}$$

where

$$\eta = 16a_{11}^2 + 81a_{20}^2,$$

$$\kappa_0 = 162a_{11}^2 r^2 - (2a_{11}^2 + 9a_{20}^2)^3 = 0,$$

$$a_{20}a_{11} \neq 0.$$

Wang (1999a) studied the relations among the above-mentioned conditions using the methods of characteristic sets, triangular systems and Gröbner bases. It has been shown that all the center conditions discovered by Jin and Wang (1990), Christopher (1991), Lloyd and Pearson (1990) for Kukles' system are already covered by those of Cherkas (1978).

Theorem 15.13: (Wang 1999a)

The set of center conditions (15.21) *holds if and only if one of the following four sets of conditions holds:* (15.17), (15.22), (15.23), *and*

$$a_{20} = a_{02} = a_{21} = a_{12} = a_{03} = 0, \quad 9a_{30} + a_{11}^2 = 0.$$

Two subclasses of the system (15.14), i.e. $a_{11}a_{03} = 0$, were investigated in Christopher and Lloyd (1990), Jin and Wang (1990), Rousseau et al. (1995) and Sadovskii (1993).

Sadovskii (1993) considered (15.14) with $a_{11} = 0$ and proved the following.

Theorem 15.14: (Sadovskii 1993)

For system (15.14) *with* $a_{11} = 0$, *the origin is a center if and only if one of the following conditions holds:*

(*i*) $a_{21} = a_{03} = 0$;

(*ii*)
$$a_{02} = -2a_{20}, \quad a_{12} = 0, \quad a_{30} = -\frac{1}{3}a_{20}^2,$$
$$a_{21} = \mp\frac{1}{3\sqrt{2}}a_{20}^2, \quad a_{03} = \pm\frac{1}{3\sqrt{2}}a_{20}.$$

It has also been shown by Sadovskii that, if E_0 in (15.14) with $a_{11} = 0$ is a focus, then the maximal order is 6 and six small-amplitude limit cycles can be constructed for its perturbation system (15.15) with $a_{11} = 0$.

The following result is established by Christopher and Lloyd (1990).

Theorem 15.15: (Christopher and Lloyd 1990)

For system (15.14) *with* $a_{03} = 0$ *the origin is a center if and only if one of the following conditions holds:*

(*i*) $a_{11} = a_{21} = 0$;

(*ii*) $a_{20} = a_{02} = a_{21} = 0$;

(*iii*) $a_{30} = a_{21} = a_{12} = 0, \quad a_{20} + a_{02} = 0$;

(*iv*)
$$a_{30} = (a_{20} + a_{02})a_{02}, \quad a_{21} = -(a_{20} + a_{02})a_{11},$$
$$a_{12} = -(a_{20} + a_{02})a_{02}^2(a_{20} + 2a_{02})^{-1}.$$

In this case, the maximal order of E_0 for (15.14) with $a_{03} = 0$ to be a focus is 5, and five small amplitude limit cycles can be constructed for system (15.15) with $a_{03} = 0$.

It is pointed out in the above papers that for Kukles' system (15.14) the necessary and sufficient conditions for E_0 to be a center still remain unknown. On the other aspect, the problem about two or more centers for cubic systems has also been considered in the literature (Conti 1988, 1990; Rousseau et al. 1995). In the next subsection, we shall be concerned with this problem for the Kukles system and give necessary and sufficient conditions for the system to have two centers.

15.4.2 Two-center conditions

Clearly, system (15.14) has at most two critical points E_{x_1} and E_{x_2} other than the origin. All these critical points are located on the x-axis, which is an isoline. By an index theorem, if $E_{x_1} = E_{x_2}$, then E_{x_1} cannot be of center-focus type; if $E_{x_1} \neq E_{x_2}$, then at most one of them, say E_{x_2}, is of center-focus type. In the sequel, we consider two cases $0 < x_1 < x_2$ and $0 > x_1 > x_2$. Note that x_1 and x_2 are distinct solutions of the quadratic equation

$$-1 + a_{20}x_2 + a_{30}x_2^2 = 0, \tag{15.24}$$

and $x_1 x_2 > 0$ if and only if

$$-a_{20}^2/4 < a_{30} < 0.$$

If $a_{20} < 0$, then

$$x_2 = \frac{1}{2a_{30}}[-a_{20} + (a_{20}^2 + 4a_{30})^{\frac{1}{2}}] < 0;$$

if $a_{20} > 0$, then

$$x_2 = \frac{1}{2a_{30}}[-a_{20} - (a_{20}^2 + 4a_{30})^{\frac{1}{2}}] > 0.$$

By using the transformation

$$\bar{x} = x - x_2, \qquad \bar{y} = y,$$

system (15.14) is reduced to

$$\begin{cases} \dot{x} = y, \\ \dot{y} = -ax + \Delta y + \bar{a}_1 x^2 + \bar{a}_2 xy + \bar{a}_3 y^2 + Q_3, \end{cases}$$

where x and y are used instead of $\bar{x}$ and $\bar{y}$, Q_3 is as in (15.16), and

$$a = a_{20}x_2 - 2, \qquad \Delta = a_{11} + a_{21}x_2,$$

$$\bar{a}_1 = a_{20} + 3a_{30}x_2, \qquad \bar{a}_2 = a_{11} + 2a_{21}x_2, \qquad \bar{a}_3 = a_{02} + a_{12}x_2.$$

Since E_{x_2} is supposed to be of center-focus type, $\Delta \equiv 0$; that is

$$a_{11} + a_{21}x_2 = 0 \tag{15.25}$$

(and $x_2 \neq 0$). It is easy to know from (15.24) and (15.25) that,

$$a = a_{20}x_2 - 2 \tag{15.26}$$

$$= \begin{cases} \dfrac{1}{2a_{30}}[a_{20} - (a_{20}^2 + 4a_{30})^{\frac{1}{2}}](a_{20}^2 + 4a_{30})^{\frac{1}{2}} > 0 & \text{if} \quad a_{20} < 0, \\ \dfrac{1}{2a_{30}}[-a_{20} - (a_{20}^2 + 4a_{30})^{\frac{1}{2}}](a_{20}^2 + 4a_{30})^{\frac{1}{2}} > 0 & \text{if} \quad a_{20} > 0. \end{cases} \tag{15.27}$$

From the above discussions, it may be easily concluded that E_{x_2} is of center-focus type if and only if (15.24) and (15.25) hold.

Theorem 15.16: (Chen et al. 2000)

Kukles' system (15.14) *has two centers if and only if* $-a_{20}^2/4 < a_{30} < 0$ *and one of the following conditions holds:*

(*i*) $a_{11} = a_{21} = a_{03} = 0;$

(*ii*) $$a_{03} = 0, \quad a_{30} = (a_{20} + a_{02})a_{02}, \quad a_{21} = -(a_{11} + a_{02})a_{11},$$
$$a_{12} = -(a_{20} + a_{02})a_{02}^2(a_{20} + 2a_{02})^{-1};$$

(*iii*) (15.17).

It is known that, if Kukles' system (15.14) has two fine foci and one of them has order 2, then the other has order at most 1. The remaining question is whether the maximal order of the other focus can be greater than 6 if one of the two foci has order 1. In fact, we have the following result.

Theorem 15.17: (Chen et al. 2000)

If Kukles' system (15.14) *has two fine foci, then the sum of their orders is less than or equal to* 7.

Chapter 16

EXACT SOLITARY WAVE SOLUTIONS OF NON-LINEAR EVOLUTION EQUATIONS

Zhibin Li

This chapter presents a mechanization method for solving non-linear evolution equations exactly and systematically. The method proceeds by reducing evolution equations to polynomial equations and solving the latter using Wu's method. We illustrate our method by solving several well-known equations, including the KdV–Burgers equation, the Kawachara equation, the Burgers–Huxley equation, and a variant of Boussinesq equations. A number of other equations and their exact solitary wave solutions found by using this method are provided in table form, which demonstrates the wide applicability of the method.

16.1 Introduction

Several methods for obtaining solitary wave solutions have been developed since the inverse scattering technique was established. Searching for exact solutions of non-linear partial differential equations (PDE) directly has become more and more attractive due to the availability of computer algebra systems. Indeed, systems like Macsyma, Maple and Mathematica allow one to perform tedious algebraic calculations.

The original idea behind the direct method goes back to Hirota (1980), who systematically solved large classes of evolution equations by bilinear transformation. Hereman et al. (1985) established a straightforward algorithm for constructing solitary wave solutions from the mixing of the real exponential traveling wave solutions of the underlying linear equation. This direct algebraic method has been used to construct solitary wave solutions of coupled systems, see Hereman (1990) for more details. Seeking traveling wave solutions, one can immediately reduce the non-linear PDE into a non-linear ordinary differential equation (ODE) and then use basic ideas from the geometrical theory of ODE. In

MATHEMATICS MECHANIZATION AND APPLICATIONS
ISBN 0-12-734760-7

particular, Painlevé analysis with experiments (see Ince 1956) helps in classifying the possible singularities of the solution of the ODE. Partly based on this type of singularity analysis, Takaoka (1989) solved the KdV equation with an extra fifth-order dispersive term (Karachwa equation). Restricting himself to hyperbolic-type solutions, Takaoka achieved his goal. With the aid of tanh functions, Lan and Wang (1990) obtained traveling wave solutions of a complicated non-linear wave equation directly. His technique is based on the *a priori* assumption that the solutions can be written in terms of such functions.

In this chapter we report a systematized approach for exactly solving non-linear evolution equations leading to mainly hyperbolic-type solitary waves. The idea of the method that can easily be used in a wider context is to reduce non-linear evolution equation(s) to polynomial equations. Solving the polynomial equations by means of the characteristic set method developed by Wu (1986a, 1994d; Chap. 1) yields exact solitary wave solutions of non-linear evolution equations.

In Sect. 16.2, we present the details of our method for the construction of solitary wave solutions of non-linear evolution equations.

In Sect. 16.3, we exemplify the technique by first applying it to the non-integrable KdV–Burgers equation; the exact solutions in the form of superposition of a soliton and a shock are obtained. We then treat the Kawachara equation which is the KdV equation with an additional fifth-order dispersive term. This example shows that we can get a solution more general than that obtained by using Hereman's method. Thirdly, we deal with the Burgers–Huxley equation and find a new kink-type solution of this non-linear diffusion equation. We consider a variant of Boussinesq equations lastly and show how an extension of the technique leads to the solitary solutions of the coupled wave equations.

In Sect. 16.4, we briefly list other equations solved by our mechanization method, providing appropriate references for comparison purposes and tabulating the results obtained so far. Finally, we discuss further possible generalization of the present method and draw some conclusions.

16.2 Solution method

In this section we discuss the details of the general method for exactly solving non-linear evolution equations leading to mainly hyperbolic-type solitary waves.

The starting point is a non-linear partial differential equation in $1 + 1$ dimensions with x and t as the space and time coordinates respectively, which describes the dynamical evolution of the wave form $u(x, t)$. The method, which is straightforwardly applicable, goes in the following five steps.

S1. A traveling wave solution requires one coordinate: $\xi = k(x - ct)$ and thus $u(x,t) = U(\xi)$, where $U(\xi)$ represents the wave solution, which travels with speed c. As a consequence, the derivatives are changed into

$$\frac{\partial}{\partial t} \rightarrow -k\,c\,\frac{d}{d\xi}, \qquad \frac{\partial}{\partial x} \rightarrow k\,\frac{d}{d\xi}. \tag{16.1}$$

This transforms the given non-linear PDE in $u(x,t)$ into an ODE in $U(\xi)$. For convenience, we usually take k and $w = k\,c$ as independent parameters.

S2. The crucial step now is the introduction of $\eta = \tanh(\xi)$ as a new independent variable. The corresponding derivatives are then changed as follows:

$$\begin{aligned}
\frac{d}{d\xi} &\rightarrow (1-\eta^2)\frac{d}{d\eta}, \\
\frac{d^2}{d\xi^2} &\rightarrow (1-\eta^2)\left[-2\eta\frac{d}{d\eta} + (1-\eta^2)\frac{d^2}{d\eta^2}\right], \\
\frac{d^3}{d\xi^3} &\rightarrow -2\eta(1-\eta^2)\left[-2\eta\frac{d}{d\eta} + (1-\eta^2)\frac{d^2}{d\eta^2}\right] \\
&\quad + (1-\eta^2)\left[-2\frac{d}{d\eta} - 2\eta\frac{d^2}{d\eta^2} + (1-\eta^2)\frac{d^3}{d\eta^3}\right].
\end{aligned} \tag{16.2}$$

Higher-order derivatives can be found accordingly. After performing this step, we get, setting $S(\eta) = U(\xi)$, another non-linear ODE for S.

S3. We now have to solve the non-linear equation in S. There is no general procedure at this stage. Often, the following series expansion is the most favorable:

$$S(\eta) = \sum_{n=0}^{m} a_n\,\eta^n. \tag{16.3}$$

The parameter m will be found by balancing the linear term(s) of highest order with the non-linear term(s). The linear term of highest order is contained in the highest derivative when applied to $S(\eta)$. This is readily seen using the set of relations (16.2). The η^2 component of the first relation in (16.2) leads to the order $(2 + m - 1) = m + 1$, the η^3 component of the first and the η^4 component of the second derivative of the second relation in (16.2) both lead to the order $m+2$, etc. On the other hand, the non-linear terms yield multiples of m. The highest-order contribution of the latter must be equal to the former result. In this way the value of m will be determined.

S4. Substituting the series expansion (16.3) into the relevant equation yields recursion relations because the coefficients of η^n $(n = 0, 1, \dots, m)$ have to vanish. The problem is now completely reduced to an algebraic one. The remaining unknowns, being the coefficients a_i, and the (non-zero) values for k and c will follow from solving a non-linear system, i.e., a set of polynomial equations.

S5. A characteristic set of the set of polynomials can be straightforwardly determined by a symbolic manipulation package based on Wu's method, and the solutions for the unknowns are then calculated. Returning to the original dependent variable u and the independent variables x and t, the exact solitary wave solution(s) of the non-linear evolution equation is/are obtained.

Remark. Normally m will be a positive integer so that a closed analytical solution can be obtained. However, we have found some examples where m takes a non-integer value. In this case, we can transform the original non-linear evolution equation in u into an equation in v by introducing a substitution $u = v^{1/m}$ and proceed then as before.

The process described above may not always work, of course, but it certainly works well for a very large class of interesting non-linear evolution equations. In the next two sections, we shall study some examples of various evolution equations, to illustrate in detail the method.

16.3 Examples

16.3.1 The Korteweg–de Vries–Burgers equation

As a first example to demonstrate the method outlined in Sect. 16.2, we have chosen a non-integrable non-linear wave equation, the Korteweg–de Vries–Burgers (KdVB) equation

$$u_t + u_x + u\,u_x - p\,u_{xx} - q\,u_{xxx} = 0, \tag{16.4}$$

where p and q are dissipative and dispersive coefficients, respectively. The equation (16.4), quoted as the simplest non-linear dissipative and dispersive system, models fluid turbulence. Numerical evaluation and qualitative investigation of (16.4) have been carried out by Canosa and Gazdag (1977).

To look for traveling waves of (16.4), the parameter m is first determined by balancing the order of the linear term of highest order with the non-linear term: $2m+1 = m+3$ or $m=2$. Next, the polynomial equations are derived using a simple program in a symbolic algebraic system such as Mathematica:

$$\begin{cases} P_1 = -wa_1 + ka_1 + 2qk^3a_1 + ka_0a - 1 - 2pk^2a_2 = 0, \\ P_2 = 2pk^2a_1 + ka_1^2 - 2wa_2 + 2ka_2 + 16qk^3a_2 + 2ka_0a_2 = 0, \\ P_3 = -wa_1 + ka_1 + 8qk^3a_1 + ka_0a_1 - 8pk^2a_2 - 3ka_1a_2 = 0, \\ P_4 = 2pk^2a_1 + ka_1^2 - 2wa_2 + 2ka_2 + 40qk^3a_2 + 2ka_0a_2 - 2ka_2^2 = 0, \\ P_5 = 2qk^2a_1 - 2pka_2 - a_1a_2 = 0, \\ P_6 = 12qk^2 - a_2 = 0. \end{cases} \tag{16.5}$$

A characteristic set $\{C_1, \ldots, C_4\}$ of the polynomial set $\{P_1, \ldots, P_6\}$ is then easily computed with the package WSolve:

$$\begin{cases} C_1 = p^2 - 100q^2k^2, \\ C_2 = -w + k + 12qk^3 + ka_0, \\ C_3 = 12pk + 5a_1, \\ C_4 = 12qk^2 - a_2. \end{cases} \tag{16.6}$$

In view of the fact that there are four equations $C_i = 0$ with five unknowns (a_0, a_1, a_2, k, w), we choose w as a free parameter and obtain

$$a_0 = -1 \pm \frac{10qw}{p} - \frac{3p^2}{25q}, \qquad a_1 = \mp\frac{6p^2}{25q}, \qquad a_2 = \frac{3p^2}{25q}, \qquad k = \pm\frac{p}{10q}.$$

Hence the exact solitary wave solutions of the KdVB equation (16.4) are

$$\begin{aligned} u(x,t) = {} & -1 \pm \frac{10qw}{p} - \frac{3p^2}{25q} \mp \frac{6p^2}{25q}\tanh\left(\pm\frac{p}{10q}x - wt\right) \\ & + \frac{3p^2}{25q}\tanh^2\left(\pm\frac{p}{10q}x - wt\right) \end{aligned} \tag{16.7}$$

with w an arbitrary constant. Taking $w = p(5q - 6p^2)/(50pq)$ yields from (16.7)

$$u^*(x,t) = -\frac{3p^2}{25q}(\operatorname{sech}^2 z \mp 2\tanh z \mp 2), \tag{16.8}$$

where $z = \pm p[x + (\frac{6p^2}{25q} - 1)t]/(10q)$. We can see easily that u^* consists of the kink-type solutions of a Burgers equation and the pulse-type solutions of a KdV equation.

16.3.2 The Kawachara equation

As an interesting example of a non-linear equation let us consider the Kawachara equation with experiments (see Kawachara 1972):

$$u_t + uu_x + pu_{xxx} + qu_{xxxxx} = 0. \tag{16.9}$$

This equation has also been derived and studied by Hunter and Scheurle (1988). They have given some analysis for the existence of solitary wave solutions and numerical calculations. Using a different method, Yamamoto and Takizawa (1981) and Hereman et al. (1986) have shown that (16.9) allows a sech^4-type solitary wave solution:

$$u(x,t) = -\frac{105}{169}\frac{p^2}{q}\,\mathrm{sech}^4\left[\sqrt{-\frac{p}{52q}}\left(x + \frac{36p^2}{169q}t\right)\right]. \tag{16.10}$$

Solving (16.9) by our method, we can obtain a more general solution. In fact, it is easy to judge that $m = 4$. Applying the same program as before, we get the following polynomial equations:

$$\left\{\begin{aligned}
P_1 =& -wa_1 - 2pk^3a_1 + 16qk^5a_1 + ka_0a_1 + 6pk^3a_3\\
&- 120qk^5a_3 = 0,\\
P_2 =& \, ka_1^2 - 2wa_2 - 16pk^3a_2 + 272qk^5a_2 + 2ka_0a_2 + 24pk^3a_4\\
&- 960qk^5a_4 = 0,\\
P_3 =& -wa_1 - 8pk^3a_1 + 136qk^5a_1 + ka_0a_1 - 3ka_1a_2 + 3wa_3\\
&+ 60pk^3a_3 - 1848qk^5a_3 - 3ka_0a_3 = 0,\\
P_4 =& \, ka_1^2 - 2wa_2 - 40pk^3a_2 + 1232qk^5a_2 + 2ka_0a_2 - 2ka_2^2\\
&- 4ka_1a_3 + 4wa_4 + 152pk^3a_4 - 7744qk^5a_4 - 4ka_0a_4 = 0,\\
P_5 =& -6pk^3a_1 + 240qk^5a_1 - 3ka_1a_2 + 3wa_3 + 114pk^3a_3\\
&- 5808qk^5a_3 - 3ka_0a_3 + 5ka_2a_3 + 5ka_1a_4 = 0,\\
P_6 =& -24pk^3a_2 + 1680qk^5a_2 - 2ka_2^2 - 4ka_0a_3 + 3ka_3^2 + 4wa_4\\
&+ 248pk^3a_4 - 19264qk^5a_4 - 4ka_0a_4 + 6ka_2a_4 = 0,\\
P_7 =& -k(120qk^4a_1 + 60pk^2a_3 - 6600qk^4a_3 + 5a_2a_3 + 5a_1a_4\\
&- 7a_3a_4) = 0,\\
P_8 =& \, k(-720qk^4a_2 - 3a_3^2 - 120pk^2a_4 + 19200qk^4a_4 - 6a_2a_4\\
&+ 4a_4^2) = 0,\\
P_9 =& -7ka_3(360qk^4 + a_4) = 0,\\
P_{10} =& \, 1680k^4q + a_4 = 0.
\end{aligned}\right. \tag{16.11}$$

Computing a characteristic set *CS* of the polynomial set $\{P_1, \ldots, P_{10}\}$ with the package WSolve, we find that *CS* consists of the following six polynomials

$$\left\{\begin{array}{l} C_1 = 31p^3 - 56784pq^2k^4 + 1406080q^3k^6, \\ C_2 = pw - 104qwk^2 + 8p^2k^3 - 1904pqk^5 + 51584q^2k^7 - pka_0 \\ \qquad + 104qk^3a_0, \\ C_3 = a_1, \\ C_4 = -280pk^2 + 29120qk^4 - 13a_2, \\ C_5 = a_3, \\ C_6 = 1680k^4q + a_4. \end{array}\right. \tag{16.12}$$

Now there are seven unknowns and six equations $C_i = 0$, so we may choose w as a free parameter. The equations are then solved for the other variables. Finally, the solitary wave solution of the Kawachara equation (16.9) reads

$$\begin{aligned} u(x,t) = &-\frac{69p^2}{169q} \mp 2\sqrt{13}\sqrt{-\frac{p}{q}w + \frac{210p^2}{169q}} \tanh^2\left(\pm\sqrt{-\frac{p}{52q}}x - wt\right) \\ &- \frac{105p^2}{169q}\tanh^4\left(\pm\sqrt{-\frac{p}{52q}}x - wt\right) \end{aligned} \tag{16.13}$$

with w an arbitrary constant.

For $w = -\sqrt{-\frac{p}{52q}\frac{36p^2}{169q}}$ we retrieve the solution (16.10) from (16.13).

16.3.3 The Burgers–Huxley equation

As an example of non-linear diffusion equations we focus on the following Burgers–Huxley equation:

$$u_t + puu_x - u_{xx} = ru(1-u)(u-s). \tag{16.14}$$

Equation (16.14) shows a prototype model for describing the interaction between reaction mechanisms, convection effects, and diffusion transport. This equation was investigated by Satsuma (1987). By use of the Hirota method in soliton theory, two solitary wave solutions of (16.14) were obtained.

Applying our method to (16.14), we expect to find more exact traveling wave solutions of (16.14). The quantity m is now determined by the relation $m + 2 = 3m$, so $m = 1$. The relevant polynomial equations are given as follows:

$$\left\{\begin{array}{l} P_1 = rsa_0 - ra_0^2 - rsa_0^2 + ra_0^3 - wa_1 + kpa_0a_1 = 0, \\ P_2 = 2k^2 + rs - 2ra_0 - 2rsa_0 + 3ra_0^2 + kpa_1 = 0, \\ P_3 = w - kpa_0 - ra_1 - rsa_1 + 3ra_0a_1 = 0, \\ P_4 = -2k^2 - 2kpa_1 + ra_1^2 = 0. \end{array}\right. \tag{16.15}$$

Computing a characteristic set of $PS = \{P_1, \ldots, P_4\}$ with the variable order $a_0 \prec k \prec a_1 \prec w$, we have $CS = \{C_1, \ldots, C_4\}$ with

$$\begin{cases} C_1 = -p^2r(-1+2a_0)(-1-s+2a_0)(-s+2a_0), \\ C_2 = 4k^4r + k^2p^2rs + 4k^2r^2s + r^3s^2 - 2k^2pra_0 - 8k^2r^2a_0 \\ \quad - 2k^2p^2rsa_0 - 8k^2r^2sa_0 - 4r^3sa_0 - 4r^3s^2a_0 + 3k^2p^2ra_0^2 \\ \quad + 12k^2r^2a_0^2 + 4r^3a_0^2 + 14r^3sa_0^2 + 4r^3s^2a_0^2 - 12r^3a_0^3 \\ \quad - 12r^3sa_0^3 + 9r^3a_0^4, \\ C_3 = 2k^2 + rs - 2ra_0 - 2rsa_0 + 3ra_0^2 + kpa_1, \\ C_4 = w - kpa_0 - ra_1 - rsa_1 + 3ra_0a_1. \end{cases} \tag{16.16}$$

Solving $CS = 0$ yields three sets of solutions of $PS = 0$:

$$\begin{cases} a_0 = \dfrac{1}{2}, \quad a_1 = \dfrac{1}{2}, \quad k = -\dfrac{p \pm \lambda}{8}, \quad w = \dfrac{-p^2 - 4r \pm p\lambda + 8rs}{16}; \\ a_0 = \dfrac{s}{2}, \quad a_1 = \dfrac{s}{2}, \quad k = -\dfrac{(p \pm \lambda)s}{8}, \quad w = \dfrac{s(8r - p^2s - 4rs \pm ps\lambda)}{16}; \\ a_0 = \dfrac{1+s}{2}, \quad a_1 = \dfrac{1-s}{2}, \quad k = \dfrac{(p \pm \lambda)(s-1)}{8}, \quad w = \dfrac{(p^2 + 4q \pm p\lambda)(1-s^2)}{16}, \end{cases}$$

where $\lambda = \sqrt{p^2 + 8r}$. So the three solitary wave solutions of the Burgers–Huxley equation (16.14) are easily obtained:

$$\begin{cases} u_1(x,t) = \dfrac{s}{2} + \dfrac{s}{2}\tanh\left[\dfrac{-(p \pm \lambda)s}{8}x - \dfrac{s(8r - p^2s - 4rs \pm ps\lambda)}{16}t\right], \\ u_2(x,t) = \dfrac{1}{2} + \dfrac{1}{2}\tanh\left[\dfrac{-(p \pm \lambda)}{8}x - \dfrac{-p^2 - 4r \pm p\lambda + 8rs}{16}t\right], \\ u_3(x,t) = \dfrac{1+s}{2} + \dfrac{1-s}{2}\tanh\left[\dfrac{(p \pm \lambda)(s-1)}{8}x \right. \\ \qquad \left. - \dfrac{(p^2 + 4r \pm p\lambda)(1-s^2)}{16}t\right]. \end{cases} \tag{16.17}$$

The last is a new solution found by our mechanization method.

For $p = 0$ and $r = 1$, we can see from (16.17) that

$$\begin{cases} u_1(x,t) = \dfrac{s}{2} + \dfrac{s}{2}\tanh\left[\pm\dfrac{s}{2\sqrt{2}}x - \dfrac{s(2-s)}{4}t\right], \\ u_2(x,t) = \dfrac{1}{2} + \dfrac{1}{2}\tanh\left[\pm\dfrac{1}{2\sqrt{2}}x - \dfrac{2s-1}{4}t\right], \\ u_3(x,t) = \dfrac{1+s}{2} + \dfrac{1-s}{2}\tanh\left[\pm\dfrac{s-1}{2\sqrt{2}}x - \dfrac{1-s^2}{4}t\right] \end{cases} \tag{16.18}$$

are three solutions of the FitzHugh–Nagumo equation

$$u_t - u_{xx} = u(1-u)(u-s); \tag{16.19}$$

see Hereman (1990) for more details.

A simple solitary wave solution of the real Newell–Whitehead equation

$$u_t - u_{xx} - u(1-u^2) = 0, \tag{16.20}$$

which was solved by Carillo and Tabor (1989), readily follows from (16.18) with $s = -1$:

$$u(x,t) = \pm\frac{1}{2}\left[1 - \tanh\left(\pm\frac{1}{2\sqrt{2}}x - \frac{3}{4}t\right)\right]. \tag{16.21}$$

16.3.4 A variant of Boussinesq equations

It is generally known that coupled non-linear equations are difficult to handle. Nevertheless, the following set of equations can be solved by our solution technique without any difficulty.

Consider

$$\begin{cases} u_t + v_x + uu_x + s\,u_{xx} + q\,u_{xxt} = 0, \\ v_t + (uv)_x + 2s\,v_{xx} + p\,u_{xxx} = 0. \end{cases} \tag{16.22}$$

This set of equations can be seen as the extended Boussinesq's equations which arise in the study of shallow water theory described by Whitham (1973). As a model for water waves, u is a velocity and v is the total depth, so physical solutions ought to have $v > 0$. Here formal properties are considered and the physical restriction $v > 0$ will be ignored.

Now, we show how our method applies to coupled system (16.22). By following the same steps, we can also make the ansätz

$$u = \sum_{i=0}^{m} a_i \tanh^i \xi, \qquad v = \sum_{j=0}^{n} b_j \tanh^j \xi \tag{16.23}$$

with $\xi = kx - wt$, where a_i $(i = 0, \ldots, m)$, b_j $(j = 0, \ldots, n)$, k, w and the integers m and n are parameters to be determined.

The requirement that the highest power of the function $\tanh z$ for the non-linear terms and that for the linear terms of highest order in each equation of (16.22) must be equal gives the following relations:

$$2m + 1 = m + 3, \qquad m + n + 1 = m + 3.$$

Thus we have $m = n = 2$. The polynomial equations are then easily obtained:

$$
\begin{cases}
P_1 = 6wk^2qa_1 + 6k^2sa_2 - 3ka_1a_2 = 0,\\
P_2 = 2ka_2(12wkq - a_2) = 0,\\
P_3 = wa_1 - 8wk^2qa_1 - ka_0a_1 - 8k^2sa_1 + 3ka_1a_2 - kb_1 = 0,\\
P_4 = -wa_1 + 2wk^2qa_1 + ka_0a_1 + 2k^2sa_2 + kb_1 = 0,\\
P_5 = 2k^2sa_1 - ka_1^2 + 2wa_2 - 40wk^2qa_2 - 2ka_0a_2 + 2ka_0a_2\\
\qquad + 2ka_2^2 - 2kb_2 = 0,\\
P_6 = -2k^2sa_1 + ka_1^2 - 2wa_2 + 16wk^2qa_2 + 2ka_0a_2\\
\qquad +2kb_2 = 0,\\
P_7 = -2k^3pa_1 + ka_1b_0 - wb_1 + ka_0b_1 + 4k^2sb_2 = 0,\\
P_8 = -16k^3pa_2 + 2ka_2b_0 - 4k^2sb_1 + 2ka_1b_1 - 2wb_2\\
\qquad + 2ka_0b_2 = 0,\\
P_9 = -2k^2pa_1 - a_2b_1 + 4ksb_2 - a_1a_2 = 0,\\
P_{10} = 8k^3pa_1 - ka_1b_0 + wb_1 - ka_0b_1 + 3ka_2b_1 - 16k^2sb_2\\
\qquad + 3ka_1b_2 = 0,\\
P_{11} = -24k^3pa_2 - 4ka_2b_2 = 0,\\
P_{12} = 40k^3pa_2 - 2ka_2b_0 + 4k^2sb_1 - 2ka_1b_1 + 2wb_2\\
\qquad - 2ka_0b_2 + 4ka_2b_2 = 0.
\end{cases}
\tag{16.24}
$$

We find that a characteristic set CS of $\{P_1, \ldots, P_{12}\}$ under the order $w \prec b_0 \prec a_0 \prec b_1 \prec a_1 \prec a_2 \prec b_2$ consists of the following polynomials:

$$
\begin{cases}
C_1 = s(s^2 - 100w^2q^2),\\
C_2 = 75k^2p^2 + 1200w^2k^2pq^2 + 16k^2ps^2 - 300w^2q^2b_0,\\
C_3 = -25k^2p - 50w^2q + 400w^2k^2q^2 + 2k^2s^2 + 50wkqa_0,\\
C_4 = 6k^2ps + 5wqb_1,\\
C_5 = 12ks - 5a_1,\\
C_6 = 12wkq - a_2,\\
C_7 = 6k^2p + b_2.
\end{cases}
\tag{16.25}
$$

Determining the zero set of CS, we get the following solitary wave solutions of the variant of Boussinesq equations (16.22):

$$
\begin{cases}
u(x,t) = \pm\left(\dfrac{5kp}{s} + \dfrac{s}{10kq} - \dfrac{12ks}{5}\right) \pm \dfrac{6ks}{5}\left[1 \pm \tanh\left(kx \mp \dfrac{s}{10q}t\right)\right]^2,\\
v(x,t) = \dfrac{25k^2p}{s^2} + 12k^2p - 6k^2p\left[1 \pm \tanh\left(kx \mp \dfrac{s}{10q}t\right)\right]^2,
\end{cases}
\tag{16.26}
$$

where k is an arbitrary constant.

If $p = 0$ is chosen in (16.26), one sees that the shocklike wave solutions of the Benjamin–Bona–Mahony–Burgers equation

$$u_t + uu_x + su_{xx} + qu_{xxt} = 0 \tag{16.27}$$

found by Li and Wang (1993) is recovered from (16.26):

$$u(x,t) = \pm\frac{s}{10kq} \mp \frac{12ks}{5} \pm \frac{6ks}{5}\left[1 \pm \tanh\left(kx \mp \frac{s}{10q}t\right)\right]^2 . \tag{16.28}$$

With $s = 0$, the set of equations (16.22) becomes

$$\begin{cases} u_t + v_x + uu_x + qu_{xxt} = 0, \\ v_t + (uv)_x + pu_{xxx} = 0. \end{cases} \tag{16.29}$$

Sachs (1988) has constructed rational solutions and the associated higher-order flows of these equations. The sech^2-type solution is readily found from (16.25):

$$\begin{cases} u(x,t) = \dfrac{w}{k} + \dfrac{kp}{2wq} - 8wkq + 12wkq\tanh^2(kx - wt), \\ v(x,t) = \dfrac{k^2p^2}{4w^2q^2} + 4k^2p - 6k^2p\tanh^2(kx - wt), \end{cases} \tag{16.30}$$

where k and w are arbitrary constants. Obviously, the choice $k = \frac{1}{2\sqrt{-2q}}$, $w = \pm\frac{\sqrt{p}}{2\sqrt{2q}}$ in (16.30) produces

$$\begin{cases} u(x,t) = \pm\dfrac{3\sqrt{p}}{2\sqrt{-q}}\,\text{sech}^2\left[\dfrac{1}{2\sqrt{-2q}}\left(x \mp \dfrac{\sqrt{p}}{\sqrt{-q}}t\right)\right], \\ v(x,t) = -\dfrac{3p}{4q}\,\text{sech}^2\left[\dfrac{1}{2\sqrt{-2q}}\left(x \mp \dfrac{\sqrt{p}}{\sqrt{-q}}t\right)\right]. \end{cases} \tag{16.31}$$

Other coupled systems of evolution equations may also be found solvable by our method. The results for a coupled system of KdV equations, non-linear B-Z reaction system, and others are listed in Table 16.4. Detailed calculations will be presented elsewhere.

16.4 Exact solutions of non-linear evolution equations

We have used the method sketched in Sect. 16.2 to find single solitary wave solutions for a variety of non-linear evolution equations. Tables 16.1–16.4 give a survey of these equations, together with the corresponding general solutions (gSol) including free parameter(s) and special solutions (sSol). Some special solutions can be found in the indicated references.

Table 16.1 Solutions of non-linear wave equations

Burgers equation: $u_t + uu_x + p\,u_{xx} = 0$
gSol: $u(x,t) = \frac{w}{k} + 2pk\tanh(kx - wt)$ $(k, w\text{: const.})$
sSol: $u(x,t) = v - v\tanh[-\frac{v}{2p}(x - vt)]$ $(v\text{: const.})$
Ref: Wahlquist and Estabrook (1975), Ablowitz et al. (1980)
Korteweg–de Vries (KdV) equation: $u_t + uu_x + pu_{xxx} = 0$
gSol: $u(x,t) = \frac{8k^3p+w}{k} - 12k^2p\tanh^2(kx - wt)$ $(k, w\text{: const.})$
sSol: $u(x,t) = 12k^2p\,\mathrm{sech}^2[k(x - 4k^2pt)]$ $(k\text{: const.})$
Ref: Whitham (1973)
Benjamin–Bona–Mahony (BBM): $u_t + u_x + uu_x + pu_{xxt} = 0$
gSol: $u(x,t) = \frac{-k+w-8k^2pw}{k} + 12kpw\tanh^2(kx - wt)$ $(k, w\text{: const.})$
sSol: $u(x,t) = -\frac{12k^2p}{1+4k^2p}\,\mathrm{sech}^2[k(x - \frac{t}{1+4k^2p})]$ $(k\text{: const.})$
Ref: Benjamin et al. (1972), Calogero (1987)
Joseph–Egri equation: $u_t + u_x + uu_x + pu_{xtt} = 0$
gSol: $u(x,t) = \frac{-k+w+8kpw^2}{k} - 12pw^2\tanh^2(kx - wt)$ $(k, w\text{: const.})$
sSol: $u(x,t) = 12w^2p\,\mathrm{sech}^2[w(\frac{x}{1+4w^2p} - t)]$ $(w\text{: const.})$
Modified KdV (mKdV) equation: $u_t + qu^2u_x + ru_{xxx} = 0$
gSol: $u(x,t) = \sqrt{\frac{-6r}{q}}\,k\tanh[k(x + 2k^2rt)]$ $(k\text{: const.})$
Ref: Wadati (1972) **Modified BBM equation**: $u_t + u_x + u^2u_x + pu_{xxt} = 0$
gSol: $u(x,t) = \pm k\sqrt{\frac{6p}{1-2k^2p}}\tanh(kx + \frac{k}{2k^2p-1}t)$ $(k\text{: const.})$
Ref: Makhakov (1978)
KdV–Burgers equation: $u_t + u_x + uu_x - pu_{xx} - qu_{xxx} = 0$
gSol: $u(x,t) = -1 \pm \frac{10qw}{p} - \frac{3p^2}{25q} \mp \frac{6p^2}{25q}\tanh(\pm\frac{p}{10q}x - wt) + \frac{3p^2}{25q}\tanh^2(\pm\frac{p}{10q}x - wt)$ $(w\text{: const.})$
sSol: $u(x,t) = -\frac{3p^2}{25q}(\mathrm{sech}^2 z + 2\tanh z + 2)$, $z = p[x + (\frac{6p^2}{25q} - 1)t]/(10q)$
Ref: Li and Wang (1993)

Table 16.1 Solutions of non-linear wave equations (continued)

BBM–Burgers equation: $u_t + u_x + uu_x - pu_{xx} - qu_{xxt} = 0$
gSol: $u(x,t) = (-1 \pm \frac{p}{10qk} \pm \frac{6pk}{5}) - \frac{12}{5}pk\tanh(kx \mp \frac{p}{10q}t) - \frac{6kp}{5}\tanh^2(kx \mp \frac{p}{10q}t)$ (k: const.)
sSol: $u(x,t) = -4[1 \pm \sqrt{1+\frac{24p^2}{25q}}](\operatorname{sech}^2 z + 2\tanh z + 2)$, $z = \frac{5\sqrt{q} \pm \sqrt{25q+24p^2}}{24p\sqrt{q}}x + \frac{p}{10q}t$
Ref: Wang (1995)
Combined KdV–mKdV equation: $u_t + (p+qu)uu_x + ru_{xxx} = 0$
gSol: $u(x,t) = -\frac{p}{2q} + \sqrt{\frac{-6r}{q}}\,k\tanh[k(x + \frac{p^2+8k^2qr}{4q}t)]$ (k: const.)
sSol: $u(x,t) = -\frac{p}{2q}\{1 - \tanh[\frac{p}{12}\sqrt{\frac{-6}{qr}}(x + \frac{p^2}{6q}t)]\}$
Ref: Coffey (1990)
Generalized KdV equation: $u_t + pu^s u_x + ru_{xxx} = 0$
gSol: $u(x,t) = \{\frac{2(1+s)(2+s)k^2r}{s^2p}\operatorname{sech}^2[k(x - \frac{4k^2r}{s^2}t)]\}^{1/s}$ (k: const.)
Ref: Schamel (1973)
Generalized combined KdV–mKdV equation: $u_t + (p+qu^s)u^s u_x + ru_{xxx} = 0$
Sol: $u(x,t) = \left\{-\frac{(1+2s)p}{2(2+s)q}\{1 \pm \tanh[\frac{s}{2}\sqrt{\frac{v}{r}}(x \pm vt)]\}\right\}^{1/s}$, $v = \frac{-(1+2s)p^2}{(1+s)(2+s)^2q}$
Ref: Dey (1986), Coffey (1990)
Unnamed equation: $u_t - u_{xxx} + p(u_{xx}u^2 + 3u_x^2u) - \frac{p^2}{3}u_xu^4 = 0$
gSol: $u(x,t) = \sqrt{-\frac{3k}{2p}\{\pm 1 + \tanh[k(x+k^2t)]\}}$ (k: const.)
Ref: Calogero (1987)

Table 16.1 Solutions of non-linear wave equations (continued)

Sharma–Tasso–Olever equation:	$u_t + 3pu_x^2 + 3pu^2u_x + 3puu_{xx} + pu_{xxx} = 0$
gSol:	$u(x,t) = \pm\sqrt{\frac{w}{3pk} - \frac{k^2}{3}} + k\tanh(kx - wt)$ $\quad$ (k, w: const.)
sSol:	$u(x,t) = k\{\pm 1 + \tanh[k(x - 4k^2pt)]\}$ $\quad$ (k: const.)
Ref:	Olever (1977)
Boussinesq equation:	$u_{tt} + pu_{xx} + q(u^2)_{xx} + ru_{xxxx} = 0$
gSol:	$u(x,t) = -\frac{w^2 + pk^2 - 8rk^4}{2qk^2} - \frac{6rk^2}{q}\tanh^2(kx - wt)$ $\quad$ (k, w: const.)
sSol:	$u(x,t) = \frac{6rk^2}{q}\operatorname{sech}^2[k(x - \sqrt{-p - 4rk^2}t)]$ $\quad$ (k: const.)
Ref:	Raadu and Chanteur (1986)
Symmetrical-regular-long-wave equation:	$u_{tt} + pu_{xx} + q(u^2)_{xt} + ru_{xxtt} = 0$
gSol:	$u(x,t) = \frac{w^2 + pk^2 - 8rw^2k^2}{2qwk} + \frac{6rwk}{q}\tanh^2(kx - wt)$ $\quad$ (k, w: const.)
sSol:	$u(x,t) = -\frac{3(c^2+p)}{2qc}\operatorname{sech}^2[\frac{1}{2}\sqrt{\frac{c^2+p}{-c^2r}}(x + ct)]$ $\quad$ (c: const.)
Ref:	Seyler and Fenstermacher (1984)
Second-order Benjamin Ono equation:	$u_{tt} + q(u^2)_{xx} + ru_{xxxx} = 0$
gSol:	$u(x,t) = \frac{-w^2 + 8rk^4}{2qk^2} - \frac{6rk^2}{q}\tanh^2(kx - wt)$ $\quad$ (k, w: const.)
sSol:	$u_1(x,t) = \frac{6rk^2}{q}\operatorname{sech}^2[k(x - \sqrt{-4rk^2}t)],$ $u_2(x,t) = -\frac{6rk^2}{q}\tanh^2[k(x \pm 2k\sqrt{2r}t)]$ $\quad$ (k: const.)
Kuramoto–Sivashinsky equation:	$u_t + uu_x + pu_{xx} - qu_{xxxx} = 0$
gSol:	$u(x,t) = \frac{w}{k} + (\frac{60kp}{19} + 120k^3q)\tanh(kx - wt)$ $-120k^3q\tanh^3(kx - wt)$ $\quad$ (w: const.; $k = \pm\frac{1}{2}\sqrt{\frac{p}{19q}}$ or $k = \pm\frac{1}{2}\sqrt{\frac{-11p}{19q}}$)
sSol:	$u(x,t) = \frac{60kp}{19}\{1 + \tanh[k(x + \frac{60kp}{19}t)]\} + 120k^3q\operatorname{sech}^2[k(x + \frac{60kp}{19}t)]$ $\tanh[k(x + \frac{60kp}{19}t)]$ $\quad$ ($k = \pm\frac{1}{2}\sqrt{\frac{p}{19q}}$ or $k = \pm\frac{1}{2}\sqrt{\frac{-11p}{19q}}$)
Ref:	Kuramoto (1980), Sivashinsky (1977)

Table 16.1 Solutions of non-linear wave equations (continued)

Fifth-order dispersive equation: $u_t + pu_x u_{xx} + qu_{xxxxx} = 0$
gSol: $u(x,t) = C - \frac{60k^2q}{p} \tanh^2[k(x - 16k^4qt)]$ (C, k: const.)
sSol: $u(x,t) = \frac{60k^2q}{p} \operatorname{sech}^2[k(x - 16k^4qt)]$ (k: const.)
Kawachara equation: $u_t + uu_x + pu_{xxx} + qu_{xxxxx} = 0$
gSol: $u(x,t) = -\frac{69p^2}{169q} \mp 2\sqrt{13}\sqrt{-\frac{p}{q}}w + \frac{210p^2}{169q} \tanh^2(\pm\sqrt{-\frac{p}{52q}}x - wt)$ $-\frac{105p^2}{169q} \tanh^4(\pm\sqrt{-\frac{p}{52q}}x - wt)$ (k: const.)
sSol: $u(x,t) = -\frac{105}{169}\frac{p^2}{q} \operatorname{sech}^4[\sqrt{-\frac{p}{52q}}(x + \frac{36p^2}{169q}t)]$
Ref: Kawachara (1972), Hunter and Scheurle (1988)
Modified Kawachara equation: $u_t + u_x + u^2u_x + pu_{xxx} + qu_{xxxxx} = 0$
gSol: $u(x,t) = \frac{p-40k^2q}{\sqrt{-10q}} - 6k^2\sqrt{-10q} \tanh^2(kx - \frac{-kp^2+10kq-240k^5q^2}{10q}t)$ (k: const.)
sSol: $u(x,t) = \frac{3p}{\sqrt{-10q}} \operatorname{sech}^2\{\sqrt{\frac{p}{-20q}}[x - (1 - \frac{8p^2}{50q})t]\}$
Ref: Hunter and Scheurle (1988)
Klein–Gordon equation: $u_{tt} - u_{xx} + pu + qu^3 = 0$
gSol: $u(x,t) = \pm\sqrt{-\frac{p}{q}} \tanh(kx \pm \sqrt{\frac{2k^2+p}{2}}t)$ (k: const.)
Ref: Dodd (1982), Narita (1984)
Hirota equation: $u_t - u_{xxt} + pu_x(1 - u_t) = 0$
gSol: $u(x,t) = C + \frac{6k}{p} \tanh[k(x - \frac{p}{1-4k^2})]$ (C, k: const.)
Ref: Hirota and Ramani (1980)
Bretherton equation: $u_{tt} + u_{xx} + u_{xxxx} = p\,u^3$
Sol: $u(x,t) = \pm\frac{\sqrt{30}}{4\sqrt{p}} \operatorname{sech}^2(\frac{1}{2\sqrt{2}}x \pm \frac{\sqrt{7}i}{4}t)$
Ref: Bretherton (1964)

Table 16.2 Solutions of non-linear diffusion equations

Fisher equation:	$u_t - u_{xx} = u(1-u)$
Sol:	$u(x,t) = \frac{1}{4}\{1 - \tanh[\frac{1}{2\sqrt{6}}(x - \frac{5}{\sqrt{6}}t)]\}^2$
Ref:	Wang (1988), Hereman (1990)
Generalized Fisher equation:	$u_t - u_{xx} - p\,\frac{1}{u}(u_x)^2 = u(1-u^c)$
Sol:	$u(x,t) = \left\{\pm\frac{1}{2}\{1 - \tanh[\pm\frac{c}{2\sqrt{4+2c+4p}}(x - \frac{4+c+4p}{\sqrt{4+2c+4p}}t)]\}\right\}^{2/c}$,
Ref:	Wang (1988), Hereman (1990)
General Burgers–Fisher equation:	$u_t + pu^c u_x - u_{xx} = qu(1-u^c)$
Sol:	$u(x,t) = \left\{\frac{1}{2} + \frac{1}{2}\tanh\{-\frac{cp}{2(1+c)}[x - (\frac{p}{1+c} + \frac{q(1+c)}{p})t]\}\right\}^{1/c}$
Ref:	Satsuma (1987)
Burgers–Huxley equation:	$u_t + puu_x - u_{xx} = ru(1-u)(u-s)$
Sol:	$u_1(x,t) = \frac{s}{2} + \frac{s}{2}\tanh[\frac{-(p\pm n)s}{8}x - \frac{s(8r-p^2s-4rs\pm psn)}{16}t]$, $u_2(x,t) = \frac{1}{2} + \frac{1}{2}\tanh[\frac{-(p\pm n)}{8}x - \frac{-p^2-4r\pm pn+8rs}{16}t]$, $u_3(x,t) = \frac{1+s}{2} + \frac{1-s}{2}\tanh[\frac{(p\pm n)(s-1)}{8}x - \frac{(p^2+4r\pm pn)(1-s^2)}{16}t]$ $\quad (n = \sqrt{p^2+8r})$
Ref:	Scott (1977), Wang (1988)
Generalized Burgers–Huxley equation:	$u_t + pu^c u_x - u_{xx} = ru(1-u^c)(u^c - s)$
Sol:	$u(x,t) = \left\{\frac{\lambda}{2} \pm \frac{\lambda}{2}\tanh\{c\lambda\mu x - c\lambda\mu[\frac{p\lambda}{1+c} - 2(1+c-\lambda)\mu\, t\}\right\}^{1/c}$ $(\lambda = 1 \text{ or } s,\ \ \mu = \frac{-p\pm\sqrt{p^2+4r+4cr}}{4(1+c)})$
Ref:	Wang (1988), Hereman (1990)
Non-linear heat conduction equation:	$u_t - (u^2)_{xx} = p\,u - q\,u^2$
Sol:	$u(x,t) = 2p\{q \pm q\tanh[\frac{\sqrt{q}}{4}(x \pm \frac{p}{\sqrt{q}}t)]\}^{-1}$
Ref:	Whilemsson (1987)

Table 16.3 Solutions of non-linear coupled wave equations

Variant of Boussinesq equations 1:	$u_t + v_x + uu_x + su_{xx} = 0, \qquad v_t + (uv)_x + rv_{xx} + pu_{xxx} = 0$
gSol:	$u(x,t) = \frac{w}{k} + k(r+s\pm d)\tanh(kx-wt)$, $v(x,t) = k^2(2p+r^2-rs\pm rd)\,\mathrm{sech}^2(kx-wt)$ $(d=\sqrt{4p+r^2-2rs+s^2};\ k,w\text{: const.})$
sSol:	$u(x,t) = -k(r+s\pm d)\left\{1-\tanh\{k[x+k(r+s\pm d)t]\}\right\}$, $v(x,t) = k^2(2p+r^2-rs\pm rd)\,\mathrm{sech}^2\{k[x+k(r+s\pm d)t]\}$ $(k\text{: const.})$
Ref:	Sachs (1988)
Variant of Boussinesq equations 2:	$u_t + v_x + uu_x + qu_{xxt} = 0, \qquad v_t + (uv)_x + pu_{xxx} = 0$
gSol:	$u(x,t) = \frac{w}{k} + \frac{kp}{2wq} - 8wkq + 12wkq\tanh^2(kx-wt)$, $v(x,t) = \frac{k^2p^2}{4w^2q^2} + 4k^2p - 6k^2p\tanh^2(kx-wt)$ $(k,w\text{: const.})$
sSol:	$u(x,t) = \pm\frac{3\sqrt{p}}{2\sqrt{-q}}\,\mathrm{sech}^2[\frac{1}{2\sqrt{-2q}}(x\mp\frac{\sqrt{p}}{\sqrt{-q}}t)], \quad v(x,t) = -\frac{3p}{4q}\,\mathrm{sech}^2[\frac{1}{2\sqrt{-2q}}(x\mp\frac{\sqrt{p}}{\sqrt{-q}}t)]$
Ref:	Sachs (1988), Whitham (1973)
Variant of Boussinesq equations 3:	$u_t + v_x + uu_x + su_{xx} + qu_{xxt} = 0, \qquad v_t + (uv)_x + 2sv_{xx} + pu_{xxx} = 0$
gSol:	$u(x,t) = \pm(\frac{5kp}{s} + \frac{s}{10kq} - \frac{12ks}{5}) \pm \frac{6ks}{5}[1\pm\tanh(kx\mp\frac{s}{10q}t)]^2$, $v(x,t) = \frac{25k^2p}{s^2} + 12k^2p - 6k^2p[1\pm\tanh(kx\mp\frac{s}{10q}t)]^2$ $(k\text{: const.})$
Ref:	Whitham (1973)
Drinfel'd–Sokolov–Wilson equations:	$u_t + pvv_x = 0, \qquad v_t + qv_{xxx} + ruv_x + su_xv = 0$
gSol:	$u(x,t) = \frac{2k^3q+w}{kr} - \frac{6k^2q}{r+2s}\tanh^2(kx-wt)$,
	$v(x,t) = \pm\sqrt{\frac{-12kqw}{(r+2s)p}}\tanh(kx-wt)$ $(k,w\text{: const.})$
sSol:	$u(x,t) = \frac{6k^2q}{r+2s}\,\mathrm{sech}^2[k(x-4k^2q\frac{r-s}{r+2s}t)]$, $v(x,t) = \frac{4k^2q}{r+2s}\sqrt{\frac{3(s-r)}{p}}\tanh[k(x-4k^2q\frac{r-s}{r+2s}t)]$ $(k\text{: const.})$
Ref:	Hirota et al. (1986)

Table 16.3 Solutions of non-linear coupled wave equations (continued)

Modified Volterra equations:	$u_t + pu_x - u + uv = 0, \qquad v_t + qv_x + v - uv = 0$
Sol:	$u(x,t) = v(x,t) = \frac{1}{2}[1 + \tanh(\frac{1}{p-q}(x - \frac{p+q}{2}t))]$
Ref:	Murray (1989)
Coupled KdV equations:	$u_t + pvv_x + quu_x + ru_{xxx} = 0, \qquad v_t + s(uv)_x + hvv_x = 0,$
gSol:	$u(x,t) = \frac{8h^2k^3r+h^2w+4psw}{k(h^2q+4ps^2)} - 12\frac{h^2k^2r}{h^2q+4ps^2}\tanh^2(kx - wt),$ $v(x,t) = \frac{2w}{kh} - \frac{2s}{h}u(x,t) \quad (k, w\text{: const.})$
sSol:	$u(x,t) = \frac{12h^2k^2r}{h^2q+4ps^2}\operatorname{sech}^2[k(x - \frac{4h^2k^2r}{h^2+4ps}t)], \quad v(x,t) = \frac{8hk^2r}{h^2+4ps} - \frac{2s}{h}u(x,t) \quad (k\text{: const.})$
Ref:	Roy et al. (1986), Kupershmidt (1985)
Coupled mKdV equations:	$u_t + pv_x + qu^2u_x + ru_{xxx} = 0, \qquad v_t + s(uv)_x + hvv_x = 0$
gSol:	$u(x,t) = \pm\sqrt{-\frac{6r}{q}}\,k\tanh[k(x + \frac{2(hk^2r+ps)}{h}t)],$ $v(x,t) = -4\frac{hk^2r+ps}{h^2} + \frac{2ks}{h}\sqrt{\frac{-6r}{q}}\tanh[k(x + \frac{2(hk^2r+ps)}{h}t)] \quad (k\text{: const.})$
Ref:	Kawamoto (1984), Roy et al. (1986)
Hirota–Satsuma equations:	$u_t - 6puu_x + rvv_x - pu_{xxx} = 0, \qquad v_t + 3puv_x + pv_{xxx} = 0$
gSol:	$u_1(x,t) = \frac{2k^3p+w}{3kp} - 2k^2\tanh^2(kx - wt),$ $v_1(x,t) = \pm 2\sqrt{\frac{k(4pk^3-3w)}{r}}\tanh(kx - wt) \quad (k, w\text{: const.})$
sSol:	$u_1(x,t) = 2k^2\operatorname{sech}^2[k(x + 8k^2pt)],$ $v_1(x,t) = \pm 2k^2\sqrt{\frac{28p}{r}}\tanh[k(x + 8k^2pt)] \quad (k\text{: const.})$
gSol:	$u_2(x,t) = \frac{8k^3p+w}{3pk} - 4k^2\tanh^2(kx - wt),$ $v_2(x,t) = \pm\frac{8k^3p+w}{k\sqrt{3pr}} \mp 4k^2\sqrt{\frac{3p}{r}}\tanh^2(kx - wt) \quad (k, w\text{: const.})$
sSol:	$u_2(x,t) = 4k^2\operatorname{sech}^2[k(x - 4k^2pt)],$ $v_2(x,t) = \pm 4k^2\sqrt{\frac{3p}{r}}\operatorname{sech}^2[k(x - 4k^2pt)] \quad (k\text{: const.})$
Ref:	Hirota and Satsuma (1981)

Table 16.4 Solutions of non-linear coupled diffusion equations

Unnamed coupled reaction-diffusion equations 1:	$u_t = du_{xx} + uv, \qquad v_t = dv_{xx} - uv$
Sol:	$u_1(x,t) = \frac{C}{4}[1 - 2\tanh(\frac{1}{2}\sqrt{\frac{C}{6d}}x - \frac{5C}{12}t) + \tanh^2(\frac{1}{2}\sqrt{\frac{C}{6d}}x - \frac{5C}{12}t)]$, $v_1(x,t) = \frac{C}{4}[3 + 2\tanh(\frac{1}{2}\sqrt{\frac{C}{6d}}x - \frac{5C}{12}t) - \tanh^2(\frac{1}{2}\sqrt{\frac{C}{6d}}x - \frac{5C}{12}t)]$ (C: const.)
	$u_2(x,t) = \frac{C}{4}[1 + 2\tanh(\frac{1}{2}\sqrt{\frac{C}{6d}}x + \frac{5C}{12}t) + \tanh^2(\frac{1}{2}\sqrt{\frac{C}{6d}}x + \frac{5C}{12}t)]$, $v_2(x,t) = \frac{C}{4}[3 - 2\tanh(\frac{1}{2}\sqrt{\frac{C}{6d}}x + \frac{5C}{12}t) - \tanh^2(\frac{1}{2}\sqrt{\frac{C}{6d}}x + \frac{5C}{12}t)]$ (C: const.)
Ref:	Malfliet (1991)
Unnamed coupled reaction-diffusion equations 2:	$u_t = du_{xx} + u^2v - pu, \qquad v_t = dv_{xx} - u^2v + pu$
Sol:	$u(x,t) = \frac{C}{2} \pm \frac{1}{2}\sqrt{C^2 - 4p}\tanh(\frac{1}{2\sqrt{2d}}\sqrt{C^2 - 4p}x \pm \frac{1}{4}\sqrt{C^2 - 4p}Ct)$, $v(x,t) = \frac{C}{2} \mp \frac{1}{2}\sqrt{C^2 - 4p}\tanh(\frac{1}{2\sqrt{2d}}\sqrt{C^2 - 4p}x \pm \frac{1}{4}\sqrt{C^2 - 4p}Ct)$ (C: const.)
Ref:	Lefever (1977)
Belousov–Zhabotinskii reaction-diffusion equations:	$u_t = u_{xx} + u(1 - u - rv), \qquad v_t = v_{xx} - suv$
sSol:	$u_1(x,t) = \frac{1}{4}[1 + \tanh(\frac{1}{2\sqrt{2}}(x + \frac{1}{\sqrt{2}}t))]^2$, $v_1(x,t) = \frac{1}{4}[1 - \tanh(\frac{1}{2\sqrt{2}}(x + \frac{1}{\sqrt{2}}t))]^2$ ($r = 2$, $s = 3$)
	$u_2(x,t) = \frac{1}{4}[1 + \tanh(\frac{1}{2\sqrt{6}}(x + \frac{1}{\sqrt{6}}t))]^2$, $v_2(x,t) = \frac{1}{2}[1 - \tanh(\frac{1}{2\sqrt{6}}(x + \frac{1}{\sqrt{6}}t))]^2$ ($r = 4/3$, $s = 1/3$)
	$u_3(x,t) = \frac{1}{4}[1 + \tanh(\frac{s}{2\sqrt{6}}(x + 5\sqrt{\frac{s}{6}}t))]^2$, $v_3(x,t) = \frac{3}{4} - \frac{1}{2}\tanh[\frac{s}{2\sqrt{6}}(x + 5\sqrt{\frac{s}{6}}t)] - \frac{1}{4}\tanh^2[\frac{s}{2\sqrt{6}}(x + 5\sqrt{\frac{s}{6}}t))]$ ($r + s = 1$)
Ref:	Troy (1980)

16.5 Concluding remarks

We have derived traveling wave solutions of solitary wave form using a relatively simple mechanization method. We believe that it can be used in a wider context. Of course we are aware of the fact that not all fundamental evolution equations can be treated by this method. For example, the non-linear Schrödinger and the sine-Gordon equations are notable exceptions, though their solutions contain hyperbolic functions as well. Nevertheless, we are convinced that our primary goal of mechanizing the solving of non-linear evolution equations has been achieved.

Finally, we should mention that the method is readily applicable to $(2+1)$-dimensional equations such as the K-P equation. We are also investigating how our method can be modified to treat complex, higher-dimensional or vector non-linear evolution equations or PDEs of a different kind.

Chapter 17

APPLICATIONS OF MECHANICAL METHODS TO PARTIAL DIFFERENTIAL EQUATIONS

Hongqing Zhang and Engui Fan

The purpose of this chapter is to report the research work on unifying and extending various special methods for solving partial differential equations (PDEs). The work described in the chapter has been done by the authors and their collaborators in the Dalian University of Technology. It includes

(1) a mechanical approach "$AC = BD$" for constructing the analytical solutions of PDEs,

(2) Wu–Ritt's characteristic set method and other computational tools for constructing analytical solutions for PDEs,

(3) applications of the differential characteristic set method to compute the symmetry groups for PDEs,

(4) automatic reasoning by the differential characteristic set method,

(5) a mechanical method for constructing and identifying the Hamiltonian structure of PDEs,

(6) applications of the differential characteristic set method to testing the Painlevé property.

(7) applications of symbolic computation for generating integrable hierarchy and for finding Lax representation for PDEs.

All the algorithms and techniques developed in this chapter can be implemented in a computer algebra system, and may have important applications in elasticity, electrodynamics, solitons, integrable systems, and other areas.

MATHEMATICS MECHANIZATION AND APPLICATIONS
ISBN 0-12-734760-7

17.1 A general scheme for solving PDEs

Our general idea is to transform a given system $Au = 0$ of PDEs to another system $Dv = 0$, which may be solved easily. The following list on possible A's and D's is given by Zhang (1997).

A	D
Any differential equations (DEs)	DEs in diagonal form
Non-linear DEs	Linear DEs
DEs with variable coefficients	DEs with constant coefficients
PDEs	Ordinary DE
DEs with separable variable	DEs without separable variable
DEs with known solutions	DEs without known solutions
Any DE	DE with nice property.

The question is how to transform $Au = 0$ to $Dv = 0$.

Theorem 17.1:

Let X be a linear space, and A, B, C, D be operators from X to X. If $AC = BD$, $B(0) = 0$, and $C \ker D \supset \ker A$, then any solution u of $Au = 0$ can be written as $u = Cv$, $Dv = 0$ for some v in X.

Proof:

From $AC = BD$ and $B(0) = 0$, we know that for any solution v of $Dv = 0$, if $u = Cv$ then $Au = ACv = BDv = 0$, i.e., $u = Cv$ is a solution of $Au = 0$. Conversely, from $C \ker D \supset \ker A$ it follows that for any solution u of $Au = 0$, there exists a solution v of $Dv = 0$ such that $u = Cv$.

Example 17.1:

For the Maxwell equation,

$$A = \begin{pmatrix} 0 & -\partial_z & \partial_y & -\frac{1}{c}\partial_t & 0 & 0 \\ \partial_z & 0 & -\partial_x & 0 & -\frac{1}{c}\partial_t & 0 \\ -\partial_y & \partial_x & 0 & 0 & 0 & \frac{1}{c}\partial_t \\ 0 & 0 & 0 & \partial_x & \partial_y & \partial_z \\ \frac{1}{c}\partial_t & 0 & 0 & 0 & -\partial_z & \partial_y \\ 0 & \frac{1}{c}\partial_t & 0 & \partial_z & 0 & -\partial_x \\ 0 & 0 & \frac{1}{c}\partial_t & -\partial_y & \partial_x & 0 \\ \partial_z & \partial_y & \partial_z & 0 & 0 & 0 \end{pmatrix},$$

$$u = (H_1\ H_2\ H_3\ E_1\ E_2\ E_3)^T, \quad f = \left(\frac{4\pi}{c}j_1\ \ \frac{4\pi}{c}j_2\ \ \frac{4\pi}{c}j_3\ \ 4\pi\rho\ \ 0\ \ 0\ \ 0\ \ 0\right),$$

$$C = \begin{pmatrix} 0 & -\partial_z & \partial_y & 0 \\ \partial_z & 0 & -\partial_x & 0 \\ -\partial_y & \partial_x & 0 & 0 \\ -\frac{1}{c}\partial_t & 0 & 0 & \partial_x \\ 0 & -\frac{1}{c}\partial_t & 0 & \partial_y \\ 0 & 0 & -\frac{1}{c}\partial_t & \partial_z \end{pmatrix}, \qquad B = \begin{pmatrix} -1 & 0 & 0 & 0 & \partial_x \\ 0 & -1 & 0 & 0 & \partial_y \\ 0 & 0 & -1 & 0 & \partial_z \\ 0 & 0 & 0 & 1 & -\frac{1}{c}\partial_t \\ 0 & 0 & 0 & 0 & 0 \\ 0 & 0 & 0 & 0 & 0 \\ 0 & 0 & 0 & 0 & 0 \\ 0 & 0 & 0 & 0 & 0 \end{pmatrix},$$

$$D = \begin{pmatrix} \Box & 0 & 0 & 0 \\ 0 & \Box & 0 & 0 \\ 0 & 0 & \Box & 0 \\ 0 & 0 & 0 & \Box \\ \partial_x & \partial_y & \partial_z & \frac{1}{c}\partial_t \end{pmatrix},$$

where

$$\Box = \partial_x^2 + \partial_y^2 + \partial_z^2 - \frac{1}{c^2}\partial_t, \qquad e_i = 0, \quad i = 1, \dots, 6;$$

$$g_i = -\frac{4\pi}{c}j_i, \quad i = 1, 2, 3, \qquad g_4 = 4\pi\rho.$$

It is easy to check that $AC = BD$ and $Ae + Bg = f$. By the Helmhotz decomposition, we can prove that $C \ker D = \ker A$, so that the general solution of $Au = f$ can be written as $u = Cv + e$, $Dv = g$, where e and g satisfy $Ae + Bg = f$.

Theorem 17.2:

With the above notation, for any $v \in X$, there exists an operator $B(v) : X \to X$ such that $ACv = B(v)D(v)$.

In fact, for any $v \in X$, there exist unique ACv and Dv respectively, so we can determine an operator $B(v)$ which maps Dv to ACv and satisfies $ACv = B(v)Dv$.

Example 17.2:

For Burgers' equation,

$$Au = \left[\partial_t + \frac{1}{2}(\partial_x)^2 - \lambda\partial_x^2\right] u = 0,$$

$$u = Cv = -2\lambda\frac{\partial}{\partial_x}\ln v, \qquad \text{(Hopf–Cole transformation)}$$

$$Bv = -\frac{2\lambda}{v}, \qquad Dv = v_t - \lambda v_{xx}.$$

It is easy to check that $ACv = B(v)D(v)$ and $B(v) \neq 0$. Because C is surjective, $C\ker = \ker A$. It follows that every solution of $Au = 0$ can be written as $u = Cv$, $Dv = 0$.

Theorem 17.1 leads to an interesting question: for a given operator A, is there an algorithm to find operators B and C? There are many approaches to tackle this problem. For instance,

- differential elimination methods,
- methods based on algebraic geometry,
- the variational method,
- methods based on Lie groups,
- methods of undetermined coefficients, and
- the characteristic set method.

These methods have applications to various fields of mathematical physics. In the following sections, we will study these applications in more detail.

17.2 Analytical solutions for system of PDEs with constant coefficients

Theorem 17.3: (Zhang 1978)

Let a_{ij} $(i, j = 1, 2, 3)$ be linear partial differential operators with constant coefficients in a convex domain Ω, and

$$A = \begin{pmatrix} a_{11} & a_{12} & a_{13} \\ a_{21} & a_{22} & a_{23} \\ a_{31} & a_{32} & a_{33} \end{pmatrix}, \qquad u = \begin{pmatrix} u_1 \\ u_2 \\ u_3 \end{pmatrix}.$$

Let A_{ij} be the algebraic cofactor of a_{ij}, $A_{3j} = B_{3j}E$ $(j = 1, 2, 3)$ and $A_{23} = B_{23}E$, where E is the greatest common divisor of A_{23} and A_{3j} $(j = 1, 2, 3)$. If a_{11} and a_{12} are relatively prime, and thus A_{32} and B_{33} are also relatively prime, then the general solution of $Au = 0$ in Ω is given by

$$\begin{pmatrix} u_1 \\ u_2 \\ u_3 \end{pmatrix} = \begin{pmatrix} 0 & a_{12} & B_{31} \\ 0 & -a_{11} & B_{32} \\ 0 & 0 & B_{33} \end{pmatrix} \begin{pmatrix} 0 \\ v_1 \\ v_2 \end{pmatrix},$$

where v_1 and v_2 satisfy

$$Ev_1 = 0, \qquad \frac{|A|}{E} v_2 = 0,$$

respectively, and $|A|$ is the determinant of A.

Example 17.3:

Consider the problem of plane stress (Zhang 1978)

$$\begin{cases} \dfrac{\partial \sigma_x}{\partial x} + \dfrac{\partial \tau_{xy}}{\partial y} = 0, \\ \dfrac{\partial \sigma_y}{\partial y} + \dfrac{\partial \tau_{xy}}{\partial x} = 0, \\ (\partial_x^2 + \partial_y^2)(\sigma_x + \sigma_y) = 0. \end{cases}$$

Let

$$A = \begin{pmatrix} \partial_x & \partial_y & 0 \\ 0 & \partial_x & \partial_y \\ \triangle & 0 & \triangle \end{pmatrix}, \qquad u = \begin{pmatrix} \sigma_x \\ \tau_{xy} \\ \sigma_y \end{pmatrix},$$

where $\triangle = \partial_x^2 + \partial_y^2$. Then $A_{31} = \partial_y^2$, $A_{32} = -\partial_{xy}^2$, $A_{33} = \partial_x^2$, $E = 1$ and $|A| = \triangle^2$. By Theorem 17.3, we can get

$$\sigma_x = \frac{\partial^2 v}{\partial y^2}, \qquad \tau_{xy} = -\frac{\partial^2 v}{\partial x \partial y}, \qquad \sigma_y = \frac{\partial^2 v}{\partial x^2}, \qquad \triangle^2 v = 0;$$

v is called the *Airy stress function.*

Example 17.4: (The three-dimensional problem of theory of elasticity of transversely isotropic bodies)

$$A = \begin{pmatrix} b_{11}\partial_x^2 + b_{66}\partial_y^2 + b_{44}\partial_z^2 & b_{12}\partial_{xy}^2 & b_{13}\partial_{xz}^2 \\ b_{12}\partial_{xy}^2 & b_{66}\partial_x^2 + b_{11}\partial_y^2 + b_{44}\partial_z^2 & b_{13}\partial_{yz}^2 \\ b_{13}\partial_{xz}^2 & b_{13}\partial_{yz}^2 & b_{44}(\partial_x^2 + \partial_y^2) + b_{33}\partial_z^2 \end{pmatrix},$$

where $b_{11}, b_{12}, b_{33}, b_{66}$ are elastic constants and $b_{66} = b_{11} - b_{12}$.

By Theorem 17.3, the solution of $Au = 0$ can be given by

$$\begin{cases} u_1 = -\dfrac{\partial^2 v_2}{\partial x \partial z} - \dfrac{\partial v_1}{\partial y}, \\ u_2 = -\dfrac{\partial^2 v_2}{\partial y \partial z} - \dfrac{\partial v_1}{\partial x}, \\ u_3 = \dfrac{b_{11}}{b_{33}} \left(\dfrac{\partial^2 v_2}{\partial x^2} + \dfrac{\partial^2 v_2}{\partial y^2} \right) + \dfrac{b_{44}}{b_{66}} \dfrac{\partial^2 v_2}{\partial z^2}. \end{cases}$$

v_1 and v_2 satisfy

$$\left(\partial_x^2 + \partial_y^2 + \frac{1}{s_0^2} \partial_z^2 \right) v_1 = 0,$$

$$\left(\partial_x^2 + \partial_y^2 + \frac{1}{s_1^2} \partial_z^2 \right) \left(\partial_x^2 + \partial_y^2 + \frac{1}{s_2^2} \partial_z^2 \right) v_2 = 0$$

respectively, where

$$\frac{1}{s_0^2} = \frac{b_{44}}{b_{66}}, \qquad \frac{1}{s_1^2} + \frac{1}{s_2^2} = b_{44}^2 + b_{11} b_{33} - \frac{b_{13}^2}{b_{11} b_{14}}.$$

Example 17.5: (Maxwell equation; Zhang 1981b)

If $j = 0$ and $\rho = 0$, the Maxwell equation can be written as

$$\begin{cases} \operatorname{rot}(E + iH) = \dfrac{1}{c} \dfrac{\partial (E + iH)}{\partial t}, \\ \operatorname{div}(E + iH) \mid_{t=0} = 0. \end{cases}$$

By Theorem 17.3, we get

$$\begin{pmatrix} E_1 \\ E_2 \\ E_3 \end{pmatrix} = \begin{pmatrix} \partial_{xz}^2 & -\dfrac{1}{c} \partial_{yt}^2 \\ \partial_{yz}^2 & \dfrac{1}{c} \partial_{xt}^2 \\ \partial z^2 - \dfrac{1}{c} \partial_t^2 & 0 \end{pmatrix} \begin{pmatrix} \phi_1 \\ \\ \phi_2 \end{pmatrix},$$

$$\begin{pmatrix} H_1 \\ H_2 \\ H_3 \end{pmatrix} = \begin{pmatrix} \dfrac{1}{c} \partial_{yt}^2 & \partial_{xz}^2 \\ -\dfrac{1}{c} \partial_{xt}^2 & \partial_{yz}^2 \\ 0 & \partial z^2 - \dfrac{1}{c} \partial_t^2 \end{pmatrix} \begin{pmatrix} \phi_1 \\ \\ \phi_2 \end{pmatrix},$$

$$\Box \phi_i = 0, \qquad i = 1, 2.$$

Theorem 17.4: (Zhang 1981a)

Let

$$A = \begin{pmatrix} a_{11} & a_{12} & \cdots & a_{1n} \\ a_{21} & a_{22} & \cdots & a_{2n} \\ \cdots & \cdots & \cdots & \cdots \\ a_{n1} & a_{n2} & \cdots & a_{nn} \end{pmatrix}, \qquad u = \begin{pmatrix} u_1 \\ u_2 \\ \vdots \\ u_n \end{pmatrix},$$

where a_{ij} *are linear partial differential operators on* $C^\infty(\Omega)$, Ω *is a convex domain and* $u_i, f_i \in C^\infty(\Omega)$, $i = 1, \ldots, n$. *Let* $\triangle^{i_1\cdots i_m}_{j_1\cdots j_m}$ *be the determinant with* i_1*th, …,* i_m*th rows and* j_1*th, …,* j_m*th columns of* A. *That is*

$$\triangle^{i_1\cdots i_m}_{j_1\cdots j_m} = \begin{vmatrix} a_{i_1j_1} & a_{i_1j_2} & \cdots & a_{i_1j_m} \\ a_{i_2j_1} & a_{i_2j_2} & \cdots & a_{i_2j_m} \\ \cdots & \cdots & \cdots & \cdots \\ a_{i_mj_1} & a_{i_mj_2} & \cdots & a_{i_mj_m} \end{vmatrix}.$$

Let D_k *be the greatest common divisor of* $\triangle^{12\cdots k-1,k}_{i_1i_2\cdots i_{k-1}i_k}$ *and* $\triangle^{i_1i_2\cdots i_{k-1}i_k}_{12\cdots k-1,k}$,

$$\triangle^{12\cdots k-1,k}_{i_1i_2\cdots i_{k-1}i_k} = D_k\delta^{12\cdots k-1,k}_{i_1i_2\cdots i_{k-1}i_k} \quad \textit{and} \quad \triangle^{i_1i_2\cdots i_{k-1}i_k}_{12\cdots k-1,k} = D_k\delta^{i_1i_2\cdots i_{k-1}i_k}_{12\cdots k-1,k},$$

where $i_1 < i_2 < \cdots < i_{k-1}$, $i_1, i_2, \ldots, i_{k-1}$ *are the numbers among* $1, \ldots, k$, *and* i_k *may be* $k, k+1, \ldots, n$. *Suppose that* $D \ker C = \ker B$, *where* B, C *and* D *will be defined below. Then the general solution to* $Au = 0$ *can be written as* $u = Cv$, $Dv = 0$, *where*

$$C = \begin{pmatrix} 1 & a_{12} & -\delta^{12}_{23} & \cdots & (-1)^k\delta^{12\cdots k-1}_{23\cdots k} & \cdots & (-1)^n\delta^{12\cdots n-1}_{23\cdots n} \\ & -a_{11} & \delta^{12}_{13} & \cdots & (-1)^{k+1}\delta^{12\cdots k-1}_{13\cdots k} & \cdots & (-1)^{n+1}\delta^{12\cdots n-1}_{13\cdots n} \\ & & -\delta^{12}_{12} & \cdots & \cdots & \cdots & \cdots \\ & & & \ddots & (-1)^{2k-1}\delta^{12\cdots k-1}_{12\cdots k-1} & \cdots & (-1)^{n+k-1}\delta^{12\cdots k-1,k+1\cdots n-1}_{12\cdots k-1,k+1\cdots n} \end{pmatrix},$$

$$D = \begin{pmatrix} 1 & & & \\ & \frac{D_2}{D_1} & & \\ & & \ddots & \\ & & & \frac{D_n}{D_{n-1}} \end{pmatrix}, \qquad B = \begin{pmatrix} a_{11} & & & & \\ a_{21} & \delta^{12}_{12} & & & \\ a_{31} & -\delta^{13}_{12} & -\delta^{123}_{123} & & \\ \vdots & \vdots & \vdots & \ddots & \\ a_{n1} & -\delta^{1n}_{12} & -\delta^{12n}_{123} & \cdots & \delta^{12\cdots n}_{12\cdots n} \end{pmatrix}.$$

Example 17.6: (Micro-elasticity for isotropic elastic body)

The operator matrix of the system of micro-elasticity for isotropic elastic body is

$$A = \begin{pmatrix} L_1 + k_1\partial_z^2 & k_1\partial_{xy}^2 & k_1\partial_{xz}^2 & 0 & -k\partial_z & k\partial_y \\ k_1\partial_{xy}^2 & L_1 + k_1\partial_y^2 & k_1\partial_{yz}^2 & k\partial_z & 0 & -k\partial_x \\ k_1\partial_{xz}^2 & k_1\partial_{yz}^2 & L_1 + k_1\partial_z^2 & -k\partial_y & k\partial_x & 0 \\ 0 & -k\partial_z & k\partial_y & L_2 + k_2\partial_z^2 & k_2\partial_{xy}^2 & k_2\partial_{xz}^2 \\ k\partial_x & 0 & -k\partial_z & k_2\partial_{xy}^2 & L_2 + k_2\partial_y^2 & k_2\partial_{yz}^2 \\ -k\partial_y & k\partial_x & 0 & k_2\partial_{xz}^2 & k_2\partial_{yz}^2 & L_2 + k_2\partial_z^2 \end{pmatrix},$$

where $L_1 = (\mu + k)\triangle - \rho\partial_t^2$, $L_2 = r\triangle - 2k - \rho l\partial_t^2$, $k_1, k_2, \mu, \lambda, l, r$ are material constants, and $\triangle$ is the Laplace operator. We can obtain that

$$D_1 = D_2 = 1, \qquad D_3 = L_1(L_1 + k_1\triangle), \qquad D_4 = L_1 + k_1\triangle,$$
$$D_5 = (L_1 + k_1\triangle)(k^2\triangle + L_1L_2),$$
$$D_6 = (L_1 + k_1\triangle)(k^2\triangle + L_1L_2)^2(L_2 + k_2\triangle).$$

Notice that $\frac{D_4}{D_3}$ is not an integral operator, but we can make $\frac{D_k}{D_{k-1}}$ integral by re-adjusting the rows and the columns of A appropriately. In the above example, we exchange the third row for the fourth and third column for the fourth in A. That is

$$A = \begin{pmatrix} L_1 + k_1\partial_z^2 & k_1\partial_{xy}^2 & 0 & k_1\partial_{xz}^2 & -k\partial_z & k\partial_y \\ k_1\partial_{xy}^2 & L_1 + k_1\partial_y^2 & k\partial_z & k_1\partial_{yz}^2 & 0 & -k\partial_x \\ 0 & -k\partial_z & L_2 + k_2\partial_z^2 & k\partial_y & k_2\partial_{xy}^2 & k_2\partial_{xz}^2 \\ k_1\partial_{xz}^2 & k_1\partial_{yz}^2 & -k\partial_y & L_1 + k_1\partial_z^2 & k\partial_x & 0 \\ k\partial_z & 0 & k_2\partial_{xy}^2 & -k\partial_x & L_2 + k_2\partial_y^2 & k_2\partial_{yz}^2 \\ -k\partial_y & k\partial_x & k_2\partial_{xz}^2 & 0 & k_2\partial_{yz}^2 & L_2 + k_2\partial_z^2 \end{pmatrix}.$$

Thus we can get

$$D_1 = D_2 = D_3 = 1, \;\; D_4 = L_1 + k_1\triangle, \;\; D_5 = (L_1 + k_1\triangle)(k^2\triangle + L_1L_2),$$
$$D_6 = (L_1 + k_1\triangle)(k^2\triangle + L_1L_2)^2(L_2 + k_2\triangle).$$

So the solution can be described as

$$u = \begin{pmatrix} \partial_x & k\partial_{xz}^2 & k\partial_y(L_2 + k_2\triangle) \\ \partial_y & k\partial_{yz}^2 & -k\partial_x(L_2 + k_2\triangle) \\ 0 & -L_1k\partial_y & \partial_{xz}^2(k_2L_1 - k^2) \\ \partial_z & -k(\partial_x^2 + \partial_y^2) & 0 \\ 0 & L_1\partial_x & k\partial_{yz}^2(k_2L_1 - k^2) \\ 0 & 0 & -[L_1L_2 + k^2\triangle + (\partial^2 + \partial^2)(k_2L_1 - k^2)] \end{pmatrix} \begin{pmatrix} \phi_1 \\ \phi_2 \\ \phi_3 \end{pmatrix},$$

where ϕ_1, ϕ_2, and ϕ_3 satisfy

$$(L_1 + k_1\triangle)\phi_1 = 0,\ (k^2\triangle + L_1L_2)\phi_2 = 0,\ (L_2 + k_2\triangle)(k^2\triangle + L_1L_2)\phi_3 = 0,$$

respectively. Some conditions on $C \ker D = \ker A$ and $\overline{C \ker D} = \ker A$ are given in Zhang and Wang (1986).

17.3 Analytical solution for systems of linear inhomogeneous PDEs with constant coefficients

The following theorem may be found in Zhang and Chao (1996).

Theorem 17.5:

Suppose that $P(D)$ and $Q(D)$ are linear partial differential operator matrices of orders $m \times q$ and $m \times l$ respectively, $G(D) = (P(D) \mid Q(D))$, and $q + l \geq m \geq 1$. If all the $m \times m$ minors of $G(D)$ have no common zero, then there exist partial differential operator matrices $R(D)$ and $S(D)$ such that

$$P(D)R(D) + Q(D)S(D) = I_m.$$

We remark that there is an algorithm for computing partial differential operator matrices $R(D)$ and $S(D)$.

From now on, R stands for the ring of linear partial differential operators with constant coefficients, and $R \mid \partial x_i$ for the ring of linear partial differential operators generated by $\partial x_1, \ldots, \partial x_{i-1}, \partial x_{i+1}, \ldots, \partial x_n$.

Theorem 17.6:

Suppose that $P(D)$ and $Q(D)$ are linear partial differential operator matrices of orders $m \times q$ and $m \times l$, respectively, and $q + l \geq m \geq 1$. Then, for each i with $1 \leq i \leq n$, there exist linear partial differential operator matrices $R_i(D)$ and $S_i(D)$ such that

$$P(D)R_i(D) + Q(D)S_i(D) = \psi_i(D)L(D)I_m$$

or

$$P(D)R(D) + Q(D)S(D) = \psi(D)L(D)I_m,$$

and there is an algorithm for finding these $R_i(D)$ and $S_i(D)$, where $\psi_i(D) \in R \mid \partial x_i$ for $i = 1, \ldots, n$, $L(D), \psi(D) \in R$, and $L(D)$ is the greatest common divisor of all the $m \times m$ minors of $(P(D) \mid Q(D))$.

Theorem 17.7: (Zhang and Feng 1994)

Let

$$A = \begin{pmatrix} a_{11} & a_{12} & \cdots & a_{1n} \\ a_{21} & a_{22} & \cdots & a_{2n} \\ \cdots & \cdots & \cdots & \cdots \\ a_{n1} & a_{n2} & \cdots & a_{nn} \end{pmatrix},$$

where a_{ij} are linear partial differential operators, and suppose that C, B, D are partial differential operator matrices such that $AC = BD$ and $C \ker D = \ker A$. Then the general solution of $Au = f$ can be written as

$$u = Cv + e, \qquad Dv = g,$$

where e, g are a solution of the equation $Ae + Bg = f$.

We may take the following steps to solve the system $Au = f$.

S1. Construct the general solution of $Au = 0$ and find D and C. Calculate matrices B from $AC = BD$.

S2. Construct matrices M and N such that $AM + BN = \psi(D)L(D)I_m$.

S3. Find a kind of solution to $Au = f$, $u = Cv + M\phi, Dv = N\phi$, where ϕ satisfies $\psi(D)L(d)I_m\phi = f$.

Example 17.7:

Consider the Maxwell equations

$$\begin{cases} \operatorname{rot} H = \dfrac{1}{c}\dfrac{\partial E}{\partial t} + \dfrac{4\pi}{c} j, \\ \operatorname{rot} E = -\dfrac{1}{c}\dfrac{\partial H}{\partial t}, \\ \operatorname{div} H = 0, \\ \operatorname{div} E = 4\pi\rho. \end{cases}$$

Let

$$A = \begin{pmatrix} 0 & -\partial_z & \partial_y & -\frac{1}{c}\partial_t & 0 & 0 \\ \partial_z & 0 & -\partial_x & 0 & -\frac{1}{c}\partial_t & 0 \\ -\partial_y & \partial_x & 0 & 0 & 0 & -\frac{1}{c}\partial_t \\ 0 & 0 & 0 & \partial_x & \partial_y & \partial_z \\ \frac{1}{c}\partial_t & 0 & 0 & 0 & -\partial_z & \partial_y \\ 0 & \frac{1}{c}\partial_t & 0 & \partial_z & 0 & -\partial_x \\ 0 & 0 & \frac{1}{c}\partial_t & -\partial_y & \partial_x & 0 \\ \partial_x & \partial_y & \partial_z & 0 & 0 & 0 \end{pmatrix},$$

$$u = \begin{pmatrix} H \\ E \end{pmatrix}, \qquad f = \begin{pmatrix} \frac{4\pi}{c}j & 4\pi\rho & 0 & 0 & 0 & 0 \end{pmatrix}^T,$$

$$C = \begin{pmatrix} 0 & -\partial_z & \partial_y & 0 \\ -\partial_x & 0 & -\partial_x & 0 \\ -\partial_y & \partial_x & 0 & 0 \\ -\frac{1}{c}\partial_t & 0 & 0 & \partial_x \\ 0 & -\frac{1}{c}\partial_t & 0 & \partial_y \\ 0 & 0 & -\frac{1}{c}\partial_t & \partial_z \end{pmatrix}, \qquad D = \begin{pmatrix} \Box & 0 & 0 & 0 \\ 0 & \Box & 0 & 0 \\ 0 & 0 & \Box & 0 \\ 0 & 0 & 0 & \Box \\ \partial_x & \partial_y & \partial_z & \frac{1}{c}\partial_t \end{pmatrix},$$

$$B = \begin{pmatrix} -1 & 0 & 0 & 0 & \partial_x \\ 0 & -1 & 0 & 0 & \partial_y \\ 0 & 0 & -1 & 0 & \partial_z \\ 0 & 0 & 0 & -1 & -\frac{1}{c}\partial_t \\ 0 & 0 & 0 & 0 & 0 \\ 0 & 0 & 0 & 0 & 0 \\ 0 & 0 & 0 & 0 & 0 \end{pmatrix},$$

and then take $Ae + Bg = f$ into consideration, where $e = (\rho_1 \ \cdots \ \rho_6)^T$ and $g = (g_1 \ \cdots \ g_5)^T$. Obviously,

$$\rho_i = 0, \quad i = 1, \ldots, 6,$$

$$g_1 = -\frac{4\pi}{c}j_1, \qquad g_2 = -\frac{4\pi}{c}j_2, \qquad g_3 = -\frac{4\pi}{c}j_3, \qquad g_4 = 4\pi\rho, \qquad g_5 = 0$$

is a solution of $Ae + Bg = f$. Thus the solution of $Au = f$ can be expressed as

$$H = \operatorname{rot}\phi, \qquad E = -\operatorname{grad}\psi - \frac{1}{c}\partial_t\phi,$$

where ϕ and ψ satisfy

$$\Box\phi = \frac{4\pi}{c}j, \qquad \Box\psi = -4\pi\rho, \qquad \operatorname{div}\phi + \frac{1}{c}\partial_t\psi = 0.$$

The above solution is the same as the classical one of the Maxwell equations.

Example 17.8:

Let $Au = 0$ be the Maxwell equation

$$C = \begin{pmatrix} -\partial_x + \frac{1}{c^2}\partial_x^{-1}\partial_t^2 & -\partial_y & -\partial_z & 0 & \frac{1}{c}\partial_x^{-1}\partial_{zt}^2 & -\frac{1}{c}\partial_x^{-1}\partial_{yt}^2 \\ -\frac{1}{c}\partial_z^{-1}\partial_{yt}^2 & \frac{1}{c}\partial_z^{-1}\partial_{xt}^2 & 0 & -\partial_x & -\partial_y & -\partial_z + \frac{1}{c^2}\partial_z^{-1}\partial_t^2 \end{pmatrix}^T,$$

$$D = \begin{pmatrix} \Box & 0 \\ 0 & \Box \end{pmatrix}, \qquad B = \begin{pmatrix} 0 & 0 & 0 & 0 & -\frac{1}{c}\partial_x^{-1}\partial_t & 0 & 0 & -1 \\ 0 & 0 & \frac{1}{c}\partial_z^{-1}\partial_t & -1 & 0 & 0 & 0 & 0 \end{pmatrix}^T.$$

It is easy to check that $AC = BD$.

Let

$$f = \left(\frac{4\pi}{c}j_1 \;\; \frac{4\pi}{c}j_2 \;\; \frac{4\pi}{c}j_3 \;\; 4\pi\rho \;\; 0 \;\; 0 \;\; 0 \;\; 0\right)^T,$$

$$g = \left(\frac{4\pi}{c}\partial_z^{-1}(\partial_x j_2 - \partial_y j_1) \;\; -4\pi\rho - \frac{4\pi}{c^2}\partial_x^{-1}\partial_z^{-1}\partial_t j_1\right)^T,$$

$$e = \left(\frac{4\pi}{c}\partial_z^{-1}j_2 \;\; -\frac{4\pi}{c}\partial_z^{-1}j_1 \;\; 0 \;\; 0 \;\; 0 \;\; \frac{4\pi}{c^2}\partial_x^{-1}\partial_z^{-1}\partial_t j_1\right)^T.$$

It is also easy to check that $Ae + Bg = f$. By Theorem 17.6, we can get

$$u = \begin{pmatrix} -\partial_x + \frac{1}{c^2}\partial_x^{-1}\partial_t^2 & -\partial_y & -\partial_z & 0 & \frac{1}{c}\partial_x^{-1}\partial_{zt}^2 & -\frac{1}{c}\partial_x^{-1}\partial_{yt}^2 \\ -\frac{1}{c}\partial_z^{-1}\partial_{yt}^2 & \frac{1}{c}\partial_z^{-1}\partial_{xt}^2 & 0 & -\partial_x & -\partial_y & -\partial_z + \frac{1}{c^2}\partial_z^{-1}\partial_t^2 \end{pmatrix}^T \begin{pmatrix} \psi_1 \\ \psi_2 \end{pmatrix} + \left(\frac{4\pi}{c}\partial_z^{-1}j_2 \;\; -\frac{4\pi}{c}\partial_z^{-1}j_1 \;\; 0 \;\; 0 \;\; 0 \;\; \frac{4\pi}{c^2}\partial_x^{-1}\partial_z^{-1}\partial_t j_1\right)^T,$$

$$Dv = \begin{pmatrix} \Box & 0 \\ 0 & \Box \end{pmatrix}\begin{pmatrix} \psi_1 \\ \psi_2 \end{pmatrix} = \begin{pmatrix} \frac{4\pi}{c}\partial_z^{-1}(\partial_x j_2 - \partial_y j_1) \\ 4\pi\rho - \frac{4\pi}{c^2}\partial_x^{-1}\partial_z^{-1}\partial_t j_1 \end{pmatrix}.$$

Example 17.9:

Now consider the equation of theory of elasticity:

$$A = \begin{pmatrix} \partial_x^2 + k_1\triangle & \partial_{xy}^2 & \partial_{xz}^2 \\ \partial_{xy}^2 & \partial_y^2 + k_1\triangle & \partial_{yz}^2 \\ \partial_{xz}^2 & \partial_{yz}^2 & \partial_z^2 + k_1\triangle \end{pmatrix}, \quad u = \begin{pmatrix} u_1 \\ u_2 \\ u_3 \end{pmatrix}, \quad f = \begin{pmatrix} f_1 \\ f_2 \\ f_3 \end{pmatrix}.$$

Let

$$C = \begin{pmatrix} \frac{1}{2} + k_1 - \frac{x}{2}\partial_x & -\frac{y}{2}\partial_x & -\frac{z}{2}\partial_x & -\frac{1}{2}\partial_x \\ -\frac{x}{2}\partial_y & \frac{1}{2} + k_1 - \frac{y}{2}\partial_y & -\frac{z}{2}\partial_y & -\frac{1}{2}\partial_y \\ -\frac{x}{2}\partial_z & -\frac{y}{2}\partial_z & -\frac{1}{2} + k_1 - \frac{z}{2}\partial_y & -\frac{1}{2}\partial_z \end{pmatrix},$$

$$B = \begin{pmatrix} -\frac{1}{2} + k_1 - \frac{x}{2}\partial_x & -\frac{y}{2}\partial_x & -\frac{z}{2}\partial_x & -\frac{1}{2}\partial_x \\ -\frac{x}{2}\partial_y & -\frac{1}{2} + k_1 - \frac{y}{2}\partial_y & -\frac{z}{2}\partial_y & -\frac{1}{2}\partial_y \\ -\frac{x}{2}\partial_z & -\frac{y}{2}\partial_z & \frac{1}{2} + k_1 - \frac{z}{2}\partial_z & -\frac{1}{2}\partial_z \end{pmatrix}.$$

Provided that $\rho = (\rho_1 \ \rho_2 \ \rho_3)^T$ and $g = (g_1 \ \cdots \ g_4)^T$, an indeterminate equation $Ae + Bg = f$ is obtained:

$$\begin{cases} (\partial_x^2 + k_1\triangle)e_1 + \partial_{xy}^2 e_2 + \partial_{xz}^2 e_3 + \left(-\frac{1}{2} + k_1 - \frac{x}{2}\partial_x\right) g_1 - \frac{y}{2}\partial_x g_2 - \frac{z}{2}\partial_x g_3 \\ \quad - \frac{1}{2}\partial_x g_4 = f_1, \\ (\partial_y^2 + k_1\triangle)e_2 + \partial_y^2 e_1 + \partial_{yz}^2 e_3 + \left(-\frac{1}{2} + k_1 - \frac{y}{2}\partial_y\right) g_2 - \frac{x}{2}\partial_y g_1 - \frac{z}{2}\partial_y g_3 \\ \quad - \frac{1}{2}\partial_y g_4 = f_2, \\ (\partial_z^2 + k_1\triangle)e_3 + \partial_{xz}^2 e_1 + \partial_{yz}^2 e_2 + \left(-\frac{1}{2} + k_1 - \frac{z}{2}\partial_z\right) g_3 - \frac{x}{2}\partial_z g_1 - \frac{y}{2}\partial_z g_2 \\ \quad - \frac{1}{2}\partial_z g_4 = f_3. \end{cases}$$

These equations have seven unknowns. We may find some of their special solutions:

$$e_1 = e_2 = e_3 = 0, \qquad g_1 = \frac{1}{k_1} f_1, \qquad g_2 = \frac{1}{k_1} f_2,$$

$$g_3 = \frac{1}{k_1} f_3, \qquad g_4 = -\frac{1}{k_1}(x f_1 + y f_2 + z f_3).$$

Then the solution of $Au = f$ can be expressed as

$$u = k_1\phi_0 - \frac{k_1}{2(1+k_1)}\text{grad}\,(\psi_0 + r\phi_0),$$

$$\triangle\phi_0 = \frac{1}{k_1}f, \qquad \triangle\psi_0 = \frac{1}{k_1}r\cdot f,$$

where $r = (x\ y\ z)^T$, $f = (f_1\ f_2\ f_3)$, and ϕ_0 is a vector function. It is just the Boussinesq–Papkovich–Neuber solution.

It is possible to solve inhomogeneous linear PDEs by Theorem 17.6, as illustrated in the above examples. We have many choices in solving the indeterminate equation $Ae + Bg = f$, and Theorem 17.7 provides us with a general method for doing so. By Theorem 17.6, we can get general solutions for systems of linear inhomogeneous partial differential equations with constant coefficients and, thus, find fundamental solutions for various problems in mathematical physics (Zhang 1991, 1994b; Zhang and Wu 1992b).

17.4 Analytical solutions for systems of linear homogeneous PDEs with variable coefficients

Let $\alpha = (\alpha_1, \ldots, \alpha_k)$ and $\beta = (\beta_1, \ldots, \beta_k)$ be two multi-indices. We write

$$|\alpha| = \alpha_1 + \cdots + \alpha_k, \qquad D^\alpha = \frac{\partial^{|\alpha|}}{\partial x_1^{\alpha_1}\cdots\partial x_k^{\alpha_k}}, \qquad \alpha! = \alpha_1!\cdots\alpha_k!,$$

$$\binom{\alpha}{\beta} = \binom{\alpha_1}{\beta_1}\cdots\binom{\alpha_k}{\beta_k}, \qquad \binom{\alpha_i}{\beta_i} = \frac{\alpha_i!}{\beta_i!(\alpha_i-\beta_i)!}, \qquad i = 1,\ldots,k.$$

Define $Au = \sum_{|\alpha|\le n} a_\alpha(x)D^\alpha u$, $u \in C^\infty(R^k)$, where $x = (x_1, \ldots, x_n) \in R^k$, R^k is a k-dimensional Euclidean space and $a_\alpha(x) \in C^\infty(R^k)$.

The operator A^* is called the *conjugate operator* of A, where

$$A^*u = \sum_{|\alpha|\le n}(-1)^n\sum_{\beta\le\alpha}\binom{\alpha}{\beta}[D^\beta a_\alpha(x)]D^{\alpha-\beta}u = \sum_{|\alpha|\le n}a_\alpha^*(x)D^\alpha u,$$

$$a_\alpha^*(x) = \sum_{\substack{|l|\le n,\,\beta\le\alpha\\ l-\beta=\alpha}}(-1)^l\binom{\alpha}{\beta}D^\beta a_l(x), \qquad |\alpha|\le n,\ u\in C^\infty(R^k)$$

and $\alpha \ge \beta$ means that $\alpha_i \ge \beta_i$ for every i $(1 \le i \le k)$.

Theorem 17.8: (Zhang and Wu 1992a)

Let

$$A = \begin{pmatrix} a_{11} & a_{12} & a_{13} \\ a_{21} & a_{22} & a_{23} \\ a_{31} & a_{32} & a_{33} \end{pmatrix}$$

be an operator matrix, where the a_{ij} *are linear partial differential operators with smooth coefficients. Then there exists a mechanical method for constructing the operator matrices* B, C*, and* D *such that* $AC = BD$ *and* D *is diagonal. If* $C \ker D = \ker A$*, then any solution of* $Au = 0$ *can be given by*

$$u = Cv, \qquad Dv = 0.$$

The result of this section has some applications in the shell theory (Zhang and Wu 1992c, 1993).

Theorem 17.9: (Inverse problem for Hamilton equation; Zhang 1994b)

Let

$$L = \partial_x^{2n} - l_{2n-1}\partial_x^{2n-1} + \cdots + (-1)^k l_k \partial_x^k + \cdots + l_0$$

be a linear self-adjoint partial differential operator, where l_k *is a linear differential operator not involving the variable* x*. Then any solution of* $L\phi = 0$ *can be expressed as a solution of the Hamilton equation* $(A - \partial_x I)u = 0$*, where* I *is a* $2n \times 2n$ *identity matrix.*

17.5 Constructing analytical solutions of non-linear PDEs

We have studied how to construct the analytical solutions of linear PDEs by the algebraic mechanical method $AC = BD$ (Zhang 1978, 1981a,b). In this section, we present some attempts in generalizing and applying this method to attack non-linear PDEs. Our main work includes the following.

(1) Transform non-linear PDEs to linear PDEs. We have constructed the transformation of linearization for Kupershmidt's equation and the Whitham–Broer–Kaup equation, and obtained their solitary solutions (Fan and Zhang 1998h; Zhang and Fan 1998). Kupershmidt's equation is

$$u_t + uu_x + h_x + \sigma u_{xx} = 0 \tag{17.1}$$

and the Whitham–Broer–Kaup equation is

$$h_t + (uh)_x - \sigma h_{xx} = 0. \tag{17.2}$$

These equations can be transformed to the heat equation

$$v_t - \sigma v_{xx} = 0$$

under transformation

$$u = -2\sigma\frac{v_x}{v}, \qquad h = 4\sigma^2\frac{2vv_{xx} - v_x^2}{v^2}.$$

The solitary solutions of (17.1) and (17.2) are

$$u = -k\sigma \tanh\frac{k}{2}(x + k\sigma t) - k\sigma, \qquad h = k^2\sigma^2\text{sech}^2\frac{k}{2}(x + k\sigma t)$$

respectively.

(2) Transform complex PDEs to simple PDEs. The famous Yang–Mills equations are a set of PDEs of second order with 12 unknown functions. For example, the Yang–Mills equations on $R^{3,1}$ are

$$\sum g^{k,k}(\partial_k F_{jk} - G_k F_{jk} + F_{jk} G_k) = 0, \quad j = 1, \ldots, 4, \tag{17.3}$$

where

$$\begin{aligned}
&g^{11} = g^{22} g^{33} = 1, \qquad g^{44} = -1,\\
&f_{jk} = \partial_k g_j - \partial_j G_k + G_j G_k - G_k G_j, \quad G_j = a_j X_1 + B_j X_2 + C_j X_3,\\
&\qquad\qquad j = 1, \ldots, 4,
\end{aligned}$$

and

$$X_1 = \begin{pmatrix} i & 0 \\ 0 & -i \end{pmatrix}, \qquad X_2 = \begin{pmatrix} 0 & 1 \\ -1 & 0 \end{pmatrix}, \qquad X_3 = \begin{pmatrix} 0 & i \\ i & 0 \end{pmatrix}.$$

Based on the idea of $AC = BD$ (Zhang 1981b), Shi (1997) introduced a kind of linear differential transformation and reduced the Yang–Mills equations (17.3) to three simple linear equations, by the Wu–Ritt's characteristic set method (Wu 1984a,b, 1986a; Chap. 1), as follows:

$$D_{31}u = D_{31}v = D_{31}w = 0, \qquad D_{31} = \partial_1^2 + \partial_2^2 + \partial_3^2 - \partial_4^2,$$

where the transformation is

$$A_i(x) = \sum_{j=1}^{4} s_{ij}\partial_j u(x), \quad B_i(x) = \sum_{j=1}^{4} s_{ij}\partial_j v(x), \quad C_i(x) = \sum_{j=1}^{4} s_{ij}\partial_j w(x)$$

with

$$S = (s_{ij}) = s \begin{pmatrix} 1 & -i & -1 & 1 \\ -i & 1 & i & i \\ 1 & i & -1 & 1 \\ 1 & -i & -1 & 1 \end{pmatrix}.$$

(3) Transform PDEs with variable coefficients to PDEs with constant coefficients. We find that the KdV equation with variable coefficients,

$$u_t + h_1(6uu_x + u_{xxx}) + 4h_2u_x - h_3(2u + xu_x) = 0, \tag{17.4}$$

can be transformed into a KdV equation with constant coefficients,

$$v_\tau + 6vv_\xi + v_{\xi\xi\xi} = 0,$$

by the transformation

$$u = g^2 v, \quad \xi = gx - 4\int h_2 g\, dt, \quad \tau = \int h_1 g^3\, dt, \quad g = \exp\big(\int h_3\, dt\big).$$

(4) Transform PDEs to ordinary differential equations (ODEs). The Clarkson–Kruskal method, which involves no group-theoretic techniques, is extended to investigate similar reductions of the variable coefficient KdV (VCKdV) equation. It is shown that its reduction equations are Painlevé I and II, respectively, which are the same as those of the KdV equations. By this property the transformation from the VCKdV equation to the KdV equation is established.

By using the transformation

$$u = \alpha + \beta w(z), \quad \alpha = -\frac{1}{6h_1\theta}\left[\left(\theta' - h_3\theta\right)x + \left(\sigma' + 4h_2\theta\right)\right], \quad \beta = \theta^2,$$

$$z = \theta x + \sigma, \; \theta = c_1 g, \; \sigma = c_2\int h_1 g^3\, dt - 4c_1\int h_2 g\, dt + Bc_1^6\left(\int h_1 g^3\, dt\right)^2,$$

$$g = \exp\left(\int h_3\, dt\right),$$

the VCKdV equation (17.4) is reduced to

$$w''' + 6ww' + Aw - \frac{1}{2}B = 0,$$

which is equivalent to either the Painlevé I or the Weierstrass elliptic function equation.

We have also found some new symmetry reductions and similar solutions for two kinds of variant Boussinesq equations. The Duffing equation and Painlevé II type equations are reduced to ODEs. Both poles and algebraic branch points for t can be entered into the solutions of variant Boussinesq equations. For details, see Fan (1998).

17.6 Further study on analytical solutions of non-linear PDEs

According to the general rules in the previous sections we have constructed various kinds of analytical solutions for non-linear PDEs. Some solitary wave solutions for a class of non-linear wave equation

$$u_{tt} - a_1 u_{xx} + a_2 u_t + a_3 u + a_4 u^3 = 0$$

are obtained by homogeneous balance method and Wu's method; see Fan and Zhang (1997).

The field equations relevant to the quantum field theory of charged solitons studied by Rajaraman are

$$\sigma_{xx} = -\sigma + \sigma^3 + d\rho^2\sigma, \tag{17.5}$$

and

$$\rho_{xx} = (f - d)\rho + \lambda\rho^3 + d\rho^2\sigma, \tag{17.6}$$

where σ and ρ are scalar fields and d, f, λ are parameters. For certain special values of d, f, and λ, three kinds of solitary wave solutions were obtained by Rajaraman and X. Wang. Using appropriate function transformations, we have found five kinds of new soliton solutions and periodic solutions of equations (17.5) and (17.6) for other special values of d, f, and λ. Applications of these results to other famous equations in physics are also given (Fan and Zhang 1998a).

The homogeneous balance method is very powerful for finding exact solutions of given PDEs. However, it involves a lot of tedious calculation, and can only find soliton-like solutions. In Fan and Zhang (1998b,c,d), we obtained other kinds of new exact solutions for the reaction-diffusion equation, KP equation, KdV and coupled KdV equation by improving some key steps of this approach. We also generalized this method to obtain Bäcklund transformation and milt-similarity solutions for non-linear PDEs (Fan and Zhang 1997, 1998b,c,d,e).

We have presented a simple and efficient method for seeking exact solutions of non-linear PDEs. Some equations were investigated by this means and new traveling wave solutions and singular traveling wave solutions were found. We now describe the method. For a given PDE in two variables

$$H(u, u_x, u_t, u_{xx}, \ldots) = 0, \tag{17.7}$$

we first consider its traveling solutions $u(x,t) = u(\xi)$, $\xi = x - \lambda t$. Then (17.7) becomes an ODE, which is integrated as long as all terms contain derivatives. The associated integration constants can be taken as zero. The next step is to find the solutions that can be expressed as

$$u(\xi) = \sum_{i=0}^{m} a_i v^i(\xi), \tag{17.8}$$

and the Riccati equation

$$v' = b_1 + b_2 v^2 \tag{17.9}$$

or Bernoulli equation

$$v' = b_1 v + b_2 v^n \tag{17.10}$$

is taken as the target equation. The parameter m is found by balancing the linear terms of highest order with the non-linear terms. Substituting (17.8) into the relevant ODE and using (17.9) or (17.10) yields a set of algebraic equations with respect to a_i $(i = 1, \ldots, m)$, k, and λ, because the coefficients of v^i have to vanish. For these relations, a_i, k, λ can be determined by Wu–Ritt's characteristic set method. The Riccati equation (17.9) has general solution

$$v = -\frac{\sqrt{\pm b_1 b_2}}{2b_2}\tanh(\sqrt{\pm b_1 b_2}\,\xi), \qquad v = -\frac{\sqrt{\pm b_1 b_2}}{2b_2}\coth(\sqrt{\pm b_1 b_2}\,\xi).$$

The Bernoulli equation (17.10) has solutions

$$v = -\left\{\frac{b_1}{2b_2}\tanh\left[\frac{b_1(n-1)}{2}\xi\right] - \frac{b_1}{2b_2}\right\}^{\frac{1}{n-1}},$$

$$v = -\left\{\frac{b_1}{2b_2}\coth\left[\frac{b_1(n-1)}{2}\xi\right] - \frac{b_1}{2b_2}\right\}^{\frac{1}{n-1}}.$$

Example 17.10:

For the Burgers–KdV equation

$$u_t + uu_x - \alpha u_{xx} + \beta u_{xxx} = 0,$$

its new solitary wave solutions and singular traveling wave solutions are obtained in Fan and Zhang (1998g).

Example 17.11:

For the seventh-order KdV equation

$$u_t + uu_x + u_{3x} - u_{5x} + eu_{7x} = 0,$$

four kinds of traveling wave solutions are obtained in Fan (1998).

With the help of Wu's method, we have solved many non-linear PDEs by taking Riccati or Bernoulli equations as target equations. Many new traveling wave solutions are found. We will study to choose other solvable equations as target equations.

17.7 Painlevé test and Bäcklund transformations for non-linear PDEs

It is well known that there exist two approaches to check the Painlevé property for non-linear PDEs. The first is the ARS algorithm (Ablowitz et al. 1980), which requires that every ODE reduction of a PDE is of the Painlevé type. An ODE has the Painlevé property if none of its solutions has movable critical points. The second one is the WTC method (Weiss et al. 1983), which directly attacks the PDEs. A PDE has the Painlevé property if its solutions are single-valued about a movable singularity manifold. According to the ARS algorithm, Zhu and Shi (1997) used Wu's method and symbolic computation to analyze singular points of the Painlevé test for PDEs. The key of this method is to get every ODE reduction of a PDE. However, it is often very difficult to do that, because there are no general methods for identifying whether we have found all reduction ODEs for a given PDE. We have applied symbolic computation to the Painlevé test. Using the WTC method, we show that two kinds of variant Boussinesq equations all possess the Painlevé property.

The first variant of Boussinesq equations is written as

$$\begin{cases} u_t + (uv)_x + v_{xx} = 0, \\ v_t + u_x + vv_x = 0. \end{cases} \tag{17.11}$$

To check the Painlevé property by means of the WTC method, we make the ansätz that u and v can be expanded about the singularity manifold $\phi(x,t)=0$ as

$$\begin{cases} u = \dfrac{1}{\phi^{\alpha}} \displaystyle\sum_{j=0}^{\infty} u_j \phi^j, \\ v = \dfrac{1}{\phi^{\beta}} \displaystyle\sum_{j=0}^{\infty} v_j \phi^j. \end{cases} \tag{17.12}$$

It is not necessary to investigate the existence of infinite u_j and v_j $(j = 0, 1, \ldots)$, because they can be determined by recursion relations except on "resonances." So it is sufficient to check the compatible condition on resonance. The algorithm takes the following steps.

W1. Substituting $u \sim u_0\phi^{-\alpha}$ and $v \sim u_0\phi^{-\beta}$ into (17.11) with leading analysis yields

$$\alpha = 2, \qquad \beta = 1, \qquad u_0 = -2\phi_x^2, \qquad v_0 = 2\phi_x.$$

W2. Substituting $u = -2\phi_x^2\phi^{-2}$ and $v = 2\phi_x\phi^{-1} + v_j\phi^{j-1}$ into the main terms $(uv)_x + v_{xxx}$ in the first equation and $u_x + vv_x$ in the second equation of (17.11), we obtain the coefficients of u_j and v_j:

$$-j(j-3)^2\phi_x^3 v_j, \qquad -(j-2)\phi_x u_j,$$

which imply that resonances are $j = 0, 2, 3$, but u_0 and v_0 have been determined in the case $j = 0$. Thus, we require that u and v be expanded to terms for $j = 3$ in (17.12).

W3. Substituting

$$u = -2\phi_x^2\phi^{-2} + u_1\phi^{-1} + u_2 + u_3\phi, \qquad v = 2\phi_x\phi^{-1} + v_1 + v_2\phi + u_3\phi^2$$

into (17.11), we obtain a set of equations by equating the coefficients of the same powers, and we know that the compatible conditions are satisfied. Therefore, the variant Boussinesq system (17.11) possesses the Painlevé property. In particular, by virtue of the truncation of the Painlevé expansion at ϕ^0 term, we get the Bäcklund transformation

$$u = 2\frac{\partial^2}{\partial x^2}\ln\phi + u_2, \qquad v = 2\frac{\partial}{\partial x}\ln\phi + v_1,$$

where ϕ, u_2, v_1 satisfy

$$\begin{aligned}
&\phi_t + v_1\phi_x + \phi_{xx} = 0, \qquad u_2 = v_{1x},\\
&u_{2t} + (u_2v_1)_x + v_{1xx} = 0,\\
&v_{1t} + u_{2x} + v_1v_{1x} = 0.
\end{aligned}$$

Using the method presented in Fan and Zhang (1998f), we have obtained Bäcklund transformations for a large class of non-linear PDEs (Fan and Zhang 1998e,h; Zhang and Fan 1997).

Example 17.12:

For the KdV equation

$$u_t + h_1(6uu_x + u_{xxx}) + 4h_2u_x - h_3(2u + xu_x) = 0,$$

its Bäcklund transformation is

$$u = 2\frac{\partial^2}{\partial x^2}\ln\phi + u_1,$$

where ϕ satisfies

$$\begin{aligned}
&\phi_{xt} + h_1\phi_{xxxx} + 4h_2\phi_{xx} - xh_3\phi_{xx} - h_3\phi_x + 6h_1u_1\phi_{xx} = 0,\\
&\phi_x\phi_t + 4h_1\phi_x\phi_{xxx} - 3h_1\phi_{xx}^2 + 4h_2\phi_x^2 - xh_3\phi_x^2 + 6h_1u_1\phi_x^2 = 0,\\
&u_{1t} + h_1(6u_1u_{1x} + u_{1xxx}) + 4h_2u_{1x} - h_3(2u_1 + xu_{1x}) = 0.
\end{aligned}$$

17.8 Symmetries and Hamiltonian structure of non-linear PDEs

Symmetries play an important role in the study of non-linear PDEs. Since Olver (1986) proved a general theorem about recursion operators (strong symmetry) for the symmetries of PDEs, the study of the bi-Hamiltonian structure, symmetries, and strong symmetries of non-linear PDEs has received much attention. An elegant geometrical theory of finite-dimensional Hamiltonian systems has been developed since the late 1960s. The well-known Liouville–Arnold theorem shows that a Hamilton system with n degrees of freedom is completely integrable if it possesses n integrals, which are involution in pairs. Much progress has been made from the 1980s onwards. However, there is still a long way to go. In recent years, the determination of symmetry groups and Hamiltonian structure of PDEs has come to prominence through the work of Wu (1984b, 1986a, 1987e, 1989b), Cartan (1952), Ritt (1950), Bluman and Kumei (1989), and Schwarz (1985, 1988, 1992).

In recent years, we have studied algorithms for constructing symmetries (Fan and Zhang 1998e; Chao 1997; Zhang and Chao 1997, 1998). Our main work includes

- applying Wu–Ritt's characteristic set method in constructing symmetries of non-linear PDEs,
- proposing the automated reasoning method based on Wu–Ritt's characteristic set method,
- developing a method for constructing and identifying the Hamiltonian structure of non-linear PDEs (Chao 1997),
- presenting a specific computation formula to construct the Hamilton structure of non-linear PDEs, and
- deriving a necessary and sufficient condition for constructing Hamiltonian structure of non-linear PDEs by a division algorithm introduced by ourselves.

A few examples are given below. For plane stress equations

$$\frac{\partial \sigma_x}{\partial x} + \frac{\partial \tau_{xy}}{\partial y} = 0, \qquad \frac{\partial \sigma_y}{\partial y} + \frac{\partial \tau_{xy}}{\partial x} = 0, \qquad \Delta(\sigma_x + \sigma_y) = 0,$$

their symmetries are

$$X_1 = -y\frac{\partial}{\partial x} + x\frac{\partial}{\partial y} - 2\tau_{xy}\frac{\partial}{\partial \sigma_x} + 2\tau_{xy}\frac{\partial}{\partial \sigma_y} + (\sigma_x - \sigma_y)\frac{\partial}{\partial \tau_{xy}}, \quad X_2 = \frac{\partial}{\partial x},$$

$$X_3 = x\frac{\partial}{\partial x} + y\frac{\partial}{\partial y}, \quad X_4 = \frac{\partial}{\partial y}, \quad X_5 = \tau_x\tau_{xy}\frac{\partial}{\partial \sigma_x} + \sigma_y\tau_{xy}\frac{\partial}{\partial \sigma_y} + \tau_x\frac{\partial}{\partial \tau_{xy}}.$$

For the vibrant equation system,

$$\rho h\left(\frac{\partial^2 u}{\partial t^2} + \frac{h^2}{12R}\frac{\partial^2 r_x}{\partial t^2}\right) = c_{11}\frac{\partial^2 u}{\partial x^2} + c_{66}\frac{\partial^2 u}{\partial y^2} + (c_{12} + c_{66})\frac{\partial^2 v}{\partial x \partial y} + \frac{c_{12}}{R}\frac{\partial \omega}{\partial x},$$

$$\rho h\left(\frac{\partial^2 u}{\partial t^2} + \frac{h^2}{12R}\frac{\partial^2 r_y}{\partial t^2}\right) = c_{66}\frac{\partial^2 v}{\partial x^2} + c_{22}\frac{\partial^2 v}{\partial y^2} + (c_{12} + c_{66})\frac{\partial^2 u}{\partial x \partial y} + \frac{c_{22}}{R}\frac{\partial \omega}{\partial y},$$

$$\rho h\frac{\partial^2 \omega}{\partial t^2} = c_{13}\frac{\partial^2 \omega}{\partial x^2} + c_{23}\frac{\partial^2 \omega}{\partial y^2} - \frac{c_{22}}{R^2}\omega - \frac{c_{12}}{R}\frac{\partial u}{\partial x} - \frac{c_{22}}{R}\frac{\partial v}{\partial x} - c_{13}\frac{\partial r_x}{\partial x} + c_{23}\frac{\partial r_x}{\partial y},$$

$$\frac{\rho h^3}{12}\left(\frac{\partial^2 r_x}{\partial t^2}+\frac{1}{R}\frac{\partial^2 u}{\partial t^2}\right)$$
$$=D_{11}\frac{\partial^2 r_x}{\partial x^2}+D_{66}\frac{\partial^2 r_x}{\partial y^2}-c_{13}r_x+(D_{12}+D_{66})\frac{\partial^2 r_x}{\partial x\partial y}-c_{13}\frac{\partial \omega}{\partial x},$$
$$\frac{\rho h^3}{12}\left(\frac{\partial^2 r_x}{\partial t^2}+\frac{1}{R}\frac{\partial^2 u}{\partial t^2}\right)$$
$$=D_{66}\frac{\partial^2 r_x}{\partial x^2}+D_{22}\frac{\partial^2 r_x}{\partial y^2}-c_{23}r_y+(D_{12}+D_{66})\frac{\partial^2 r_x}{\partial x\partial y}-c_{23}\frac{\partial \omega}{\partial y},$$

which have Hamilton structure by our mechanical method

$$\dot{P}=-\frac{\delta H}{\delta Q},\qquad \dot{Q}=-\frac{\delta H}{\delta P},$$

where

$$H=H(\dot{P},\dot{Q})=\int[PQ-\frac{1}{2}L(QQ)]\,dx\,dy,$$

$$\dot{Q}=\begin{pmatrix}\dot{q}_1\\ \dot{q}_2\\ \dot{q}_3\\ \dot{q}_4\\ \dot{q}_5\end{pmatrix}=\begin{pmatrix}\frac{1}{c_{66}}\left(P_1-\frac{c_{12}+c_{66}}{2}\frac{\partial v}{\partial x}\right)\\ \frac{1}{c_{22}}\left(P_2-\frac{c_{12}+c_{66}}{2}\frac{\partial v}{\partial x}-\frac{c_{22}}{2R}\right)\\ \frac{1}{c_{23}}\left(P_3-\frac{c_{22}}{2R}v-\frac{c_{23}}{2}r_y\right)\\ \frac{1}{D_{66}}\left(P_4-\frac{D_{12}+D_{66}}{2}\frac{\partial r_y}{\partial x}\right)\\ \frac{1}{D_{22}}\left(P_5-\frac{D_{12}+D_{66}}{2}\frac{\partial r_x}{\partial x}+c_{23}w\right)\end{pmatrix}.$$

Acknowledgments. The authors thank Ziming Li for his help in improving the presentation of this chapter.

Chapter 18

EQUATION SOLVING IN ROBOTICS AND MECHANISMS

Qizheng Liao

Wu's method and ideas were introduced into the fields of robotics and mechanisms in about 1990 and have been successfully used to solve a number of problems such as in inverse kinematics of serial chain manipulators, forward kinematics of parallel chain manipulators, synthesis of planar four-bar mechanisms, and reverse force analysis of spring system compliant mechanisms. This chapter reviews some of the problems solved in this way: the inverse kinematics of several kinds of manipulators, the forward kinematics of the 3-TPS parallel manipulator, the forward kinematics of the general 6-6 Stewart platform, the five position synthesis of the planar four-bar linkage, and the reverse force analysis of two kinds of three-spring systems. In particular, the Stewart platform is the foundation for the so-called *NC machine tools* of the 21st century. The results obtained with these methods may have potential applications to enhancing the accuracy of this kind of NC machines tool.

18.1 Introduction

In the research of robotics and mechanisms there are two kinds of problems. One is kinematic analysis, and the other is synthesis. The first one consists of three steps: displacement, velocity, and acceleration analyses. Among them the displacement analysis is the most difficult one, because it deals with non-linear equations; the velocity and acceleration analyses normally deal with linear equations, so they are relatively easy. In this chapter only displacement analysis is considered for the PUMA-type and the Intelledex660T industrial manipulators, the 3-TPS, and the 6-6 Stewart platform mechanisms.

Robot manipulators are normally driven by actuators. The position and orientation (or the displacement) of the hand are determined by the displacement of the actuators. The purpose of the displacement analysis is to find the

MATHEMATICS MECHANIZATION AND APPLICATIONS
ISBN 0-12-734760-7

relations between the two kinds of displacement, the hand's and the actuator's displacement. The forward problem is to find the position and orientation of the manipulator when the displacement of the actuator is given, while the inverse one is just the opposite. Traditionally robot manipulators, like the PUMA560 that is widely used in industry, are series of chain mechanisms. The forward displacement analysis of this kind of manipulator is relatively easy but the inverse one is difficult. Attention has been focused on parallel manipulators such as 3-TPS and the Stewart platform, for they have many new advantages such as high stiffness and more accuracy. But, unlike for serial chains, the forward displacement analysis is difficult while the inverse one is easy. In this chapter the two difficult problems, the inverse displacement analysis of the serial chains (such as the PUMA-type and the Intelledex660T) and the forward displacement analysis of the parallel chains (such as 3-TPS and the 6-6 Stewart platform) are discussed.

Except for the analysis of mechanisms, the synthesis is also discussed in this chapter. This is finding the dimensions of the mechanisms when some of the properties of it are given. In fact this is the first step in the design of a machine. Because of the complexity of this problem, in this chapter only the simplest case — the five position synthesis of the planar four-bar mechanism — is considered. The purpose of this problem is to find a group of dimensions of a planar four-bar mechanism in order that the link can pass through five given positions; usually it is called *Burmester's problem*, and here it is solved by using Wu's method.

Spring systems can be used in manipulators as a compliant wrist to help machine part assembly or as a key part when using the so-called *kinestatic control*, that is, the simultaneous regulation of force and displacement solely via the control of displacement. There are two problems in the analysis of spring systems. One is the forward static force analysis to find the total force when the position of the spring system is given. The reverse static force analysis is to find the position of the system when all the forces are given; this problem is difficult. In this chapter we consider the problem of the reverse force analysis for two systems: one is a planar three-spring system, which is based on the analysis of planar two-spring systems, and the other is a spatial three-spring system, which is the simplest case of three dimensions.

18.2 Inverse kinematics of serial chain manipulators

Serial chain mechanisms, such as manipulators, are widely used in industry. The inverse kinematics of serial chain mechanisms is difficult but necessary for the control. Most manipulators used in industry are designed to be of special dimensions, such as parallel of axis or of zero link length; this can greatly reduce the complexity of the inverse kinematics. The general 6R manipulator is among the most difficult cases; it was solved by Liao et al. (1986). At that time Wu's method had not been introduced into this field, and researchers were busy working on the parallel chain mechanisms.

Since Wu's method was introduced into the area of robotics in the 1990s, most of the work has been done on the parallel chains using the method and Wu's ideas. Wu (1989e) tried the PUMA-type manipulator and got some results. He also used his method in a simpler case, the PUMA560 manipulator, and the results obtained agree with those of Paul (1981a).

Yang, Fu, and Zeng (1997) did some work in the more general case and got results for several industrial manipulators, such as PUMA-type and Intelledex660T. It seems that their method can be used in more complicate cases. This section reviews the work of Wu and Yang and his co-workers.

18.2.1 Inverse kinematics of PUMA type and PUMA560 manipulators

The following is Wu's work.

By a PUMA-type robot we mean a 6R-robot for which the joint-axes J_i are such that

$$J_1 \perp J_2 \| J_3 \perp J_4 \perp J_5 \perp J_6$$

(where $\|, \perp$ mean *parallel to* and *orthogonal to* respectively, and J_2, J_3 are non-coincident). The usual PUMA560, for example, is such a PUMA-type robot that has some further geometrical restrictions: J_2 intersects $J_1, J_5 \| J_2$, and J_4, J_5, J_6 are concurrent.

We use the standard notations as in Paul (1981a) or Paul (1981b) with only slight modifications. The link lengths are A_i and the link distances are D_i. In particular, we have $A_6 = 0$ and $D_6 = 0$. For the PUMA-type manipulators the twist angles α_i are taken to be

$$\frac{\pi}{2}, 0, \frac{\pi}{2}, \frac{\pi}{2}, \frac{\pi}{2}, 0.$$

The 4×4 transformation matrix of frame FR_j to frame FR_i will be denoted by A_{ij} and that of the frame FR_6 associated to the end-effector to the frame FR_0 associated to the base by

$$E = (E_{ji})$$

with E_{ji} the element in the ith row and jth column for $i = 1, 2, 3$ and $j = 1, \ldots, 4$, and the fourth row $(0, 0, 0, 1)$. The inverse kinematic equation (IKE) of the robot is given by

$$A_{01}A_{12}A_{23}A_{34}A_{45}A_{56} = E. \tag{18.1}$$

To solve the equation we introduce the matrices $U = A_{32}A_{21}A_{10}$, $V = A_{34}A_{45}A_{56}$, and $W = V - UE$. Let W_{ji} be the element of W in the ith row and the jth column and set $W_{ji} = P_k$, with $k = 3(j-1) + i$ for $i = 1, 2, 3$ and $j = 1, \ldots, 4$. Then the IKE reduces to a set of equations $P_k = 0, k = 1, \ldots, 12$, as (18.2).

For simplicity of notation, let us introduce the following polynomials F_{ij}:

$$\begin{aligned}
&F_{i1} = S_1E_{i2} + C_1E_{i1}, \qquad F_{i2} = -S_1E_{i1} + C_1E_{i2}, \quad (i = 1, 2, 3)\\
&F_{41} = S_1E_{42} + C_1E_{41} - A_1,\\
&F_{42} = -S_1E_{41} + C_1E_{42} + D_{23}, \qquad (D_{23} = D_2 + D_3)\\
&F_{43} = E_{43} - D_1.
\end{aligned}$$

Let S_{23} and C_{23} denote the sine and cosine of the angle-sum $\theta_2 + \theta_3$. We consider C_i, S_i $(i = 1, 3, \ldots, 6)$, and C_{23}, S_{23} as variables connected by the relations $C_i^2 + S_i^2 = 1$, etc. Let the order of the variables be

$$S_1 \prec C_1 \prec S_{23} \prec C_{23} \prec S_4 \prec C_4 \prec S_5 \prec C_5 \prec S_6 \prec C_6 \prec S_3 \prec C_3.$$

With respect to this order we may then write the polynomials P_k $(k = 1, \ldots, 12)$ in the form below:

$$\begin{aligned}
P_1 &= C_6C_5C_4 + S_6S_4 - C_{23}F_{11} - S_{23}E_{13},\\
P_2 &= C_6C_sS_4 - S_6C_4 + F_{12},\\
P_3 &= C_6S_5 + C_{23}E_{13} - S_{23}F_{11},\\
P_4 &= C_6S_4 - S_6C_5C_4 - C_{23}F_{21} - S_{23}E_{23},\\
P_5 &= -C_6C_4 - S_6C_5S_4 + F_{22},\\
P_6 &= -S_6S_5 + C_{23}E_{23} - S_{23}F_{21},\\
P_7 &= S_5C_4 - C_{23}F_{31} - S_{23}E_{33},\\
P_8 &= S_5S_4 + F_{32},\\
P_9 &= -C_5 + C_{23}E_{33} - S_{23}F_{31},\\
P_{10} &= C_3A_2 + C_5C_4A_5 + C_4A_4 + S_4D_5 - C_{23}F_{41} - S_{23}F_{43} + A_3,\\
P_{11} &= C_5S_4A_5 - C_4D_5 + S_4A_4 + F_{42},\\
P_{12} &= S_3A_2 + S_5A_5 + C_{23}F_{43} - S_{23}F_{41} + D_4.
\end{aligned} \tag{18.2}$$

In addition, we have

$$\begin{aligned} P_{13} &= C_1^2 + S_1^2 - 1, \\ P_{14} &= C_{23}^2 + S_{23}^2 - 1, \\ P_{11+i} &= C_i^2 + S_i^2 - 1, \quad (i = 4, 5, 6) \\ P_{18} &= C_3^2 + S_3^2 - 1. \end{aligned} \tag{18.3}$$

Our problem of solving the IKE amounts now to the determination of the zero set Zero(PS), where PS is the set consisting of the above 18 polynomials P_k, $k = 1, \ldots, 18$, in (18.2) and (18.3). Applying the general algorithm (somewhat modified) for the determination of a characteristic set of PS, we find during the procedure a polynomial set PS' consisting of 12 polynomials P'_k, $k = 1, \ldots, 12$, such that

$$\mathrm{Zero}(PS) = \mathrm{Zero}(PS'/S_4F_{32}) \cup \mathrm{Zero}(PS \cup \{S_4\}) \cup \mathrm{Zero}(PS \cup \{F_{32}\}).$$

The polynomials P'_k are as follows:

$$\begin{aligned} P'_1 &= P_{13}, \qquad P'_2 = P_{14}, \\ P'_3 &= C_4F_{32} + S_4C_{23}F_{31} + S_4S_{23}E_{33}, \\ P'_4 &= S_5 - C_4C_{23}F_{31} - C_4S_{23}E_{33} + S_4F_{32}, \\ P'_5 &= C_5 - C_{23}E_{33} + S_{23}F_{31}, \\ P'_6 &= S_6 - C_4F_{12} - S_4C_{23}F_{11} - S_4S_{23}E_{13}, \\ P'_7 &= C_6 - C_4F_{22} - S_4C_{23}F_{21} - S_4S_{23}E_{23}, \\ P'_8 &= S_3A_2 + S_5A_5 + C_{23}F_{43} - S_{23}F_{41} + D_4, \\ P'_9 &= C_3A_2 + C_5C_4A_5 + C_4A_4 + S_4D_5 - C_{23}F_{41} - S_{23}F_{43} + A_3, \\ P'_{10} &= S_4(C_{23}F_{32}E_{33}A_5 + C_{23}F_{31}D_5 - S_{23}F_{32}F_{31}A_5 \\ &\quad +S_{23}E_{33}D_5 + F_{32}A_4) + F_{42}F_{32}, \\ P'_{11} &= S_4^2[(C_{23}F_{31} + S_{23}E_{33})^2 + F_{32}^2] - F_{32}^2, \\ P'_{12} &= S_4^2(C_{23}^4F_{31}^4A_5^2 + \cdots) + \cdots. \end{aligned} \tag{18.4}$$

The last polynomial P'_{12} in (18.4) has 69 terms in all.

Consider the particular PUMA-type robot, the PUMA560, where

$$D_5 = A_5 = A_4 = 0.$$

Then P'_{10} and P'_{12} in (18.4) are simplified to

$$\begin{aligned} P'_{10} &= F_{42}F_{32}, \\ P'_{12} &= F_{32}^2[2C_{23}(F_{43}D_4 - F_{41}A_3) - 2S_{23}(F_{43}A_3 + F_{41}D_4) \\ &\quad +(F_{43}^2 + F_{41}^2 + 2D_4 + A_3^2 - A_2^2)]. \end{aligned}$$

For $F_{32} \neq 0$, we have $F_{42} = 0$ or $-S_1E_{41} + C_1E_{42} = -D_{23}$, which may then be solved for θ_1 unless $E_{41} = E_{42} = 0$ (in case $D_{23} = 0$); the excluded case corresponds to some so-called singular position of the robot. Suppose that this case does not occur. Then from P'_{12} we may solve for θ_{23} unless $F_{43}D_4 - F_{41}A_3$ and $F_{43}A_3 + F_{41}D_4$ are both 0; the excluded case is again supposed not to occur. Then P'_3 will give θ_4, P'_4 and P'_5 give θ_5, P'_6 and P'_7 give θ_6, and finally P'_8 and P'_9 give θ_3 and hence also θ_2. We have thus determined the part $\text{Zero}(PS'/S_4F_{32})$ which corresponds to the solutions given by Paul (1981b) for the particular PUMA-type robot PUMA560. On the other hand, the parts $\text{Zero}(PS \cup \{S_4\})$ and $\text{Zero}(PS \cup \{F_{32}\})$ will give solutions corresponding to the singular points or dead points of the robot in question.

18.2.2 Inverse kinematics of 6R manipulators with simple geometry

Yang, Fu, and Zeng (1997) solved the inverse kinematics problem for several industrial manipulators, such as Puma 560, Puma 762, and Intelledex660T. Their method is based on the solution of general 6R manipulators. It can be divided into seven steps Y1–Y7.

Y1. For a special manipulator, express all non-zero link lengths and offset distances as symbols a_i and d_i with twist angles $\alpha_i = 0$ or $\pm 90°$, and then construct the matrices A_i, $i = 1, \ldots, 6$.

Y2. Given arbitrary position $[x, y, z]$ and orientation $[\mathbf{u}, \mathbf{v}, \mathbf{w}]$ of the hand relative to the manipulator's base, construct A_h. Then, like (18.1), the closure equation can be rewritten as the following matrix equations:

$$\begin{aligned}
&A_3A_4A_5 = A_2^{-1}A_1^{-1}A_hA_6^{-1},\\
&A_2A_3A_4 = A_1^{-1}A_hA_6^{-1}A_5^{-1},\\
&A_1A_2A_3 = A_hA_6^{-1}A_5^{-1}A_4^{-1},\\
&A_6A_h^{-1}A_1A_2 = A_5^{-1}A_4^{-1}A_3^{-1},\\
&A_5A_6A_h^{-1}A_1 = A_4^{-1}A_3^{-1}A_2^{-1},\\
&A_4A_5A_6A_h^{-1} = A_3^{-1}A_2^{-1}A_1^{-1}.
\end{aligned}$$

Y3. For the above six matrix equations, write the first three elements of the third columns as a vector $\mathbf{L}_k$ and form the first three elements of the fourth columns into a vector $\mathbf{P}_k$, where $k = 1, \ldots, 6$. Hence every pair of $\mathbf{P}_k$ and $\mathbf{L}_k$ involves at most five variables $(\theta_{k1}, \theta_{k2}, \theta_{k3}, \theta_{k4}, \theta_{k5})$.

Y4. For each pair of $\mathbf{L}_k$ and $\mathbf{P}_k$, construct the following vector and scalar products to obtain fourteen scalar equations:

$$\begin{aligned}
&\mathbf{P}_k = 0,\\
&\mathbf{L}_k = 0,\\
&\mathbf{P}_k \cdot \mathbf{P}_k = 0,\\
&\mathbf{P}_k \cdot \mathbf{L}_k = 0,\\
&\mathbf{P}_k \times \mathbf{L}_k = 0,\\
&(\mathbf{P}_k \cdot \mathbf{P}_k)\mathbf{L}_k - (2\mathbf{P}_k \cdot \mathbf{L}_k)\mathbf{P}_k = 0,
\end{aligned}$$

where $k = 1, \ldots, 6$. Then each set of equations above is written in the following matrix form:

$$Q(s_{k1}s_{k2} \;\; s_{k1}c_{k2} \;\; c_{k1}s_{k2} \;\; c_{k1}c_{k2} \;\; s_{k1} \;\; c_{k1} \;\; s_{k2} \;\; c_{k2})^T = b_k,$$

where Q is a 14×8 matrix whose elements are all parameters, θ_{k1} and θ_{k2} are the two variables on the left-hand side of the above kth matrix equation, and vector b_k involves the other three variables $\theta_{k3}, \theta_{k4}, \theta_{k5}$.

Y5. For the above six sets of linear equations, by Gaussian elimination one can get six sets of condition equations, of which each set involves at most three variables $(\theta_{k3}, \theta_{k4}, \theta_{k5})$ and holds the zeros of the original closure equation. To simplify symbolic computation, extract a set of equations from the six sets of condition equations whose symbolic expression is the simplest. Generally one can assume that the set of equations is

$$\begin{aligned}
&f_1(c_{k3}, c_{k4}, c_{k5}, s_{k3}, s_{k4}, s_{k5}) = 0,\\
&f_2(c_{k3}, c_{k4}, c_{k5}, s_{k3}, s_{k4}, s_{k5}) = 0,\\
&f_3(c_{k3}, c_{k4}, c_{k5}, s_{k3}, s_{k4}, s_{k5}) = 0.
\end{aligned}$$

Solving the linear equations $f_1 = 0, f_2 = 0$ for c_{k3}, s_{k3} and then substituting the solutions into $f_3 = 0$ and $c_{k3}^2 + s_{k3}^2 - 1 = 0$, one gets two equations:

$$\begin{aligned}
&f_4(c_{k4}, c_{k5}, s_{k4}, s_{k5}) = 0,\\
&f_5(c_{k4}, c_{k5}, s_{k4}, s_{k5}) = 0.
\end{aligned}$$

Using the substitutions

$$c_{k4} = \frac{1-t^2}{1+t^2}, \qquad s_{k4} = \frac{2t}{1+t^2}, \qquad c_{k5} = \frac{1-r^2}{1+r^2}, \qquad s_{k5} = \frac{2r}{1+r^2},$$

one obtains two polynomial equations in two variables r, t:

$$f(t,r) = 0,$$
$$g(t,r) = 0.$$

Y6. Solve the above simultaneous equations in two unknowns by the resultant algorithm. Here the Bézout resultant may be used instead of the usual Sylvester resultant to reduce the symbolic computations greatly.

Y7. After $c_{k3}, c_{k4}, c_{k5}, s_{k3}, s_{k4}, s_{k5}$ have been solved from a set of condition equations, the other three variables can be linearly solved out from the corresponding closure equation.

The above method was used to solve the equations for PUMA-type, PUMA560, PUMA762, and Intelledex660T robotic manipulators, and the desired results were obtained. The running times are 31, 39, 33, and 27 seconds, respectively.

18.3 Forward kinematics of parallel manipulators

After the inverse kinematics of the general serial chain manipulator was solved, researchers began to direct their attention to parallel manipulators. In the research of parallel manipulators the difficult problem becomes solving the forward kinematics. Especially for the general case, so far the general 6-6 in-parallel platform as shown in Fig. 18.1 is still open. Various simple cases are discussed by many researchers. The simplest case is the 3-TPS manipulator (Figs. 18.8–18.10), which has been solved by Liao and Liang (1990) using the resultant elimination. Later it was also solved by Han et al. (1998a) using Wu's method (Chap. 1). The resulting univariate equation is given by a polynomial of degree 8. Some other cases are also discussed, e.g., in Yin and Liang (1994) and Xiao and Liang (1992). A more complex case — the 6-6 planar shaped Stewart platform with some special dimensions as shown in Fig. 18.11 — was discussed by Huang and Wu (1991, 1992); a polynomial equation of degree 14 was obtained. After that Wen and Liang (1994) solved the general 6-6 Stewart platform with coplanar shaped platforms (Fig. 18.12) by using the ball-joint coordinates as variables. They set up 18 equations in 18 unknowns, eliminated 17 of them in three steps, and finally obtained a resultant as the determinant of a 21×21 matrix. Then, by numerical calculation, this determinant is reduced to a 10×10 determinant that can be expanded to a polynomial of degree

20. Afterwards, Wu and Huang (1996) improved Wen and Liang's method using symbolic calculation and obtained the univariate equation that is the same as Wen and Liang's. Later they used this result and solved the so-called Florida manipulator (Wu and Huang 1998; Fig. 18.7), which is produced by the University of Florida, USA. It deals with a special case of the general Stewart platform, and the result is a univariate equation of degree 8.

Although various kinds of parallel mechanisms have been solved, the general case — the general 6-6 in-parallel mechanism which has no planar shaped restrictions — remains unsolved. So far the general case, Fig. 18.1, together with a somewhat simpler case in which one of the base and the moving platform is planar shaped and the other has no restrictions, and Fig. 18.2 have yet to be solved.

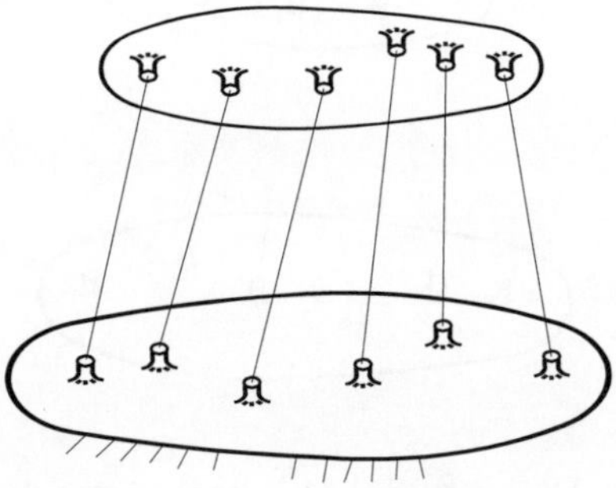

Figure 18.1 A 6-6 in-parallel mechanism

Besides the planar shaped platforms that have been widely studied, another special case, using combined ball joints, is also considered by many researchers. The 6-3 platform, which was solved by Rong and Liang (1991), results in a

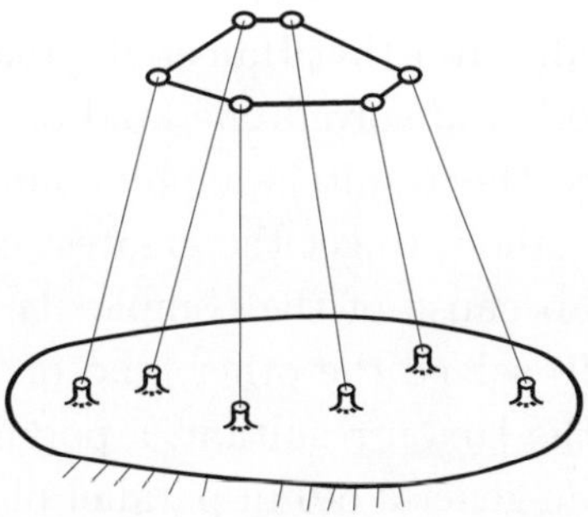

Figure 18.2 A 6-6 in-parallel mechanism with one planar shaped platform.

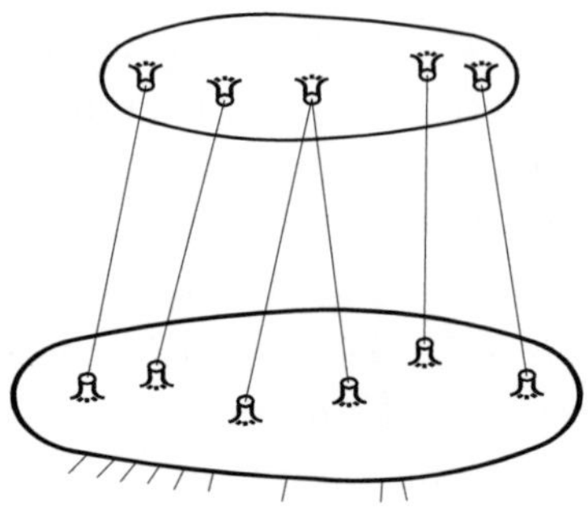

Figure 18.3 A 6-5 in-parallel mechanism.

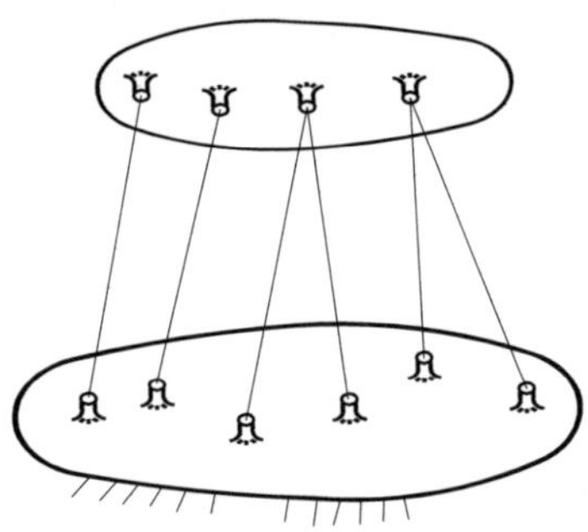

Figure 18.4 A 6-4 in-parallel mechanism.

polynomial equation of degree 16. A more complex case — the general 4-6 in-parallel platform (Fig. 18.4) — was solved by Liao et al. (1995), where the vector elimination method was used. The process is to set up 13 equations in 12 unknowns. After using vector elimination, two equations with only two unknowns are obtained. By computing the resultant, a polynomial equation of degree 32 is finally obtained. After that Han et al. (1998b) used a similar method together with Wu's method, and solved one kind of 5-5 in-parallel platform as shown in Fig. 18.6, where the result is a polynomial equation of degree 24. However, in Han et al.'s method, to get the greatest common divisor, numerical calculation must be used because of the complexity of the equations. At the same time, Su et al. (1999) solved the other kind of 5-5 in-parallel platform as shown in Fig. 18.5 using the Dixon resultant; a polynomial equations of degree 40 was obtained. So far the general 6-6 in-parallel platform (Fig. 18.1) and the 6-5 in-parallel platform (Fig. 18.3) remain unsolved. This is obviously a difficult task.

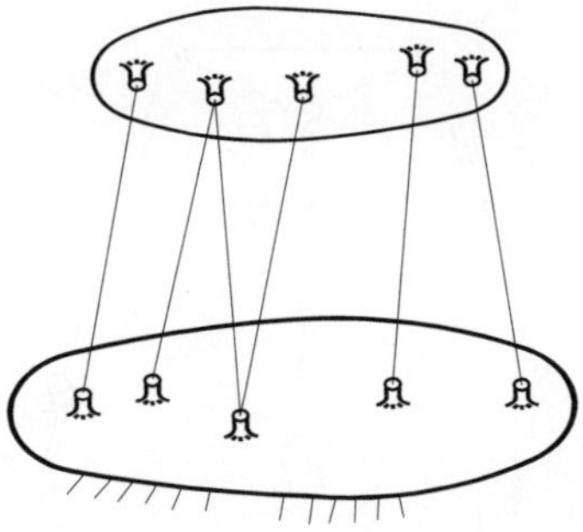

Figure 18.5 A 5-5a in-parallel mechanism.

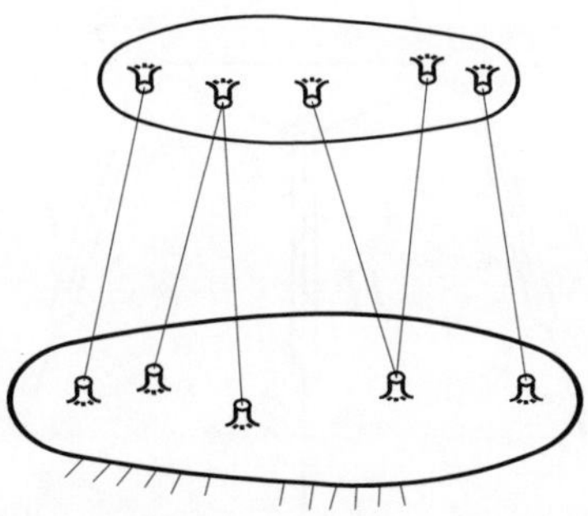

Figure 18.6 A 5-5a in-parallel mechanism.

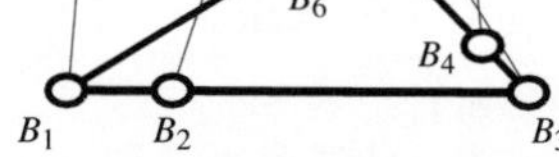

Figure 18.7 A 6-6 Florida manipulator.

In what follows we will review some of the important parts of this topic.

18.3.1 Forward kinematics of 3-TPS parallel manipulators

A *3-TPS* manipulator is a parallel mechanism with two platforms connected by three chains as shown in Fig. 18.8. Each chain consists of a universal joint

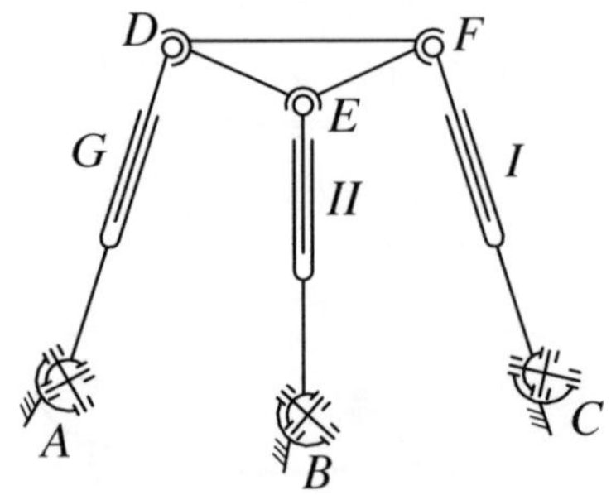

Figure 18.8 A 3-TPS manipulator.

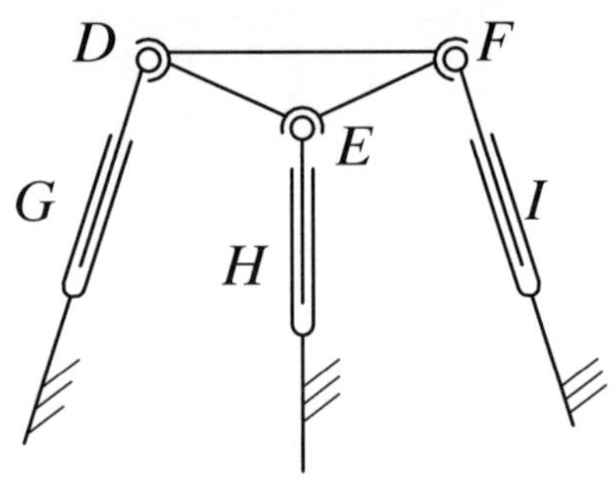

Figure 18.9 A 3-PS structure.

(A, B, C), a prismatic pair (G, H, I), and a spherical joint (D, E, F). The six input drives are at the universal joint, and correspond to the six degrees of freedom of the moving platform. When the six inputs are given, the whole mechanism behaves like the structure as shown in Fig. 18.9. So the forward positional analysis of the 3-TPS mechanism is the same as finding the positions of the 3-PS structure as in Fig. 18.9, where all the six inputs are absorbed by the dimensions of the structure.

As shown in Fig. 18.10, A, B, C, are three fixed points in the three-dimensional space. Their positions are expressed as vectors $\mathbf{A}, \mathbf{B}, \mathbf{C}$. The other three unit vectors $\mathbf{l}$, $\mathbf{m}$, $\mathbf{n}$ are attached along the three legs. We have

$$\mathbf{l} \cdot \mathbf{l} = \mathbf{m} \cdot \mathbf{m} = \mathbf{n} \cdot \mathbf{n} = 1.$$

When the input is given, all the six vectors $\mathbf{A}, \mathbf{B}, \mathbf{C}, \mathbf{l}, \mathbf{m}, \mathbf{n}$ are known parameters. Assume that the three unknowns are

$$x = |AD|, \qquad y = |BE|, \qquad z = |CF|,$$

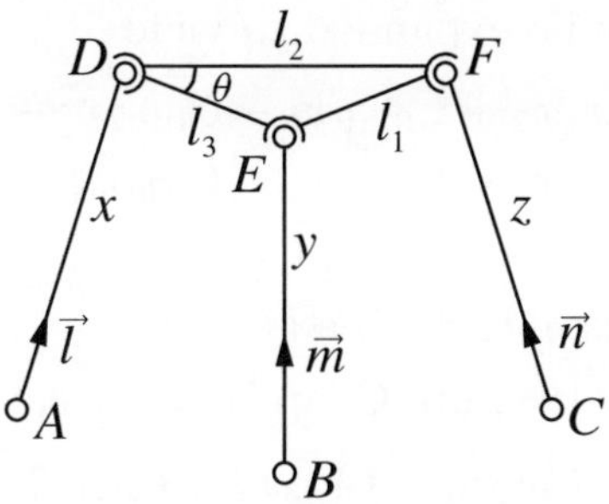

Figure 18.10 The dimension of a 3-PS structure.

where $|AD|, |BE|, |CF|$ are the distances between the pairs of points. Then the vectors **D, E, F** can be expressed as

$$\mathbf{D} = x\mathbf{l} + \mathbf{A}, \tag{18.5}$$

$$\mathbf{E} = y\mathbf{m} + \mathbf{B}, \tag{18.6}$$

$$\mathbf{F} = z\mathbf{n} + \mathbf{C}. \tag{18.7}$$

Because the moving platform is a fixed triangle, the lengths of the three sides are all constants. Thus

$$(\mathbf{D} - \mathbf{F}) \cdot (\mathbf{D} - \mathbf{F}) = l_2^2, \tag{18.8}$$

$$(\mathbf{D} - \mathbf{E}) \cdot (\mathbf{D} - \mathbf{E}) = l_3^2, \tag{18.9}$$

$$2(\mathbf{D} - \mathbf{E}) \cdot (\mathbf{D} - \mathbf{F}) = l_2^2 + l_3^2 - l_1^2. \tag{18.10}$$

Substituting (18.5)–(18.7) into equations (18.8)–(18.10), we get

$$(x\mathbf{l} + \mathbf{A} - z\mathbf{n} - \mathbf{C}) \cdot (x\mathbf{l} + \mathbf{A} - z\mathbf{n} - \mathbf{C}) = l_2^2, \tag{18.11}$$

$$(x\mathbf{l} + \mathbf{A} - y\mathbf{m} - \mathbf{B}) \cdot (x\mathbf{l} + \mathbf{A} - y\mathbf{l} - \mathbf{B}) = l_3^2, \tag{18.12}$$

$$(x\mathbf{l} + \mathbf{A} - y\mathbf{m} - \mathbf{B}) \cdot (x\mathbf{l} + \mathbf{A} - z\mathbf{n} - \mathbf{C}) = l_2^2 + l_3^3 - l_1^2. \tag{18.13}$$

There are only three unknowns x, y, z in the above equations (18.11)–(18.13). Let the unknowns be renamed as

$$x_1 = x, \quad x_2 = z, \quad x_3 = y.$$

Then (18.11)–(18.13) may be expanded to yield

$$\begin{aligned}
P_1 &= C_{0120}x_1^2 + C_{0111}x_1x_2 + C_{0110}x_1 + C_{0102}x_2^2 + C_{0101}x_2 + C_{0100},\\
P_2 &= C_{0220}x_1^2 + C_{0211}x_1x_3 + C_{0210}x_1 + C_{0202}x_3^2 + C_{0201}x_3 + C_{0200},\\
P_3 &= C_{03200}x_1^2 + C_{03110}x_1x_3 + C_{03101}x_1x_2 + C_{03100}x_1 + C_{03011}x_2x_3\\
&\quad + C_{03010}x_3 + C_{03001}x_2 + C_{03000}.
\end{aligned}$$

Using the method of Wu (1984a,b; Chap. 1) they can be triangularized as CS:

$$\begin{aligned}
C_1 &= C_{050} + C_{051}x_1 + C_{052}x_1^2 + C_{053}x_1^3 + C_{054}x_1^4 + C_{055}x_5\\
&\quad + C_{056}x_1^6 + C_{057}x_1^7 + C_{058}x_1^8,\\
C_2 &= C_{0400} + C_{0410}x_1 + C_{0420}x_1^2 + C_{0430}x_1^3 + C_{0440}x_1^4\\
&\quad + (C_{0401} + C_{0411}x_1 + C_{0421}x_1^2 + C_{0431}x_1^3)x_2,\\
C_3 &= C_{0600} + C_{0610}x_1 + C_{0620}x_1^2 + C_{0630}x_1^3 + C_{0640}x_1^4 + C_{0650}x^5\\
&\quad + (C_{0601} + C_{0611}x_1 + C_{0621}x_1^2 + C_{0631}x_1^3 + C_{0641}x_1^4)x_3.
\end{aligned} \tag{18.14}$$

According to the polynomials in (18.14) there are eight sets of solutions, corresponding to the eight configurations or possible positions of the moving platform. An example was given in Han et al. (1998a)

18.3.2 Forward kinematics of the 6-6 Stewart platform

We now discuss the forward kinematics of a 6-6 Stewart platform in three parts. The first is the work of Huang and Wu (1991, 1992); they solved a special case of the 6-6 Stewart platform. The second part is the work of Wen and Liang (1994), who solved the general case of the 6-6 Stewart platform by using the coordinates of the ball joints as variables. The third part is the work of Wu and Huang (1996); they solved the general case of the 6-6 Stewart platform as Wen and Liang did, and they improved Wen and Liang's method by using symbolic calculation. Some other mechanisms, such as 6-4 (Liao et al. 1995), 5-5a (Su et al. 1999) and 5-5b (Han et al. 1998b) in-parallel manipulators, are not discussed here.

18.3.3 Forward kinematics of a special 6-6 Stewart platform

Huang and Wu (1991, 1992) solved a special case of the Stewart platform (Fig. 18.11), where the six ball joints of the base and moving platforms are located on a circle with the following restrictions

$$\begin{aligned}
&B_1B_3 = B_3B_5 = B_5B_1,\\
&B_1B_2 = B_3B_4 = B_5B_6,\\
&P_1P_3 = P_3P_5 = P_5P_1,\\
&P_1P_2 = P_3P_4 = P_5P_6.
\end{aligned}$$

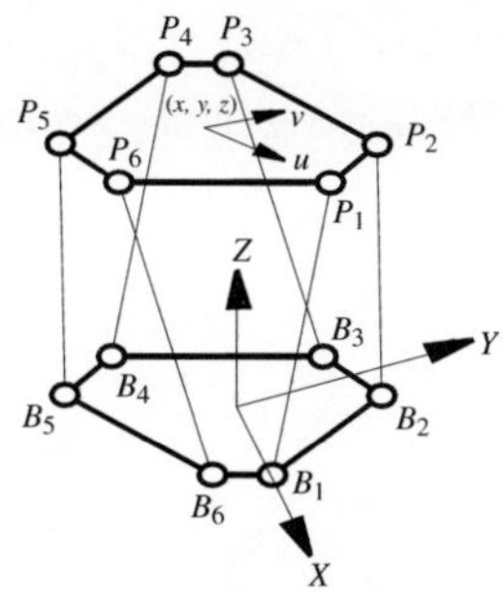

Figure 18.11 6-6 Stewart platform with special dimensions

First they set up nine equations in nine unknowns:

$$\begin{aligned} &u_1^2 + u_2^2 + u_3^2 = 1, \\ &v_1^2 + v_2^2 + v_3^2 = 1, \\ &u_1 v_1 + u_2 v_2 + u_3 v_3 = 0 \end{aligned} \tag{18.15}$$

and six other equations, each for one leg. For example, the equation for the fourth leg is

$$\begin{aligned} &(-c + c_1 u_1 r + s_1 v_1 r - x)^2 + (-s + c_1 u_2 r + s_1 v_2 r - y)^2 \\ &\quad + (-c + c_1 u_3 r + s_1 v_3 r - z)^2 - l_4^2 = 0. \end{aligned} \tag{18.16}$$

The nine unknowns are $u_1, u_2, u_3, v_1, v_2, v_3, x, y, z$. All the other variables are constants.

Now introduce three new variables:

$$\begin{aligned} &u = u_1 x + u_2 y + u_3 z, \\ &v = v_1 z + v_2 y + v_3 z, \\ &w = x^2 + y^2 + z^2. \end{aligned} \tag{18.17}$$

Using (18.15) and (18.17), equation (18.16) can be changed into six linear equations in nine unknowns $u_1, u_2, u, v_1, v_2, v, x, y, w$. By solving these six equations, the six unknowns u_1, u_2, u, v_1, v_2, v can be expressed linearly in x, y, w.

The other six equations can be written according to (18.15) and (18.17) and by substitution of the six solutions of the linear equations:

$$\begin{aligned} &u_3^2 = 1 - u_1^2 - u_2^2 = p_1(x, y, w), \\ &v_3^2 = 1 - v_1^2 - v_2^2 = p_2(x, y, w), \\ &u_3 v_3 = -u_1 v_1 - u_2 v_2 = p_3(x, y, w), \\ &u_3 z = u - u_1 x - u_2 y = p_4(x, y, w), \\ &v_3 z = v - v_1 x - v_2 y = p_5(x, y, w), \\ &z^2 = w - x^2 - y^2 = p_6(x, y, w). \end{aligned} \tag{18.18}$$

Equation (18.18) has six variables u_3, v_3, z, x, y, w. Eliminating three of them from the left-hand side of (18.18), we get six equations in x, y, w as follows:

$$
\begin{aligned}
&r^2(p_6p_1 - p_4^2) = 0,\\
&r^2(p_6p_2 - p_5^2) = 0,\\
&r^2(p_6p_3 - p_4p_5) = 0,\\
&r^3(p_1p_5 - p_3p_4) = 0,\\
&r^3(p_2p_4 - p_3p_5) = 0,\\
&r^4(p_1p_2 - p_3^2) = 0.
\end{aligned}
\tag{18.19}
$$

After reducing the degree of (18.19), some new equations of the same degrees in x, y, w can be obtained. Using these equations, Huang and Wu constructed a resultant as the determinant of a 10×10 matrix and got an equation of degree 14 in w. The (triangularized) equations for the other variables are not difficult to obtain.

18.3.4 Forward kinematics of the general 6-6 Stewart platform: Method 1

After Huang and Wu had solved the special case of the Stewart platform, Wen and Liang (1994) solved the general 6-6 Stewart platform as in Fig. 18.12.

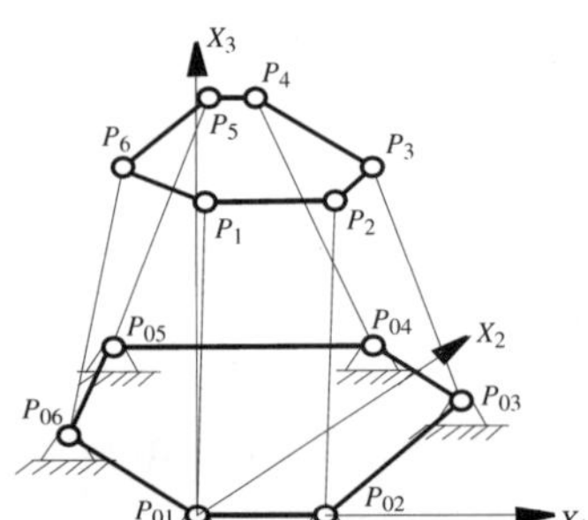

Figure 18.12 6-6 general Stewart platform

Some part of Wen and Liang's method is motivated by the work of Huang and Wu (1991). In their method, Wen and Liang set up 18 equations in 18 unknowns:

$$
\begin{aligned}
&(x_{11} - x_{21})^2 + (x_{12} - x_{22})^2 + (x_{13} - x_{23})^2 - D_{12}^2 = 0,\\
&(x_{11} - x_{31})^2 + (x_{12} - x_{32})^2 + (x_{13} - x_{33})^2 - D_{13}^2 = 0,\\
&(x_{21} - x_{31})^2 + (x_{22} - x_{32})^2 + (x_{23} - x_{33})^2 - D_{23}^2 = 0,\\
&x_{kj} = V_{k1}x_{1j} + V_{k2}x_{2j} + V_{k3}x_{3j} \quad (k = 4, 5, 6;\ j = 1, 2, 3)
\end{aligned}
$$

and

$$\begin{aligned}
&(x_{11})^2 + (x_{12})^2 + (x_{13})^2 - L_1^2 = 0,\\
&(x_{21} - X_{021})^2 + (x_{22})^2 + (x_{23})^2 - L_2^2 = 0,\\
&(x_{31} - X_{031})^2 + (x_{32} - X_{032})^2 + (x_{33})^2 - L_3^2 = 0,\\
&(x_{41} - X_{041})^2 + (x_{42} - X_{042})^2 + (x_{43})^2 - L_4^2 = 0,\\
&(x_{51} - X_{051})^2 + (x_{52} - X_{052})^2 + (x_{53})^2 - L_5^2 = 0,\\
&(x_{61} - X_{061})^2 + (x_{62} - X_{062})^2 + (x_{63})^2 - L_6^2 = 0.
\end{aligned}$$

The 18 unknowns are $x_{i,j}$, $i = 1, \ldots, 6$ and $j = 1, 2, 3$. After linear elimination of nine unknowns, Wen and Liang get nine equations in nine unknowns; they can be changed as

$$\begin{aligned}
x_{11} &= x_{12}A_{441} + x_{22}A_{442} + x_{32}A_{443} + A_{444},\\
x_{21} &= x_{12}A_{451} + x_{22}A_{452} + x_{32}A_{453} + A_{454},\\
x_{31} &= x_{12}A_{461} + x_{22}A_{462} + x_{32}A_{463} + A_{464};
\end{aligned} \tag{18.20}$$

$$x_{13}^2 = -(x_{11})^2 - (x_{12})^2 + L_1^2, \tag{18.21}$$

$$x_{23}^2 = -(x_{21} - X_{021})^2 - (x_{22})^2 + L_2^2, \tag{18.22}$$

$$x_{33}^2 = -(x_{31} - X_{031})^2 - (x_{32} - X_{032})^2 + L_3^2; \tag{18.23}$$

$$x_{13}x_{23} = -x_{11}x_{21} - x_{12}x_{22} - x_{21}X_{021} - A_{37}, \tag{18.24}$$

$$x_{13}x_{33} = -x_{11}x_{31} - x_{12}x_{32} - x_{31}X_{031} + x_{32}X_{032} - A_{38}, \tag{18.25}$$

$$\begin{aligned}
x_{33}x_{23} = &- x_{31}x_{21} - x_{32}x_{22} - x_{21}X_{021} + x_{31}X_{031}\\
&+ x_{32}X_{032} - A_{39}.
\end{aligned} \tag{18.26}$$

Using a method such as Huang and Wu's to eliminate the variables from the left-hand side of (18.21)–(18.26) yields six equations in six unknowns:

$$\begin{aligned}
&(18.21) \cdot (18.26) - (18.24) \cdot (18.25) = 0,\\
&(18.22) \cdot (18.25) - (18.24) \cdot (18.26) = 0,\\
&(18.23) \cdot (18.24) - (18.25) \cdot (18.26) = 0,\\
&(18.21) \cdot (18.22) - (18.24)^2 = 0,\\
&(18.21) \cdot (18.23) - (18.25)^2 = 0,\\
&(18.22) \cdot (18.23) - (18.26)^2 = 0.
\end{aligned} \tag{18.27}$$

Now substitute (18.20) into (18.27): the six resulting equations have only three unknowns: x_{12}, x_{22}, x_{32}.

Using these six equations to eliminate x_{12} and x_{22}, one can obtain a resultant as the determinant of a 21×21 matrix. Its expansion is a univariate polynomial equation. But before the expansion, the highest-degree terms of the polynomial in most elements may be made zero by using elementary row operations. Then the expansion of the determinant is a polynomial of degree 20. Here the row operations are performed by numerical calculation.

18.3.5 Forward kinematics of the general 6-6 Stewart platform: Method 2

After Wen and Liang had solved the general 6-6 Stewart platform, Wu and Huang (1996) improved Wen and Liang's method by symbolic calculation and obtained the same results. Wu and Huang's work presented below is based on the method of Huang and Wu (1991).

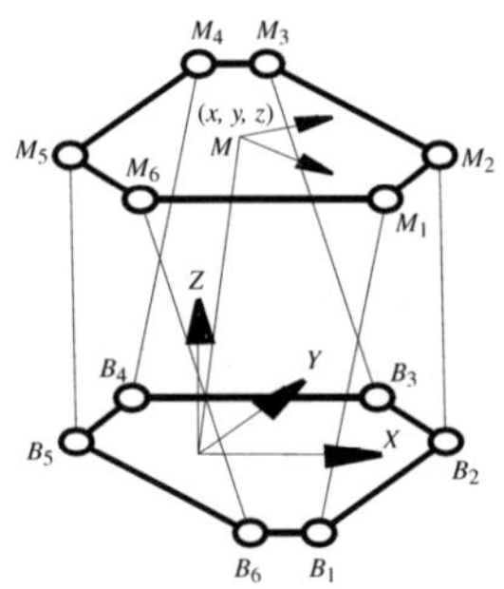

Figure 18.13 6-6 general Stewart platform

The polynomial system for the problem. In Fig. 18.13 where the legs are fixed on the ground, the coordinates of base points B_i are given by $(x_i, y_i, 0)$, $i = 1, \ldots, 6$. Since the platform is planar, we denote the coordinates of points M_i of fixation of the legs by (p_i, q_i) with respect to any rectangular coordinate system given in the platform plane. Let its origin be M with coordinates (x, y, z), and the direction cosines of its axes MP and MQ be (u_1, u_2, u_3) and (v_1, v_2, v_3) respectively. Thus the coordinates of M_i can be expressed as

$$(p_i u_1 + q_i v_1 + x, p_i u_2 + q_i v_2 + y, p_i u_3 + q_i v_3 + z).$$

Let the lengths of $B_i M_i$ be l_i. We have six equations

$$\begin{aligned} f_i = {} & (p_i u_1 + q_i v_1 + x - x_i)^2 + (p_i u_2 + q_i v_2 + y - y_i)^2 + \\ & (p_i u_3 + qiv_3 + z)^2 - l_i^2 = 0, \quad i = 1, \ldots, 6. \end{aligned} \tag{18.28}$$

Three other obvious equations are

$$\begin{aligned} f_7 &= u_1^2 + u_2^2 + u_3^2 - 1 = 0, \\ f_8 &= v_1^2 + v_2^2 + v_3^2 - 1 = 0, \\ f_9 &= u_1 v_1 + u_2 v_2 + u_3 v_3 = 0. \end{aligned} \tag{18.29}$$

These nine equations in nine unknowns $u_1, u_2, u_3, v_1, v_2, v_3, x, y, z$ form the fundamental system describing the problem.

The transformed system. By using f_7, f_8, f_9 in (18.29) and introducing

$$\begin{aligned} u &= u_1 x + u_2 y + u_3 z, \\ v &= v_1 x + v_2 y + v_3 z, \\ w &= x^2 + y^2 + z^2, \end{aligned} \tag{18.30}$$

the first six equations (18.28) can be written as

$$\begin{aligned} f_{10+i} = {} & p_i x_i u_1 + p_i y_i u_2 - p_i u + q_i x_i v_1 + q_i y_i v_2 \\ & - q_i v + x_i x + y_i y - 1/2w + m_i = 0, \qquad i = 1, \ldots, 6, \end{aligned} \tag{18.31}$$

where

$$m_i = \frac{1}{2}(l_i^2 - x_i^2 - y_i^2 - p_i^2 - q_i^2).$$

Solving (18.31) for u_1, u_2, u, v_1, v_2, v, we get the linear expressions of these variables in terms of x, y and w:

$$\begin{aligned} f_{21} &= a_0 u_1 + a_{11} x + a_{12} y + a_{13} w + a_{14} = 0, \\ f_{22} &= a_0 u_2 + a_{21} x + a_{22} y + a_{23} w + a_{24} = 0, \\ f_{23} &= a_0 u + a_{31} x + a_{32} y + a_{33} w + a_{34} = 0, \\ f_{24} &= a_0 v_1 + a_{41} x + a_{42} y + a_{43} w + a_{44} = 0, \\ f_{25} &= a_0 v_2 + a_{51} x + a_{52} y + a_{53} w + a_{54} = 0, \\ f_{26} &= a_0 v + a_{61} x + a_{62} y + a_{63} w + a_{64} = 0. \end{aligned}$$

By f_7, f_8, f_9 in (18.29) and the definitions of u, v, and w in (18.30), we get the expressions of $u_3^2, v_3^2, u_3 v_3, u_3 z, v_3 z, z^2$ in terms of x, y, and w, denoted by $f_{31}, \ldots, f_{36}$ respectively:

$$\begin{aligned} f_{31} &= 1 - u_1^2 - u_2^2 = 1 - a_0^{-2}(A^2 + B^2), \\ f_{32} &= 1 - v_1^2 - v_2^2 = 1 - a_0^{-2}(F^2 + G^2), \\ f_{33} &= -u_1 v_1 - u_2 v_2 = -a_0^{-2}(AF + BG), \\ f_{34} &= u - u_1 x - u_2 y = a_0^{-1}(-C + Ax + By), \\ f_{35} &= v - v_1^2 - v_2 y = a_0^{-1}(-H + Fx + Gy), \\ f_{36} &= w - x^2 - y^2. \end{aligned}$$

From these expressions the following six equalities obviously hold true:

$$h_1 = a_0^2(f_{31}f_{36} - f_{34}^2) = 0,$$
$$h_2 = a_0^2(f_{32}f_{36} - f_{35}^2) = 0,$$
$$h_3 = a_0^2(f_{33}f_{36} - f_{34}f_{35}) = 0,$$
$$h_4 = a_0^3(f_{31}f_{35} - f_{33}f_{34}) = 0,$$
$$h_5 = a_0^3(f_{32}f_{34} - f_{33}f_{35}) = 0,$$
$$h_6 = a_0^4(f_{31}f_{32} - f_{33}^2) = 0.$$

For any j the total degree of h_j in x, y, and w is 4.

We will solve this polynomial system in the next section. Once the values of x, y, and w are obtained, the values for the other unknowns may easily be computed.

Elimination. From the six polynomials $h_1, \ldots, h_6$, we construct the following nine polynomials:

$$h_7 = Fh_1 - Ah_3 + xh_4,$$
$$h_8 = Gh_1 - Bh_3 + yh_4,$$
$$h_9 = Ah_2 - Fh_3 + xh_5,$$
$$h_{10} = Bh_2 - Gh_3 + yh_5,$$
$$h_{11} = Fh_4 + Ah_5 + xh_6,$$
$$h_{12} = Gh_4 + Bh_5 + yh_6,$$
$$h_{13} = Gh_7 + Bh_9 + yh_{11} \equiv Fh_8 + Ah_{110} + xh_{12},$$
$$h_{14} = (1/2)(a_0^2h_1 + a_0^2h_2 + Hh_4 + Ch_5 + Fh_7 - Gh_8, \\ + Ah_9 - Bh_{10} + 2xh_1),$$
$$h_{15} = (1/2)(a_0^2h_1 + a_0^2h_2 + Hh_4 + Ch_5 - Fh_7 + Gh_8, \\ - Ah_9 + Bh_{10} + 2yh_1).$$

It is easy to check that, for $j = 7, \ldots, 15$, the total degree of h_j in x, y and w is still 4, not 5.

Now consider h_j as polynomials in x and y with polynomial coefficients in w. We can write the 15 equations $h_j = 0$ in matrix form

$$M_{15\times 15}T = 0,$$

where

$$T = (x^4 \;\; x^3y \;\; x^2y^2 \;\; xy^3 \;\; y^4 \;\; x^3 \;\; x^2y \;\; xy^2 \;\; y^3 \;\; x^2 \;\; xy \;\; y^2 \;\; x \;\; y \;\; 1)^T$$

and the ith row of $M_{15\times 15}$ corresponds to h_i.

For any solution of the system, we have

$$g_1(w) = \det(M_{15\times 15}) = 0.$$

It is not difficult to check that the degree of g_1 in w is at most 20 when $g_1(w) \not\equiv 0$. The equations for the other unknowns can be obtained easily.

18.4 Synthesis of planar four-bar mechanisms

Synthesis of mechanisms is the first step in the design of a machine. Here the synthesis is called *five position rigid body guidance*: given the five positions and orientations of a plane (where the link of a four-bar mechanism is fixed), try to find a point B on the plane such that the corresponding five points B_i will locate on a circle when the plane passes through the five given positions and orientations. If one can find such a point, then a revolute joint can be put there, and so can another revolute joint on the frame at the center A of the circle. Then one can connect the joints by a link; this is called an $R-R$ *dyad*. Furthermore, one tries to find another point C and the center D, and connect them by another $R-R$ dyad. In this way, we can form a four-bar mechanism $ABCD$, such that the plane of the link BC can pass through the five given positions and orientations.

This problem is called *Burmester's problem*, and was solved once by drawing a series of graphs. Here we present the work done by Guan and Liu (1995) using Wu's method.

Mathematical modeling. Let Q_0 be the reference point on the moving plane p' that moves to Q_j when p' moves from the position p_0 to p_j. The displacement of the moving plane is shown in Fig. 18.14.

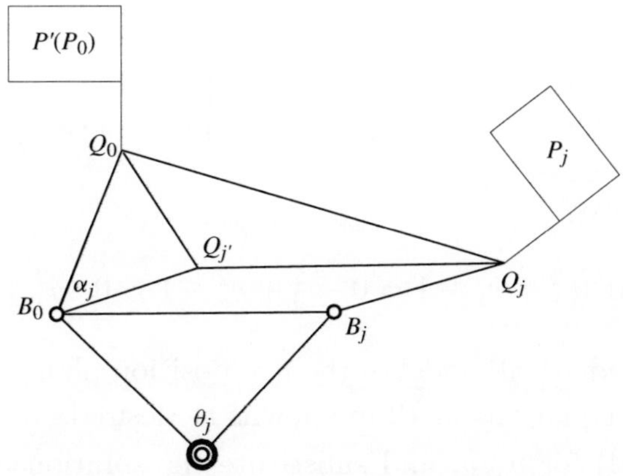

Figure 18.14 A plane p_0 moves to p_j

Since $B_0Q'_jQ_jB_j$ is a parallelogram, from $\triangle Q_0Q'_jQ_j$ we have

$$\mathbf{r}_{01}\cdot(m_j)+\mathbf{r}_{02}\cdot(n_j)-\mathbf{w}_j=0,\quad j=1,\ldots,4,$$

where $\mathbf{r}_{01}$ is the vector from point A to point B_0, whose components are denoted by x_1, x_2, $\mathbf{r}_{02}$ is the vector from point B_0 to the reference point Q_0, whose components are denoted by x_3, x_4, $\mathbf{w}_j$ is the displacement vector of the reference point Q_0, whose components are c_j, d_j, and $(m_j), (n_j)$ are displacement matrices which represent rotations about points A, B_0 respectively.

The matrix form of the above equation is

$$\begin{aligned}&(x_1\ \ x_2)\begin{pmatrix}u_j-1 & v_j\\ -v_j & u_j-1\end{pmatrix}+(x_3\ \ x_4)\begin{pmatrix}a_j-1 & b_j\\ -b_j & a_j-1\end{pmatrix}\\ &\quad-(c_j\ \ d_j)=0,\quad j=1,\ldots,4,\end{aligned}\tag{18.32}$$

where a_j, b_j, c_j, d_j are given and $u_j=\cos\theta_j, v_j=\sin\theta_j, \alpha_j=\cos\alpha_j, b_j=\sin\alpha_j$. Obviously

$$u_j^2+v_j^2=1,\quad a_j^2+b_j^2=1,\quad j=1,\ldots,4.\tag{18.33}$$

Design equations. By expanding (18.32), eight equations are obtained as follows:

$$\begin{array}{llllllll}
-c_1 & -x_1 & -x_3 & +a_1x_3 & -b_1x_4 & +x_1u_1 & -x_2v_1 & =0,\\
-c_2 & -x_1 & -x_3 & +a_2x_3 & -b_2x_4 & +x_1u_2 & -x_2v_2 & =0,\\
-c_3 & -x_1 & -x_3 & +a_3x_3 & -b_3x_4 & +x_1u_3 & -x_2v_3 & =0,\\
-c_4 & -x_1 & -x_3 & +a_4x_3 & -b_4x_4 & +x_1u_4 & -x_2v_4 & =0,\\
-d_1 & -x_2 & +b_1x_3 & -x_4 & +a_1x_4 & +x_2u_1 & +x_1v_1 & =0,\\
-d_2 & -x_2 & +b_2x_3 & -x_4 & +a_2x_4 & +x_2u_2 & +x_1v_2 & =0,\\
-d_3 & -x_2 & +b_3x_3 & -x_4 & +a_3x_4 & +x_2u_3 & +x_1v_3 & =0,\\
-d_4 & -x_2 & +b_4x_3 & -x_4 & +a_4x_4 & +x_2u_4 & +x_1v_4 & =0.
\end{array}\tag{18.34}$$

From (18.33) we have

$$u_1^2+v_1^2-1=0,\ u_2^2+v_2^2-1=0,\ u_3^2+v_3^2-1=0,\ u_4^2+v_4^2-1=0.\tag{18.35}$$

Thus the mathematical model of the five position plane rigid body guidance now consists of 12 equations in 12 unknowns as desired.

We solve (18.34) for u_j, v_j and substitute the solutions into (18.35). Then four basic polynomial equations are obtained. Let these polynomials be denoted

P_i; we have

$$\begin{aligned}
P_1 &= e_1 + 2c_1x_1 + 2d_1x_2 + 2c_1x_3 - 2a_1c_1x_3 - 2b_1d_1x_3 + f_1x_1x_3 - 2b_1x_2x_3 \\
&\quad + f_1x_3^2 + 2b_1c_1x_4 + 2d_1x_4 - 2a_1d_1x_4 + 2b_1x_1x_4 + f_1x_2x_4 + f_1x_4^2, \\
P_2 &= e_2 + 2c_2x_1 + 2d_2x_2 + 2c_2x_3 - 2a_2c_2x_3 - 2b_2d_2x_3 + f_2x_1x_3 - 2b_2x_2x_3 \\
&\quad + f_2x_3^2 + 2b_2c_2x_4 + 2d_2x_4 - 2a_2d_2x_4 + 2b_2x_1x_4 + f_2x_2x_4 + f_2x_4^2, \\
P_3 &= e_3 + 2c_3x_1 + 2d_3x_2 + 2c_3x_3 - 2a_3c_3x_3 - 2b_3d_3x_3 + f_3x_1x_3 - 2b_3x_2x_3 \\
&\quad + f_3x_3^2 + 2b_3c_3x_4 + 2d_3x_4 - 2a_3d_3x_4 + 2b_3x_1x_4 + f_3x_2x_4 + f_3x_4^2, \\
P_4 &= e_4 + 2c_4x_1 + 2d_4x_2 + 2c_4x_3 - 2a_4c_4x_3 - 2b_4d_4x_3 + f_4x_1x_3 - 2b_4x_2x_3 \\
&\quad + f_4x_3^2 + 2b_4c_4x_4 + 2d_4x_4 - 2a_4d_4x_4 + 2b_4x_1x_4 + f_4x_2x_4 + f_4x_4^2,
\end{aligned}$$

where e_j, f_j depend on given data. Let $PS = \{P_1, \ldots, P_4\}$. By the elimination method of Wu (1984a, 1987b), we may get a characteristic set CS of PS according to the scheme (S) in Chap. 1. In the symbolic computation system Mathematica V.2.0, the obtained CS consists of the following four polynomials:

$$\begin{aligned}
C_1 &= a_{31} + a_{32}x_1 + a_{33}x_2 + a_{34}x_3 + a_{35}x_4, \\
C_2 &= a_{35}a_{4l} - a_{3l}a_{45} + (a_{35}a_{42} - a_{32}a_{45})x_1 + (a_{35}a_{43} - a_{33}a_{45})x_2 + i_{44}x_3, \\
C_3 &= -b_{35}^2b_{4l}^2 + b_{3l}b_{45} + (-b_{35}b_{42} + b_{32}b_{45})x_1 + (-b_{35}b_{43} + b_{33}b_{45})x_1^2 + i_{64}x_2, \\
C_4 &= b_{35}^2b_{41}^2b_{45} - 2b_{3l}b_{35}b_{4l}b_{45}^2 + b_{3l}^2b_{45}^2 + b_{35}b_{4l}b_{441}i_{64} - b_{31}b_{441}b_{45}i_{64} \\
&\quad + b_{41}i_{64}^2 + (2b_{35}^2b_{41}b_{42}b_{45} - 2b_{32}b_{35}b_{41}b_{45}^2 - 2b_{31}b_{35}b_{42}b_{45}^2 + 2b_{31}b_{32}b_{45}^3 \\
&\quad + b_{35}b_{42}b_{441}i_{64} + b_{35}b_{4l}b_{442}i_{64} - b_{32}b_{442}b_{45}i_{64} - b_{3l}b_{442}b_{45}i_{64} \\
&\quad + b_{42}i_{64}^2)x_1 + (b_{35}^2b_{42}^2b_{45} + b_{35}^2b_{41}b_{43}b_{45} - 2b_{33}b_{35}b_{41}b_{45}^2 - 2b_{32}b_{35}b_{42}b_{45}^2 \\
&\quad - 2b_{3l}b_{35}b_{43}b_{45}^2 + b_{32}^2b_{45}^3 + 2b_{31}b_{33}b_{45}^2 + b_{35}b_{43}b_{441}i_{64} + b_{35}b_{42}b_{442}i_{64} \\
&\quad - b_{33}b_{441}b_{45}i_{64} - b_{32}b_{442}b_{45}i_{64} + b_{43}i_{64}^2)x_1^2 + (2b_{35}^2b_{42}b_{43}b_{45} \\
&\quad - 2b_{33}b_{35}b_{42}b_{45}^2 - 2b_{32}b_{35}b_{43}b_{45}^2 + 2b_{32}b_{33}b_{45}^3 + b_{35}b_{43}b_{442}i_{64} \\
&\quad - b_{33}b_{442}b_{45}i_{64})x_1^3 + (b_{35}^2b_{43}^2b_{45} - 2b_{33}b_{35}b_{43}b_{45}^2 + b_{33}^2b_{45}^3)x_1^4,
\end{aligned}$$

where all the newly introduced symbols are determined by the given data. From CS, one can get at most four real solutions for x_1 and thus for x_2, x_3, x_4.

18.5 Reverse force analysis of spring systems

The simplest case of a spring system is the planar two-spring system, which was solved by Pigoski and Duffy (1995), resulting in a univariate equation of degree 6. After that Zhang et al. (1997) solved the spatial three-spring system and obtained a univariate equation of degree 22. Sun et al. (1997) solved the planar three-spring system, where the result is a univariate equation of degree 54. This section reviews the two kinds of three-spring systems.

18.5.1 The planar three-spring system

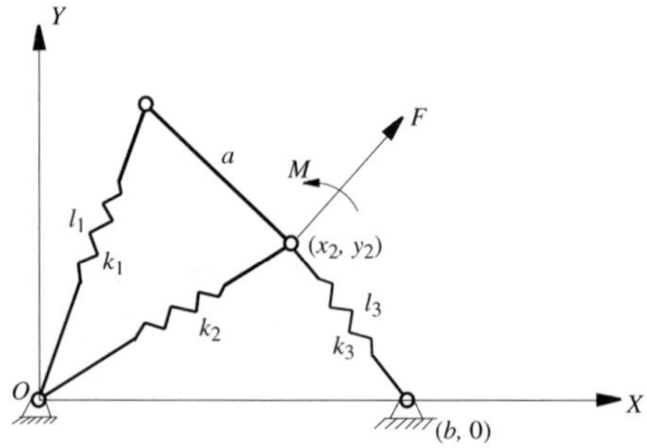

Figure 18.15 A planar three-spring system

The planar three-spring system is shown in Fig. 18.15. Under a planar Cartesian coordinate system, the coordinates of the two basic pivots on the frame are $(0, 0)$ and $(b, 0)$. Two joints on the platform are (x_1, y_1) and (x_2, y_2), and the distance between them is a. The two joints are connected with the two basic pivots by three spring rods. The free lengths of the three springs are l_{01}, l_{02}, l_{03}, and k_1, k_2, k_3 are constants. A known external force $\mathbf{F}$ acts on the point (x_2, y_2) of the platform, and a known external couple M acts on the platform. When the platform is in a static equilibrium, the lengths of the three springs are supposed to be l_1, l_2, l_3. According to the geometric relationship the following four equations can be established:

$$\begin{aligned}
&x_1^2 + y_1^2 - l_1^2 = 0,\\
&x_2^2 + y_2^2 - l_2^2 = 0,\\
&(x_2 - b)^2 + y_2^2 - l_3^2 = 0,\\
&(x_1 - x_2)^2 + (y_1 - y_2)^2 - a^2 = 0.
\end{aligned} \tag{18.36}$$

According to the equilibrium of the force and couple, we have the following equations:

$$\mathbf{V}_1 F_1 + \mathbf{V}_2 F_2 + \mathbf{V}_3 F_3 = \mathbf{F}, \tag{18.37}$$

$$k_3(l_3 - l_{03})by_2/l_3 = M + F_y x_2 - F_x y_2 \tag{18.38}$$

where

$$\mathbf{V}_1 = \begin{pmatrix} x_1/l_1 \\ y_1/l_1 \end{pmatrix}, \ \mathbf{V}_2 = \begin{pmatrix} x_2/l_2 \\ y_2/l_2 \end{pmatrix}, \ \mathbf{V}_3 = \begin{pmatrix} (x_2 - b)/l_3 \\ y_2/l_3 \end{pmatrix}, \ \mathbf{F} = \begin{pmatrix} F_x \\ F_y \end{pmatrix}$$

and

$$F_1 = k_1(l_1 - l_{01}), \qquad F_2 = k_2(l_2 - l_{02}), \qquad F_3 = k_3(l_3 - l_{03}). \tag{18.39}$$

Note that the vector equation (18.37) leads to two scalar equations. In the seven equations (18.36)–(18.38), there are seven unknowns $x_1, y_1, x_2, y_2, l_1, l_2, l_3$.

After eliminating the four unknowns x_1, y_1, x_2, y_2, we get three equations:

$$\begin{aligned} &A_{1000} + A_{1001}l_3 + A_{1002}l_3^2 + A_{1003}l_3^3 + A_{1004}l_3^4 + A_{1010}l_2 + A_{1011}l_2l_3 \\ &\quad + A_{1012}l_2l_3^2 + A_{1013}l_2l_3^3 + A_{1014}l_2l_3^4 + A_{1020}l_2^2 + A_{1021}l_2^2l_3 \\ &\quad + A_{1022}l_2^2l_3^2 + A_{1030}l_2^3 + A_{1031}l_2^3l_3 + A_{1032}l_2^3l_3^2 + A_{1111}l_1l_2l_3 \\ &\quad + A_{1112}l_1l_2l_3^2 + A_{1211}l_1^2l_2l_3 + A_{1212}l_1^2l_2l_3^2 = 0, \end{aligned} \tag{18.40}$$

$$\begin{aligned} &A_{2000} + A_{2001}l_3 + A_{2002}l_3^2 + A_{2003}l_3^3 + A_{2004}l_3^4 + A_{2005}l_3^5 + A_{2006}l_3^6 \\ &\quad + A_{2020}l_2^2 + A_{2021}l_2^2l_3 + A_{2022}l_2^2l_3^2 + A_{2023}l_2^2l_3^3 + A_{2024}l_2^2l_3^4 \\ &\quad + A_{2040}l_2^4 + A_{2041}l_2^4l_3 + A_{2042}l_2^4l_3^2 = 0, \end{aligned} \tag{18.41}$$

$$\begin{aligned} &A_{3001}l_3 + A_{3002}l_3^2 + A_{3021}l_2^2l_3 + A_{3022}l_2^2l_3^2 + A_{3100}l_1 + A_{3101}l_1l_3 \\ &\quad + A_{3102012}l_1l_3^2 + A_{3103}l_1l_3^3 + A_{3104}l_1l_3^4 + A_{3111}l_1l_2l_3 \\ &\quad + A_{3112}l_1l_2l_3^2 + A_{3120}l_1l_2^2 + A_{3121}l_1l_2^2l_3 + A_{3122}l_1l_2^2l_3^2 + A_{3201}l_1^2l_3 \\ &\quad + A_{3202}l_1^2l_3^2 + A_{3301}l_1^3l_3 + A_{3302}l_1^3l_3^2 = 0. \end{aligned} \tag{18.42}$$

In these equations all the coefficient A_{ijkl} contain no unknowns and can be obtained directly from the known parameters.

Equation (18.41) does not contain l_1, so we can proceed with the following three steps of elimination.

Firstly, eliminate l_1 from (18.40) and (18.42) by using the resultant given as a 5×5 determinant. An equation containing only l_2 and l_3 can be obtained and has the following form:

$$\begin{aligned} &B_{1000} + B_{1001}l_3 + B_{1002}l_3^2 + B_{1003}l_3^3 + B_{1004}l_3^4 + B_{1005}l_3^5 + B_{1006}l_3^6 \\ &\quad + B_{1007}l_3^7 + B_{1008}l_3^8 + B_{1009}l_3^9 + B_{10010}l_3^{10} + B_{10011}l_3^{11} \\ &\quad + B_{10012}l_3^{12} + B_{10013}l_3^{13} + B_{10014}l_3^{14} = 0, \end{aligned} \tag{18.43}$$

where the B_{ijkl} contain only the unknown l_2 and the known parameters. Secondly, eliminate l_3 from (18.41) and (18.43) by using a resultant. Then a 20×20 determinant containing only l_2 is obtained. The last step is to reduce the degree of the obtained determinant column by column. In the process two polynomials of degree 4 and two polynomials of degree 2 are drawn out. Solving them separately, we can get 12 roots. Through a proper test, they are all proved to be extraneous roots, and should be abandoned. After that we can obtain a 10×10 determinant. Expanding this determinant, we can get a polynomial equation of degree 54 in one unknown l_2, which contains no extraneous roots. Solving this equation we can get 54 roots for l_2; the solutions depending on l_2 for all the other unknowns can be obtained without difficulties.

18.5.2 Reverse force analysis of a spatial three-spring system

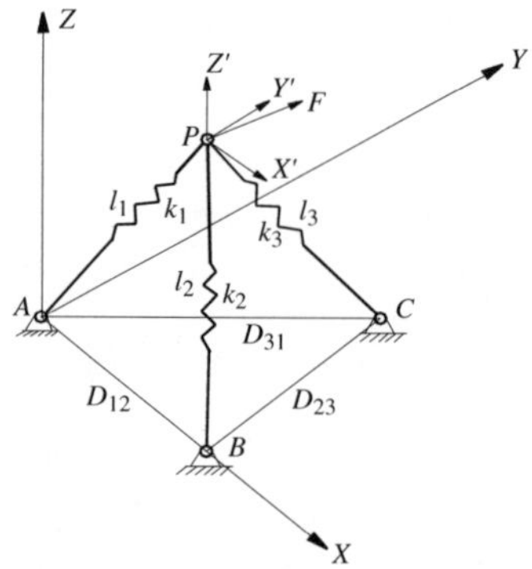

Figure 18.16 A spatial three-spring system

Now set up a rectangular coordinate system as shown in Fig. 18.16. The coordinates of the joint centers A, B, and C in the base $\triangle ABC$ are $(0,0,0)$, $(X_2,0,0)$, and $(X_3,Y_3,0)$, respectively. Here, we assume that the coordinates of P are (x,y,z) and the spring lengths in equilibrium positions are l_1, l_2, and l_3, which are all unknown system variables.

The geometric constraint equations for the spring system can be written as follows:

$$x^2+y^2+z^2-l^2=0, \tag{18.44}$$

$$(x-X_2)^2+y^2+z^2-l_2^2=0, \tag{18.45}$$

$$(x-X_3)^2+(y-Y_3)^2+z^2-l_3^2=0. \tag{18.46}$$

The spring elasticity constants k_1, k_2, and k_3 and free lengths l_{01}, l_{02}, and l_{03} are considered known. The internal forces of the three springs are

$$f_1 = k_1(l_1 - l_{01}),$$
$$f_2 = k_2(l_2 - l_{02}),$$
$$f_3 = k_3(l_3 - l_{03}).$$

The force balance equation can be expressed in the form

$$f_1\mathbf{v}_1 + f_2\mathbf{v}_2 + f_3\mathbf{v}_3 = -\mathbf{F}, \tag{18.47}$$

where the known externally applied force $\mathbf{F} = F_x\mathbf{i} + F_y\mathbf{j} + F_z\mathbf{k}$ and $\mathbf{v}_1, \mathbf{v}_2, \mathbf{v}_3$ are unit vectors specifying the orientations of the lines of action of the three springs, and

$$\mathbf{v}_1 = \begin{pmatrix} -x/l_1 \\ -y/l_1 \\ -z/l_1 \end{pmatrix}, \quad \mathbf{v}_2 = \begin{pmatrix} (X_2 - x/l_2 \\ -y/l_2 \\ -z/l_2 \end{pmatrix}, \quad \mathbf{v}_3 = \begin{pmatrix} (X_3 - x/l_3 \\ (Y_3 - y)/l_3 \\ -z/l_3 \end{pmatrix}. \tag{18.48}$$

Rearranging the vector equation (18.47) with substitution (18.48) produces three scalar equations:

$$\begin{aligned} F_x - k_1x - k_2X - k_3x + k_1l_{01}x/l_1 + k_2l_{02}x/l_2 + k_3l_{03}x/l_3 \\ + k_2X_2 - k_2l_{02}X_2/l_2 + k_3X_3 - k_3l_{03}X_3/l_3 = 0, \end{aligned} \tag{18.49}$$

$$\begin{aligned} F_y - k_1y - k_2y - k_3y + k_1l_{01}y/l_1 + k_2l_{02}y/l_2 + k_3l_{03}/l_3 + k_3Y_3 \\ - k_3l_{03}Y_3/l_3 = 0, \end{aligned} \tag{18.50}$$

$$F_z - k_1z - k_2z - k_3z + k_1l_{01}z/l_1 + k_2l_{02}z/l_2 + k_3l_{03}z/l_3 = 0. \tag{18.51}$$

Equations (18.44), (18.45), (18.46), (18.49), (18.50), and (18.51) are considered as the fundamental equations, in which there are six unknowns l_1, l_2, l_3, x, y, z. The key to the problem in this formulation is to eliminate five variables from the six equations.

Equations (18.49), (18.50), and (18.51) are linear in x, y, and z, respectively. Solving them for x, y, z and substituting the solution into (18.44), (18.45), and (18.46) yield three equations:

$$\sum_{j,k,m=0}^{2} T_{1jkm}l_1^j l_2^k l_3^m = 0, \tag{18.52}$$

$$\sum_{j,k,m=0}^{2} T_{2jkm}l_1^j l_2^k l_3^m = 0, \tag{18.53}$$

$$\sum_{j,k,m=0}^{2} T_{3jkm}l_1^j l_2^k l_3^m = 0. \tag{18.54}$$

It is important to note that two additional equations are needed in the elimination. By forming the self-scalar products of the left and right sides of (18.47) and properly eliminating x, y, z, this pair of closure equations can be expressed in the abbreviated form

$$\sum_{j,k=0,\ m=0,3}^{3} A_{4jkm}l_1^j l_2^k l_3^m = 0, \tag{18.55}$$

$$\sum_{j,m=0,\ k=0,3}^{3} A_{5jkm}l_1^j l_2^k l_3^m = 0. \tag{18.56}$$

Equations (18.55) and (18.56) together with (18.52), (18.53), and (18.54) form a set of five elements from which one can construct a resultant as the determinant of a 16×16 matrix to eliminate l_2, l_3 and get a polynomial equation that contains only l_1.

After factoring out some extraneous roots, the resultant is of degree 22. The solutions depending on these roots for the other unknowns can be obtained successively.

Chapter 19

BLENDING SEVERAL IMPLICIT ALGEBRAIC SURFACES

Guochen Feng, Hongju Ren, and Yunshi Zhou

One of the central problems in computer-aided geometric design (CAGD) is blending surfaces, which provides the theoretical basis for the design technology of space surfaces. This chapter summarizes some basic results achieved in recent years on blending implicit algebraic surfaces. Our presentation focuses mainly on the work done in the research groups of the authors and J. Warren, in which Wu's method and polynomial ideal theory are used.

19.1 Introduction

Due to the development of the theory of multi-spline functions since the 1970s (see Wang et al. 1994), the theory of parametric and explicit surfaces has been studied extensively. However, its usefulness in the design of space surfaces is sometimes limited. Implicit algebraic surfaces must be carefully studied, which requires the heavy machinery of algebraic geometry. But the constructive theory of classic algebraic geometry is so restricted that it cannot be used efficiently to resolve many practical problems. Recently, many results in commutative algebra and algebraic geometry have been achieved by means of the Gröbner basis method developed by Buchberger (1985) and Wu's method (Chap. 1), which has in turn helped the research on implicit algebraic surfaces.

In the list of references, one may find some dissertations on the blending of implicit algebraic surfaces. Hoffmann and Hopcroff (1990) presented a general algorithm for blending surfaces by means of potential method. Warren (1989) pointed out that blending surfaces are generally quartic, and gave some examples of cubic blending surfaces. Wu and Wang (1994) presented a general algorithm for blending surfaces by means of Wu's method, but the algorithm is complicated, and no existence conditions of cubic blending surfaces were given. Since 1994, the authors and their collaborators have studied the theory of

MATHEMATICS MECHANIZATION AND APPLICATIONS
ISBN 0-12-734760-7

blending surfaces of two or more algebraic surfaces with plane intersections and the feasible algorithms for finding blending surfaces with the lowest degree; see Feng et al. (1998), Guo and Ren (1998), Wu et al. (1995, 1996), Zhou and Wu (1995), and Zhang and Yu (1998). We first consider, for several definite surfaces, the construction of a surface with the same degree (and call it *ruled surface*) such that it intersects with these surfaces along their plane intersections, and then construct the GC^1 blending surface with it. The effort to find the ruled surfaces with our method is so small that one can even do it by hand.

In what follows, we will give the basic results in recent years about blending implicit algebraic surfaces, including mainly the use of Wu's method and the work of Warren and ours.

We assume that the problem is considered in $\boldsymbol{R}[x, y, z]$ and $S(f)$ always denotes the algebraic surface determined by the polynomial equation $f(x, y, z) = 0$, where $\boldsymbol{R}$ is the field of real numbers. If F is a polynomial set, we denote by $S(F)$ the affine variety determined by F, i.e., the set of common zeros of all the polynomials in F. For arbitrary $f_1, \ldots, f_s \in \boldsymbol{R}[x, y, z]$, the ideal generated by $f_1, \ldots, f_s$ is denoted as $\mathrm{Ideal}(f_1, \ldots, f_s)$. The problem we will deal with can be expressed as follows.

Problem:

Given three sets of irreducible algebraic curves $S(g_i, h_i)$, $S(g_j, h_j)$, and $S(g_k, h_k)$ in the real 3-space $\boldsymbol{R}^3$ with $i \in I$, $j \in J$, $k \in K$, and I, J, K all being finite sets of indices, and given also two sets of irreducible algebraic surfaces $S(g_j)$ and $S(g_k)$ $(j \in J, k \in K)$ containing $S(g_j, h_j)$ and $S(g_k, h_k)$ respectively, determine an irreducible algebraic surface $S(f)$ of given degree m verifying the following conditions:

(a) $S(f)$ contains all the curves $S(g_i, h_i)$, $S(g_j, h_j)$, and $S(g_k, h_k)$ for $i \in I$, $j \in J$, and $k \in K$.

(b) $S(f)$ touches smoothly each of $S(g_j)$ and $S(g_k)$ along the curves $S(g_j, h_j)$ and $S(g_k, h_k)$, respectively, for $j \in J$ and $k \in K$. More precisely, for each point on $S(g_j, h_j)$ or $S(g_k, h_k)$ which is regular for $S(g_j, h_j)$, $S(f)$, and $S(g_j)$ or for $S(g_j, h_k)$, $S(f)$, and $S(g_k)$, $S(f)$ and $S(g_j)$ or $S(f)$ and $S(g_k)$ have the same tangent plane at that point.

(c) $S(f)$ possesses the same curvature as $S(g_k)$ along the curve $S(g_k, h_k)$ for each $k \in K$. More precisely, for each point on $S(g_k, h_k)$ which is regular for $S(g_k, h_k)$, $S(f)$ and $S(g_k)$, $S(f)$ and $S(g_k)$ have the same (Gaussian) curvature at that point.

19.2 Blending surfaces by Wu's method and ideal theory

19.2.1 Wu's method

Warren (1989) introduced a new notion of geometric continuity that he called *recalling* continuity.

Let $S(g)$ and $S(f)$ be two algebraic surfaces that meet along an irreducible algebraic space curve W. $S(g)$ and $S(f)$ meet with *order k recalling continuity* along W (here it is denoted GC^k), if

(1) $S(g)$ and $S(f)$ are smooth (and non-singular) at all but a finite number of points on W, and

(2) there exist $A, B \in \boldsymbol{C}[x, y, z]$, not identically zero on W, such that all the partial derivatives of Ag and Bf up to order k agree on W (where $\boldsymbol{C}$ denotes the field of complex numbers).

Wu (1993d) (see also Wu and Wang 1994) gave a general algorithm for the above problem using his characteristic set method. The reader is referred to Chap. 1 for the details of Wu's method of characteristic sets and the relevant concepts and notations including *initial*, *ascending set*, *irreduciblility* of an ascending set and *pseudo-remainder* $\mathrm{prem}(P, AS)$ of a polynomial P with respect to an ascending set AS.

An irreducible algebraic surface in $\boldsymbol{R}^3$ can be expressed as $P = 0$, where $P \in \boldsymbol{R}[x, y, z]$ is irreducible. An irreducible curve in $\boldsymbol{R}^3$ can be defined by an ascending set.

Let c be an irreducible algebraic curve in space, and defined by an irreducible ascending set $IRR = \{P_1, P_2\}$. S and S' are irreducible algebraic surfaces defined by polynomials P and P', respectively. We have the following theorems.

Theorem 19.1:

The surface S contains c wholly iff (if and only if) $\mathrm{prem}(P, IRR) = 0$.

Theorem 19.2:

Suppose that c is contained wholly in S and in S', but not contained wholly in the singularity part of S or S'. Set $P_x = \frac{\partial P}{\partial x}$, $P'_x = \frac{\partial P'}{\partial x}$, etc., and form the polynomials

$$D_1 = P_x P'_y - P_y P'_x, \qquad D_2 = P_x P'_z - P_z P'_x, \qquad D_3 = P_y P'_z - P_z P'_y.$$

Then S and S' touch smoothly along c iff

$$\mathrm{prem}(D_i, IRR) = 0 \quad \text{for} \quad i = 1, 2, 3.$$

Let

$$\begin{aligned} H &= P_x^2 + P_y^2 + P_z^2, \\ V &= P_{xx}P_{yy}P_z^2 + P_{xx}P_{zz}P_y^2 + P_{yy}P_{zz}P_x^2 - 2P_xP_yP_{xy}P_{zz} \\ &\quad -2P_xP_yP_{xz}P_{yz} - 2P_yP_zP_{xz}P_{yz} + 2P_xP_yP_{xz}P_{yz} \\ &\quad +2P_xP_zP_{xy}P_{yz} + 2P_yP_zP_{xy}P_{xz} - P_x^2P_{yz}^2 - P_y^2P_{xx}^2 - P_y^2P_{xy}^2. \end{aligned}$$

Theorem 19.3:

Let c, S and S' be as before and satisfy the conditions in Theorem 19.2. Then S and S' have the same curvatures at the points regular to c, S and S' iff

$$\mathrm{prem}(H^2V' - H'^2V, IRR) = 0.$$

With the above theorems, we transform the problem of surface-fitting into that of solving polynomial systems; that is, if a polynomial P which defines the space surface S satisfies the conditions in the theorems, then the coefficients of P must satisfy a set of polynomial equations.

The method of solving the problem may now be described as follows.

For any irreducible algebraic surface $S(f)$ of degree m defined by $P = 0$, let us write P in the form

$$P = \sum_m u_{ijk} x^i y^j z^k,$$

in which $\sum_m$ is to be extended over triples (i, j, k) verifying

$$i \geq 0, \qquad j \geq 0, \qquad k \geq 0, \qquad i + j + k \leq m. \tag{19.1}$$

We shall denote by U_m the set of all u_{ijk} verifying (19.1).

Consider the condition that $S(f)$ contains the whole irreducible algebraic curve defined by an irreducible ascending set IRR. By Theorem 19.1 the condition becomes

$$R = \mathrm{prem}(P, IRR) = 0. \tag{19.2}$$

Let the polynomial set formed of the coefficients of various power-products of x, y, z in R be $US_a \subset \boldsymbol{R}[U_m]$. Then the condition (19.2) is identical to

$$US_a = 0. \tag{19.3}$$

For the requirement (a) in the problem of Sect. 19.1, let us form, in accordance with US_a of (19.3), polynomial sets US_{ai}, US_{aj}, and US_{ak} for each of the curves $S(g_i, h_i)$, $S(g_j, h_j)$, and $S(g_k, h_k)$. Similarly, for the requirement (b) in the problem, Theorem 19.2 gives polynomial sets US_{bj} and US_{bk}; for the requirement (c), Theorem 19.3 gives polynomial sets US_{ck}, whose vanishing is the corresponding condition to be verified. Remark that all the polynomials in US_{ai}, US_{aj}, US_{ak}, US_{bj}, and US_{bk} are linear in the u's, while those in US_{ck} are in general non-linear in the u's. Combining all the above polynomial sets into a single one $US \subset \boldsymbol{R}[U_m]$, we see that all the possible solutions $S(f)$ verifying the requirements of the problem are furnished by the real zeros of US.

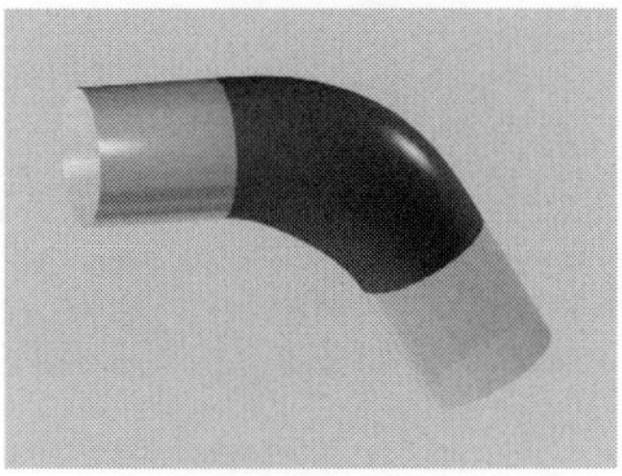

Figure 19.1

Figure 19.2

Example 19.1: (Fig. 19.1)

A general cubic surface S in $\boldsymbol{R}^3$ is of the form

$$\begin{aligned}f(x,y,z) = {} & a_1x^3 + a_2y^3 + a_3z^3 + a_4x^2y + a_5x^2z + a_6y^2x + a_7y^2z + a_8z^2x \\ & + a_9z^2y + a_{10}xyz + a_{11}x^2 + a_{12}y^2 + a_{13}z^2 + a_{14}xy + a_{15}xz \\ & + a_{16}yz + z_{17}x + z_{18}y + z_{19}z + a_{20}\end{aligned}$$

In $\boldsymbol{R}^3$, consider two circular cylinders CYL_1 and CYL_2 with the x-axis and y-axis as their axes and two circles C_1 and C_2 of section by the planes orthogonal to these axes respectively. The two circles C_1 and C_2 are defined by the irreducible ascending sets $AS_1 = \{h_1, g_1\}$ and $AS_2 = \{h_2, g_2\}$ with

$$h_1 = x - d_1,\ g_1 = y^2 + z^2 - r_1^2; \qquad h_2 = y - d_2,\ g_2 = x^2 + z^2 - r_2^2.$$

We want to determine such cubic surfaces that contain the circles C_1, C_2 and touch the cylinders CYL_1, CYL_2 smoothly along these circles. Based on the above theorems and using the package Wsolve, D.-K. Wang proved that such surfaces exist only if $r_1^2 + d_1^2 = r_2^2 + d_2^2$, and in that case there is only one possible cubic surface, given by

$$\begin{aligned}& z^2(yd_2 + xd_1 - d_2^2 - d_1^1) + (y^3d_2 + x^3d_1) \\ & \quad + yx(yd_1 + xd_2 - 2d_2d_1) - (y^2 + x^2)(d_2^2 + d_1^2) \\ & \quad + yd_2(d_1^2 - r_1^2) + xd_1(d_2^2 - r_2^2) + (r_1^2 + d_1^2)(r_2^2 + d_2^2) - 2d_1^2d_2^2 = 0.\end{aligned}$$

When $d_1 = d_2 = 2$ and $r_1 = r_2 = 1$, we have

$$\begin{aligned}f(x,y,z) = {} & 2x^2y + 2y^2x + 2x^3 + 6y + 2y^3 - 8x^2 - 8y^2 - 8z^2 + 8 \\ & -8xy + 6x + 2z^2y + 2z^2x = 0.\end{aligned}$$

Example 19.2: (Fig. 19.2)

Consider a circular cylinder CYL_1 with x-axis as its axis, circular sections C_1 and C_2 by two planes orthogonal to the x-axis and symmetric with respect to the yz plane, a circular cylinder CYL_2 with the y-axis as its axis, and circular sections C_3 and C_4 by two planes orthogonal to the y-axis and symmetric with respect to the xz plane. The irreducible ascending sets defining them are $AS_i = \{h_i, g_i\}$, $i = 1, \ldots, 4$, with

$$h_1 = x - d_1,\ g_1 = y^2 + z^2 - r_1^2; \qquad h_2 = x + d_1,\ g_2 = y^2 + z^2 - r_1^2;$$
$$h_3 = y - d_2,\ g_3 = x^2 + z^2 - r_2^2; \qquad h_4 = y + d_2,\ g_4 = x^2 + z^2 - r_2^2.$$

Now seek for a quartic surface in $\boldsymbol{R}^3$ such that it touches the four cylinders smoothly along $C_1, \ldots, C_4$ respectively.

When $b_1 = -1$, $b_3 = 1$, $d_1 = d_2 = 2$, and $r_1 = r_2 = 1$, the surface is defined by

$$f(x, y, z) = -23 + 8x^2 + 8y^2 + 6z^2 - x^4 - y^4 + z^4.$$

19.2.2 Blending implicit algebraic surfaces by ideal theory

Warren (1989) studied problems of blending surfaces by means of ideal theory.

Basic hypothesis. Let g_1, g_2 be two given irreducible polynomials with degree m, and h_1, h_2 be linear polynomials. They define the surfaces $S(g_i)$ and $S(h_i)$ $(i = 1, 2)$ respectively. Assume that $S(h_i)$ intersects transversely with $S(g_i)$, and $S(g_i, h_i)$ is an irreducible curve of degree m.

Under the above hypothesis, we want to construct an algebraic surface $S(f)$ of degree n such that $S(f)$ meets $S(g_i)$ along $S(g_i, h_i)$ with GC^1 continuity respectively.

We have discussed an algorithm for seeking the polynomial f with Wu's method. Warren (1989) transformed the problem of seeking f into the determination of ideal membership.

Theorem 19.4:

If $S(f)$ meets $S(g_i)$ along $S(g_i, h_i)$ with GC^k continuity respectively, then

$$f \in \mathrm{Ideal}(g_1, h_1^{k+1}) \cap \mathrm{Ideal}(g_2, h_2^{k+1}), \quad k = 0, 1, \ldots.$$

Otherwise, if there exists a polynomial a_i not identically zero on $S(g_i, h_i)$ such that

$$f = a_i g_i + b_i h_i^{k+1}, \quad i = 1, 2,$$

then $S(f)$ meets $S(g_i)$ along $S(g_i, h_i)$ with GC^k continuity respectively.

According to this and with the intersections of ideals, we can have the general expression of f.

In particular, if $m = 2$, $k = 1$ and f can be written as follows,

$$f \in \text{Ideal}(g_1 g_2, g_1 h_2^2, g_2 h_1^2, h_1^2 h_2^2),$$

then f will be quartic. Warren suggested that f can be re-written as the following two expressions:

$$f = g_1 g_2 + \lambda h_1^2 h_2^2, \qquad f = g_1 h_2^2 + \mu g_2 h_1^2,$$

where λ and μ are constants. For some special cases, he obtained cubic blending surfaces with GC^1 continuity.

With Theorem 19.4, the problem of determining the blending surface $S(f)$ can be transformed into the computation of syzygy. Therefore we can talk about the problem of seeking blending surfaces with the lowest degree.

Proposition 19.5: (Wu et al. 1995)

Under the basic hypothesis above, for $f \in \text{Ideal}(g_i, h_i^{k+1})$, *there exist polynomials* a_i *and* b_i *such that*

$$\deg(a_i) \le \deg(f) - m, \qquad \deg(b_i) \le \deg(f) - (k+1).$$

Therefore, we have the following.

Theorem 19.6: (Wu et al. 1995)

Under the basic hypothesis, there exists an $S(g)$ *of degree* m *which passes through the curve* $S(g_i, h_i)$ *of degree* m $(i = 1, 2)$ *iff there exist non-zero constants* λ_1 *and* λ_2 *such that*

$$g = \lambda_1 g_1 + a_1 h_1 = \lambda_2 g_2 + a_2 h_2$$

and $\deg(a_i) \le m - 1$.

There exists an $S(f)$ of degree $n \ge m$ which meets $S(g_i)$ along $S(g_i, h_i)$ with GC^k continuity respectively iff there exist a non-zero polynomial u_i of degree $n - m$ and a polynomial v_i of degree $n - (k+1)$ such that

$$f = u_i g_i + v_i h_i^{k+1}, \quad i = 1, 2.$$

Conclusion. There exists an $S(f)$ of degree $n > m$ which meets $S(g_i)$ along $S(g_i, h_i)$ with GC^k continuity respectively iff there exists a non-zero polynomial u_i of degree $n - m$ such that

$$(u_1 g_1 - u_2 g_2) = 0 \mod \text{Ideal}(h_1^{k+1}, h_2^{k+1}). \tag{19.4}$$

The above equation is the syzygy equation with respect to u_i.

In order to determine u_i, we can expand u_i and g_i with respect to h_i and then compare the coefficients of (19.4). Thus, we can obtain a system of linear algebraic equations with respect to u_i and construct f by means of solving this system.

This method may be applied universally. But for a concrete problem, we need to solve it step by step.

In the next part, we will solve the problem by means of some special skills and give some practical results. It is obvious that the above method can be applied to problems of blending several surfaces.

19.3 Blending surfaces with ruled surface

We will construct a general blending surface with so-called *ruled surfaces*, give the concrete expressions of the blending surface for blending quadratic surfaces, and obtain the conditions for the existence of surfaces with lowest degree and its expressions. These results can be found in Feng et al. (1998), Guo and Ren (1998), Wu et al. (1995), Zhou and Wu (1995), and Zhang and Yu (1998).

19.3.1 A fundamental principle

Definition 19.1:

Let g_i be quadratic polynomials and h_i be linear polynomials. If there exists a quadratic polynomial g such that $S(g)$ passes through $S(g_i, h_i)$, then we call $S(g)$ a ruled surface *with respect to $S(g_i)$.*

Theorem 19.6 gives the existence condition of ruled surfaces.

Proposition 19.7: (Wu et al. 1995)

If a ruled surface $S(g)$ exists, then

$$\text{Ideal}(g_1, h_1) \cap \text{Ideal}(g_2, h_2) = \text{Ideal}(g, h_1 h_2).$$

Proposition 19.8: (Warren 1989)

For any $f \in \text{Ideal}(g, h_1 h_2)$, there exist polynomials u and v such that

$$f = ug + vh_1 h_2 \tag{19.5}$$

and $\max\{\deg(ug), \deg(vh_1h_2)\} = \deg(f)$.

Theorem 19.9: (Wu et al. 1995)

Let a ruled surface $S(g)$ exist. Then, there is a cubic polynomial $f \in \text{Ideal}(g_1, h_1^2)$ $\cap\text{Ideal}(g_2, h_2^2)$, which is stated as (19.5), iff there exist linear polynomials u and v such that

$$ua_1 + vh_2 \in \text{Ideal}(g, h_1), \qquad ua_2 + vh_1 \in \text{Ideal}(g, h_2). \tag{19.6}$$

If we restrict f to be irreducible, then u and v must be linearly independent and, especially, $v \neq 0$.

It is not easy to determine u and v by solving the above syzygy equation. We will discuss a special expression for u; in this case, its solution can be obtained easily.

Theorem 19.10: (Wu et al. 1995)

In Theorem 19.9, if we make the restriction that

$$u = \alpha_1 h_1 + \alpha_2 h_2, \tag{19.7}$$

where α_1 and α_2 are constants not all zero, then the condition

$$\begin{cases} ua_1 + vh_2 \in \text{Ideal}(g_1, h_1), \\ ua_2 + vh_1 \in \text{Ideal}(g_2, h_2) \end{cases}$$

is equivalent to

$$\alpha a_2 + \gamma h_1 = \beta a_1 + \delta h_2, \tag{19.8}$$

where $\alpha, \beta, \gamma, \delta$ are constants and $\alpha\beta \neq 0$.

If (19.7) and (19.8) hold, then the cubic $f = ug + vh_1h_2$ belonging to $\text{Ideal}(g_1, h_1^2)$ $\cap\text{Ideal}(g_2, h_2^2)$ can be expressed as

$$f = ug + (\alpha a_2 + \gamma h_1)h_1h_2$$

and is irreducible provided that $v = \alpha a_2 + \gamma h_1 \neq 0$.

Similarly to the blending of two surfaces, it is not difficult to prove the following.

Lemma 19.11:

$S(f)$ contains $S(g_i, h_i)$ $(i = 1, 2, 3)$ iff

$$f \in \text{Ideal}(g_1, h_1) \cap \text{Ideal}(g_2, h_2) \cap \text{Ideal}(g_3, h_3).$$

Lemma 19.12:

An irreducible quadric $S(g)$ contains $S(g_i, h_i)$ $(i = 1, 2, 3)$ iff

$$g = \lambda_1 g_1 + a_1 h_1 = \lambda_2 g_2 + a_2 h_2 = \lambda_3 g_3 + a_3 h_3, \tag{19.9}$$

where the λ_i are real numbers and a_i are linear polynomials $(i = 1, 2, 3)$.

Lemma 19.13:

Let an irreducible quadric $S(g)$ contain $S(g_i, h_i)$ $(i = 1, 2, 3)$, then

$$\text{Ideal}(g_1, h_1) \cap \text{Ideal}(g_2, h_2) \cap \text{Ideal}(g_3, h_3) = \text{Ideal}(g, h_1 h_2 h_3).$$

Lemma 19.14:

For any $f \in \text{Ideal}(g, h_1 h_2 h_3)$, there exist polynomials c and d such that

$$f = cg + dh_1 h_2 h_3, \qquad \deg(c) \leq \deg(f) - 2, \qquad \deg(d) \leq \deg(f) - 3.$$

Lemma 19.15:

If $S(f)$ meets $S(g_i)$ with GC^1 continuity along $S(g_i, h_i)$ respectively, then

$$f \in \text{Ideal}(g_1, h_1^2) \cap \text{Ideal}(g_2, h_2^2) \cap \text{Ideal}(g_3, h_3^2)$$

and f can be written as

$$f = a_1 g_1 + b_2 h_3^2 = a_2 g_2 + b_2 h_2^2 = a_3 g_3 + b_3 h_3^2.$$

If a_i is not identically zero on $S(g_i, h_i)$ $(i = 1, 2, 3)$, then $S(f)$ meets $S(g_i)$ with GC^1 continuity along $S(g_i, h_i)$.

From Lemma 19.13, if the ruled surface $S(g)$ exists, it is not difficult to conclude that

$$\text{Ideal}(g_1, h_1) \cap \text{Ideal}(g_2, h_2) \cap \text{Ideal}(g_3, h_3) \subseteq \text{Ideal}(g, h_1 h_2 h_3).$$

Therefore, from Lemma 19.14 in the construction of GC^1 blending surfaces, f can be expressed as

$$f = ug + vh_1 h_2 h_3, \tag{19.10}$$

$$\deg(u) \leq \deg(f) - 2, \qquad \deg(v) \leq \deg(f) - 3. \tag{19.11}$$

Theorem 19.16:

If the ruled surface $S(g)$ exists, then $f \in \text{Ideal}(g_1, h_1^2) \cap \text{Ideal}(g_2, h_2^2) \cap \text{Ideal}(g_3, h_3^2)$, where f is defined by (19.10) and (19.11), iff there exist polynomials u and v such that

$$ua_i + vh_j h_k \in \text{Ideal}(g, h_i), \quad i, j, k = 1, 2, 3, \tag{19.12}$$

and they are pairwise different, where u and v satisfy (19.11).

In other words, the necessary and sufficient condition is that there exist polynomials u, v and b_i, c_i that satisfy (19.11) and such that

$$ua_i + vh_jh_k = b_ig + c_ih_i, \quad i, j, k = 1, 2, 3; \tag{19.13}$$

$$\deg(b_i) \leq \deg(f) - 2, \qquad \deg(c_i) \leq \deg(f) - 3.$$

We make the restriction that

$$u = \alpha_1 h_2 h_3 + \alpha_2 h_1 h_3 + \alpha_3 h_1 h_2, \tag{19.14}$$

where α_i $(i = 1, 2, 3)$ are non-zero real numbers. Thus, we get the following constructive theorem.

Theorem 19.17:

Under the basic hypothesis and the assumption of the existence of a quadric ruled surface and that u satisfies (19.14), *there exits an irreducible quartic polynomial*

$$f = ug + vh_1h_2h_3 \in \bigcap_{i=1}^{3} \mathrm{Ideal}(g_i, h_i^2)$$

iff there exist non-zero real numbers α_i and β_i $(i = 1, 2, 3)$ such that

$$v = \beta_1 h_1 - \alpha_1 a_1 = \beta_2 h_2 - \alpha_2 a_2 = \beta_3 h_3 - \alpha_3 a_3$$

holds, where $v = \beta_1 h_1 - \alpha_1 a_1 \neq 0$.

Although finding the solution is very complicated, it can be done in many cases.

For the same problem as in Example 19.1, we can prove by our method that there exists a cubic blending surface $S(f)$ with GC^1 continuity iff $r_1^2 + d_1^2 = r_2^2 + d_2^2$ and $d_1, d_2 \neq 0$.

Figure 19.3

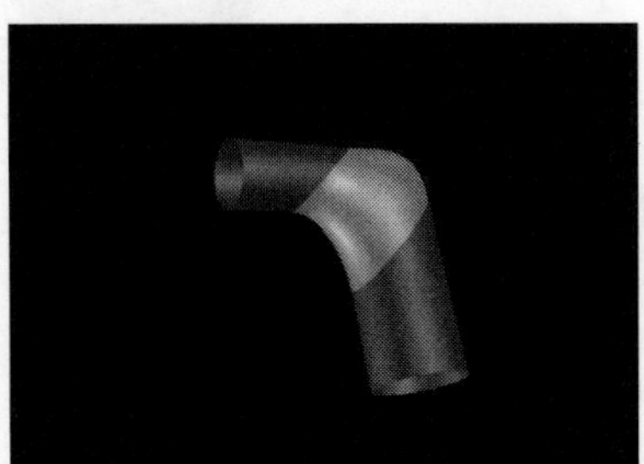

Figure 19.4

Example 19.3:

Consider

$$\begin{cases} g_1 = x^2 + y^2 - 1, \\ g_2 = x^2 + y^2 + z^2 - 4, \end{cases} \qquad \begin{cases} h_1 = z - 2, \\ h_2 = z - 1. \end{cases}$$

There exists a cubic polynomial f of the form (Fig. 19.3)

$$\begin{aligned} f = &[-(\beta_1 + 2\beta_2)(\beta_1 + \beta_2)z][x^2 + y^2 + (-2 + z)(1 + z) - 3] \\ &+(-2 + z)(-1 + z)(-4 - 3\beta_1 + 2z) \end{aligned}$$

such that the surfaces $S(f)$ and $S(g_i)$ meet with GC^1 continuity along the intersection $S(g_i, h_i)$. It should be pointed out that Warren's method can only produce quartic surfaces.

Example 19.4: (Fig. 19.4)

Let

$$\begin{cases} g_1 = y^2 + z^2 - r_1^2, \\ g_2 = x^2 + z^2 - r_2^2, \end{cases} \qquad \begin{cases} h_1 = x + a_0 y + b_0, \\ h_2 = c_0 x + y + d_0. \end{cases}$$

This example is a generalization of Example 19.1.

(i) $S(h_1) \not\parallel S(h_2)$, i.e. $a_0 c_0 \neq 1$. There exists a cubic GC^1 blending surface iff $q^2 - (a_0 q - b_0 p)^2 = r_1^2 - r_2^2$, where $p = d_0 - b_0 c_0$ and $q = 1 - a_0 c_0$.

(ii) $S(h_1) \parallel S(h_2)$, i.e. $a_0 c_0 \neq 1$. If $a_0 = c_0 = 1$ or -1 and $b_0, d_0 \neq 0$, then there exists a cubic GC^1 blending surface.

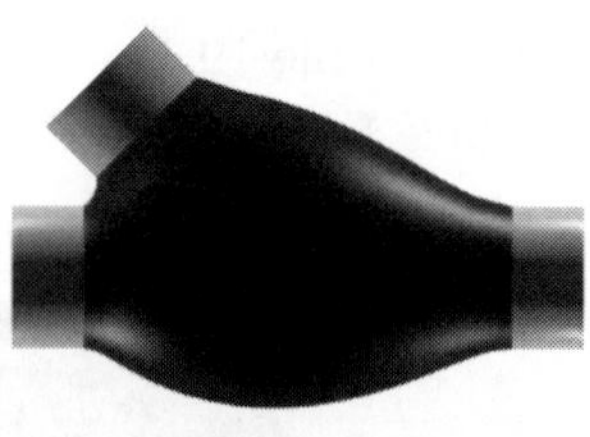

Figure 19.5

Example 19.5: (Fig. 19.5)

Let two cylinders and two clipping surfaces be of the form

$$\begin{cases} g_1 = y^2 + z^2 - r_1^2, \\ g_2 = y^2 + z^2 - r_2^2, \end{cases} \qquad \begin{cases} h_1 = x - d_1, \\ h_2 = x - d_2, \end{cases}$$

let the axis of the third cylinder intersect with the x-axis at the origin in the xy plane, and let their intersectional angle be θ. We obtain the blending surface

$$f = (x^2 + 4x\sin\theta + 4y\cos\theta - 12)(y^2 + z^2 - 1) + (x^2 - 4)^2.$$

When taking $\theta = \frac{\pi}{4}$, we have

$$(x^2 + 2\sqrt{2}x + 2\sqrt{2}y - 12)(y^2 + z^2 - 1) + (x^2 - 4)^2 = 0.$$

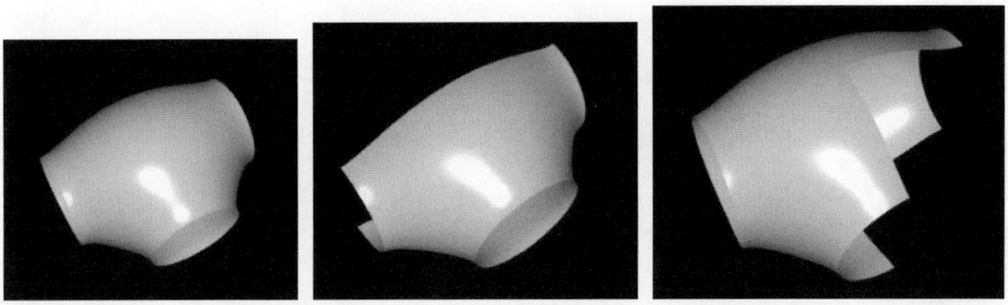

(a) (b) (c)

Figure 19.6

Example 19.6: (Figs. 19.6a–c)

Let three cylinders and three clipping surfaces be of the form

$$\begin{cases} g_1 = y^2 + z^2 - r_1^2, \\ g_2 = y^2 + z^2 - r_2^2, \\ g_3 = x^2 + z^2 - r_3^2, \end{cases} \qquad \begin{cases} h_1 = x - d_1, \\ h_2 = x - d_2, \\ h_3 = y - d_3. \end{cases}$$

We obtain a sufficient and necessary condition for the existence of a quadric ruled surface, which is $r_1^2 + d_1^2 = r_2^2 + d_2^2 = r_3^2 + d_3^2$.

As an example, we take $r_1 = r_2 = r_3 = 1, d_1 = d_3 = 2, d_2 = -2$, and then obtain a blending surface

$$f = -4(x^2 + 4y - 12)(x^2 + y^2 + z^2 - 5) + 16(x - 2)(x + 2)(y - 2).$$

19.3.2 Constructing blending surfaces of low degree in general cases

A. Two surfaces

Based on Theorems 19.9 and 19.16, constructing a blending surface can be transformed into solving a syzygy equation (19.6) or (19.9). When

$$u = \alpha_1 h_1 + \alpha_2 h_2 \quad \text{or} \quad u = \alpha_1 h_2 h_3 + \alpha_2 h_1 h_3 + \alpha_3 h_1 h_2,$$

we have obtained all expressions of blending surfaces of low degree. It is difficult to solve the syzygy equation (19.6) or (19.9) when u does not satisfy the above conditions. In what follows, we will solve this problem using the expression of the polynomial f with respect to clipping planes, the so-called *canonical expansion*. With the expressions of u, v, a_i and (19.12), the problem can be changed into seeking the solutions of the linear algebraic equations derived from the coefficients in the expressions.

We will discuss the problem of blending two surfaces in two cases and give necessary and sufficient conditions of existence of cubic blending surfaces and their expressions respectively.

Case 1. $S(h_1)$ is not parallel to $S(h_2)$. Let

$$h_1 = \xi_1 x + \eta_1 y + \zeta_1 z + \delta_1, \qquad h_2 = \xi_2 x + \eta_2 y + \zeta_2 z + \delta_2. \tag{19.15}$$

We choose h_3 such that $S(h_3) \perp S(h_i)$ $(i = 1, 2)$ and $S(h_3)$ passes through the origin; then h_3 can be defined by

$$h_3 = \begin{vmatrix} x & y & z \\ \xi_1 & \eta_1 & \zeta_1 \\ \xi_2 & \eta_2 & \zeta_2 \end{vmatrix}. \tag{19.16}$$

From (19.15) and (19.16), we can express x, y, z in terms of h_1, h_2, h_3. Thus an arbitrary polynomial g_i can be expressed with respect to h_k $(k = 1, 2, 3)$. In the following, we will use the expression of g_i with respect to h_1, h_2, h_3 directly, and call this expression the *canonical expression* of g_i.

Provided that the ruled surface $S(g)$ exists, by Proposition 19.7, there exist non-zero constants λ_1, λ_2 and linear polynomials a_1, a_2 such that

$$\lambda_1 g_1 - \lambda_2 g_2 = a_1 h_1 - a_2 h_2. \tag{19.17}$$

Substituting the canonical expressions of g_i and a_i,

$$g_i = \gamma_{00}^i + \sum_{l=1}^{2}\sum_{k=1}^{3} \gamma_{lk}^i h_k^l + \sum_{k\neq j\neq l\neq k} \gamma_{mk}^i h_j h_l, \qquad a_i = \alpha_0^i + \sum_{k=1}^{3} \alpha_k^i h_k,$$

into (19.17) and comparing the coefficients, we have

$$\gamma_{00}^\lambda = 0, \qquad \gamma_{13}^\lambda = 0, \qquad \gamma_{23}^\lambda = 0, \tag{19.18}$$

$$\begin{cases} \gamma_{11}^\lambda = \alpha_0^1, & \gamma_{21}^\lambda = \alpha_1^1, & \gamma_{m2}^\lambda = \alpha_3^1, \\ -\gamma_{12}^\lambda = \alpha_0^2, & -\gamma_{22}^\lambda = \alpha_2^2, & -\gamma_{m1}^\lambda = \alpha_3^2, \\ \alpha_2^1 = \theta + \dfrac{\gamma_{m3}}{2}, & \alpha_1^2 = \theta - \dfrac{\gamma_{m3}^\lambda}{2}, \end{cases}$$

where $\gamma_{ij}^\lambda = \lambda_1 \gamma_{ij}^1 - \lambda_2 \gamma_{ij}^2$ and θ is an arbitrary constant.

Therefore, if the ruled surface exists, we have

$$\begin{cases} a_1 = \gamma_{11}^{\lambda} + \gamma_{21}^{\lambda} h_1 + \left(\theta + \dfrac{\gamma_{m3}^{\lambda}}{2}\right) h_2 + \gamma_{m2}^{\lambda} h_3, \\ a_2 = -\gamma_{12}^{\lambda} + \left(\theta - \dfrac{\gamma_{m3}^{\lambda}}{2}\right) h_1 - \gamma_{22}^{\lambda} h_2 - \gamma_{m1}^{\lambda} h_3. \end{cases} \tag{19.19}$$

Theorem 19.18: (Wu et al. 1995)

A necessary and sufficient condition for the existence of the ruled surface $S(g)$ is that the coefficients of g_i satisfy (19.18). *In this case, a_1 and a_2 are defined by* (19.19) *and θ is an arbitrary parameter.*

For an arbitrary polynomial set W, we will use $(W)_{12}$ to express the residual of W modulo h_1 and h_2 and $(W)_i$ to express the residual of W modulo h_i $(i = 1, 2)$.

In the following, we always assume that g exists and has been calculated. For convenience, we will use g_i instead of $\lambda_i g_i$; then

$$\gamma_{ij}^{\lambda} = \gamma_{ij}^{1} - \gamma_{ij}^{2}, \qquad g = g_1 - a_1 h_1 = g_2 - a_2 h_2.$$

Let

$$u = \mu_0 + \sum_{k=1}^{3} \mu_k h_k, \qquad v = \nu_0 + \sum_{k=1}^{3} \nu_k h_k.$$

We have proved the following results.

Theorem 19.19:

Under the existence condition of the ruled surface and the assumption that $\gamma_{11}^{\lambda} = \gamma_{12}^{\lambda} = \gamma_{m1}^{\lambda} = \gamma_{m2}^{\lambda} = 0$, there exists an $f \in \mathrm{Ideal}(g_1, h_1^2) \cap \mathrm{Ideal}(g_2, h_2^2)$ which takes form as (19.5) *and of which the degree is not greater than 3 and $(u)_{12} \neq 0$ iff $\gamma_{m3}^{\lambda} = 0$. In this case, the ruled surface is a GC^1 blending surface.*

Theorem 19.20:

Under the existence condition of the ruled surface and $u_{12} \neq 0$, there exists a cubic polynomial $f \in \mathrm{Ideal}(g_1, h_1^2) \cap \mathrm{Ideal}(g_2, h_2^2)$ such that it takes form as (19.5) *iff one of the following conditions holds:*

(1) $\sigma = \dfrac{\gamma_{m2}^{\lambda}}{\gamma_{11}^{\lambda}} = -\dfrac{\gamma_{m1}^{\lambda}}{\gamma_{12}^{\lambda}}$ *satisfy*

$$\sigma^2 \gamma_{00} - \sigma \gamma_{13} + \gamma_{23} = 0,$$
$$\sigma(2\gamma_{00}\gamma_{m3}^{\lambda} - \gamma_{11}\gamma_{12}^{\lambda} + \gamma_{12}\gamma_{11}^{\lambda}) = \gamma_{m1}\gamma_{11}^{\lambda} - \gamma_{m2}\gamma_{12}^{\lambda} + \gamma_{13}\gamma_{m3}^{\lambda}$$

when $\gamma_{11}^{\lambda}\gamma_{12}^{\lambda} \neq 0$;

(2) $\gamma_{00} = 0, \qquad \gamma_{m1}^{\lambda}\gamma_{11} + \gamma_{m2}^{\lambda}\gamma_{12} - \gamma_{m3}^{\lambda}\gamma_{13} = 0$

when $\gamma_{11}^{\lambda} = \gamma_{12}^{\lambda} = 0$ *and* $\gamma_{m1}^{\lambda}\gamma_{m2}^{\lambda} \neq 0$;

(3) $$\alpha_0^2\gamma_{m2} - \alpha_3^2\gamma_{11} - \left(\gamma_{13} - \frac{2\alpha_3^2}{\alpha_0^2}\gamma_{00}\right)(\alpha_2^1 - \alpha_1^2) = 0$$

when $\gamma_{11}^{\lambda} = \gamma_{m2}^{\lambda} = 0$ *and* $\gamma_{12}^{\lambda} \neq 0$;

(4) $\gamma_{13}\gamma_{m3}^{\lambda} - \gamma_{11}\gamma_{m1}^{\lambda} = 0$ *when* $\gamma_{11}^{\lambda} = \gamma_{m2}^{\lambda} = 0$, $\gamma_{12}^{\lambda} = 0$ *and* $\gamma_{00} = 0$.

For (1)–(4) above, the coefficients of u and v are expressed in order as follows:

$$\mu_0 = \gamma_{00}\kappa, \qquad \mu_2\gamma_{11}^{\lambda} - \mu_1\gamma_{12}^{\lambda} = (\gamma_{00}\gamma_{m3}^{\lambda} + \gamma_{11}^{\lambda}\gamma_{12} - \gamma_{12}^{\lambda}\gamma_{11})\kappa,$$

$$\mu_3 = (\gamma_{13} - \sigma\gamma_{00})\kappa, \qquad \nu_0 = \left[\gamma_{11}^{\lambda}\gamma_{12} - \left(\theta + \frac{\gamma_{m3}^{\lambda}}{2}\right)\right]\kappa - \nu_2\gamma_{11}^{\lambda},$$

$$\nu_2 = \gamma_{11}^{\lambda}\gamma_{22}\kappa - \mu_2\left(\theta + \frac{\gamma_{m3}^{\lambda}}{2}\right), \qquad \nu_1 = -\gamma_{12}^{\lambda}\gamma_{21}\kappa - \mu_1\left(\theta - \frac{\gamma_{m3}^{\lambda}}{2}\right),$$

$$\nu_3 = \left[\gamma_{11}^{\lambda}\gamma_{m1} - (\gamma_{13} - \sigma\gamma_{00})\left(\theta - \frac{\gamma_{m3}^{\lambda}}{2}\right)\right]\kappa + \mu_1\gamma_{m1}^{\lambda},$$

where κ and one of μ_1 and μ_2 are free parameters;

$$\mu_0 = \gamma_{13}\bar{\kappa}, \quad \mu_3 = \gamma_{23}\bar{\kappa}, \quad \mu_2\gamma_{m2}^{\lambda} - \mu_1\gamma_{m1}^{\lambda} = (\gamma_{m2}^{\lambda}\gamma_{m1} + \gamma_{m1}^{\lambda}\gamma_{m2} - \gamma_{m3}^{\lambda}\gamma_{23})\bar{\kappa},$$

$$\nu_0 = \left[\gamma_{m2}^{\lambda}\gamma_{12} - \left(\theta + \frac{\gamma_{m3}^{\lambda}}{2}\right)\right]\bar{\kappa}, \quad \nu_1 = -\gamma_{m1}^{\lambda}\gamma_{21}\bar{\kappa} - \left(\theta - \frac{\gamma_{m3}^{\lambda}}{2}\right)\mu_1,$$

$$\nu_2 = \gamma_{m2}^{\lambda}\gamma_{22}\bar{\kappa} - \left(\theta + \frac{\gamma_{m3}^{\lambda}}{2}\right)\mu_2, \quad \nu_3 = \left[\gamma_{m2}^{\lambda}\gamma_{m1} - \left(\theta + \frac{\gamma_{m3}^{\lambda}}{2}\right)\gamma_{23}\right]\bar{\kappa} - \gamma_{m2}^{\lambda}\mu_2,$$

where one of μ_1 and μ_2 is a free parameter;

$$\mu_0 = \gamma_{00}\kappa, \quad \mu_3 = \left(\gamma_{13} - \frac{\gamma_{m1}^{\lambda}}{\gamma_{12}^{\lambda}}\gamma_{00}\right)\kappa, \quad \mu_1 = \left(\gamma_{11} - \frac{1}{\gamma_{12}^{\lambda}}\gamma_{00}\gamma_{m3}^{\lambda}\right)\kappa,$$

$$\nu_0 = -\gamma_{00}\left(\theta + \frac{\gamma_{m3}^{\lambda}}{2}\right)\kappa, \quad \nu_1 = \left[\gamma_{12}^{\lambda}\gamma_{21} - \left(\theta - \frac{\gamma_{m3}^{\lambda}}{2}\right)\left(\gamma_{11} + \frac{1}{\gamma_{12}^{\lambda}}\gamma_{00}\gamma_{m3}^{\lambda}\right)\right]\kappa,$$

$$\nu_2 = -\left(\theta + \frac{\gamma_{m3}^{\lambda}}{2}\right)\mu_2, \quad \nu_3 = \left(\theta + \frac{\gamma_{m3}^{\lambda}}{2}\right)\left(\frac{\gamma_{m1}^{\lambda}}{\gamma_{12}^{\lambda}}\gamma_{00} - \gamma_{13}\right)\kappa,$$

where κ, θ, μ_2 are free parameters and $\kappa \neq 0$;

$$\mu_0 = \gamma_{13}\bar{\kappa}, \qquad \mu_3 = \gamma_{23}\bar{\kappa}, \qquad \mu_1 = \left(\gamma_{m2} - \frac{1}{\gamma_{m1}^{\lambda}}\gamma_{23}\gamma_{m3}^{\lambda}\right)\bar{\kappa},$$

$$\nu_0 = -\left(\theta + \frac{\gamma_{m3}^{\lambda}}{2}\right)\gamma_{13}\bar{\kappa}, \quad \nu_1 = \left[\left(\frac{\gamma_{m3}^{\lambda}}{2} - \theta\right)\left(\gamma_{m2} - \frac{1}{\gamma_{m1}^{\lambda}}\gamma_{23}\gamma_{m3}^{\lambda}\right) - \gamma_{m1}^{\lambda}\gamma_{21}\right]\bar{\kappa},$$

$$\nu_2 = -\left(\theta + \frac{\gamma_{m3}^{\lambda}}{2}\right)\mu_2, \quad \nu_3 = -\left(\theta + \frac{\gamma_{m3}^{\lambda}}{2}\right)\gamma_{23}\bar{\kappa}.$$

Case 2. $S(h_1)$ is parallel to $S(h_2)$. In this case, $h_2 = h_1 + \delta$, and we denote it by $\bar{h}_1$. For convenience, we assume that h_1 passes through the origin, and the coefficient of x is 1; then

$$h_1 = x + \eta y + \zeta z.$$

For our purpose, let us construct other two planes h_2 and h_3 such that $S(h_2) \perp S(h_3)$ and their intersection is the normal line of h_1. It is not difficult to calculate that

$$\begin{cases} h_2 = -(\eta\eta_2 + \zeta\zeta_2)x + \eta_2 y + \zeta_2 z, \\ h_3 = (\eta\zeta_2 + \zeta\eta_2)x - [\zeta_2 + \zeta(\zeta\zeta_2 + \eta\eta_2)]y + [\eta_2 + \eta(\eta\eta_2 + \zeta\zeta_2)]z, \end{cases}$$

where η_2 and ζ_2 are arbitrary constants. If η_2 and ζ_2 are fixed, then x, y, z can be expressed uniquely in terms of h_1, h_2, h_3.

For the definite quadric g_i, we can also express it as a polynomial in h_k $(k = 1, 2, 3)$.

$$g_i = \gamma_{00}^i + \sum_{l=1}^{3}\sum_{k=1}^{3}\gamma_{lk}^i h_k^l + \sum_{j\neq k\neq l\neq j}\gamma_{mj}^i h_k h_l.$$

Proposition 19.21:

If the two clipping planes are parallel, then, under the basic hypothesis, a necessary and sufficient condition for the existence of the quadric ruled surface is

$$\gamma_{22}^{\lambda} = \gamma_{23}^{\lambda} = \gamma_{m1}^{\lambda} = 0.$$

From $g = \lambda_1 g_1 - a h_1 = \lambda_2 g_2 - \bar{a}(h_1 + \delta)$, we obtain

$$\lambda_1 g_1 - \lambda_2 g_2 = a h_1 - \bar{a}(h_1 + \delta). \tag{19.20}$$

Substituting the expressions of $a, \bar{a}, g_1, g_2$ into (19.20), we have

$$\bar{\alpha}_0 = -\frac{1}{\delta}\gamma_{00}^{\lambda}, \qquad \alpha_0 = \gamma_{11}^{\lambda} - \frac{1}{\delta}\gamma_{00}^{\lambda} + \alpha_1\delta, \qquad \bar{\alpha}_2 = -\frac{1}{\delta}\gamma_{12}^{\lambda},$$
$$\alpha_2 = \gamma_{m3}^{\lambda} - \frac{1}{\delta}\gamma_{12}^{\lambda}, \qquad \bar{\alpha}_3 = -\frac{1}{\delta}\gamma_{13}^{\lambda}, \qquad \alpha_3 = \gamma_{m2}^{\lambda} - \frac{1}{\delta}\gamma_{13}^{\lambda}, \qquad \bar{\alpha}_1 = \alpha_1 - \gamma_{21}^{\lambda},$$

where α_1 is an arbitrary real number.

Define $\Delta, \omega_1, \omega_2$ by

$$\begin{cases} \Delta = \alpha_2\bar{\alpha}_3 - \alpha_3\bar{\alpha}_2, \\ \omega_1 = \alpha_2^2\gamma_{23} + \alpha_3^2\gamma_{22} - \alpha_2\alpha_3\gamma_{m1}, \\ \omega_2 = \bar{\alpha}_2^2\gamma_{23} + \bar{\alpha}_3^2\gamma_{22} - \bar{\alpha}_2\bar{\alpha}_3\gamma_{m1}. \end{cases}$$

Proposition 19.22:

Under the existence condition of the ruled surface and the assumption that one of $\Delta, \omega_1, \omega_2$ is non-zero, there exists an $f \in$ Ideal$(g_1, h_1^2) \cap$ Ideal$(g_2, (h_1 + \delta)^2)$ iff

$$u = \rho_1 h_1 + \rho_2(h_1 + \delta). \tag{19.21}$$

Theorem 19.23:

Let the two clipping planes h_1 and h_2 be parallel. Then, under the existence condition of the ruled surface, there exist linear polynomials u, v with u not satisfying (19.21) *such that*

$$f = ug + vh_1(h_1 + \delta) \in \text{Ideal}(g_1, h_1^2) \cap \text{Ideal}(g_2, (h_1 + \delta)^2)$$

iff all the following conditions hold:

(1) $\Delta = \omega_1 = \omega_2 = 0$;

(2) α_2 *and* $\bar{\alpha}_2$ *are not both zero; and*

(*a*) *if* $\bar{\alpha}_2 \neq 0$, *then*

$$\eta = \alpha_0 + \bar{\alpha}_0 - \bar{\alpha}_1\delta,$$
$$\bar{\theta} = \frac{1}{\gamma_{22}}[(\alpha_2 + \bar{\alpha}_2)(\bar{\alpha}_2\gamma_{13} - \bar{\alpha}_3\gamma_{12}) + \bar{\alpha}_2\delta(\bar{\alpha}_3\gamma_{m3} - \bar{\alpha}_2\gamma_{m2})]$$

satisfy

$$\bar{\theta} + \eta\left(2\bar{\alpha}_3 - \bar{\alpha}_2\frac{\gamma_{m1}}{\gamma_{22}}\right) = 0;$$

(*b*) *if* $\bar{\alpha}_2 = 0$ *and* $\alpha_2 \neq 0$, *then*

$$\eta, \quad \theta = \frac{1}{\gamma_{22}}[(\alpha_2 + \bar{\alpha}_2)(\alpha_2\gamma_{13} - \alpha_3\gamma_{12}) + \bar{\alpha}_2\delta(\alpha_3\gamma_{m3} - \alpha_2\gamma_{m2})]$$

satisfy

$$\theta + \eta\left(2\alpha_3 - \alpha_2\frac{\gamma_{m1}}{\gamma_{22}}\right) = 0.$$

If $\bar{\alpha}_2 \neq 0$, then u and v are defined by

$$\mu_3 = \left(\frac{\gamma_{m1}}{\gamma_{22}} - \frac{\bar{\alpha}_3}{\bar{\alpha}_2}\right)\mu_2,$$

$$\mu_1 = \frac{1}{\bar{\alpha}_2\delta}\left\{\mu_0(\alpha_2 + \bar{\alpha}_2) - \left[\frac{\gamma_{12}}{\gamma_{22}}(\alpha_2 + \bar{\alpha}_2) + \frac{\bar{\alpha}_2}{\gamma_{22}}\delta\gamma_{m3} + \eta\right]\mu_2\right\},$$

$$\begin{cases} \nu_0 = \dfrac{1}{\delta}\left(\dfrac{\alpha_2}{\gamma_{22}}\gamma_{00}\mu_2 - \alpha_0\mu_0\right), \\ \nu_1 = \dfrac{1}{\gamma_{22}}\left[\bar{\alpha}_2(\gamma_{00} - \gamma_{11}\delta + \gamma_{21}\delta^2) + \alpha_2\gamma_{00} - \left(1 + \dfrac{\alpha_2}{\bar{\alpha}_2}\right)\gamma_{12} - \delta\gamma_{m3}\right]\mu_2 \\ \qquad -\dfrac{1}{\bar{\alpha}_2}\eta\mu_2 + \left(1 - \eta + \dfrac{\alpha_2}{\bar{\alpha}_2}\right)\mu_0, \\ \nu_2 = \dfrac{1}{\delta}\left[\left(\dfrac{\alpha_2}{\gamma_{22}}\right)\mu_2 - \alpha_2\mu_0\right], \\ \nu_3 = \dfrac{1}{\delta}\left[\left(\dfrac{\alpha_2}{\gamma_{22}}\gamma_{13} - \dfrac{\gamma_{m1}\alpha_0}{\gamma_{22}} + \dfrac{\alpha_0\bar{\alpha}_3}{\bar{\alpha}_2}\right)\mu_2 - \alpha_3\mu_0\right]; \end{cases}$$

if $\bar{\alpha}_2 = 0$ and $\alpha_2 \neq 0$, then u and v are defined by

$$\mu_3 = \left(\frac{\gamma_{m1}}{\gamma_{22}} - \frac{\alpha_3}{\alpha_2}\right)\mu_2, \qquad \mu_0 = \left(\frac{\gamma_{12}}{\gamma_{22}} - \frac{\eta}{\alpha_2}\right)\mu_2,$$

$$\begin{cases} \nu_0 = \dfrac{1}{\delta}\left[\dfrac{1}{\gamma_{22}}(\alpha_2\gamma_{00} - \alpha_0\gamma_{12}) + \dfrac{\alpha_0}{\alpha_2}\eta\right]\mu_2, \\ \nu_1 = \dfrac{1}{\delta^2}\left[\dfrac{1}{\gamma_{22}}(\alpha_2\gamma_{00} - \eta\gamma_{12}) + \dfrac{\eta_2}{\alpha_2}\eta\right]\mu_2 + \dfrac{1}{\delta}(\bar{\alpha}_0 - \bar{\alpha}_1\delta)\mu_1, \\ \nu_2 = \dfrac{1}{\delta}(\bar{\alpha}_0 - \bar{\alpha}_1\delta)\mu_2, \qquad \nu_3 = \dfrac{1}{\delta}(\bar{\alpha}_0 - \bar{\alpha}_1\delta)\left(\dfrac{\gamma_{m1}}{\gamma_{22}} - \dfrac{\alpha_3}{\alpha_2}\right)\mu_2, \end{cases}$$

where μ_1, μ_2 are arbitrary real parameters.

Notice that we have given the existence condition and its concrete polynomial of the form

$$f = ug + vh_1h_2 \in \text{Ideal}(g_1, h_1^2) \cap \text{Ideal}(g_2, h_2^2),$$

and we can judge whether f is the GC^1 blending surface using Proposition 19.7.

We should mention that the case in which g_i is a cubic polynomial has been discussed in Guo and Ren (1998).

B. THREE SURFACES.

Suppose that h_1, h_2, h_3 can be written as

$$h_i = \xi_{i1}x + \xi_{i2}y + \xi_{i3}z + \xi_{i0}, \quad i = 1, 2, 3.$$

In this section, we discuss the case in which $S(h_i)$, $i = 1, 2, 3$, intersect at a point. In this case, x, y, z can be expressed linearly with respect to h_i. So any polynomial can be rewritten as one in h_i, and such an expression is unique. In the following, we always express polynomials in this way. In paricular, we express g_i — the defined quadric polynomial — as

$$g_i = \gamma_{00}^i + \sum_{\substack{k=1 \\ k \neq j \neq l}}^{3} (\gamma_{1k}^i h_k + \gamma_{2k}^i h_k^2 + \gamma_{mk}^i h_j h_l).$$

First we find the quadric $S(g)$ which should meet $S(g_i)$ along $S(g_i, h_i)$ with GC^0 continuity. Such a surface is called a *ruled surface*. Then under the assumption that the ruled surface exists, we seek the smooth blending surface.

From Lemma 19.12, a necessary and sufficient condition for the existence of the quadric ruled surface is that there exist non-zero constants λ_i $(i = 1, 2, 3)$ and linear polynomials a_i $(i = 1, 2, 3)$ such that

$$g = \lambda_1 g_1 + a_1 h_1 = \lambda_2 g_2 + a_2 h_2 = \lambda_3 g_3 + a_3 h_3. \tag{19.22}$$

Another statement of this condition may be given as follows.

Theorem 19.24:

Under the basic hypothesis, a necessary and sufficient condition for the existence of the quadric ruled surface is the existence of non-zero real numbers λ_i $(i = 1, 2, 3)$ *such that*

$$\begin{matrix} \lambda_i \gamma_{00}^i = \lambda_j \gamma_{00}^j, \\ \lambda_i \gamma_{lk}^i = \lambda_j \gamma_{lk}^j, \end{matrix} \quad 1 \leq i < j \leq 3, \ l = 1, 2, \ k = 1, 2, 3, \ k \neq i, j. \tag{19.23}$$

If the ruled surface exists, according to (19.23) we may write

$$\begin{cases} r_{00} = \lambda_i r_{00}^i, & 1 \leq i < j \leq 3, \\ r_{lk} = \lambda_i r_{lk}^i = \lambda_j r_{lk}^j, & l = 1, 2, \\ r_{mi} = \lambda_i r_{mi}^i, & k = 1, 2, 3; \ k \neq i, j. \end{cases} \tag{19.24}$$

From (19.22), we have the following result.

Proposition 19.25:

If the ruled surface exists, then the coefficients of

$$a_i \equiv \alpha_0^i + \sum_{j=1}^{3} \alpha_j^i h_j$$

in the expression (19.22) *can be defined by the following equations:*

$$\alpha_0^i = r_{1i} - \lambda_i r_{1i}^i, \qquad \alpha_i^i = r_{2i} - \lambda_i r_{2i}^i, \qquad \alpha_j^i = r_{mk} - \lambda_i r_{mk}^i, \quad k \neq i, j,$$

where r_{00}, r_{lk}, r_{mi} *are defined by* (19.24), *and* g *can be written as*

$$g = r_{00} + \sum_{\substack{i=1 \\ i \neq j \neq k}}^{3} (r_{1i}h_i + r_{2i}h_i^2 + r_{mi}h_j h_k). \tag{19.25}$$

Since $\lambda_i \neq 0$, $i = 1, 2, 3$, if the ruled surface exists then either the γ_{00}^i are all zero, or they are all not zero. If $\gamma_{00}^i \neq 0$, we may normalize g_i such that $\gamma_{00}^i \equiv 1$. In this case,

$$g_i = 1 + \sum_{l=1}^{2} \gamma_{li}^i h_i^l + \sum_{i \neq j \neq k} \gamma_{mj}^i h_i h_k + \sum_{\substack{j=1 \\ j \neq i \neq k}}^{3} (r_{1j}h_j + r_{2j}h_j^2, +r_{mi}h_j h_k).$$

Correspondingly, $r_{00} = 1$ and $\lambda_i \equiv 1$; we will only discuss this case.

By Theorem 19.17, constructing a cubic $S(f)$ such that it meets $S(g_i)$ along $S(g_i, h_i)$ with GC^1 continuity is equivalent to determining all possible linear polynomial u and constant v according to (19.12). By means of the expressions of a_i, g and

$$u = \mu_0 + \mu_1 h_1 + \mu_2 h_2 + \mu_3 h_3,$$

this problem is reduced to seeking the solutions of the linear algebraic equations with respect to the coefficients of u and v. Thus, we have the following theorem.

Theorem 19.26:

If the ruled surface $S(g)$ *defined by* (19.10) *exists, there is an irreducible cubic polynomial* $f \in \bigcap_{i=1}^{3} \text{Ideal}(g_i, h_i^2)$ *iff the following equalities all hold:*

$$\begin{aligned}
&\alpha_0^k \neq 0, \qquad \Delta_k \equiv \alpha_0^i \alpha_k^j - \alpha_k^i \alpha_0^j = 0, \qquad \omega_k \equiv r_{2k} - \eta_k r_{1k} + \eta_k^2, \\
&\alpha_0^1(r_{m1} + 2\eta_2\eta_3 - \eta_3 r_{12} - \eta_2 r_{13}) = \alpha_0^2(r_{m2} + 2\eta_1\eta_3 - \eta_1 r_{13} - \eta_3 r_{11}) \\
&\qquad = \alpha_0^3(r_{m3} + 2\eta_1\eta_2 - \eta_1 r_{12} - \eta_2 r_{11}), \qquad k = 1, 2, 3,
\end{aligned}$$

where

$$\eta_k \equiv \frac{\alpha_k^i}{\alpha_0^i} = \frac{\alpha_k^j}{\alpha_0^j}, \quad i, j, k = 1, 2, 3, \ i \neq j \neq k.$$

In this case, we have

$$\begin{aligned} f &= ug + vh_1h_2h_3, \\ u &= 1 + (r_{11} - \eta_1)h_1 + (r_{12} - \eta_2)h_2 + (r_{13} - \eta_3), \\ v &= \alpha_0^1(r_{m1} + 2\eta_2\eta_3 - \eta_2 r_{13} - \eta_3 r_{12}). \end{aligned}$$

From Theorem 19.17, constructing a quartic $S(f)$ such that it meets $S(g_i)$ along $S(g_i, h_i)$ with GC^1 continuity is equivalent to determining quadric polynomials

$$u = \mu_{00} + \sum_{\substack{k=1 \\ k \neq i \neq j}}^{3} (\mu_{1k}h_k + \mu_{2k}h_k^2 + \mu_{mk}h_ih_j),$$

$$v = \nu_0 + \nu_1 h_1 + \nu_2 h_2 + \nu_3 h_3,$$

such that they satisfy (19.12).

Theorem 19.27:

Let $\Delta_k \equiv \alpha_0^i\alpha_k^j - \alpha_k^i\alpha_0^j$. If $\Delta_1\Delta_2\Delta_3 \neq 0$, then there does not exist an irreducible quartic GC^1 transition polynomial.

If $\Delta_k = 0$ and $\alpha_0^i\alpha_0^j \neq 0$, then

$$\begin{cases} \omega_k = r_{2k} - \eta_k r_{1k} + \eta_k^2, \\ \xi_k = r_{1k} - \eta_k, \\ \rho_k = \kappa_k - \eta_k\kappa, \end{cases}$$

where

$$\eta_k \equiv \frac{\alpha_k^i}{\alpha_0^i} = \frac{\alpha_k^j}{\alpha_0^j}$$

and κ, κ_k are real numbers.

Theorem 19.28:

Under the existence condition of the ruled surface $S(g)$ defined by (19.25) and $\gamma_{00} = 1$, there exists an irreducible quartic polynomial

$$f = ug + vh_1h_2h_3 \in \bigcap_{k=1}^{3} \mathrm{Ideal}(g_k, h_k^2)$$

iff one of the following conditions holds:

(1) (a) $\omega_k \rho_k = 0, \quad k = 1,2,3,$

(b) $$\begin{aligned}\omega_k(\alpha_0^i \rho_j - \alpha_0^j \rho_i) = \rho_k[\alpha_0^i(r_{1j}\eta_k - r_{1k}\eta_j) - \alpha_0^j(r_{1i}\eta_k - r_{1k}\eta_i) \\ -2(\alpha_0^i \eta_j \eta_k - \alpha_0^j \eta_i \eta_k) - (\alpha_0^i r_{mi} - \alpha_0^j r_{mj})], \\ i,j,k = 1,2,3, \quad i \neq j \neq k \neq i,\end{aligned}$$

when $\Delta_k = 0$ *and* $a_0^k \neq 0$, $k = 1,2,3$;

(2) $\alpha_2^1 \alpha_3^2 \alpha_1^3 - \alpha_3^1 \alpha_1^2 \alpha_2^3 = 0$

when $\Delta_k = 0$ *and* $\alpha_0^i = 0, \alpha_k^i \neq 0$ $(i,k = 1,2,3,\ i \neq k)$;

(3) (a) $\omega_i = \omega_j = 0,$

(b) $r_{1i}\rho_j - r_{1j}\rho_i = 0,$

(c) $$\begin{aligned}\rho_i[\alpha_0^i(r_{mi} - \eta_i r_{1k}) - \alpha_k^i(r_{1j} - 2\eta_j) - \alpha_0^k(r_{mn} - r_{1i}\eta_j - r_{1j}\eta_i \\ + 2\eta_i\eta_j)] - \rho_i\alpha_0^k(r_{2j} - r_{1j}\eta_j + \eta_j^2) = 0, \quad i,j = 1,2,3,\ i,j \neq k,\end{aligned}$$

when $\Delta_i = \Delta_j = 0$ *and* $\Delta_k \neq 0$, $i,j,k = 1,2,3$, $i \neq j \neq k \neq i$;

(4) (a) $\alpha_0^i - \alpha_0^j \neq 0, \quad i \neq k \neq j,$

(b) $\omega_k = 0,$

(c) $\Delta_i \omega_{jj}^i - \Delta_j \omega_{ii}^j = 0,$

(d) $$\frac{\alpha_0^j \omega_{jj}^i}{\Delta_i} - \frac{\alpha_0^i \omega_{ii}^j}{\Delta_j} = \alpha_0^i r_{1j} - \alpha_0^j r_{1i} - (\alpha_j^i - \alpha_i^j),$$

(e) $$\begin{aligned}(\alpha_k^i r_{mi} - \alpha_k^j r_{mj})(\kappa_k - \kappa) - [(\alpha_j^i - \alpha_i^j)(\xi_k - \eta_k) + (\alpha_0^i r_{1j} \\ - \alpha_0^j r_{1i})\eta_k]\rho_k = 0\end{aligned}$$

when $\Delta_i\Delta_j \neq 0$ *and* $\Delta_k = 0$, $i,j,k = 1,2,3$, $i \neq j \neq k \neq i$.

For (1)–(4), the coefficients of μ and ν can be expressed in order as follows:

$$\begin{cases} \mu_{00} = \kappa, \qquad \mu_{1k} = r_{1k}\kappa + \rho_k, \qquad \mu_{2k} = r_{2k}\kappa + \xi_k\rho_k \\ \mu_{mj} = r_{mj}\kappa + \alpha_0^i \dfrac{(\mu_{mj} - r_{mj}\kappa)}{\alpha_0^j} - (\dfrac{\alpha_0^1 \xi_j}{\alpha_0^j - \xi_j})\rho_k - (\alpha_0^i \rho_k - \rho_i)\xi_k, \end{cases}$$

$$j,k = 2,3, \quad k \neq j,$$

$$\begin{cases} \nu_0 = \alpha_0^1(r_{m1}\kappa - \mu_{m1} + \xi_3\rho_2 + \xi_2\rho_3), \\ \nu_1 = \alpha_2^1 r_{m2}(\rho_1 + \eta_1\kappa) - \alpha_1^2\mu_{m2} + \alpha_0^2 r_{21}\rho_3 - \alpha_1^2\rho_1\xi_3, \\ \nu_k = \alpha_0^1 r_{m1}(\rho_k + \eta_k\kappa) - \alpha_k^1\mu_{m1} + \alpha_0^1 r_{2k}\rho_j - \alpha_j^1\rho_k\xi_k, \end{cases}$$

where μ_{ml}, κ are arbitrary constants;

$$f = \left[g\kappa - \left(h_2h_3 + \frac{\alpha_3^1 h_1 h_3}{\alpha_3^2} + \frac{\alpha_2^1 h_1 h_2}{\alpha_2^3}\right)\chi\right] g$$
$$+ \left(\frac{\alpha_1^2\alpha_3^1 h_1}{\alpha_3^2} + \alpha_2^1 h_2 + \alpha_3^1 h_3\right)\chi h_1 h_2 h_3,$$

where κ, χ are arbitrary constants;

$$\mu_{00} = \kappa, \quad \mu_{1k} = r_{1k}\kappa, \quad \mu_{2k} = r_{2k}\kappa, \quad \mu_{1i} = r_{1i} + \rho_i, \quad \mu_{2i} = r_{2i}\kappa + \xi_i\rho_i,$$

$$\mu_{mk} = r_{mk}\kappa + \left(\frac{\xi_i + \alpha_0^j\omega_{ij}^k}{\alpha_0^k\Delta_k}\right)\rho_j + \left(\frac{\xi_j - \alpha_0^i\omega_{ii}^k}{\alpha_0^k\Delta_k}\right)\rho_i - \frac{(r_{1k}\alpha_0^i - \alpha_k^i)\rho_j}{\alpha_0^k},$$

$$\mu_{mi} = r_{mi}\kappa + \frac{(\omega_{ij}^k\rho_i - \omega_{ii}^k\rho_j)}{\Delta_k}, \qquad \nu_0 = \alpha_0^i(r_{mi} - \mu_{mi}) + \xi_k^i\rho_j,$$

$$\nu_j = \alpha_0^i r_{mi}\kappa_i - \alpha_j^i\mu_{mi} - \alpha_k^i\xi_j\rho_j, \quad \nu_k = \alpha_k^i(r_{mi}\kappa - \mu_{mi}) + \alpha_0^i r_{2k}\rho_j,$$
$$i, j, k = 1, 2, 3, \quad i \neq j \neq k,$$

where

$$\omega_{ij}^k = \alpha_0^i\alpha_0^j r_{2k} - \alpha_0^i\alpha_k^j r_{1k} + \alpha_k^i\alpha_k^j, k \neq i, j, \qquad \rho_i = \kappa_i - \eta_i\kappa,$$

and κ_i, κ are real numbers;

$$\mu_{00} = \kappa, \qquad \mu_{li} = r_{li}\kappa, \quad l = 1, 2, \; i = 1, 2, 3, \; i \neq k,$$
$$r_{1k}\kappa + \rho_k, \qquad \mu_{2k} = r_{2k}\kappa + \xi_k\rho_k,$$

$$\mu_{mi} = r_{mi}\kappa + \frac{\rho_k\omega_{ik}^j}{\Delta_j}, \quad i = 1, 2, 3, \; i \neq k, \quad \mu_{mk} = r_{mk}\kappa + \frac{\rho_k\omega_{ii}^j}{\Delta_j}, \quad i \neq k \neq j,$$

$$\nu_0 = \frac{\alpha_0^i\omega_{ii}^j\rho_k}{\Delta_j}, \qquad \nu_j = \left(\frac{\alpha_j^i\omega_{ik}^j}{\Delta_j} + \alpha_0^i r_{2j}\right)\rho_k, \quad i \neq k \neq j,$$

$$\nu_k = (\alpha_0^i r_{mi} + \alpha_k^i\omega_{ii}^j - \alpha_j^i\xi_k)\rho_k.$$

In the following, we discuss the case: $S(h_i) \parallel S(h_j, h_k)$ and $S(h_j, h_k) \not\subseteq S(h_i)$ $(i \neq j \neq k)$. Let $h_i = \xi_{i1}x + \xi_{i2}y + \xi_{i3}z + \xi_{i0}$, $i = 1, 2, 3$. In this case,

$$\begin{vmatrix} \xi_{11} & \xi_{12} & \xi_{13} \\ \xi_{21} & \xi_{22} & \xi_{23} \\ \xi_{31} & \xi_{32} & \xi_{33} \end{vmatrix} = 0.$$

Then there exist non-zero constants $\sigma_1, \sigma_2, \sigma_3$ such that

$$\sum_{i=1}^{3}\sigma_i h_i = \sum_{i=1}^{3}\sigma_i\xi_{i0} \equiv \sigma_0 \neq 0.$$

So, if we let $\bar{h}_i = \dfrac{\sigma_i h_i}{\sigma_0}$ then $\sum_{i=1}^{3} \bar{h}_i = 1$. In the following discussion, we suppose that

$$h_1 + h_2 + h_3 = 1.$$

Otherwise, let

$$h_t = \begin{vmatrix} \xi_{i1} & \xi_{i2} & \xi_{i3} \\ \xi_{j1} & \xi_{j2} & \xi_{j3} \\ x & y & z \end{vmatrix};$$

then h_j, h_k, h_t form a new coordinate system. So any polynomial can be expressed uniquely as one in h_i, h_j, h_t. In particular, we have

$$g_l = \gamma_{00}^{lk} + \sum_{s=1}^{2} (\gamma_{si}^{lk} h_i^s + \gamma_{sj}^{lk} h_j^s) + \gamma_{mi}^{lk} h_j h_t + \gamma_{mj}^{lk} h_i h_t + \gamma_{mt}^{lk} h_i h_j,$$
$$i, j, k, l = 1, 2, 3, \quad i \neq j \neq k \neq i.$$

In what follows, we first expand the variables considered in terms of h_t, h_i, h_j $(i \neq j)$. Also, we may expand f in terms of h_t, h_i, h_k and h_k, h_j. The transformation formulas between two coordinate systems are important to get our results. Now we analyze the relationship between the coefficients of the expansions and the conditions for the existence of the blending surfaces.

We have the following.

Theorem 19.29:

Under the basic hypothesis, a necessary and sufficient condition for the existence of a quadric ruled surface $S(g)$ is the existence of non-zero real numbers λ_i $(i = 1, 2, 3)$ such that

$$\lambda_i \gamma_{00}^{ik} = \lambda_j \gamma_{00}^{jk}, \qquad \lambda_i \gamma_{1t}^{ik} = \lambda_j \gamma_{1t}^{jk}, \qquad \lambda_i \gamma_{2t}^{ik} = \lambda_j \gamma_{2t}^{jk}$$
$$(1 \leq i \neq j \neq k \leq 3). \tag{19.26}$$

And if (19.26) *holds, then the solution of*

$$g = \lambda_1 g_1 + a_1 h_1 = \lambda_2 g_2 + a_2 h_2 = \lambda_3 g_3 + a_3 h_3,$$

in which a_i $(i = 1, 2, 3)$ are linear polynomials, is of the form

$$\gamma_{00}^{k} = \lambda_i \gamma_{00}^{i}, \qquad \gamma_{1t}^{k} = \lambda_i \gamma_{1t}^{i}, \qquad \gamma_{2t}^{k} = \lambda_i \gamma_{2t}^{i},$$

with $i, k = 1, 2, 3, \quad i \neq k$.

Proposition 19.30:

If the ruled surface exists, then the coefficients of

$$a_i = a_0^{ik} + a_i^{ik} h_i + a_j^{ik} h_j + a_t^{ik} h_t$$

in the expression (19.22) *are defined by the following equations:*

$$a_0^{ik} = \lambda_j \gamma_{1i}^{jk} - \lambda_i \gamma_{1i}^{ik}, \qquad a_i^{ik} = \lambda_j \gamma_{2i}^{jk} - \lambda_i \gamma_{2i}^{ik},$$
$$a_j^{ik} = (\lambda_j \gamma_{2i}^{jk} - \lambda_i \gamma_{2i}^{ik}) - (\lambda_k \gamma_{2i}^{kj} - \lambda_i \gamma_{2i}^{ij}), \qquad a_t^{ik} = \lambda_j \gamma_{mj}^{jk} - \lambda_i \gamma_{mi}^{ik}.$$

In this case g can be written as

$$g = \gamma_{00}^k + \sum_{s=1}^{2} (\gamma_{si}^k h_i + \gamma_{sj}^k h_j) + \gamma_{mi}^k h_j h_t + \gamma_{mj}^k h_i h_t + \gamma_{mt}^k h_i h_j,$$

$$\gamma_{1i}^k = \lambda_i \gamma_{1i}^{ik} + \alpha_0^{it}, \qquad \gamma_{1j}^k = \lambda_j \gamma_{1j}^{jk} + \alpha_0^{jk},$$
$$\gamma_{2i}^k = \lambda_i \gamma_{2i}^{ik} + \alpha_i^{ik}, \qquad \gamma_{2j}^k = \lambda_j \gamma_{2j}^{jk} + \alpha_j^{jk},$$
$$\gamma_{mi}^k = \lambda_j \gamma_{mi}^{jk} + \alpha_t^{jk}, \qquad \gamma_{mj}^k = \lambda_i \gamma_{mj}^{ik} + \alpha_t^{ik}, \qquad \gamma_{mt}^k = \lambda_i \gamma_{mt}^{ik} + \alpha_j^{ik},$$

with $i, j = 1, 2, 3$, $i \neq j \neq k \neq i$.

In the following, we put

$$\begin{cases} u = \mu_{00}^k + \mu_{1i}^k h_i + \mu_{1j}^k h_j + \mu_{1t}^k h_t + \mu_{2i}^k h_i^2 + \mu_{2j}^k h_j^2 \\ \qquad + \mu_{2t}^k h_t^2 + \mu_{mi}^k h_j h_t + \mu_{mj}^k h_i h_t + \mu_{mt}^k h_i h_j, \\ v = \nu_0^k + \nu_i^k h_i + \nu_j^k h_j + \nu_t^k h_t, \\ \Delta_i = \alpha_0^{ji} \alpha_t^{ki} - \alpha_t^{ji} \alpha_0^{ki}, \qquad \eta_i = \dfrac{\alpha_0^{ji}}{\alpha_t^{ji}} = \dfrac{\alpha_0^{ki}}{\alpha_t^{ki}}, \\ \omega_i = \gamma_{00}^i - \eta_i \gamma_{1t}^i + \eta_i^2 \gamma_{2t}^i, \qquad \xi_i = \gamma_{1t}^i - \eta_i \gamma_{2t}^i, \\ \theta_{jk} = (\alpha_t^{ki} - \alpha_t^{ji}) \omega_i + (\alpha_t^{ji} \gamma_{1k}^i - \alpha_t^{ki} \gamma_{1j}^i) + (\alpha_t^{ki} \gamma_{mk}^i - \alpha_t^{ji} \gamma_{mj}^i) \\ \qquad - (\alpha_k^{ji} - \alpha_j^{ki})(\xi_i - \eta_i \gamma_{2t}^i), \\ \phi_{jk} = \left(\dfrac{\alpha_t^{ki} \omega_j^{ki}}{\Delta_j} - \dfrac{\alpha_t^{ji} \omega_k^{ji}}{\Delta_k} \right) + (\alpha_t^{ki} - \alpha_t^{ji}) \xi_i \\ \qquad + (\alpha_k^{ji} - \alpha_j^{ki}) \gamma_{2t}^i + (\alpha_t^{ki} \gamma_{mk}^i - \alpha_t^{ji} \gamma_{mj}^i), \\ \omega_k^{ij} = \alpha_t^{ik} \alpha_t^{jk} \gamma_{00}^k - \alpha_t^{ik} \alpha_0^{jk} \gamma_{1t}^k + \alpha_0^{ik} \alpha_0^{jk} \gamma_{2t}^k, \qquad \psi_{jk} = \Delta_j \omega_k^{jj} - \Delta_k \omega_j^{kk}, \end{cases}$$

where $k = 1, 2, 3$ and $1 \leq i \neq j \neq k \leq 3$.

Theorem 19.31:

Suppose that the ruled surface $S(g)$ exists and $(a_i)_{I_{ij}} \neq 0$, $1 \leq i \neq j \neq k \leq 3$. An irreducible cubic blending surface $S(f)$ exists iff

$$\begin{cases} \Delta_1 = \Delta_2 = \Delta_3 = 0, \qquad \omega_1 = \omega_2 = \omega_3 = 0, \\ \alpha_j^{ik}(\xi_j - \xi_k) - \alpha_t^{ik}\gamma_{2j}^k = \alpha_k^{ji}(\xi_k - \xi_i) - \alpha_t^{ji}\gamma_{2k}^i \\ \quad = \alpha_i^{kj}(\xi_i - \xi_j) - \alpha_t^{kj}\gamma_{2i}^j \ (= v). \end{cases} \tag{19.27}$$

If (19.27) holds, then $S(f)$ is determined by the following

$$f = [\xi_k + g_{2t}^k h_t + (\xi_j - \xi_k)h_j + (\xi_k - \xi_i)h_i]g + [\alpha_i^{kj}(\xi_i - \xi_j) - \alpha_t^{kj}\gamma_{2i}^k]h_1h_2h_3,$$

$v \neq 0$, and $S(f)$ meets $S(g_i)$ with GC^1 continuity along $S(g_i, h_i)$ $(i = 1, 2, 3)$.

Theorem 19.32:

When $\Delta_1\Delta_2\Delta_3 \neq 0$, even if the ruled surface $S(g)$ exists, there does not exist any irreducible polynomial f such that u and v satisfy (19.13).

Theorem 19.33:

Under the existence condition of the ruled surface $S(g)$, there exists a quartic polynomial f with $\nu \neq 0$ iff one of the following conditions holds:

(1) $\omega_i = \theta_{jk} = \phi_{jk} = \psi_{jk} = 0$
when $\Delta_1\Delta_2\Delta_3 \neq 0$, $(a_i)_{I_{ij}} \neq 0$, $1 \leq i \neq j \neq k \leq 3$, and $\alpha_t^{ik}\alpha_t^{jk} \neq 0$;

(2) $\omega_i = \omega_j = 0, \qquad \theta_{ik}^i\theta_{jk}^j = 0$
when $\Delta_i = \Delta_j = 0$, $\Delta_k \neq 0$ $(1 \leq i \neq j \neq k \leq 3)$ and $\alpha_t^{12}\alpha_t^{23}\alpha_t^{31} \neq 0$;

(3) *there exists ρ_k such that*

$$\omega_k\rho_k = 0, \qquad \omega_k[\alpha_t^{ik}\rho_j - \alpha_0^{jk}\rho_i] + \theta_{ij}\rho_k = 0 \qquad (1 \leq i \neq j \neq k \leq 3)$$

when $\Delta_1 = \Delta_2 = \Delta_3 = 0$ and $\alpha_t^{12}\alpha_t^{23}\alpha_t^{31} \neq 0$.

For (1)–(3), the coefficients of μ and ν can be expressed in order as follows:

$$\begin{cases} u = \kappa g - \rho_i \left[\dfrac{\omega_k^{ji}}{\Delta_k}h_i - \dfrac{\omega_k^{jj}}{\Delta_k}h_j + \left(\xi_i + \dfrac{\omega_k^{ji}}{\Delta_k}\right)h_i^2 + \dfrac{\omega_k^{jj}}{\Delta_k}h_j^2 + \gamma_{2t}^i h_i h_t \right. \\ \qquad \left. + \left(\dfrac{\omega_k^{ji}}{\Delta_k} + \dfrac{\omega_j^{kk} - \omega_j^{ki}}{\Delta_j}\right)h_ih_j\right], \\ v = \dfrac{\rho_i}{\Delta_k}[\alpha_0^{ik}\omega_k^{jj} + (a_i^{jk}\xi_i\Delta_k - a_t^{ji}\gamma_{2i}^k\Delta_k + a_i^{jk}\omega_k^{ji})h_i + a_j^{ik}\omega_k^{jj}h_j + a_j^{ik}\omega_k^{jj}h_t], \end{cases}$$

where κ and ρ_i are parameters;

$$\begin{cases} \mu_{00}^k = \gamma_{00}^k \kappa, \quad \mu_{1i}^k = \gamma_{1i}^k \kappa + \dfrac{\omega_k^{ii}\rho_j - \omega_k^{ji}\rho_i}{\Delta_k}, \quad \mu_{1j}^k = \gamma_{1j}^k \kappa + \dfrac{\omega_k^{ij}\rho_j - \omega_k^{jj}\rho_i}{\Delta_k}, \\ \mu_{1t}^k = \gamma_{1t}^k \kappa, \qquad \mu_{2i}^k = \gamma_{2i}^k \kappa + \xi_i \rho_i - \dfrac{\omega_k^{ii}\rho_j - \omega_k^{ji}\rho_i}{\Delta_k}, \\ \mu_{2j}^k = \gamma_{2j}^k \kappa + \xi_j \rho_j - \dfrac{\omega_k^{ij}\rho_j - \omega_k^{jj}\rho_i}{\Delta_k}, \qquad \mu_{2t}^k = \gamma_{2t}^k \kappa, \\ \mu_{mi}^k = \gamma_{mi}^k \kappa + \gamma_{2t}^j \rho_j, \qquad \mu_{mj}^k = \gamma_{mj}^k \kappa + \gamma_{2t}^i \rho_i, \\ \mu_{mt}^k = \left(\dfrac{\alpha_t^{ij}}{\alpha_t^{kj}} - 1\right)\mu_{1j}^k - \mu_{1i}^k + \dfrac{2\alpha_t^{ij}}{\alpha_t^{kj}}(\gamma_{00}^k - \gamma_{00}^j)\kappa - \dfrac{R_j}{\alpha_t^{kj}}, \end{cases}$$

$$\begin{cases} \nu_0^k = \left(\alpha_t^{ij}\gamma_{00}^k + \dfrac{\alpha_0^{ik}\omega_k^{ji}}{\Delta_k}\right)\rho_i - \dfrac{\alpha_0^{ik}\omega_k^{ij}}{\Delta_k}\rho_j, \\ \nu_i^k = \left[\alpha_t^{ij}\gamma_{2i}^k + \alpha_i^{jk}\left(\xi_i + \dfrac{\omega_k^{ji}}{\Delta_k}\right)\right]\rho_i - \dfrac{\alpha_i^{jk}\omega_k^{ii}}{\Delta_k}\rho_j, \\ \nu_j^k = \left[\alpha_t^{ij}\gamma_{2j}^k + \alpha_j^{ik}\left(\xi_j - \dfrac{\omega_k^{ij}}{\Delta_k}\right)\right]\rho_j + \dfrac{\alpha_j^{ik}\omega_k^{jj}}{\Delta_k}\rho_i, \\ \nu_t^k = \dfrac{a_t^{ik}\omega_k^{jj}}{\Delta_k}\rho_i + \left(\alpha_t^{ij}\gamma_{1t}^k - \alpha_0^{ik}\gamma_{2t}^j - \dfrac{\alpha_t^{ik}\omega_k^{ij}}{\Delta_k}\right)\rho_j, \end{cases}$$

where κ, ρ_i and ρ_j are parameters;

$$\begin{cases} \mu_{00}^k = \gamma_{00}^k + \xi_k \rho_k, \qquad \mu_{1t}^k = \rho_{2t}^k \rho_k, \qquad \mu_{2t}^k = \gamma_{2t}^k, \\ \mu_{1i}^k = \dfrac{1}{\alpha_t^{jk}}(\alpha_t^{ik}\mu_{1j}^k + R_k), \qquad \mu_{2i}^k = \gamma_{00}^i \kappa - \mu_{00}^k - \mu_{1t}^k, \\ \mu_{mi}^k = \gamma_{mi}^k \kappa + \gamma_{2t}^j(\rho_j - \rho_k), \\ \mu_{mt}^k = \dfrac{1}{\alpha_t^{kj}}(\alpha_t^{ij}\mu_{1k}^j + R_j) - \mu_{1i}^k + 2[(\gamma_{00}^j - \gamma_{00}^k)\kappa + \xi_j\rho_j - \xi_k\rho_k]\dfrac{\alpha_t^{ij} - \alpha_t^{kj}}{\alpha_t^{kj}}, \\ \nu_0^k = \alpha_t^{ik}[(\kappa_k\gamma_{1j}^k - \eta_k\mu_{1j}^k) + \gamma_{00}^k\rho_j - (\omega_k + \eta_j\xi_k)\rho_k], \\ \nu_i^k = \alpha_t^{ik}[(\kappa\gamma_{1j}^k - \mu_{1j}^k) + \xi_k\rho_j + (\xi_j - 2\xi_k)\rho_k], \\ \nu_j^k = \alpha_t^{jk}\{(\eta_j - \eta_k)[(\gamma_{00}^j - \gamma_{00}^k)\kappa + \xi_j\rho_j - \xi_k\rho_k - \mu_{1j}^k] - (\kappa_j - \kappa_k)\gamma_{2j}^k\}, \end{cases}$$

where μ_{1j}^k, κ, ρ_i, ρ_j and ρ_k $(1 \le i \ne j \ne k \le 3)$ are parameters.

Example 19.7:

Let

$$\begin{aligned} g_1 &= -1 + x + xy + xz + yz + x^2, \\ g_2 &= -1 + y + xy + xz + yz + y^2, \\ g_3 &= -1 + z + xy + xz + yz + z^2, \\ h_1 &= x, \qquad h_2 = y, \qquad h_3 = z; \end{aligned}$$

we have

$$\begin{aligned} g &= g_1 + a_1 x = -1 + xy + xz + yz, \\ f &= (-1 + xy + xz + yz) + xyz. \end{aligned}$$

Example 19.8:

Let

$$\begin{aligned} g_1 &= 2 + y + z - y^2 - z^2 + yz, \\ g_2 &= 2 + x + z - x^2 - z^2 + xz, \\ g_3 &= 2 + x + z - x^2 - y^2 + xy, \\ h_1 &= x, \qquad h_2 = y, \qquad h_3 = z; \end{aligned}$$

we have

$$\begin{aligned} g &= 2 + x + y + z - x^2 - y^2 - z^2 + yz + xz + xy, \\ f &= (2 - x - y - z)g + 3xyz. \end{aligned}$$

Example 19.9:

Suppose that there are three elliptical cylinders and three clipping planes:

$$\begin{cases} g_1 = \dfrac{y^2}{b^2} + \dfrac{z^2}{c^2} - r_1^2, \\ g_2 = \dfrac{x^2}{a^2} + \dfrac{z^2}{c^2} - r_2^2, \\ g_3 = \dfrac{x^2}{a^2} + \dfrac{y^2}{b^2} - r_3^2, \end{cases} \qquad \begin{cases} h_1 = x - d_1, \\ h_2 = y - d_2, \\ h_3 = z - d_3. \end{cases}$$

Then there exists a quadric $S(g)$ containing the planar quadratic curves $S(g_i, h_i)$ $(i = 1, 2, 3)$ iff

$$r_1^2 + \frac{d_1^2}{a^2} = r_2^2 + \frac{d_2^2}{b^2} = r_3^2 + \frac{d_3^2}{c^2}$$

and the blending surface $S(f)$ is defined by

$$\begin{aligned} f = {} & (c^2 d_1 d_2 h_1 h_2 + b^2 d_1 d_3 h_1 h_3 + a^2 d_2 d_3 h_2 h_3) \left(\frac{x^2}{a^2} + \frac{y^2}{b^2} + \frac{z^2}{c^2} - \frac{d_1^2}{a^2} - r_1^2 \right) \\ & - 2 d_1 d_2 d_3 h_1 h_2 h_3. \end{aligned}$$

Chapter 20

GEOMETRIC THEOREM PROVERS AND ALGEBRAIC EQUATION SOLVERS

Shang-Ching Chou, Xiao-Shan Gao, Zhuo-Jun Liu, Ding-Kang Wang, and Dongming Wang

The methods of mechanizing mathematics are realized by means of computer software for solving scientific and engineering problems via symbolic and hybrid computation. This chapter provides a collection of geometric theorem provers and algebraic equation solvers that are pieces of mathematical software based mostly on Wu's method and were developed mainly by members of the extended Wu's group. The early theorem provers, though efficient, were written in basic programming languages and on primitive computers. Now there exist more powerful and mature geometric theorem provers of which some have already been published as commercial software. On the other hand, building effective equation solvers is still at the experimental stage and remains for further research and development.

20.1 Introduction

The mechanization of mathematics consists of three layers: theory and algorithms, software, and applications. The value and significance of algorithmic developments are rendered visible by means of their successful applications. Moreover, computational problems from practical applications often serve as key sources for motivating further development and improvement of theory and algorithms. The role of software development lies in bridging algorithms and applications. Therefore, the design and implementation of powerful software tools are always a central issue to the development of mathematics mechanization. Only with effective implementation on computers can good algorithms be put to use for solving practical problems from science and engineering.

When Wu proposed his method of automated geometry theorem proving (AGTP) in the late 1970s, there was no computer available to him. The computation for the first few examples in his papers was carried out with pencil

MATHEMATICS MECHANIZATION AND APPLICATIONS
ISBN 0-12-734760-7

and paper, involving polynomials of hundreds of terms (Wu 1978, 1979). In the early 1980s, Wu was able to use an HP personal computer and wrote his first geometric theorem prover in Fortran 77. This prover is capable of proving very difficult geometric theorems (including Feuerbach's and Morley's theorems) and was also used to discover several new ones. During 1983–1984, Chou developed a geometric theorem prover at the University of Texas based on Wu's method (with modifications), first using the PASCAL language and then using Macsyma on Lisp Machines (see Chou 1984, 1988). With this prover Chou was able later to prove about 600 non-trivial geometric theorems, which exerted a widespread influence in the field of automated reasoning.

Stimulated by the work and provers of Wu (1978, 1984a) and Chou (1984), several students of Wu in Beijing (Gao 1987; Wang and Hu 1987; Wang and Gao 1987) and other groups of researchers in the USA and Europe (Ko 1986; Kusche et al. 1989) implemented efficient geometric theorem provers in the middle 1980s. These early provers were developed mainly for proving and discovering theorems in elementary geometries. A number of other software systems have been developed since then not only for automated theorem proving and discovering and formula derivation in elementary as well as differential geometries but also for solving systems of polynomial equations with applications to various problems in computer vision, solid modeling, linkage design, robotics, chemical computation, mechanics, theoretical physics and CAD as we have seen partially in the previous chapters.

A side remark: although computer algebra systems had already been available in the West, Wu and his students did not have access to such systems in the early and middle 1980s. Each of them wrote his own Fortran programs, starting with representations and operations of large integers and multivariate polynomials. In some way they undertook the painful tasks of re-inventing the basic techniques for designing and implementing polynomial algebra systems.

In this chapter, we focus on the development of effective geometric theorem provers and polynomial equation solvers by researchers from the extended Wu's group. A short description for each software package or system is provided. Related work by other researchers is mentioned only briefly; the reader should consult the indicated references and sources for details.

As some of the provers for geometric theorems and the solvers for polynomial equations are based on the same algebraic method of Wu, such provers may also be used with minor modifications as solvers, and vice versa. Therefore, the classification between provers and solvers is made mainly according to their primary purposes.

20.2 Geometric theorem provers

Most of the early provers implemented only a simple version of Wu's method that is based on the well-ordering principle and successive pseudo-reduction. A complete version of Wu's method involves zero and variety decompositions, and provides an effective method for AGTP and a symbolic approach for solving systems of polynomial equations and hence has much wider scope of application. The complete version of Wu's method for AGTP has been implemented in some of the later provers. A few special yet more powerful modules have also been designed for certain restricted classes of theorems. The provers described in this section are listed chronologically, roughly according to the periods when they were developed.

20.2.1 The China-Prover: A historical theorem prover for geometry

This historical China-Prover was written by Wu during 1980–1983 in Fortran 77 on HP9835A and HP1000 micro-computers (Wu 1984c). It is based on the method proposed by Wu (1978) himself; this method is called the simple version of Wu's method in Chap. 5. The prover consists of three parts which perform the three steps of the proving:

- translating the geometric relations into polynomial equations;
- transforming the hypothesis polynomial set into a triangular form using the well-ordering principle; and
- determining the successive pseudo-remainders of the conclusion polynomial with respect to the hypothesis polynomials in triangular form.

If the final remainder is zero then the theorem is valid under some non-degeneracy conditions. Otherwise, it is generally false.

This version of Wu's method, though quite simple, is powerful enough to prove many difficult theorems and to discover some new results. It is the first geometric theorem prover based on Wu's algebraic method that was written by Wu to show the effectiveness of the method. The detailed implementation and extensive experimental study of Wu's method of AGTP were later carried out by his students and other researchers.

20.2.2 GEO: A Macsyma theorem prover at UT

The geometry theorem prover GEO developed by Chou (1984) at the University of Texas (UT) is based on Wu's method (Wu 1978, 1984a). It was implemented first in PASCAL and then in Macsyma on Lisp Machines during 1983–1985. This prover has been used to prove a large number of difficult geometric theorems and is widely known in the area of AGTP. There are several distinctive features in the prover of that time.

- The input is purely geometric, i.e., the user does not have to select coordinates manually – this job is done by the prover automatically and optimally.
- It has graphic displays: the geometric input also generates the graphic diagram of the theorem.
- It decomposes a triangular form into irreducible components in the case of quadratic successive algebraic extension fields over the base field $\boldsymbol{Q}$ (for theorems that have been run into this prover so far, there is no need of decomposition over algebraic extensions with degree > 2).
- For a class of constructive geometric theorems (i.e., geometric statements about the configurations that can be drawn by rulers and compasses), the output is also geometric, i.e., non-degeneracy conditions are given in geometric form such as "Points A, B, C are not collinear."

The prover was later used to prove over 600 theorems, of which 512 are reported in the book by Chou (1988). Some "new" theorems were also discovered with this prover (Chou 1988).

20.2.3 CPS/PS/DPMS: Early Fortran theorem provers at AC

These provers were written by Gao and D. Wang in Fortran 77 on an HP-1000 computer during 1985–1987 at Academia Sinica (AC). They are based on Wu's method with modifications for various reasons and were used to prove more than one hundred geometric theorems, of which 95 are included in Wang and Gao (1987).

The prover CPS implemented by Wang is based on Hilbert's mechanical method for pure intersection point theorems described in Wu (1982). In CPS, Hilbert's method is extended to cover theorems involving metric relations such as circles and angles (Wang 1989b). Wang also wrote a general theorem prover PS on the basis of Wu's well-ordering principle. The provers CPS and PS were used to prove a number of quite difficult theorems; several "new" theorems were also discovered by them (Wang 1989a).

DPMS (Differential Polynomial Manipulation System) was implemented by Gao (see Wang and Gao 1987). It was used to prove geometric theorems of constructive type, to prove identities of trigonometric functions and hyperbolic functions (Gao 1987, 1990), and to find geometric loci automatically. In DPMS, some techniques are used to transform reducible geometric theorems into irreducible ones, so that the prover is complete for this kind of constructive geometric theorems. DPMS also includes an implementation of Wu–Ritt's well-ordering principle in differential case and the Gröbner basis method of Buchberger (1985). Gao tried to use the Gröbner basis program to prove the Jacobi conjecture for certain concrete cases. However, this program was quickly abandoned by Gao, because of the availability of more convenient programming languages such as Lisp and C/Xwindows.

20.2.4 GEOTHER: A theorem proving environment

GEOTHER (GEOmetry THeorem provER) is an environment implemented in Maple with drawing routines and interface written in C for manipulating and proving geometric theorems (Wang 1996b). In GEOTHER a theorem is specified by means of predicates of the form `Theorem(H,C,X)` asserting that `H` implies `C`, where `H` and `C` are lists or sets of predicates that correspond to the geometric hypotheses and the conclusion of the theorem, and the optional `X` is a list of variables for the internal computation. The information contained in the specification may be all that is needed in order to manipulate and prove the theorem. From the specification GEOTHER can *automatically*

- assign coordinates to each point in some optimal manner;
- translate the predicate representation of the theorem into an English statement, into a first-order logical formula, or into algebraic expressions;
- draw one or several diagrams for the theorem;
- prove the theorem using any of the six algebraic provers;
- translate the generated algebraic non-degeneracy conditions into geometric/predicate form,
- generate a LaTeX and/or a PostScript file documenting the theorem and its proof.

The assignment of coordinates to points can be done optionally by the user. The algebraic provers are based on Wu's method (Wu 1984a), the method of

Gröbner bases (Buchberger 1985), and a method proposed in Wang (1995b,c). GEOTHER can run with an ATINF graphic interface using menu-driver and contains a collection of theorems in both elementary and differential geometries with sample specifications that have been proved. These proved theorems include those named after Steiner, Morley, Thébault, MacLane, and Bertrand. The package also has functions for handling the collection of theorems and on-line help for all functions with examples. A new module for proving geometric theorems using Clifford algebra and rewrite rules is being added.

20.2.5 Euclid: A program for producing readable proofs

Euclid[1] is written by Chou, Gao, and Zhang in Lisp on a Next workstation based on the methods of producing readable proofs (Chou et al. 1994a). It uses a polynomial factorization package written by P. S. Wang from Kent State University, USA.

Euclid implements the following methods of AGTP: *the area method* for Euclidean geometry, *the full-angle method* for Euclidean geometry, *the volume method* for solid geometry, *the vector method* for metric geometries, and *the argument method* for non-Euclidean geometries. Detailed discussions of these methods may be found in Chap. 7 or Chou et al. (1994a).

Most of the work related to the area method was originally developed and experimented with this software; more than 400 theorems have been proved (Chou et al. 1994a). The program has also been used to prove 110 theorems with the full-angle method (Chou et al. 1994b), 90 theorems with the argument method (Chou, Gao, Yang, and Zhang 1994), and 135 theorems from the American Mathematical Monthly (Chou et al. 1994c).

20.3 Algebraic equation solvers

20.3.1 SOLVER: A package based on the well-ordering principle

SOLVER was developed by Wu (1986e,f) on a Dual System 83/20 using Fortran 77. It implements the well-ordering principle (Theorem 1.4) for solving systems of algebraic equations. As mentioned by Wu, the zero decomposition algorithm (Theorem 1.5) could be carried out by repeated use of the program with input provided by the user. Wu proposed several improvements on his original algorithm to avoid the occurrence of large polynomials:

[1]The prover is available via ftp at emcity.cs.twsu.edu: pub/geometry/software/euc.tar.Z.

- The concept of *weak characteristic set* was introduced. A characteristic set *CS* is said to be *weak* if only the initials of the polynomials in *CS* are reduced with respect to *CS*.
- The well-ordering principle was modified to avoid the appearance of too many polynomials.
- Since polynomial factorization (even implemented in Fortran) is expensive, Wu used primitive polynomials instead of irreducible ones.

SOLVER aims at polynomial equation solving and its diverse applications. Wu (1986f) discussed specifically how to use SOLVER to solve problems arising from eigenvalue and eigenvector calculation, theory of automatic control, design of mechanisms, mechanical derivation of geometric formulas, and AGTP.

20.3.2 DEC: An implementation of Wu–Ritt's decomposition method

DEC (decomposition) was implemented by Chou and Gao (1990b) in Lisp on both SUN workstations and Lisp machines. With DEC a detailed experimental study of techniques in the effective implementation of Wu–Ritt's zero decomposition theorem was carried out. These techniques are summarized in Chou and Gao (1992a). DEC has the following modules:

- Wu–Ritt's zero decomposition method for polynomial systems and ordinary differential polynomial systems. The decomposition may be affected by the following parameters.
  - DEC-DIM: decide whether to use the dimension theorem.
  - DEC-VER: decide the type of ascending chains in the decomposition — weak form (Chou and Gao 1990b), Ritt's sense, Wu's sense, or triangular form.
  - DEC-TRI: decide the type of elimination procedures — bottom-up triangulation, top-down triangulation, Gröbner bases reduction or multiple-p.r.s. (Li 1987).
  - DEC-FAC: decide when to perform factorization over $\boldsymbol{Q}$ (DEC uses a polynomial factorization package written by P. S. Wang).
  - DEC-AFAC: decide whether to find irreducible ascending chains (the factorization program was written by the authors of DEC).

- DEC-PAR: this is a set of variables which are treated like parameters (i.e., the decomposition is performed over $\boldsymbol{Q}$(DEC-PAR)$[x_1, \ldots, x_n]$).

- Quantifier elimination over algebraically closed fields. This part has two functions: projection of a quasi-algebraic set and quantifier elimination over an algebraically closed field. These functions are based on an improved version (Gao and Chou 1992c) of an algorithm proposed by Wu (1990a).

- Gröbner bases computation. The program can find the Gröbner basis of a polynomial set in $\boldsymbol{Q}$(DEC-PAR)$[x_1, \ldots, x_n]$, where DEC-PAR is a set of variables treated like parameters.

DEC has been used to prove more than 400 geometric theorems with given non-degeneracy conditions (Chou and Gao 1990b), to derive geometric formulas for 120 problems (Chou and Gao 1989a), and to prove about 100 theorems from differential geometry and mechanics (Chou and Gao 1991).

20.3.3 CharSets: A characteristic sets package in Maple

CharSets is a Maple package based on Wu–Ritt's method of characteristic sets (Ritt 1950; Wu 1984a,b, 1986a, 1989b; Wang 1989–97) implemented by Wang (1995d, 1996e). The first version CharSets 1.0 was made publicly available with Maple's share library in early 1991. Two subsequent versions 1.1 and 1.2 were also distributed. The current version CharSets 2.0 is available on the World Wide Web as:

http://www-leibniz.imag.fr/ATINF/Dongming.Wang/charsets-2.0.tar.Z

CharSets has been used by a number of individuals for different purposes. Some of the routines have been adapted into other computer algebra systems including Singular and Risa/Asir.

The early versions of CharSets were implemented for computing characteristic sets and various zero decompositions of polynomial systems and other related polynomial calculations. Their capabilities include

- computing (modified) characteristic sets of polynomial sets (in different senses),

- decomposing polynomial sets into ascending sets and into irreducible and quasi-irreducible ascending sets,

- decomposing algebraic varieties into irreducible components,

- decomposing polynomial ideals into primary components,
- solving systems of polynomial equations, and
- factorizing multivariate polynomials over algebraic extension fields.

A new module has been added to CharSets version 2.0 that contains eight functions for computing *differential* characteristic sets and the corresponding zero decompositions for systems of *ordinary differential* polynomials. See Wang (1996e) for more details.

Some of the routines from the CharSets package are used by the provers in GEOTHER reviewed in Sect. 20.2.4. Several selected functions from CharSets and algebraic provers from GEOTHER were parallelized via workstation networks (Wang 1991b).

20.3.4 WSOLVE: An implementation of Wu–Ritt's decomposition algorithm for solving polynomial systems

WSOLVE was implemented by D.-K. Wang in 1993 in the Maple system on both SUN workstations and PCs. WSOLVE has two main functions: *wsolve* and *e_val*. The former is an implementation of Wu–Ritt's zero decomposition algorithm, and its main aim is for solving systems of polynomial equations. The function *e_val* is an implementation of Wu's finite kernel theorem that computes the extremal value of a polynomial under some polynomial constraints (see Chap. 1). It can be used to prove theorems involving inequalities.

WSOLVE uses the following techniques for improving the practical efficiency.

- Factorize every polynomial which is given in the initial set or produced in the process of computation.
- Use a parameter to decide which kinds of triangular form — ascending set, weak ascending set or triangular set — to compute.
- Check the redundant branches which are produced from the initials during the computations of pseudo-remainders.

The program WSOLVE has been used to solve problems involving polynomial equations and to prove trigonometric inequalities (Wang 1993).

20.3.5 SACCS/CS: Implementations of the characteristic set method in C

The SACCS library was implemented by Zhi (see Wang and Zhi 1995; Zhi 1996) in the C language, making use of routines from SACLIB, a publicly available library for symbolic and algebraic computation. SACCS has functions for most of the computations (except for primary decomposition) via characteristic sets that are listed in Sect. 20.3.3 for ordinary polynomials. For polynomial factorization over algebraic extension fields, an algorithm suggested in Wang (1992d; see also Chap. 13) was implemented in SACCS; an implementation of an algorithm proposed by Zhi (1997; Chap. 13) using modular techniques was not finished.

SACCS was developed as a first step towards a parallel implementation of characteristic sets algorithms with PARSAC, a parallel version of SACLIB. The parallel implementation has been done later by I. A. Ajwa from Ashland University, USA on the basis of the sequential SACCS.

Zhi also made a simplified version of the CharSets package in Maple which has been named CSET.

Independently, Li (1995a) implemented the characteristic set method based on SACLIB for the European project PoSSo (Polynomial System Solving). The implementation called CS is able to compute a zero decomposition of coarse form for any given system of polynomials. Two distinct features of CS are

- using the algebraic subresultant algorithm to perform top-down elimination, and
- using modular techniques to decide whether pseudo-remainders with respect to a simple ascending set are zero.

More details about CS are described in Li (1995a).

20.4 New systems

The solvers and provers reviewed in Sects. 20.2 and 20.3 were implemented either in the classical programming languages like Fortran and Lisp, or on the top of popular computer algebra systems like Maple. Some of these solvers and provers are also quite preliminary and have limited capabilities. An attempt of the extended Wu's group has been to develop more comprehensive and advanced software systems for Wu-type algorithms that are independent of the existing symbolic computation software and that take advantage of the new generation of computer hardware, operating systems, and programming languages. The design and implementation of such systems may take into account the special features of the algorithms in question and various technical considerations in order to achieve ideal performance. The two systems described in this section are some of the results from our effort and trials.

20.4.1 Geometry Expert

A prototype of this software was first written in 1992 by Gao and Lin with C/Xwindow/Xview on a Sun workstation. The program consists of two parts: a prover and a graphic editor. The prover is basically a C transformation of that of DPMS (see Sect. 20.2.3). The editor is one for *dynamic geometry* that allows free dragging. From 1993 to 1996, Gao added many new functions based on new AGTP methods introduced by Chou et al. (1994a) and developed an Xwindow version of Geometry Expert (GEX)[2] (Chou et al. 1996a). Several new features for dynamic diagram construction were also added. In 1997 Gao re-wrote the program with C++ on Windows 95 platform, and a Chinese version of GEX was then published as a commercial software product (Gao et al. 1998).

GEX is a software tool for *dynamic geometric drawing* and *automated geometric theorem proving and discovering.* The *dynamic geometric editor* mainly has the following functions:

- geometric input of diagrams and statements for the prover,
- dynamic geometric transformations and measurements,
- animation and loci generation.

With the loci generation function, one may generate diagrams for "elementary functions." One may also generate loci with linkages. As proved by Kempe (1876), the class of loci includes all those defined by algebraic curves of the form $f(x, y) = 0$, where $f(x, y)$ is a polynomial of x and y.

The dynamic geometry part of GEX has been used for CAI (computer-aided instruction; see Gao et al. 1998a), geometry research, mechanics design (Gao et al. 1998b), and computer vision (Gao and Cheng 1998).

GEX is also an efficient computer program for geometric reasoning. Within its domain, it invites comparison with the best of human geometry provers. It implements six AGTP methods: Wu's method, the area method, the deductive database method, Gröbner bases method, the vector method, and the method based on full-angles. This part is basically a C transformation of the AGTP algorithms in Euclid (Sect. 20.2.5) and DPMS (Sect. 20.2.3). The prover of GEX has been used by many researchers to deal with various kinds of geometric problems.

[2]The binary codes for GEX can be obtained via ftp at emcity.cs.twsu.edu: /pub/geometry/tsoftware/(ge_sparc.tar.Z, ge_linux.tar.Z).

20.4.2 ELIMINO: A realization for Wu's method

ELIMINO derives from a Latin word meaning *eliminate*. It has been taken as the name of an ongoing software project (Lin et al. 1998) whose objective is to implement a software system that has its own packages for long integer computation, polynomial manipulation, and Wu's method of polynomial equation solving and geometric theorem proving; so ELIMINO also stands for the system. As a project, ELIMINO may be considered as a continuation of the previous work described in Ju et al. (1997), Liu and Liu (1999), and Liu and Wu (1996).

As a symbolic computation system, ELIMINO has very general capabilities for polynomial operations and computation of characteristic sets in Wu's method. It also has an interactive programming environment which allows the user to add new facilities.

From the object-oriented point of view, a system can be considered as a collection of objects. In the system ELIMINO, every object has an associated datatype, which can be thought as a property of the object. The datatype of an object determines the operations that are applicable to the object and clearly separates the various kinds of objects in the system. The following are some of implemented features of ELIMINO.

- Numbers are the most fundamental objects to be manipulated in the implementation of ELIMINO. ELIMINO provides several different kinds of numbers such as integers, fractions, and floating numbers. When appropriate, they can be combined in the same computation because the system knows how to convert among these numbers automatically.

  Integers can be as large as desired, under the only limitation of total storage available to the system. Fractions are quotients of integers. Cancellation between numerators and denominators occurs automatically.

  For approximations, floating point calculations can be performed with any desired number of digits. The function `precision` or `PREC` can be used to set the number to be used.

- Polynomial is another fundamental object treated by ELIMINO. It plays a central role in the system. Some basic operators for polynomial computation including arithmetic, GCD calculation, and factorization have been implemented for the purpose of realizing Wu's method.

- As a software tool for mathematical research, ELIMINO provides an embedded C-like programming language. The language allows the user to program his/her ideas and algorithms using procedures and then run the procedures to verify the ideas or solve the problems.

- Functions can be as important as the values on which they act. In ELIMINO, there are two kinds of functions: system internal functions and user defined functions. The former are implemented in the kernel part, and they provide the common and fundamental facilities to the user. The user defined functions are those defined by the user himself/herself using the programming language and internal functions. In fact, they are part of the User API.

- Implementation of Wu's method is one of the primary goals for this system. Although it can be realized as a group of user defined functions, we implemented it in the kernel part of the system.

20.5 Related packages and systems

The software packages and systems presented in the previous sections were developed mainly by members of Wu's group at large. There are a number of relevant systems and packages developed by other researchers based on Wu's method and other similar methods. We are unable to give a comprehensive and fair survey on all such systems and packages. The information and pointers in this section are provided informally and to help the reader locate the desired literature and sources.

20.5.1 Provers and systems for geometry

First of all the reader is advised to look at Hong et al. (1995) for another collection of early geometric theorem provers. A few other provers implemented using Wu–Ritt's decomposition method are mentioned in the next subsection.

Different pieces of software have been developed by L. Yang, J.-Z. Zhang and their collaborators over the years. See Chaps. 7 and 9, Yang (1998) and other relevant publications by these authors.

There are several methods for quantifier elimination over real closed fields including the general method of cylindrical algebraic decomposition introduced by G. E. Collins and improved by H. Hong (from North Carolina State University, USA) and others, and some specialized methods proposed by V. Weispfenning (from Universität Passau, Germany; Sturm and Weispfenning 1997). These methods have been used for geometric theorem proving. The reader may contact these authors or consult the relevant publications by them and their co-workers for more information.

A few systems that have been developed for doing geometry (research and education) in general and for theorem proving in some special cases include the well-known

- Cabri-Géomètre (http://www-cabri.imag.fr/) initiated by J.-M. Laborde (from Université Joseph Fourier de Grenoble and CNRS, France);
- The Geometer's Sketchpad (http://www.keypress.com/sketchpad/) developed by N. Jackiw (from Key Curriculum Press, USA); the new systems
- Cinderella (http://www.cinderella.de/) developed by J. Richter-Gebert and U. Kortenkamp (from ETH Zurich, Switzerland), Dr. Geo (http://members.xoom.com/FeYiLai/dr_geo/doctor_geo.html) and Geometry Inventor.

20.5.2 Other implementations of Wu–Ritt's decomposition method

An implementation of Wu–Ritt's zero decomposition method in Lisp was made by Ko (1988) for the purpose of proving geometric theorems. Kapur and Wan (1990) implemented a geometric theorem prover based on a refutation approach using a modified version of Wu–Ritt's zero decomposition method. McPhee et al. (1994) implemented a prover for geometric theorems involving inequalities based on a modified version of Wu–Ritt's zero decomposition method and Hong's implementation of Collins' CAD method.

Part of the CharSets package was re-implemented in Risa/Asir by T. Shimoyama (from Fujitsu Laboratories Ltd., Japan) for the purpose of primary decomposition and in Singular (http://www.singular.uni-kl.de/) by M. W. Messollen (from Universität des Saarlandes, Germany).

A recent implementation of Wu's decomposition method in Axiom/Aldor has been done by Aubry and Moreno Maza (1999).

20.5.3 Implementations of other triangular sets methods

Besides Wu's method there are three known methods based on triangular sets proposed by Lazard (1991), Kalkbrenber (1993), and Wang (1993a,b). The two of 1993 have been implemented and experimented by their proposers (see also Wang 1996a). Recently, a comparative implementation of the four methods including Wu's in Axiom/Aldor has been made by Aubry and Moreno Maza (1999) for solving polynomial systems.

Some methods presented in Yang et al. (1995) based on resultants and in Wang (1998a, 2000a,b) based on subresultant regular subchains have also been implemented by the authors themselves.

20.5.4 Gröbner bases packages

The method of Gröbner bases (Buchberger 1985) is one of the most relevant to Wu's method. This method may address many of the problems, in particular geometric theorem proving and algebraic equation solving, to which Wu's method can be applied. The Gröbner basis method has been implemented in most of the popular computer algebra systems including Mathematica, Maple, Macsyma, and Axiom. There are also special-purpose implementations, of which we may mention the one in CoCoA (http://cocoa.dima.unige.it/) and the Gb package developed by J.-C. Faugère (from Université Paris VI, France). The two mentioned have been considered among some of the best implementations for Gröbner bases. The software Gb has been combined with a RealSolving package by F. Rouillier (from INRIA Lorraine, France) for wider applications.

Acknowledgments. The authors thank Ziming Li and Lihong Zhi for their contributions to Sect. 20.3.5.

REFERENCES

Abbott, J. A. (1989): On the factorization of polynomials over algebraic fields. Ph.D thesis, School of Math. Sci., Univ. of Bath, UK.

Abhyankar, S. S., Bajaj, C. (1988): Automatic parametrization of rational curves and surfaces III: Algebraic plane curves. Comput. Aided Geom. Des. **5**: 309–321.

Abhyankar, S. S., Bajaj, C. (1989): Automatic parametrization of rational curves and surfaces IV: Algebraic space curves. ACM Trans. Graph. **8**/4: 325–333.

Ablowitz, M. J., Ramani, A., Segur, H. (1980): A connection between non-linear evolution equations and ordinary differential equations of P-type I, II. J. Math. Phys. **21**: 715–721, 1006–1015.

Alonso, M. E., Mora, T., Niesi, G., Raimondo, M. (1992): Local parametrization of space curves at singular points. In: Proc. Int. Workshop Math. Mech. (Wu, W.-t., Cheng, M., eds.), Int. Academic Publ., Beijing, pp. 135–166.

Anderson, J. R. (1981): Tuning of search of the problem space for geometry proofs. In: Proc. Int. Joint Conf. Artif. Intell. (Vancouver, August 24-28, 1981), pp. 165–170.

Arnon, D. (1988): Geometric reasoning with logic and algebra. Artif. Intell. **37**: 37–60.

Arnon, D., Mignotte, M. (1988): On mechanical quantifier elimination for elementary algebra and geometry. J. Symb. Comput. **5**: 237–259.

Arnon, D. S., Sederberg, T. W. (1984): Implicit equation for a parametric surface by Gröbner bases. In: Proc. 1984 MACSYMA User's Conf. (Golden, V. E., ed.), General Electric, Schenectady, New York, pp. 431–436.

Aubry, P. (1999): Ensembles triangulaires de polynômes et résolution de systèmes algébriques. Implantation en Axiom. Ph.D thesis, Univ. Paris VI, France.

Aubry, P., Lazard, D., Moreno Maza, M. (1999): On the theories of triangular sets. J. Symb. Comput. **28**: 105–124.

Aubry, P., Moreno Maza, M. (1999): Triangular sets for solving polynomial systems: A comparative implementation of four methods. J. Symb. Comput. **28**: 125–154.

Auzinger, W., Stetter, H. J. (1988): An elimination algorithm for the computation of all zeros of a system of multivariate polynomial equations. Int. Ser. Numer. Math. **86**: 11–30.

Bachmair, L., Dershowitz, N. (1987): Inference rules for rewrite-based first order theorem proving. In: Proc. 2nd Symp. Logic Comput. Sci. (Ithaca, New York, June 1987), IEEE Comput. Soc. Press, pp. 331–337.

Bachmair, L., Ganzinger, H. (1997): A theory of resolution. Tech. Rep. MPI-I-97-2-005, Max-Planck-Institut für Informatik, Germany.

Balbiani, P., Dugat, V., Fariñas del Cerro, L., Lopez, A. (1994): Eléments de géométrie mécanique. Hermès, Paris.

Bancilhon, F., Ramakrishnan, R. (1986): An amateur's introduction to recursive query processing strategies. In: Proc. ACM SIGMOD Conf. (Zanioilo, C., ed.), pp. 16–52.

Bareiss, E. H. (1968): Sylvester's identity and multi-step integer-preserving Gaussian elimination. Math. Comput. **22**: 565–578.

Baxter, R. J. (1972): Partition function of the eight vertex model. Ann. Phys. **70**: 193–228.

Baxter, R.J. (1982): Exactly solved models in statistical mechanics. Academic Press, London.

Bazhanov, V. V., Stroganov, Y. G. (1985): Hidden symmetry of free-Fermion model. Theor. Math. Fiz. **62**: 253–260.

Bayro-Corrochano, E., Buchholz, S., Sommer, G. (1996): Self-organizing Clifford neural network. In: Proc. IEEE ICNN '96 (Washington D. C., 1996), pp. 120–125.

Bayro-Corrochano, E., Lasenby, J. (1995): Object modeling and motion analysis using Clifford algebra. In: Proc. Europe-China Workshop GMICV '95 (Xi'an, China, April 27-29, 1995), pp. 143–149.

Benjamin, T. B., Bona, J. L., Mahony, J. J. (1972): Equations for long waves in non-linear dispersive systems. Philos. Trans. R. Soc. Lond. (Ser. A) **272**: 47–48.

Berkovich, L. M., Tsirulik, V. G. (1986): Differential resultants and some of their applications. Diff. Uravn. **22**/5: 750–757.

Bluman, G. W., Kumei, S. (1989): Symmetries and differential equations. Springer, New York.

Boy de la Tour, T., Fèvre, S., Wang, D. (1999): Clifford term rewriting for geometric reasoning in 3D. In: Automated deduction in geometry (Gao, X.-S., Wang, D., Yang, L., eds.), LNAI[3] **1669**, Springer, Berlin, Heidelberg, pp. 130–155.

Bretherton, F. P. (1964): Slow viscous motion round a cylinder in a simple shear. J. Fluid Mech. **12**: 591–614.

Brown, W. S., Traub, J. F. (1971): On Euclid's algorithm and the theory of subresultants. J. ACM **18**: 505–514.

Boyer, R. S. (1971): Locking: A restriction of resolution. Ph.D thesis, The Univ. of Texas at Austin, USA.

Bronstein, M., Petkovšek, M. (1994): On Ore rings, linear operators and factorization. Prog. Comput. Software **20**: 14–26.

Bronstein, M., Petkovšek, M. (1996): An introduction to pseudo-linear algebra. Theor. Comput. Sci. **157**: 3–33.

Brown, W. S. (1971): On Euclid's algorithm and the computation of polynomial greatest common divisors. J. ACM **18**: 478–504.

Brown, W. S., Traub, J. F. (1971): On Euclid's algorithm and the theory of subresultants. J. ACM **18**: 505–514.

Buchberger, B. (1985): Gröbner bases: An algorithmic method in polynomial ideal theory. In: Multidimensional systems theory (Bose, N. K., ed.), D. Reidel, Dordrecht, Boston, pp. 184–232.

[3]Lecture Notes in Artificial Intelligence

Buchberger, B. (1987): Applications of Gröbner bases in non-linear computational geometry. In: Mathematical aspects of scientific software (Rice, J. R., ed.), Springer, New York, pp. 59–87 [also in LNCS[4] **296** (Janßen, R., ed.), Springer, Berlin, Heidelberg, pp. 52–80].

Buntine, W. (1988): Generalized subsumption and its application to induction and redundancy. Artif. Intell. **36**/2: 149–179.

Calogero, F. J. (1987): The evolution partial differential equation. J. Math. Phys. **28**: 538–555.

Canosa, J., Gazdag, J. J. (1977): The Korteweg–de Vries–Burgers equation. J. Comput. Phys. **23**: 393–403.

Carillo, F., Tabor, M. (1989): Painleve expansions for nonintegrable evolution equation. Phys. D **39**: 77–94.

Carrà Ferro, G., Gallo, G., Gennaro, R. (1997): Probabilistic verification of elementary geometry statements. In: Automated deduction in geometry (Wang, D., ed.), LNAI **1360**, Springer, Berlin, Heidelberg, pp. 85–101.

Cartan, E. (1952): La topologie des groupes de Lie. Exp. de Géom. **8**: 1307–1330.

Castelnuvo, (1894): Sulla rationalita della involuzioni Pinae. Math. Ann. **44**: 125–155.

Chang, C.-L., Lee, R. C. (1973): Symbolic logic and mechanical theorem proving. Academic Press, New York.

Chao, L. (1997): Wu-differential characteristic method and its applications to PDEs symmetries and mechanics. Ph.D thesis, Dalian Univ. of Tech., China.

Char, B. W., Geddes, K. O., Gonnet, G. H., Leong, B. L., Monagan, M. B., Watt, S. M. (1991): Maple V — Language reference manual. Springer, New York.

Chardin, M. (1991): Differential resultants and subresultants. In: Proc. FCT '91 (Gosen, Germany, September 9–13, 1991), LNCS **529**, Springer, Berlin, Heidelberg, pp. 180–189.

Chazarain, J., Riscos, A., Alonso, J. A., Briales, E. (1991): Multi-valued logic and Gröbner bases with applications to modal logic. J. Symb. Comput. **11**: 181–194.

Chen, L., Lu, Z., Wang, D. (2000): A class of cubic systems with two centers or two foci. J. Math. Anal. Appl. **242**: 154–163.

Chen, L., Wang, M. (1979): The relative position and number of limit cycles of a quadratic differential system. Acta Math. Sinica **22**: 751–758.

Cherkas, L. A. (1978): Conditions for the equation $yy' = \sum_{i=0}^{3} p_i(x)y^i$ to have a center. Differentsial'nye Uravneniya **14**: 1594–1600.

Chevalley, C. (1954): The algebraic theory of spinors. Columbia Univ. Press, New York.

Chionh, E. W., (1990): Base points, resultants, and the implicit representation of rational surfaces. Ph.D thesis, The Univ. of Waterloo, Canada.

Chou, S.-C. (1984): Proving elementary geometry theorems using Wu's algorithm. In: Automated theorem proving: After 25 years (Bledsoe, W. W., Loveland, D. W., eds.), Contemp. Math. **29**, Am. Math. Sci., Providence, pp. 243–286.

[4]Lecture Notes in Computer Science

Chou, S.-C. (1985): Proving and discovering theorems in elementary geometries using Wu's method. Ph.D thesis, Dept. of Math., The Univ. of Texas at Austin, USA.

Chou, S.-C. (1987): A method for mechanical derivation of formulae in elementary geometry. J. Automat. Reason. **3**: 291–299.

Chou, S.-C. (1988): Mechanical geometry theorem proving. D. Reidel, Dordrecht, Boston.

Chou, S.-C. (1990): Automated reasoning using the characteristic set method and the Gröbner basis method. In: Proc. ISSAC[5] '90 (Tokyo, Japan, August 20–24, 1990), ACM Press, New York, pp. 255–260.

Chou, S.-C., Gao, X.-S. (1989a): A collection of 120 computer solved geometry problems in mechanical formula derivation. Tech. Rep. TR-89-22, Dept. of Comput. Sci., The Univ. of Texas at Austin, USA.

Chou, S.-C., Gao, X.-S. (1989b): A class of geometry statements of constructive type and geometry theorem proving, Tech. Rep. TR-89-37, Dept. of Comput. Sci., The Univ. of Texas at Austin, USA.

Chou, S.-C., Gao, X.-S. (1990a): Methods for mechanical geometry formula deriving. In: Proc. ISSAC '90 (Tokyo, Japan, August 20–24, 1990), ACM Press, New York, pp. 265–270.

Chou, S.-C., Gao, X.-S. (1990b): Ritt–Wu's decomposition algorithm and geometry theorem proving. In: Proc. CADE[6]-10 (Kaiserslautern, Germany, July 24–27, 1990), LNCS **449**, Springer, Berlin, Heidelberg, pp. 207–220 [also as Tech. Rep. TR-89-09, Dept. of Comput. Sci., The Univ. of Texas at Austin, USA].

Chou, S.-C., Gao, X.-S. (1991): Theorems proved automatically using Wu's method — Part on differential geometry (space curves) and mechanics. Math. Mech. Res. Preprints[7] 6: 37–55.

Chou, S.-C., Gao, X.-S. (1992a): An algebraic system based on the characteristic set method. In: Proc. Int. Workshop Math. Mech. (Wu, W.-t., Cheng, M., eds.), Int. Academic Publ., Beijing, pp. 1–18.

Chou, S.-C., Gao, X.-S. (1992b): Proving geometry statements of constructive type. In: Proc. CADE-11 (Saratoga Springs, USA, June 15–18, 1992), LNCS **607**, Springer, Berlin, Heidelberg, pp. 20–34.

Chou, S.-C., Gao, X.-S. (1992c): Automated reasoning in differential geometry and mechanics using characteristic method III: Mechanical formula derivation. In: Automated reasoning (Shi, Z., ed.), Elsevier, North-Holland, pp. 1–12.

Chou, S.-C., Gao, X.-S. (1993a): Automated reasoning in differential geometry and mechanics using the characteristic set method I: An improved version of Ritt–Wu's decomposition algorithm. J. Automat. Reason. **10**: 161–172.

Chou, S.-C., Gao, X.-S. (1993b): Automated reasoning in differential geometry and mechanics using the characteristic set method II: Mechanical theorem proving. J. Automat. Reason. **10**: 173–189.

[5]International Symposium on Symbolic and Algebraic Computation

[6]International Conference on Automated Deduction

[7]Edited and distributed by Mathematics Mechanization Research Center, Institute of Systems Science, Academia Sinica, Beijing 100080, China

Chou, S.-C., Gao, X.-S.: (1993c): Automated reasoning in differential geometry and mechanics using the characteristic method IV: Bertrand curves. Syst. Sci. Math. Sci. **6**: 186–192.

Chou, S.-C., Gao, X.-S. (1994): Mechanical theorem proving in Riemann geometry using Wu's method. In: Computer mathematics (Tianjin, China, 1991), Nankai Ser. Pure Appl. Math. Theor. Phys. **5**, World Scientific, River Edge, pp. 136–157.

Chou, S.-C., Gao X.-S. (1995): The computer searches for Pascal conics. Int. J. Comput. Math. Appl. **29**/2: 63–71.

Chou, S.-C., Gao, X.-S., Arnon, D. (1992): On the mechanical proof of geometry theorems involving inequalities. In: Issues in robotics and nonlinear geometry (Hoffmann, C. M., ed.), JAI Press, Greenwich, pp. 139–181.

Chou, S.-C., Gao, X.-S., Li, Z. (1994): Computations with rational parametric equations. In: Computer mathematics (Tianjin, China, 1991), Nankai Ser. Pure Appl. Math. Theor. Phys. **5**, World Scientific, River Edge, pp. 86–111.

Chou, S.-C., Gao, X.-S., McPhee, N. (1989): A combination of Ritt–Wu's method and Collins' method. Tech. Rep. TR-89-28, Dept. of Comput. Sci., The Univ. of Texas at Austin, USA.

Chou, S.-C., Gao, X.-S., Yang, L., Zhang, J.-Z. (1994): A collection of 90 mechanically solved geometry problems from non-Euclidean geometries. Tech. Rep. TR-94-10, Dept. of Comput. Sci., Wichita State Univ., USA.

Chou, S.-C., Gao, X.-S., Zhang, J.-Z. (1992): Automated production of traditional proofs for theorems in Euclidean geometry IV: A collection of 400 geometry theorems. Tech. Rep. TR-92-7, Dept. of Comput. Sci., Wichita State Univ., USA.

Chou, S.-C., Gao, X.-S., Zhang, J.-Z. (1993a): Automated production of traditional proofs for constructive geometry theorems. In: Proc. 8th IEEE Symp. Logic Comput. Sci. (Montreal, Canada, June 19–23, 1993), IEEE Comput. Soc. Press, pp. 48–56.

Chou, S.-C., Gao, X.-S., Zhang, J.-Z. (1993b): Mechanical geometry theorem proving by vector calculation. In: Proc. ISSAC '93 (Kiev, Ukraine, July 6–8, 1993), ACM Press, New York, pp. 284–291.

Chou, S.-C., Gao, X.-S., Zhang, J.-Z. (1994a): Machine proofs in geometry. World Scientific, Singapore.

Chou, S.-C., Gao, X.-S., Zhang, J.-Z. (1994b): A collection of 110 geometry theorems and their machine produced proofs using full-angles. Tech. Rep. TR-94-4, Dept. of Comput. Sci., Wichita State Univ., USA.

Chou, S.-C., Gao, X.-S., Zhang, J.-Z. (1994c): Automated solution of 135 geometry problems from American mathematical monthly. Tech. Rep. TR-94-8, Dept. of Comput. Sci., Wichita State Univ., USA.

Chou, S.-C., Gao, X.-S., Zhang, J.-Z. (1995a): Automated production of traditional proofs in solid geometry. J. Automat. Reason. **14**: 257–291.

Chou, S.-C., Gao, X.-S., Zhang, J.-Z. (1995b): Automated geometry theorem proving and geometry education. In: Proc. ATCM[8] '95 (Singapore, December 18–21, 1995), Assoc. of Math. Educators, Singapore, pp. 319–328.

[8]Asian Technology Conference in Mathematics

Chou, S.-C., Gao, X.-S., Zhang, J.-Z. (1995c): A deductive database approach to automated geometry theorem proving and discovering. Tech. Rep. WSUCS-95-2, Dept. of Comput. Sci., Wichita State Univ., USA [also J. Automat. Reason. (in press)].

Chou, S.-C., Gao, X.-S., Zhang, J.-Z. (1996a): An introduction to Geometry Expert. In: Proc. CADE-13 (McRobbie, M. A., Slaney, J. K., eds.), LNAI **1104**, Springer, Berlin, Heidelberg, pp. 235–239.

Chou, S.-C., Gao, X.-S., Zhang, J.-Z. (1996b): Automated generation of readable proofs with geometric invariants I: Multiple and shortest proof generation. J. Automat. Reason. **17**: 325–347.

Chou, S.-C., Gao, X.-S., Zhang, J.-Z. (1996c): Automated generation of readable proofs with geometric invariants II: Proving theorems with full-angles. J. Automat. Reason. **17**: 349–370.

Chou, S.-C., Ko, H.-P. (1986): On mechanical theorem proving in Minkowskian plane geometry. In: Proc. IEEE Symp. Logic Comput. Sci. (Cambridge, UK, June 16–18, 1986), IEEE Comput. Soc. Press, pp. 187–192.

Chou, S.-C., Schelter, W. F. (1986): Proving geometry theorems with rewrite rules. J. Automat. Reason. **2**: 253–273.

Chou, S.-C., Schelter, W. F., Yang, J.-G. (1989): Characteristic sets and Gröbner bases in geometry theorem proving. In: Resolution of equations in algebraic structures (Aït-Kaaci, H., Nivat, M., eds.), Academic Press, San Diego, pp. 33–92.

Chou, S.-C., Schelter, W. F., Yang, J.-G. (1990): An algorithm for constructing Gröbner bases from characteristic sets and its application to geometry. Algorithmica **5**: 147–154.

Chou, S.-C., Yang, G. (1989): On the algebraic formulation of certain geometry statements and mechanical geometry theorem proving. Algorithmica **4**: 237–262.

Christopher, C. J. (1994): Invariant algebraic curves and conditions for a centre. Proc. R. Soc. Edinb. (Sect. A) **124**/6: 1209–1229.

Christopher, C. J., Lloyd, N. G. (1990): On the paper of Jin and Wang concerning the conditions for a centre in certain cubic systems. Bull. Lond. Math. Soc. **22**: 5–12.

Christopher, C. J., Lloyd, N. G. (1995): Polynomial systems: A lower bound for the Hilbert numbers. Proc. R. Soc. Lond. A **450**: 219–224.

Chuang, J. H., Hoffman, C. M. (1989): On local implicit approximation and its applications. ACM Trans. Graph. **8**/4: 298–324.

Chyzak, F., Salvy, B. (1998): Non-commutative elimination in Ore algebras proves multivariate identities. J. Symb. Comput. **26**/2: 173–187.

Clifford, W. K. (1876): On the classification of geometric algebras. In: Mathematical papers (Tucher, R., ed.), MacMillan, 1982.

Clifford, W. K. (1878): Application of Grassmann's extensive algebra. Am. J. Math. **I**: 350–358.

Coelho, H., Pereira, L. M. (1986): Automated reasoning in geometry theorem proving with Prolog. J. Automat. Reason. **2**: 329–390.

Coffey, M. W. (1990): On series expansion giving closed form solutions of KdV-equations. SIAM. J. Appl. Math. **50**: 1580–1592.

Collins, G. E. (1966): Polynomial remainder sequences and determinants. Am. Math. Mon. **73**/7: 708–712.

Collins, G. E. (1967): Subresultants and reduced polynomial remainder sequences. J. ACM **14**: 128–142.

Collins, G. E. (1971): The calculation of multivariate polynomial resultants. J. ACM **19**: 525-532.

Collins, G. E. (1975): Quantifier elimination for real closed fields by cylindrical algebraic decomposition. In: LNCS **33**, Springer, Berlin, Heidelberg, pp. 134–183.

Collins, G. E., Encarnación, M. J. (1995): Efficient rational number reconstruction. J. Symb. Comput. **20**: 299–313.

Collins, G. E., Hong, H. (1991): Partial cylindrical algebraic decomposition for quantifier elimination. J. Symb. Comput. **12**: 299–328.

Collins, G. E., Loos, R. (1982): Real zeros of polynomials. In: Computer algebra: Symbolic and algebraic computation (Buchberger, B., Collins, G. E., Loos, R., eds.), Springer, Wien, New York, pp. 83–94.

Conti, P., Traverso, C. (1995): A case of automatic theorem proving in Euclidean geometry. In: Proc. AAECC-11 (Paris, France, July 17–22, 1995), LNCS **948**, Springer, Berlin, Heidelberg, pp. 183–193.

Conti, R. (1988): On a class of cubic systems with two centers. Results Math. **14**: 30–37.

Conti, R. (1990): On centers of cubic systems. Annales Polonici Math. **5**: 123–128.

Corless, R. M., Gianni, P. M., Trager, B. M., Watt, S. M. (1995): The singular value decomposition for polynomial systems. In: Proc. ISSAC '95 (Montreal, Canada, July 10–12, 1995), ACM Press, New York, pp. 195–207.

Cox, D., Little, J., O'Shea, D. (1992): Ideals, varieties, and algorithms. Springer, New York, Berlin [2nd ed. (1996)].

Coxeter, H. S. M. (1968): Non-Euclidean geometry. Univ. of Toronto Press, Toronto.

Crapo, H. (1991): Invariant theoretic methods in scene analysis and structural mechanics. J. Symb. Comput. **11**: 523–548.

Crapo, H., Richter-Gebert, J. (1995): Automatic proving of geometric theorems. In: Invariant methods in discrete and computational geometry (White, N. L., ed.), Kluwer, Dordrecht, pp. 167–196.

Crumeyrolle, A. (1990): Orthogonal and symplectic Clifford algebras. D. Reidel, Dordrecht, Boston.

Das, S. K. (1992): Deductive databases and logic programming. Addison-Wesley Publ. Co., Wokingham.

Davis, M., Putnam, H. (1960): A computing procedure for quantification theory. J. ACM **7**: 201–215.

Davis, P. J., Cerutti, E. (1969): Formac meets Pappus. Am. Math. Mon. **76**: 895–905.

Demidovich, B. P., Maron, I. A. (1981): Computational mathematics [translated from the Russian by G. Yankovsky]. Mir Publ., Moscow.

Dershowitz, N., Jouannaud, J. P., Klop, J. W. (1991): Open problems in rewriting. In: Proc. RTA-91 (Como, Italy, April 10–12, 1991), LNCS **488**, Springer, Berlin, Heidelberg, pp. 445–456.

Dey, B. J. (1986): Domain wall solution of KdV like equations with higher order nonlinearity. J. Phys. A: Math. Gen. **19**: 9–12.

Dietrich, R. (1986): Relation resolution and algebraic completion for Horn logic. In: Proc. CADE-8 (Siekmann, J. H., ed.), LNCS **230**, Springer, Berlin, Heidelberg, pp. 62–78.

Dodd, R. K. (1982): Solitons and nonlinear wave equations. Academic Press, London.

Doran, C., Hestenes, D., Sommen, F., Acker, N. V. (1993): Lie groups as spin groups. J. Math. Phys. **34**/8: 3642–3669.

Drinfel'd, V. G. (1987): Quantum groups. In: Proc. Int. Congr. Math. (Berkeley, USA, August 3–11, 1986), AMS Press, Providence, pp. 798–820.

Ecalle, J., Martinet, J., Moussu, R., Ramis, J.-P. (1987): Non-accumulation des cycles-limites I. C. R. Acad. Sci. Paris (Ser. I) **304**: 375–377.

Encarnación, M. J. (1995): Computing GCDs of polynomials over algebraic number fields. J. Symb. Comput. **20**: 287–297.

Encarnación, M. J. (1997): Factoring polynomials over algebraic number fields via norms. In: Proc. ISSAC '97 (Hawaii, USA, July 21–23, 1997), ACM Press, New York, pp. 265–270.

Eisenbud, D. (1993): Open problems in computational algebraic geometry. In: Computational algebraic geometry and commutative algebra (Eisenbud, D., Robbiano, L., eds.) — Proc. Conf. (Cortona, Italy, June 17–21, 1991), Symp. Math. **34**, Cambridge Univ. Press, Cambridge, pp. 49–70.

Evans, T. G. (1969): A heuristic program to solve geometry analogy problems. In: Semantic information processing (Minsky, M. L., ed.), MIT Press, Cambridge.

Fan, C., Wu, F. Y. (1970): General lattice model of phase transition. Phys. Rev. B**2**: 723–738.

Fan, E. (1998): Soliton and integrable system: Exact solution, similarity reduction, Hamiltonian structure, zero curvature representation and applications of Wu's method and computer algebra. Ph.D thesis, Dalian Univ. of Tech., China.

Fan, E., Zhang, H. (1997): The solitary wave solutions for a class of nonlinear wave equation. Acta Phys. Sinica **46**: 1254–1258.

Fan, E., Zhang, H. (1998a): Exact solutions to the coupled scalar field equations. Acta Phys. Sinica **47**: 1064–1070

Fan, E., Zhang, H. (1998b): New exact solutions for a system of coupled KdV equation. Phys. Lett. A **245**: 389–392.

Fan, E., Zhang, H. (1998c): Exact solutions for nonlinear wave equations. Commun. Theor. Phys. **30**: 309–312.

Fan, E., Zhang, H. (1998d): Some exact explicit solutions for nonlinear heat equation in mathematics and physics. In: Proc. ASCM[9] '98 (Li Z., ed.), Lanzhou Univ. Press, Lanzhou, pp. 159–164.

Fan, E., Zhang, H. (1998e): Backlund transformation, Lax pairs, symmetries and exact solutions for variable coefficient KdV equations. Acta Phys. Sinica **7**: 649–654.

[9]Asian Symposium on Computer Mathematics

Fan, E., Zhang, H. (1998f): A new approach to Backlund transformation of nonlinear evolution equation. Appl. Math. Mech. **19**: 667–670.

Fan, E., Zhang, H. (1998g): A note on homogeneous balance method. Phys. Lett. A **246**: 403–406.

Fan, E., Zhang, H. (1998h): Backlund transformation formation and exact solutions for Whitham–Broer–Kaup equations in shallow water. Appl. Math. Mech. **19**: 667–670.

Fearnley-Sander, D. (1999): Plane Euclidean reasoning. In: Automated deduction in geometry (Gao, X.-S., Wang, D., Yang, L., eds.), LNAI **1669**, Springer, Berlin, Heidelberg, pp. 86–110.

Fearnley-Sander, D., Stokes, T. (1997): Area in Grassmann geometry. In: Automated deduction in geometry (Wang, D., ed.), LNAI **1360**, Springer, Berlin, Heidelberg, pp. 141–170.

Feigenbaum, E. A., Feldman, J. (eds., 1963): Computers and thought. McGraw Hill, New York.

Feng, G., Wu, W., Huang, K. (1993a): Second equivalence theorem of multivariate polynomial system and eigenvalue-problem. Adv. Math. (Beijing) **22**/5: 476–477.

Feng, G., Wu, W., Huang, K. (1993b): The computation of eigenvalue problem equivalent to multivariate polynomial system and the Gröbner bases. Adv. Math. (Beijing) **22**/3: 282–284.

Feng, G., Wu, T., Yu, K., Zhou, Y. (1998): Three-way blends of implicit algebraic surfaces with canonical expansion. In: Proc. ASCM '98 (Li, Z., ed.), Lanzhou Univ. Press, Lanzhou, pp. 201–208.

Feng, G., Wu, W., Zhang, S. (1993c): The eigenvalue problem equivalent to multivariate polynomial system. Numer. Math. J. Chinese Univ. (English Ser.) **2**/2: 234–241.

Feng, G., Zhang, S. (1992): Reducing the multivariate polynomial system to eigenvalue problem. In: Proc. Int. Workshop Math. Mech. (Wu, W.-t., Cheng, M., eds.), Int. Academic Publ., Beijing, pp. 19–27.

Feng, G., Zhang, S., Huang, K. (1995a): Converting multipolynomial systems into eigenvalue-problems via well arranged bases. In: Proc. ATCM '95 (Singapore, December 18–21, 1995), Assoc. of Math. Educators, Singapore, pp. 21–29.

Feng, G., Zhang, S., Liu, Y. (1995b): The structure of solutions to algebraic system and matrices in eigenvalue method. In: Proc. ASCM (Shi, H, Kobayashi, H., eds.), Scientists Inc., Tokyo, pp. 43–53.

Feng, G., Zhang, S., Zhang, C. (1996): The eigenvalue method for computing high-dimensional varieties. Northeastern Math. J. **12**/1: 1–4.

Fèvre, S. (1997): Integration of reasoning and algebraic calculus in geometry. In: Automated deduction in geometry (Wang, D., ed.), LNAI **1360**, Springer, Berlin, Heidelberg, pp. 218–234.

Fèvre, S., Wang, D. (1998a): Proving geometric theorems using Clifford algebra and rewrite rules. In: Proc. CADE-15 (Lindau, Germany, July 5–10, 1998), LNAI **1421**, Springer, Berlin, Heidelberg, pp. 17–32.

Fèvre, S., Wang, D. (1998b): Combining algebraic computing and term-rewriting for geometry theorem proving. In: Proc. AISC '98 (Plattsburgh, USA, September 16–18, 1998), LNAI **1476**, Springer, Berlin, Heidelberg, pp. 145–156.

Fèvre, S., Wang, D. (1999): Combining Clifford algebraic computing and term-rewriting for geometric theorem proving. Fundamenta Informaticae **39**/1-2: 85–104.

Fisher, M. J., Rabin, M. O. (1974): Super-exponential complexity of Presburger arithmetic. In: Proc. Symp. Appl. Math. (New York City, USA, 1973), pp. 27–41.

Frenkel, I. B., Reshetikhin, N. Y. (1992): Quantum affine algebras and holonomic difference equations. Commun. Math. Phys. **146**/1: 1–60.

Gallaire, H., Minker, J., Nicola, J. M. (1984): Logic and databases: A deductive approach. ACM Comput. Surveys **16**/2: 153–185.

Gantmacher, F. R. (1953): Theorie der Matrizen. Staatsverlag für technisch-theoretische Literatur, Moskau [in Russian].

Gao, X.-S. (1987): Trigonometric identities and mechanical theorem proving in elementary geometries. J. Syst. Sci. Math. Sci. **7**: 264–272 [in Chinese].

Gao, X.-S. (1989): The minimal characteristic basis of a polynomial ideal. J. Syst. Sci. Math. Sci. **2**: 236–242.

Gao, X.-S. (1990): Transcendental functions and mechanical theorem proving in elementary geometries. J. Automat. Reason. **6**: 403–417.

Gao, X.-S. (1991): On the theory of resolvents and its applications. Math. Mech. Res. Preprints 6: 79–93.

Gao, X.-S., Cheng, H.-F. (1998): On the solution classification of the "P3P" problem. In: Proc. ASCM '98 (Li, Z., ed,), Lanzhou Univ. Press, Lanzhou, pp. 185-200.

Gao, X.-S., Chou, S.-C. (1991a): Computations with parametric equations. In: Proc. ISSAC '91 (Bonn, Germany, July 15–17, 1991), ACM Press, New York, pp. 122–127.

Gao, X.-S., Chou, S.-C. (1991b): On the normal parametrization of curves and surfaces. Int. J. Comput. Geom. Appl. **1**: 125–136.

Gao, X.-S., Chou, S.-C. (1992a): Implicitization of rational parametric equations. J. Symb. Comput. **14**: 459–470.

Gao, X.-S., Chou, S.-C. (1992b): On the parametrization of algebraic curves. Appl. Algebra Eng. Commun. Comput. **3**/1: 27–38.

Gao, X.-S., Chou, S.-C. (1992c): Solving parametric algebraic systems. In: Proc. ISSAC '92 (Berkeley, USA, July 27–29, 1992), ACM Press, New York, pp. 335–341 [also in Math. Mech. Res. Preprints 7: 14–30].

Gao, X.-S., Chou, S.-C. (1993a): On the dimension of an arbitrary ascending chain. Chinese Sci. Bull. **38**/10: 799–804.

Gao, X.-S., Chou, S.-C. (1993b): A zero structure theorem for differential parametric systems. J. Symb. Comput. **16**: 585–593.

Gao, X.-S., Chou, S.-C. (1998a): Solving geometric constraint systems I: A global propagation approach. Comput.-Aided Des. **30**/1: 47–54.

Gao, X.-S., Chou, S.-C. (1998b): Solving geometric constraint systems II: A symbolic computational approach. Comput.-Aided Des. **30**/2: 115–122.

Gao, X.-S., Zhang, J.-Z., Chou, S.-C. (1998): Geometry Expert. Nine Chapters Publ., Taipei, Taiwan [in Chinese].

Gao, X.-S., Zhu, C., Huang, Y. (1998a): Building dynamic mathematical models with Geometry Expert I: Geometric transformations, functions and plane curves. In: Proc. ATCM '98 (Yang, W.-C., Shirayanagi, K., Chu, S.-C., Fitz-Gerald, G., eds.), Springer, Singapore, pp. 216–224.

Gao, X.-S., Zhu, C., Huang, Y. (1998b): Building dynamic mathematical models with geometry expert II: Linkages. In: Proc. ASCM '98 (Li, Z., ed.), Lanzhou Univ. Press, Lanzhou, pp. 15–22.

Garrity, T., Warren, J. (1989): On computing the intersection of a pair of algebraic surfaces. Comput. Aided Geom. Des. **6**: 137–153.

Geddes, K. O., Czapor, S. R., Labahn, G. (1992): Algorithms for computer algebra. Kluwer, Boston.

Gelernter, H. (1959): Realization of a geometry theorem proving machine. In: Proc. Int. Conf. Info. Process. (Paris, France, June 15–20, 1959), pp. 273–282 [also in Feigenbaum and Feldman (1963), pp. 134–152].

Gelernter, H., Hanson, J. R., Loveland, D. W. (1960): Empirical explorations of the geometry-theorem proving machine. In: Proc. Western Joint Comput. Conf. (San Francisco, USA, May 3–5, 1960), pp. 143–147.

Gianni, P., Trager, B. M., Zacharias, G. (1988): Gröbner bases and primary decomposition of polynomial ideals. J. Symb. Comput. **6**: 149–167.

Gilmore, P. (1970): An examination of the geometry theorem machine. Artif. Intell. **1**: 171–187.

Goldstein, H. (1973): Elementary geometry theorem proving. MIT AI Memo no. 28, MIT, USA.

Gonzàlez-Vega, L. (1991): Determinantal formula for the solution set of zero-dimensional ideals. J. Pure Appl. Algebra **76**: 57–80.

Greenberg, M. J. (1980): Euclidean and non-Euclidean geometries — Development and history (2nd ed.). Freeman, San Francisco.

Griffiths, P., Harris, J. (1978): Principles of algebraic geometry. Wiley, New York.

Grünbaum B., Shephard, G. C. (1993): From Menelaus to computer assisted proofs in geometry. Preprint, Dept. of Math., Washington State Univ., USA.

Guan, A.-w., Liu, H.-l. (1995): The machine proof of five positions plane rigid body guidance mechanism synthesis. In: Proc. ASCM (Beijing, China, August 18–20, 1995), Scientists Inc., Tokyo, pp. 145–152.

Guo, W., Ren, H. (1998): Blending of two implicit cubic algebraic surfaces. Acta Sci. Naturalium Universitatis Jilinensis **2**: 20–22.

Habicht, W. (1948): Zur inhomogenen Eliminationstheorie. Commun. Math. Helvetici **21**: 99–116.

Han, L., Liao, Q., Liang, C.-G. (1998a): Displacement analysis of the 3-TPS spatial parallel manipulator by using Wu–Ritt's method. Mech. Sci. Tech. **17**/4: 520–522.

Han, L., Liao, Q., Liang, C.-G. (1998b): Forward displacement analysis of one kind of general 5-5 parallel manipulator. Math. Mech. Res. Preprints 16: 140–152.

Hartshorne, R. (1977): Algebraic geometry. Springer, New York.

Havel, T. (1991): Some examples of the use of distances as coordinates for Euclidean geometry. J. Symb. Comput. **11**: 579–593.

Havel, T., Dress, A. (1992): Distance geometry and geometric algebra. Found. Phys. **23**: 1357–1374.

Heintz, J. (1983): Definability and fast quantifier elimination in algebraically closed fields. Theor. Comput. Sci. **24**: 239–278.

Helm, R. (1990): On the elimination of redundant derivations during execution. In: Proc. 1990 North Am. Conf. Logic Prog., MIT Press, Cambridge, pp. 551–568.

Hereman, W. (1990): Partial integrable evolution equations in physics. In: Proc. SSTP '90 (Conte, P., ed.), Kluwer, Dordrecht, pp. 585–586.

Hereman, W., Banerjee, P. P., Korpel, A. (1986): Exact solitary wave solutions of non-linear evolution and wave equations using a direct algebraic method. J. Phys. A: Math. Gen. **19**: 607–628.

Hereman, W., Korpel, A., Banerjee, P. P. (1985): Exact wave solutions of non-linear evolution and wave equations using a direct algebraic method. Wave Motion **7**: 283–290.

Hestenes, D. (1966): Space-time algebra. Gordon and Breach, New York.

Hestenes, D. (1987): New foundations for classical mechanics. D. Reidel, Dordrecht, Boston.

Hestenes, D. (1991): The design of linear algebra and geometry. Acta Appl. Math. **23**: 65–93.

Hestenes, D., Sobczyk, G. (1984): Clifford algebra to geometric calculus. D. Reidel, Dordrecht, Boston.

Hestenes, D., Ziegler, R. (1991): Projective geometry with Clifford algebra. Acta Appl. Math. **23**: 25–63.

Hilbert, D. (1971): Foundations of geometry (2nd ed.). Open Court Publ., La Salla.

Hirota, R. (1980): Direct methods in soliton theory. In: Solitons (Topics in Current Phys. 17) (Bullough, R. K., Caudry, P. J., eds.), Springer, Berlin, pp. 157–175.

Hirota, R., Grammaticos, B., Ramani, A. (1986): Soliton structure of the Drinfel'd–Sokolov–Wilson equation. J. Math. Phys. **27**: 1499–1505.

Hirota, R., Ramani, A. (1980): The Miura transformations of Kaup's equation and Mikhailov's equation. Phys. Lett. **76**A: 95–96.

Hirota, R., Satsuma, J. (1981): Soliton solutions of a coupled Korteweg–de Vries equation. Phys. Lett. **85**A: 407–408.

Hodge, W. V. D., Pedoe, D. (1952): Methods of algebraic geometry (vol. 2). Cambridge Univ. Press, Cambridge.

Hoe, J. (1977): Les systemes d'équations polynômes dans le Siyuan Yujian (1303) par Chu Shih-Chieh. Inst. Hautes Études Chinoises, Collège de, France.

Hoffmann, C. M. (1990): Algebraic and numeric techniques for offsets and blends. In: Computation of curves and surfaces (Dahmen, W., Gasca, M., Micchelli, C. A., eds.), Kluwer, Dordrecht, pp. 499–529.

Hong, H., Wang, D., Winkler, F. (eds., 1995): Short description of existing provers. Ann. Math. Artif. Intell. **13**: 195–202.

Hong, J. (1986a): How fast the algebraic approximation can be? Sci. Sinica (Ser. A) **29**: 813–823.

Hong, J. (1986b): Can we prove geometry theorem by computing an example? Sci. Sinica (Ser. A) **29**: 824–834.

Hong, J. (1986c): Proving by example and gap theorems. In: Proc. 27th Ann. Symp. Found. Comput. Sci. (Toronto, Canada, October 27–29, 1986), IEEE Comput. Soc. Press, pp. 107–116.

Hou, X. (1995): A new single-example experimental algorithm to automated geometry theorem proving and its computer implementation. Master thesis, Chengdu Inst. of Comput. Appl., Academia Sinica, China [in Chinese].

Hsiang, J. (1982): Topics in automated theorem proving and program generation. Ph.D thesis, UIUCDCS-R-82-1113, Univ. of Illinois at Urbana-Champaign, USA.

Hsiang, J. (1985): Refutational theorem proving using term-rewriting systems. Artif. Intell. **25**: 255–300.

Hsiang, J. (1987): Rewrite method for theorem proving in first order theory with equality. J. Symb. Comput. **3**: 133–152.

Hu, S., Wang, D. (1986): Fast factorization of polynomials over rational number field or its extension fields. Kexue Tongbao **31**: 150–156.

Huang, Y., Wu, W. (1989): A modified version of an algorithm for solving multivariate polynomial system. Math. Mech. Res. Preprints 5: 23–29.

Huang, Y., Wu, W. (1991): The kinematic solutions of the Stewart platform. Math. Mech. Res. Preprints 6: 94–101.

Huang, Y., Wu, W. (1992): Kinematic solution of a Stewart platform. In: Proc. Int. Workshop Math. Mech. (Wu, W.-t., Cheng, M., eds.), Int. Academic Publ., Beijing, pp. 181–188.

Hunter, J. K., Scheurle, J. (1988): Existence of perturbed solitary wave solutions to a model equation for water waves. Phys. D **32**: 253–163.

Il'yashenko, Yu. S. (1991): Finiteness theorems for limit cycles. Transl. Am. Math. Soc. (Ser. 2) **148**: 55–64.

Ince, E. L. (1956): Ordinary differential equations. Dover, New York.

Jacobson, N. (1974): Basic algebra I. Freeman, San Francisco.

James, E. M., Lloyd, N. G. (1991): A cubic system with eight small-amplitude limit cycles. IMA J. Appl. Math. **47**: 163–171.

Jimbo, M. (1989): Introduction to the Yang–Baxter equation. In: Braid group, knot theory and statistical mechanics (Yang, C. N., Ge, M. L., eds.), World Scientific, London, pp. 111–134.

Jin, X., Wang, D. (1990): On the conditions of Kukles for the existence of a centre. Bull. Lond. Math. Soc. **22**: 1–4.

Ju, Y., Liu, J., Liu, Z.-J. (1997): On the interactive & programming environment for STAR. Math. Mech. Res. Preprints 15: 99–112.

Kalkbrener, M. (1990a): Implicitization of rational parametric curves and surfaces. In: Proc. AAECC-8 (Tokyo, Japan, August 20–24, 1990), LNCS **508**, Springer, Berlin, Heidelberg, pp. 249–259.

Kalkbrener, M. (1990b): Implicitization by using Gröbner bases. Tech. Rep. no. 90–27, RISC-Linz, Johannes Kepler Univ., Austria.

Kalkbrener, M. (1993): A generalized Euclidean algorithm for computing triangular representations of algebraic varieties. J. Symb. Comput. **15**: 143–167.

Kalkbrener, M. (1994): Prime decompositions of radicals in polynomial rings. J. Symb. Comput. **18**: 365–372.

Kalkbrener, M. (1995): A generalized Euclidean algorithm for geometry theorem proving. Ann. Math. Artif. Intell. **13**: 73–95.

Kandri-Rody, A., Weispfenning, V. (1990): Non-commutative Gröbner bases in algebras of solvable type. J. Symb. Comput. **9**: 1–26.

Kapur, D. (1986a): Geometry theorem proving using Hilbert's Nullstellensatz. In: Proc. SYMSAC '86 (Waterloo, Canada, July 21–23, 1986), ACM Press, New York, pp. 202–208.

Kapur, D. (1986b): Using Gröbner base to reason about geometry problems. J. Symb. Comput. **2**: 399–408.

Kapur, D., Mundy, J. L. (1988): Wu's method and its application to perspective viewing. Artif. Intell. **37**: 15–36.

Kapur, D., Narendran, P. (1985): An equational approach to theorem proving in first-order predicate calculus. In: Proc. 9th Int. Joint Conf. Artif. Intell. (Los Angeles, USA, August 1985), Morgan Kaufmann Publ., San Francisco, pp. 1146–1153.

Kapur, D., Saxena, T., Yang, L. (1994): Algebraic and geometric reasoning with Dixon resultants. In: Proc. ISSAC '94 (Oxford, UK, July 20–22, 1994), ACM Press, New York, pp. 99–107.

Kapur, D., Wan, H. K. (1990): Refutational proofs of geometry theorems via characteristic sets. In: Proc. ISSAC '90 (Tokyo, Japan, August 20–24, 1990), ACM Press, New York, pp. 277–284.

Kapur, D., Zhang, H. (1989): An overview of RRL: Rewrite rule laboratory. In: Proc. RTA-89 (Chapel Hill, USA, April 3–5, 1989), LNCS **355**, Springer, Berlin, Heidelberg, pp. 513–529.

Karmarkar, N., Lakshman, Y. N. (1996): Approximate polynomial greatest common divisors and nearest singular polynomials. In: Proc. ISSAC '96 (Zurich, Switzerland, July 24–26, 1996), ACM Press, New York, pp. 35–39.

Kawachara, T. J. (1972): Oscillatory soliton waves in dispersive media. J. Phys. Soc. Japan **33**: 260–269.

Kawamoto, S. (1984): Cusp soliton of the I+0-type coupled nonlinear wave equation. J. Phys. Soc. Japan **53**: 1203–1209.

Kempe, A. B. (1876): On a general method of describing plane curves of the nth degree by linkwork. Proc. Lond. Math. Soc. **VII**: 213–216.

Knuth, D. E. (1997): The art of computer programming (vol. 2, 3rd ed.). Addison-Wesley, Reading, Massachusetts.

Knuth, D. E., Bendix, P. B. (1970): Simple word problems in universal algebras. In: Computational problems in abstract algebras (Leech, J., ed.), Pergamon Press, Oxford, pp. 263–297.

Ko, H.-P. (1986): ALGE-prover II: A new edition of ALGE-prover. Tech. Rep. 86CRD081, General Electric Co., Schenectady, USA.

Ko, H.-P. (1988): Geometry theorem proving by decomposition of quasi-algebraic sets: An application of the Ritt–Wu principle. Artif. Intell. **37**: 95–122.

Kobayashi, H., Moritsugu, S., Hogan, R. W. (1988): Solving systems of algebraic equations. In: Proc. ISSAC '88 (Rome, Italy, July 4–8, 1988), LNCS **358**, Springer, Berlin, Heidelberg, pp. 139–149.

Koedinger, K. R., Anderson, J. R. (1990): Abstract planning and perceptual chunks: Elements of expertise in geometry. Cognitive Sci. **14**: 511–550.

Korf, R. E. (1985): Depth-first interactive-deepening: An optimal admissible tree search. Artif. Intell. **27**/1: 97–109.

Kronecker, L. (1882): Grundzüge einer arithmetischen Theorie der algebraischen Größen. J. f. d. reine u. angew. Math. **92**: 1–122.

Kukles, I. S. (1944): Sur quelques cas de distinction entre un foyer et un centre. Dokl. Akad. Nauk SSSR **42**: 208–211.

Kupershmidt, B.-A. (1985): A coupled Korteweg–de Vries equation with dispersion. J. Phys. A: Math. Gen. **18**: L571–573.

Kuramoto, Y. (1980): Instability and turbulence of wavefronts in reaction-diffusion systems. Progr. Theor. Phys. **63**: 1885–1903.

Kusche, K., Kutzler, B., Stifter, S. (1989): Implementation of a geometry theorem proving package in Scratchpad II. In: Proc. EUROCAL '87 (Leipzig, GDR, June 2–5, 1987), LNCS **378**, Springer, Berlin, Heidelberg, pp. 246–257.

Kutzler, B., Stifter, S. (1986a): Automated geometry theorem proving using Buchberger's algorithm. In: Proc. SYMSAC '86 (Waterloo, Canada, July 21–23, 1986), ACM Press, New York, pp. 209–214.

Kutzler, B., Stifter, S. (1986b): On the application of Buchberger's algorithm to automated geometry theorem proving. J. Symb. Comput. **2**: 389–397.

Landis, E. M., Petrovskii, I. G. (1957): On the number of limit cycles of the equation $dy/dx = P(x,y)/Q(x,y)$ where P and Q are polynomials. Math. Sb. N. S. **43**: 149–168.

Lan, H., Wang, K. (1990): Exact solutions for coupled nonlinear equations II. J. Phys. A: Math. Gen. **23**/18: 4097–4105.

Landau, S. (1985): Factoring polynomials over algebraic number fields. SIAM J. Comput. **14**: 184–195.

Langemyr, L., McCallum, S. (1989): The computation of polynomial greatest common divisors over an algebraic number field. J. Symb. Comput. **8**: 429–444.

Lawson, H. B., Michelsohn, M.-L. (1989): Spin geometry. Princeton Univ. Press, Princeton.

Lazard, D. (1991): A new method for solving algebraic systems of positive dimension. Discrete Appl. Math. **33**: 147–160.

Lefever, R. (1977): Dissipative structures in a soluble non-linear reaction-diffusion system. Phys. Lett. **60**A: 389–390.

Lenstra, A. K. (1982): Lattices and factorization of polynomials over algebraic number fields. In: Proc. EUROCAM '82 (Marseille, France, April 5–7, 1982), LNCS **144**, Springer, Berlin, Heidelberg, pp. 32–39.

Lenstra, A. K. (1987): Factoring multivariate polynomials over algebraic number fields. SIAM J. Comput. **16**: 591–598.

Li, H. (1994): New explorations in automated theorem proving in geometries. Ph.D thesis, Peking Univ., China.

Li, H. (1995): Automated reasoning with differential forms. In: Proc. ASCM (Beijing, China, August 18–20, 1995), Scientists Inc., Tokyo, pp. 29–32.

Li, H. (1996a): Clifford algebra and area method. Math. Mech. Res. Preprints 14: 37–69.

Li, H. (1996b): Clifford algebra and Lobachevski geometry. In: Clifford algebras and their application in mathematical physics — Proc. 4th Conf. (Aachen, Germany, May 28–31, 1996), Kluwer, Dordrecht, pp. 239–245 (1998).

Li, H. (1997a): Ordering in mechanical geometry theorem proving. Sci. China (Ser. A) **40**/3: 225–233.

Li, H. (1997b): On mechanical theorem proving in differential geometry: Local theory of surfaces. Sci. China (Ser. A) **40**/4: 350–356.

Li, H. (1997c): Hyperbolic geometry with Clifford algebra. Acta Appl. Math. **48**/3: 317–358.

Li, H. (2000): Vectorial equation-solving for mechanical geometry theorem proving. J. Automat. Reason. (in press).

Li, H., Cheng, M. (1997): Proving theorems in elementary geometry with Clifford algebraic method. Adv. Math. (Beijing) **26**/4: 357–371.

Li, H., Cheng, M. (1998a): Clifford algebraic reduction method for automated theorem proving in differential geometry. J. Automat. Reason. **21**/1: 1–21.

Li, H., Cheng, M. (1998b): Ordering in mechanical theorem proving in differential geometry. Acta Math. Appl. Sinica **14**/4: 358–362.

Li, H., Shi, H. (1997): On Erdos' ten-point problem. Acta Math. Sinica (New Ser.) **13**/2: 221–230.

Li, J., Li, C. (1995): Global bifurcations of planar disturbed Hamiltonian systems and distributions of limit cycles of cubic systems. Acta Math. Sinica **28**: 509–521.

Li, L. (1996): Approximate parametrization of prime ideals in neighborhood of a given point. In: Proc. 2nd ASCM (Kobe, Japan, August 20–22, 1996), Scientists Inc., Tokyo, pp. 19–30.

Li, S. (1993a): Algebraic differential geometric theory of nonlinear control systems. Ph.D thesis, Academia Sinica, China.

Li, S. (1993b): Polynomial equation solving and its application to control theory. In: Proc. World Congr. IFAC '93 (Syndey, Australia, 1993), vol. 2, pp. 209–213.

Li, S. (1994): Wu–Ritt's method for nonlinear control systems. Syst. Sci. Math. Sci. **7**/3: 223–228.

Li, S. (1998): Minimal realization of nonlinear control systems. J. Syst. Sci. Math. Sci. **18**/3: 303–308.

Li, Z. (1987): On the triangulation for any polynomial set. Math. Mech. Res. Preprints 2: 48–54.

Li, Z. (1989): Automatic implicitization of parametric objects. Math. Mech. Res. Preprints 4: 54–62.

Li, Z. (1995a): An implementation of the characteristic set method for solving algebraic equations. In: Proc. PoSSo Workshop (Paris, France, March 1–4, 1995), pp. 107–122.

Li, Z. (1995b): Mechanical theorem proving of the local theory of surfaces. Ann. Math. Artif. Intell. **13**: 25–46 [also in Math. Mech. Res. Preprints 6: 102–120 (1991)].

Li, Z. (1996): A subresultant theory for linear differential, linear difference and Ore polynomials. Tech. Rep. no. 96-14, RISC-Linz, Johannes Kepler Univ., Austria.

Li, Z. (1998): A subresultant theory for Ore polynomials with applications. In: Proc. ISSAC '98 (Gloor, O., ed.), ACM Press, New York, pp. 132–139.

Li, Z., Nemes, I. (1997): A modular algorithm for computing greatest common right divisors of Ore polynomials. In: Proc. ISSAC '97 (Küchlin, W. W., ed.), ACM Press, New York, pp. 282–298.

Li, Z., Wang, M. (1993): Travelling wave solutions to the two-dimensional KdV–Burgers equation. J. Phys. A: Math. Gen. **26**: 6027–6029.

Liao, Q., Liang, C.-G. (1990): Displacement analysis of spatial 3-TPS mechanism. In: Proc. 7th Nat. Conf. Mech. (J. Tianjin Univ.), pp. 257–260.

Liao, Q., Liang, C.-G., Zhang, Q. (1986): A novel approach to the displacement analysis of general spatial 7R mechanism. Chinese J. Mech. Eng. **22**/3: 1–9.

Liao, Q., Seneviratne, L. D., Earles, S. W. E. (1995): Forward positional analysis for the general 4-6 in-parallel platform. J. Mech. Eng. Sci. — Proc. Inst. Mech. Eng. **209**: 55–67.

Lin, D., Liu, Z.-J. (1992): Some results on theorem proving in finite geometry. In: Proc. Int. Workshop Math. Mech. (Wu, W.-t., Cheng, M., eds.), Int. Academic Publ., Beijing, pp. 222–235.

Lin, D., Liu, J., Liu, Z.-J. (1998): Mathematical research software: ELIMINO. In: Proc. ASCM '98 (Lanzhou, China, August 6–8, 1998), Lanzhou Univ. Press, Lanzhou, pp. 107–114.

Liu, J., Liu, Z.-J. (1999): Upgrading & using lex and yacc to develop message-driven GUI syntax analyzing software. J. Comput. Eng. Des. **20**/1: 17–23.

Liu, X., Jiang, Y. (1987): Mechanical theorem proving. Science Press, Beijing [in Chinese].

Liu, X., Sun, J. (1994): Generalized resolution and NC-resolution. J. Comput. Tech. **9**/2: 160–167.

Liu, Z.-J., Wu, W. (1996): STAR: Small tool for algebraic research. In: Proc. 2nd ASCM (Kobe, Japan, August 20–22, 1996), Scientists Inc., Tokyo, pp. 121–129.

Lloyd, J. W. (1984): Foundations of logic programming. Springer, Berlin.

Lloyd, N. G., Blows, T. R., Kalenge, M. C. (1988): Some cubic systems with several limit cycles. Nonlinearity **1**: 653–669.

Lloyd, N. G., Pearson, J. M. (1990): REDUCE and the bifurcation of limit cycles. J. Symb. Comput. **9**: 215–224.

Lloyd, N. G., Pearson, J. M. (1992): Computing center conditions for certain cubic systems. J. Comput. Appl. Math. **40**: 323–336.

Loos, R. (1982): Generalized polynomial remainder sequences. In: Computer algebra: Symbolic and algebraic computation (Buchberger, B., Collins, G. E., Loos, R., eds.), Springer, Wien, New York, pp. 115–137.

Loveland, D. W. (1969): A simplified format for the model-elimination theorem-proving procedure. J. ACM **16**: 349–363.

Loveland, D. W. (1978): Automated theorem proving: A logical basis. North-Holland, New York.

Ma, S., Ning, S. (1997): Deriving some new conditions on the existence of eight limit cycles for a cubic system. Comput. Math. Appl. **33**: 59–84.

Ma, S., Ning, S. (1998): Practically solving some problems expressed in the first order theory of real closed field. Int. J. Comput. Math. **69**: 265–282.

Ma, Z.-Q. (1993): Yang–Baxter equation and quantum algebra. Science Press, Beijing.

Makhakov, V. G. (1978): Dynamics of classical solitons (in non-integrable systems). Phys. Rep. **35**: 1–128.

Malfliet, W. (1991): Series solution of nonlinear coupled reaction-diffusion equations. J. Phys. A: Math. Gen. **24**: 5499–5503.

Mandache, A. (1994): The Gröbner basis algorithm and subresultant theory. In: Proc. ISSAC '94 (Giesbrecht, M., ed.), ACM Press, New York, pp. 20–23.

Manocha, D., Canny, J. F. (1991): Efficient techniques for multipolynomial resultant algorithms. In: Proc. ISSAC '91 (Bonn, Germany, July 15–17, 1991), ACM Press, New York, pp. 86–95.

Manocha, D., Canny, J. (1992): Implicit representations of rational parametric surfaces. J. Symb. Comput. **13**/5: 485–510.

Marinari, M. G., Möller, H. M., Mora, T. (1991): Göbner bases of ideals given by dual bases. In: Proc. ISSAC '91 (Bonn, Germany, July 15–17, 1991), ACM Press, New York, pp. 55–63.

McCharen, J. D., Overbeek, R. A., Wos, L. A. (1976): Problems and experiments for and with automated theorem-proving programs. IEEE Trans. Comput. C-**25**: 773–782.

McPhee, N., Chou, S.-C., Gao, X.-S. (1994): A combination of Ritt–Wu's method and Collins' method. In: Proc. CADE-12, (Nancy, France, June 28 – July 1, 1994), LNAI **814**, Springer, Berlin, Heidelberg, pp. 401–415.

Mishra, B. (1993): Algorithmic algebra. Springer, New York.

Möller, H. M., Stetter, H. J. (1995): Multivariate polynomial equations with multiple zeros solved by matrix eigenproblems. Numer. Math. **70**(3):311–329

Mourrain, B. (1992): New aspects of geometrical calculus with invariants. Preprint from http://www-sop.inria.fr/safir/whoswho/Bernard.Mourrain/publi.html.

Mourrain, B., Stolfi, N. (1995a): Computational symbolic geometry. In: Invariant methods in discrete and computational geometry (White, N. L., ed.), Kluwer, Dordrecht, pp. 107–139.

Mourrain, B., Stolfi, N. (1995b): Applications of Clifford algebras in robotics. In: Computational kinematics '95 — Proc. 2nd Workshop (Merlet, J.-P., Ravani, B., eds.), D. Reidel, Dordrecht, Boston, pp. 141–150.

Muller, J., Socher-Ambrosius, R. (1988a): Topics in completion theorem proving. SEKI-Report SR-88-13, SEKI, Germany.

Muller, J., Socher-Ambrosius, R. (1988b): On the unnecessity of multiple overlaps in completion theorem proving. In: Proc. 12th German Workshop Artif. Intell. (Eringerfeld, Germany, September 1988), Springer, Berlin, Heidelberg, pp. 169–178.

Murakami, J. (1992): A state model for the multi-variable Alexander polynomial. Int. J. Mod. Phys. A **7**: 765–782.

Murray, J. (1989): Mathematical biology. Biomathematics **19**, Springer, Berlin.

Murray, N. V. (1982): Completely non-clausal theorem proving. Artif. Intell. **18**/1: 67–85.

Narita, K. J. (1984): A discrete model for wave equation with cubic nonlinearity. J. Phys. Soc. Japan **53**: 2472–2478.

Nemyskii, V. V., Stepannov, V. V. (1960): Qualitative theory of differential equations. Princeton Univ. Press, Princeton.

Nevins, A. J. (1975): Plane geometry theorem proving using forward chaining. Artif. Intell. **6**: 1–23.

Newell, A., Shaw, J. C., Simon, H. A. (1963): Empirical explorations with the logic theory machine: A case study in heuristics. In: Feigenbaum and Feldman (1963), pp. 109–133.

Ning, S., Ma, S., Kwek, K. H., Zheng, Z. (1994): A cubic system with eight small-amplitude limit cycles. Appl. Math. Lett. **7**: 23–27.

Noro, M., Yokoyama, K. (1997): Factoring polynomials over algebraic extension fields. Preprint, Fujitsu Laboratories Ltd., Japan.

Olever, P. J. (1977): Evolution equations possessing infinitely many symmetries. J. Math. Phys. **18**: 1212–1215.

Olver, P. J. (1986): Applications of Lie groups to differential equations. GTM **107**, Springer, New York.

Ollivier, F. (1989): Inversibility of rational mappings and structural identifiability in automatics. In: Proc. ISSAC '89 (Portland, USA, July 17–19, 1989), ACM Press, New York, pp. 43–54.

Ore, O. (1931): Linear equations in non-commutative fields. Ann. Math. **32**: 463–477.

Ore, O. (1933): Theory of non-commutative polynomials. Ann. Math. **34**: 480–508.

Ostrowski, A. M. (1960): Solution of equations and systems of equations. Academic Press, New York, London.

Otrokov, N. F. (1954): On the number of limit cycles of a differential equation in the neighborhood of a critical point. Mat. Sbornik **34**: 127–144.

Paul, R. P. (1981a): Robot manipulators: Mathematics, programming, and control. MIT Press, Cambridge.

Paul, R. P. (1981b): Kinematic control equations for simple manipulators. IEEE Trans. SMC **11**: 449–460.

Peano, G. (1888): Geometric calculus according to the Ausdehnungslehre of H. Grassmann. Translation copyright 1997 by L. Kannenberg, who can be contacted at kannenbel@woods.uml.edu.

Pearson, J. M., Lloyd, N. G., Christopher, C. J. (1996): Algorithmic derivation of centre conditions. SIAM Review **38**: 619–636.

Perron, O. (1951): Algebra I die Grundlagen. de Gruyter & Co., Berlin.

Pigoski, T., Duffy, J. (1995): An inverse force analysis of a planar two-spring system. Trans. ASME. J. Mech. Des. **117**/4: 548–553.

Raadu, M., Chanteur, G. (1986): Formation of double layers: Shocklike solutions of an mKdV-equation. Phys. Scripta. **33**: 240–245.

Rabin, M. O. (1977): Decidable theories. In: Handbook of mathematical logic (Barwise J., ed.), Horth-Holland, Amsterdam, New York, pp. 595–630.

Reiter, R. (1977): A semantically guided deductive system for automatic theorem proving. IEEE Trans. Comput. C-**25**/4: 328–334.

Riesz, M. (1958): Clifford numbers and spinors. Inst. for Fluid Dynamics and Appl. Math. Lecture Ser. **38**, Univ. of Maryland, College Park.

Ritt, J. F. (1932/1938): Differential equations from the algebraic standpoint. Am. Math. Soc., New York.

Ritt, J. F. (1950): Differential algebra. Am. Math. Soc., New York.

Robinson, A. (1983): Proving a theorem (as done by man, logician, or machine). In: Automation of reasoning (Siekmann, J., Wrightson, G., eds.), Springer, Berlin, pp. 74–78.

Robinson, J. A. (1965): A machine-oriented logic based on the resolution principle. J. ACM **12**/1: 23–41.

Rong, H., Liang, C.-G. (1991): A direct displacement solution to the triangle-platform 6-SPS parallel manipulator. In: Proc. 8th World Congr. Theory Mach. Mech. (Prague, Czechoslovakia, 1991), vol. 4, Inst. of Thermol. Mech., Czechoslovak Academy of Sci., pp. 1237–1240.

Rousseau, C., Schlomiuk, D., Thibaudeau, P. (1995): The centres in the reduced Kukles system. Nonlinearity **8**: 541–569.

Roy, C. G., Bagchi, B., Sinha, D. K. (1986): Traveling-wave solutions and the coupled Korteweg–de Vries equation. J. Math. Phys. **27**: 2558–2561.

Rubald, C. M. (1973): Algorithms for polynomials over a real algebraic number field. Ph.D thesis, Dept. of Comp. Sci., Univ. of Wisconsin, USA.

Ruiz-Altaba, M. (1992): Solution of Yang–Baxter equation and statistical model. Phys. Lett. **277**: 326–351.

Sachs, R. L. (1988): On the integrable variant of the Boussinesq system: Painleve property, rational solutions, a related many-body system, and equivalence with the akns hierarchy. Phys. D **30**: 1–27.

Sadovskii, A. P. (1993): On the limiting cycles of a cubic system describing nonlinear oscillations. Diff. Uravn. **29**: 226–231.

Sagiv, Y. (1988): Optimizing datalog programs. In: Foundations of deductive databases and logic programming (Minker, J., ed.), Morgan Kaufmann Publ., Los Altos, pp. 659–698.

Sasaki, T., Suzuki, M., Kolar, M., Sasaki, M. (1991): Approximate factorization of multivariate polynomials and absolute irreducibility testing. Japan J. Ind. Appl. Math. **8**/3: 357–375.

Satsuma, J. (1987): Exact solutions of Burgers' equation with reaction terms. In: Topics in soliton theory and exactly solvable nonlinear equations (Ablowitz, M., Fuchssteiner, B., Kruskal, M., eds.), World Scientific, Singapore, pp. 157–168.

Schamel, H. (1973): A modified Korteweg–de Vries equation for icon-acoustic waves due to resonant electrons. J. Plasma Phys. **9**: 377–387.

Schicho, J. (1995): Rational parametrization of algebraic surfaces. Ph.D thesis, Johannes Kepler Univ., Austria.

Schwartz, J. T. (1980): Fast probabilistic algorithms for verification of polynomial identities. J. ACM **27**: 701–717.

Schwarz, F. (1985): Automatically determining symmetries of partial differential equation. Comput. **34**: 91–106.

Schwarz, F. (1988): Symmetries of differential equations: From Sophus Lie to computer algebra. SIAM Review **30**: 450–481.

Schwarz, F., Augstin, S. (1992): An algorithm for determining the size of symmetry groups. Comput. **49**: 95–115.

Scott, A. C. (1977): Neuro physics. Wiley, New York.

Sederberg, T. W. (1986): Improperly parametrized rational curves. Comput. Aided Geom. Des. **3**: 67–75.

Sederberg, T. W., Anderson, D. C., Goldman, R. N. (1984): Implicit representation of parametric curves and surfaces. Comput. Vision Graph. Image Process. **28**: 72–84.

Seidenberg, A. (1954): A new decision method for elementary algebra. Ann. Math. **60**: 365–371.

Seidenberg, A. (1956a): Some remarks on Hilbert's Nullstellensatz. Arch. Math. **7**: 235–240.

Seidenberg, A. (1956b): An elimination theory for differential algebra. Univ. California Publ. Math. (N.S.) **3**/2: 31–66.

Sendra, J. R., Winkler, F. (1991): Symbolic parametrization of curves. J. Symb. Comput. **12**: 607–632.

Seyler, C. E., Fenstermacher, D. L. (1984): A symmetric regularized-long-wave equation. Phys. Fluids **27**: 4–7.

Shannon, S., Sweedler, M. (1988): Using Gröbner bases to determine algebraic membership. J. Symb. Comput. **6**: 267–273.

Shen, B. Z. (1992): A subresultant sequence and its application. J. Symb. Comput. **14**/5: 505–522.

Shen, Z. (1986): On the number of limit cycles of polynomials. J. Grad. Sch. Academia Sinica **3**: 1–4.

Shi, H. (1992): Solving of the Yang–Baxter equation. In: Proc. Int. Workshop Math. Mech. (Wu, W.-t., Cheng, M., eds.), Int. Academic Publ., Beijing, pp. 79–88.

Shi, H. (1993): Solving the Yang–Baxter equation. In: Proc. XXI DGMTP (Yang, C. N., Ge, M. L., Zhou, X. W., eds.), World Scientific, Singapore, pp. 305–308.

Shi, H. (1997): Reduced form of Yang–Mills equation of SU(2) on $R^{2,2}$, $R^{3,1}$ and $R^{4,0}$. Math. Mech. Res. Preprints 15: 1–12.

Shi, H., Zhou, F. (1997): Flat central configurations of four planet motions. In: Automated deduction in geometry (Wang, D., ed.), LNAI **1360**, Springer, Berlin, Heidelberg, pp. 208–217.

Shi, S. (1980): A concrete example of the existence of four limit cycles for plane quadratic systems. Sci. Sinica A**23**: 153–158.

Shimoyama, T., Yokoyama, K. (1996): Localization and primary decomposition of polynomial ideals. J. Symb. Comput. **22**: 247–277.

Shoenfield, J. R. (1967): Mathematical logic. Addison-Wesley, Reading, Massachusetts.

Sivashinsky, G. (1977): Nonlinear analysis of hydrodynamic instability in laminar flames I: Derivation of basic equations. Acta Astronautica **4**: 1117–1206.

Slagle, J. R. (1967): Automatic theorem proving with renamable and semantic resolution. J. ACM **14**: 687–697.

Slagle, J. R. (1974): Automated theorem proving for theories with simplifiers, commutativity and associativity. J. ACM **21**: 622–642.

Smale, S. (1970a): Topology and mechanics I. Invent. Math. **10**: 305–331; II: The planar n-body problem. Invent. Math. **11**: 45–64.

Smale, S. (1970b): Problems on the nature of relative equilibria in celestial mechanics. In: Manifolds - Amsterdam 1970 — Proc. NUFFIC Summer School Manifolds 1970, LNM **197**, Springer, Berlin, Heidelberg, pp. 194–198 (1971).

Stickel, M. (1985): A Prolog technology theorem prover: Implementation by an extended Prolog compiler. J. Automat. Reason. **4**/4: 353–380.

Stifter, S. (1993): Geometry theorem proving in vector spaces by means of Gröbner bases. In: Proc. ISSAC '93 (Kiev, Ukraine, July 6–8, 1993), ACM Press, New York, pp. 301–310.

Sturm, T., Weispfenning, V. (1997): Computational geometry problems in REDLOG. In: Automated deduction in geometry (Wang, D., ed.), LNAI **1360**, Springer, Berlin, Heidelberg, pp. 58–86.

Sturmfels, B. (1993a): Computational algebraic geometry of projective configurations. J. Symb. Comput. **11**: 595–618.

Sturmfels, B. (1993b): Algorithms in invariant theory. Springer, Wien, New York.

Su, H., Liao, Q., Liang, C.-G. (1999): Direct positional analysis for a kind of 5-5 in-parallel robotic mechanism. Mech. Mach. Theory **34**: 285–301.

Sun, L., Liang, C.-G., Liao, Q. (1997): A reverse static force analysis of a special planar three-spring system. Mech. Mach. Theory **32**/5: 609–615.

Sun, X.-d., Wang, S.-k. (1995): Solutions of Yang–Baxter equation with color parameters. Sci. China (Ser. A) **38**: 1105–1116.

Sun, X.-d., Wang, S.-k., Wu, K. (1995a): Classification of six vertex solutions of color Yang–Baxter equation. J. Math. Phys. **10**: 6043–6063.

Sun, X.-d., Wang, S.-k., Wu, K. (1995b): Wu algorithm and solutions of Yang–Baxter equation for eight vertex model. Commun. Theor. Phys. **23**: 125–128.

Sun, X.-d., Wang, S.-k, Wu, K. (1996): Wu algorithm and eight vertex solutions of color Yang–Baxter equation. Commun. Theor. Phys. **26**: 213–218.

Takaoka, M. (1989): Pole distribution and steady pulse solution of the fifth order Korteweg–de Vries equation. J. Phys. Soc. Japan **58**: 73–81.

Tarski, A. (1951): A decision method for elementary algebra and geometry (2nd revised ed.). Univ. California Press, Berkeley, Los Angeles [also as Tech. Rep. R-109, The Rand Corporation, Santa Monica, USA (1948)].

Thomas, J. M. (1937): Differential systems. Am. Math. Soc., New York.

Trager, B. M. (1976): Algebraic factoring and rational function integration. In: Proc. SYMSAC '76 (Yorktown Heights, USA, August 10–12, 1976), ACM Press, New York, pp. 219–226.

Troy, W. C. (1980): Explosive instability of reaction-diffusion equations. J. Diff. Equ. **36**: 89–98.

Turaev, V. G. (1988): The Yang–Baxter equation and invariants of links. Invent. Math. **92**: 527–553.

van der Waerden, B. L. (1930): Morderne Algebra (Aufl. 1). Springer, Berlin.

van der Waerden, B. L. (1945): Einführung in die algebraische Geometrie. Dover Publ., New York.

van der Waerden, B. L. (1949): Modern algebra (vol. 1) [translated from the 2nd revised German edition by F. Blum]. Frederick Vngar Publ. Co., New York.

Wadati, M. (1972): The exact solution of the modified Korteweg–de Vries equation. J. Phys. Soc. Japan **32**: 1681–1724.

Wahlquist, H. D., Estabrook, F. B. (1975): Prolongation structures of non-linear evolution equations. J. Math. Phys. **16**: 1–7.

Walker, R. (1950): Algebraic curves. Princeton Univ. Press, Princeton.

Wang, D. (1988): Proving-by-examples method and inclusion of varieties. Kexue Tongbao **33**: 1121–1123 [extended Chinese version in J. Syst. Sci. Math. Sci. **9**: 40–46 (1989)].

Wang, D. (1989a): A new theorem discovered by computer prover. J. Geom. **36**: 173–182.

Wang, D. (1989b): On Wu's method for proving constructive geometric theorems. In: Proc. 11th Int. Joint Conf. Artif. Intell. (Detroit, USA, August 20–25, 1989), Morgan Kaufmann Publ., San Francisco, pp. 419-424.

Wang, D. (1989c): Computer algebraic methods for investigating plane differential systems of center and focus type. In: Computers & mathematics (Kaltofen, E., Watt, S. M., eds.), Springer, New York, Berlin, pp. 91–99.

Wang, D. (1989d): A program for computing the Liapunov functions and Liapunov constants in Scratchpad II. ACM SIGSAM Bull. **23**/4: 25–31.

Wang, D. (1989e): A method for determining the finite basis of an ideal from its characteristic set with application to irreducible decomposition of algebraic varieties. Math. Mech. Res. Preprints 4: 124–140 [also available as Tech. Rep. no. 89-50.0, RISC-Linz, Johannes Kepler Univ., Austria].

Wang, D. (1989–97): Characteristic sets and zero structure of polynomial sets. Lecture notes, RISC-Linz, Johannes Kepler Univ., Austria.

Wang, D. (1990): A class of cubic differential systems with 6-tuple focus. J. Diff. Equ. **87**: 305–315.

Wang, D. (1991a): Mechanical manipulation for a class of differential systems. J. Symb. Comput. **12**: 233–254 [preliminary version in Math. Mech. Res. Preprints 1: 38–52 (1987)].

Wang, D. (1991b): On the parallelization of characteristic-set-based algorithms. In: Proc. 1st Int. ACPC Conf. (Salzburg, Austria, September 30 – October 2, 1991), LNCS **591**, Springer, Berlin, Heidelberg, pp. 338–349.

Wang, D. (1991c): Reasoning about geometric problems using algebraic methods. In: Medlar 24-month deliverables, DOC, Imperial College, Univ. of London, UK.

Wang, D. (1991d): On Wu's method for solving systems of algebraic equations. Tech. Rep. no. 91-52.0, RISC-Linz, Johannes Kepler Univ., Austria.

Wang, D. (1992a): Irreducible decomposition of algebraic varieties via characteristic sets and Gröbner bases. Comput. Aided Geom. Des. **9**: 471–484.

Wang, D. (1992b): A strategy for speeding-up the computation of characteristic sets. In: Proc. 17th Int. Symp. Math. Found. Comput. Sci. (Prague, Czechoslovakia, August 24–28, 1992), LNCS **629**, Springer, Berlin, Heidelberg, pp. 504–510.

Wang, D. (1992c): Some improvements on Wu's method for solving systems of algebraic equations. In: Proc. Int. Workshop Math. Mech. (Beijing, China, July 16–18, 1992), Int. Academic Publ., Beijing, pp. 89–100.

Wang, D. (1992d): A method for factorizing multivariate polynomials over successive algebraic extension fields. Preprint, RISC-Linz, Johannes Kepler Univ., Austria.

Wang, D. (1993a): An elimination method based on Seidenberg's theory and its applications. In: Computational algebraic geometry (Eyssette, F., Galligo, A., eds.), Progr. Math. **109**, Birkhäuser, Boston, pp. 301–328.

Wang, D. (1993b): An elimination method for polynomial systems. J. Symb. Comput. **16**: 83–114.

Wang, D. (1994): Algebraic factoring and geometry theorem proving. In: Proc. CADE-12 (Nancy, France, June 28 – July 1, 1994), LNAI **814**, Springer, Berlin, Heidelberg, pp. 386–400.

Wang, D. (1995a): Reasoning about geometric problems using an elimination method. In: Automated practical reasoning: Algebraic approaches (Pfalzgraf, J., Wang, D., eds.), Springer, Wien, New York, pp. 147–185.

Wang, D. (1995b): Elimination procedures for mechanical theorem proving in geometry. Ann. Math. Artif. Intell. **13**: 1–24.

Wang, D. (1995c): A method for proving theorems in differential geometry and mechanics. J. Univ. Comput. Sci. **1**/9: 658–673.

Wang, D. (1995d): An implementation of the characteristic set method in Maple. In: Automated practical reasoning: Algebraic approaches (Pfalzgraf, J., Wang, D., eds.), Springer, Wien, New York, pp. 187–201.

Wang, D. (1996a): Solving polynomial equations: Characteristic sets and triangular systems. Math. Comput. Simulation **42**: 339–351.

Wang, D. (1996b): GEOTHER: A geometry theorem prover. In: Proc. CADE-13 (New Brunswick, USA, July 30 – August 3, 1996), LNAI **1104**, Springer, Berlin, Heidelberg, pp. 166–170.

Wang, D. (1996c): Geometry machines: From AI to SMC. In: Proc. AISMC-3 (Steyr, Austria, September 23–25, 1996), LNCS **1138**, Springer, Berlin, Heidelberg, pp. 213–239.

Wang, D. (1996d): An elimination method for differential polynomial systems I. Syst. Sci. Math. Sci. **9**/3: 216–228.

Wang, D. (1996e): CharSets 2.0 — A package for decomposing (differential) polynomial systems. Preprint, LEIBNIZ – Institut IMAG, France [also available from http://www-leibniz.imag.fr/ATINF/Dongming.Wang/charsets-2.0.tar.Z].

Wang, D. (1997): Clifford algebraic calculus for geometric reasoning with application to computer vision. In: Automated deduction in geometry (Wang, D., ed.), LNAI **1360**, Springer, Berlin, Heidelberg, pp. 115–140.

Wang, D. (1998a): Decomposing polynomial systems into simple systems. J. Symb. Comput. **25**: 295–314.

Wang, D. (1998b): Gröbner bases applied to geometric theorem proving and discovering. In: Gröbner bases and applications (Buchberger, B., Winker, F., eds.), Cambridge Univ. Press, Cambridge, pp. 281–301.

Wang, D. (1998c): Unmixed and prime decomposition of radicals of polynomial ideals. ACM SIGSAM Bull. **32**/4: 2–9.

Wang, D. (1999a): Polynomial systems from certain differential equations. J. Symb. Comput. **28**: 301–313.

Wang, D. (1999b): Decomposing algebraic varieties. In: Automated deduction in geometry (Gao, X.-S., Wang, D., Yang, L., eds.), LNAI **1669**, Springer, Berlin, Heidelberg, pp. 180–206.

Wang, D. (2000a): Elimination methods. Springer, Wien, New York.

Wang, D. (2000b): Computing triangular systems and regular systems. J. Symb. Comput. (to appear).

Wang, D., Gao, X.-S. (1987): Geometry theorems proved mechanically using Wu's method — Part on Euclidean geometry. Math. Mech. Res. Preprints 2: 75–106.

Wang, D., Hu, S. (1987): A mechanical proving system for constructible theorems in elementary geometry. J. Syst. Sci. Math. Sci. **7**: 163–172 [in Chinese].

Wang, D., Zhi, L. (1998): Algebraic factorization applied to geometric problems. In: Proc. ASCM '98 (Lanzhou, China, August 6–8, 1998), Lanzhou Univ. Press, Lanzhou, pp. 23–36.

Wang, D.-K. (1993): Polynomial equations solving and mechanical geometric theorem proving. Ph.D thesis, Academia Sinica, China.

Wang, D.-K., Zhi, L. (1995) Software development in MMRC. In: Proc. ATCM '95 (Singapore, December 18–21, 1995), Assoc. of Math. Educators, Singapore, pp. 234–243.

Wang, H. (1960), Toward mechanical mathematics. IBM J. Res. Dev. **4**: 2–22.

Wang, M.-L. (1995): Exact solutions for the RLW–Burgers equation. Math. Appl. (China) **8**: 51–55.

Wang, P. S. (1976): Factoring multivariate polynomials over algebraic number fields. Math. Comput. **30**: 324–336.

Wang, P. S. (1981): A p-adic algorithm for univariate partial fractions. In: Symbolic and algebraic computation — Proc. ACM Symp. (Wang, P. S., ed.), ACM Press, New York, pp. 212–217.

Wang, P. S., Guy, M. J. T., Davenport, J. H. (1992): p-adic reconstruction of rational numbers. ACM SIGSAM Bull. **16**/2: 2–3.

Wang, R., Shi, X., Luo, Z., Su, Z. (1994): Multivariate spline and application. Science Press, Beijing.

Wang, S.-k. (1996): Classification of eight vertex solutions of the color Yang–Baxter equation. J. Phys. A: Math. Gen. **29**: 2259–2277.

Wang, X., Liu, X. (1982): Generalized resolution. Chinese J. Comput. **5**/2: 81–92 [in Chinese].

Wang, X.-Y. (1988): Exact and explicit solitary wave solutions for the generalised Fisher equation. Phys. Lett. A **131**: 277–279.

Warren, J. (1989): Blending algebraic surfaces. ACM Trans. Graph. **8**/4: 263–278.

Weber, H. (1898): Lehrbuch der Algebra (Aufl. 2). Vieweg, Braunschweig.

Weinberger, P. J., Rothschild, L. P. (1976): Factoring polynomials over algebraic number fields. ACM Trans. Math. Software **2**: 335–350.

Weiss, J., Tabor, M., Carnevale, G. (1983): The Painleve property for partial differential equations. J. Math. Phys. **24**: 522–526.

Wen, F., Liang, C.-G. (1994): Displacement analysis of the 6-6 Stewart platform mechanisms. Mech. Mach. Theory **29**/4: 547–557.

Whilemsson, H. (1987): Explosive instabilities of reaction-diffusion equations. Phy. Rev. A **36**: 965–979.

White, N. L. (1991): Multilinear Cayley factorization. J. Symb. Comput. **11**: 421–438.

Whitehead, A. N. (1898): A treatise on universal algebra with applications. Cambridge Univ. Press, Cambridge [Reprint: Hafner, New York (1960)].

Whiteley, W. (1991): Invariant computations for analytic projective geometry. J. Symb. Comput. **11**: 549–578.

Whitham, G. B. (1973): Linear and nonlinear waves. North-Holland, New York.

Winkler, F. (1996): Polynomial algorithms in computer algebra. Springer, Wien, New York.

Wos, L. (1988): Automated reasoning: 33 basic problems. Prentice Hall, Englewood Cliffs.

Wos, L., Robinson, G., Carson, D. (1965): Efficiency and completeness of the set of support strategy in theorem proving. J. ACM **12**: 536–541.

Wu, J. (1996): First order polynomial based theorem proving. Res. Rep. no. 23, Inst. of Math. and School of Math. Sci., Peking Univ., China.

Wu, J., Liu, Z.-J. (1996a): Remainder method for the mechanical theorem proving in first-order predicate calculus. Chinese J. Comput. **19**/10: 728–734 [in Chinese].

Wu, J., Liu, Z.-J. (1996b): On theorem proving using generalized odd-superposition II. Sci. China (Ser. E) **39**/6: 608–619 [Chinese version in Sci. China (Ser. E) **26**/5: 442–449].

Wu, J., Liu, Z.-J. (1998): Well-behaved inference rules for first-order theorem proving. J. Automat. Reason. **21**/3: 381–400.

Wu, J., Tan, H. (1994): An algebraic method to decide the deduction problem in propositional many-valued logics. In: Proc. Int. Symp. Multiple-Valued Logics (Boston, USA, May 1994), IEEE Comput. Soc. Press, pp. 272–275.

Wu, T., Gao, W., Feng, G. (1995): Blending of implicit algebraic surfaces. In: Proc. ASCM (Shi, H., Kobayashi, H., eds.), Scientists Inc., Tokyo, pp. 125–131.

Wu, T., Zhou, Y., Feng, G. (1996): Blending two quadratics algebraic surfaces. In: Proc. 2nd ASCM (Kobe, Japan, August 20–22, 1996), Scientists Inc., Tokyo, pp. 73–79.

Wu, T.-J. (1994): Some test problems on applications of Wu's method in nonlinear programming problems. Chinese Quart. J. Math. **9**: 8–17.

Wu, T.-J. (1995): On a collision problem. Acta Math. Sci. **15** (Suppl.): 32–38.

Wu, W., Huang, Y. (1996): The direct kinematic solution of the planar Stewart platform with coplanar ground points. J. Comput. Math. **14**/3: 263–272.

Wu, W., Huang, Y. (1998): An investigation of the polynomial for forward analysis of a Florida platform. Syst. Sci. and Math. Sci. **11**/2:181–183.

Wu, W.-t. (1978): On the decision problem and the mechanization of theorem-proving in elementary geometry. Sci. Sinica **21**: 159–172 [also in Automated theorem proving: After 25 years (Bledsoe, W. W., Loveland, D. W., eds.), Contemp. Math. 29, Am. Math. Soc., Providence, pp. 235–242 (1984)].

Wu, W.-t. (1979): On the mechanization of theorem-proving in elementary differential geometry. Sci. Sinica (Math. Suppl. I): 94–102.

Wu, W.-t. (1980): Mechanical theorem proving in elementary geometry and differential geometry. In: Proc. 1980 Beijing Symp. Diff. Geom. Diff. Equ. (Beijing, China, August 18 – September 21, 1980), vol. 2, Science Press, Beijing, pp. 1073–1092 (1982).

Wu, W.-t. (1982): Toward mechanization of geometry — Some comments on Hilbert's "Grundlagen der Geometrie." Acta Math. Sci. **2**: 125–138.

Wu, W.-t. (1983): Some remarks on mechanical theorem-proving in elementary geometry. Acta Math. Sci. **3**: 357–360.

Wu, W.-t. (1984a): Basic principles of mechanical theorem proving in elementary geometries. J. Syst. Sci. Math. Sci. **4**: 207–235 [also in J. Automat. Reason. **2**: 221–252 (1996)].

Wu, W.-t. (1984b): Basic principles of mechanical theorem proving in geometries (part on elementary geometries). Science Press, Beijing [in Chinese, English translation as Wu (1994a)].

Wu, W.-t. (1984c): Some recent advance in mechanical theorem-proving of geometries. In: Automated theorem proving: After 25 years (Bledsoe, W. W., Loveland, D. W., eds.), Contemp. Math. **29**, Am. Math. Soc., Providence, pp. 235–242.

Wu, W.-t. (1986a): On zeros of algebraic equations — An application of Ritt principle. Kexue Tongbao **31**: 1–5.

Wu, W.-t. (1986b): A mechanization method of geometry I: Elementary geometry. Chinese Quart. J. Math. **1**: 1–14.

Wu, W.-t. (1986c): A mechanization method of geometry and its applications I: Distances, areas, and volumes. J. Syst. Sci. Math. Sci. **6**: 204–216.

Wu, W.-t. (1986d): A mechanization method of geometry and its applications, I: Distances, areas, and volumes in Euclidean and non-Euclidean geometries. Kexue Tongbao **32**: 436–440.

Wu, W.-t. (1986e): A general description of the SOLVER package. Math. Practice Theory 2: 32–39 [in Chinese].

Wu, W.-t. (1986f): Examples of applications of the SOLVER package. Math. Practice Theory 3: 1–11 [in Chinese].

Wu, W.-t. (1987a): On reducibility problem in mechanical theorem proving of elementary geometries. Chinese Quart. J. Math. **2**: 1–20.

Wu, W.-t. (1987b): A zero structure theorem for polynomial-equations-solving and its applications. Math. Mech. Res. Preprints 1: 2–12.

Wu, W.-t. (1987c): Mechanical derivation of Newton's gravitational laws from Kepler's laws. Math. Mech. Res. Preprints 2: 53–61.

Wu, W.-t. (1987d): A mechanization method of geometry and its applications II: Curve pairs of Bertrand type. Kexue Tongbao **32**/9: 535–538 [Chinese version in **17**: 1281–1284 (1986)].

Wu, W.-t. (1987e): A constructive theory of differential algebraic geometry. In: Differential geometry and differential equations (Gu, C. H., Beyer, M., Bryant, R. L., eds.), LNM **1255**, Springer, Berlin, Heidelberg, pp. 173–189.

Wu, W.-t. (1988): A mechanization method of geometry and its applications III: Mechanical proving of polynomial inequalities and equations-solving. Syst. Sci. Math. Sci. **1**: 1-17 [also in Math. Mech. Res. Preprints 2: 1–17 (1987)].

Wu, W.-t. (1989a): A mechanization method of geometry and its applications IV: Some theorems in planar kinematics. Syst. Sci. Math. Sci. **2**: 97–109.

Wu, W.-t. (1989b): On the foundation of algebraic differential geometry. Syst. Sci. Math. Sci. **2**: 289–312 [also in Math. Mech. Res. Preprints 3: 1–26].

Wu, W.-t. (1989c): Some remarks on characteristic-set formation. Math. Mech. Res. Preprints 3: 27–29.

Wu, W.-t. (1989d): On the generic zero and Chow basis of an irreducible ascending set. Math. Mech. Res. Preprints 4: 1–21.

Wu, W.-t. (1989e): A mechanization method of geometry and its applications VI: Solving inverse kinematic equations of Puma-type robots. Math. Mech. Res. Preprints 4: 49–53.

Wu, W.-t. (1990a): On a projection theorem of quasi-varieties in elimination theory. Chinese Ann. Math. (Ser. B) **11**: 220–226 [also in Math. Mech. Res. Preprints 4: 40–48 (1989)].

Wu, W.-t. (1990b): On the chemical equilibrium problem and equations-solving. Acta Math. Sci. **10**: 361–374.

Wu, W.-t. (1990c): Automation of theorem-proving. Math. Mech. Res. Preprints 5: 5–22 [also in Proc. Workshop Adv. Comput. Artif. Intell. (Beijing, China, May 15–17, 1990; Wang, C. W., Hwang, K., eds.), pp. 11–15].

Wu, W.-t. (1990d): Equations-solving and theorem-proving: Zero-set formulation and ideal formulation. In: Proc. Asian Math. Conf. (Hong Kong, August 14-18, 1990; Li, Z., Shum, K. P., Yang, C. C., Yang, L., eds.), pp. 1–10.

Wu, W.-t. (1991a): On the construction of Gröbner basis of a polynomial ideal based on Riquier–Janet theory. Syst. Sci. Math. Sci. **4**: 193–207.

Wu, W.-t. (1991b): A survey of developments of mathematics mechanization in China. In: Chinese mathematics into the 21st century (Wu, W.-t., Cheng, M., eds.), Peking Univ. Press, Beijing, pp. 15–40.

Wu, W.-t. (1991c): Mechanical theorem proving of differential geometries and some of its applications in mechanics. J. Automat. Reason. **7**: 171–191.

Wu, W.-t. (1992a): A mechanization method of equations-solving and theorem-proving. Adv. Comp. Res. **6**: 103–138.

Wu, W.-t. (1992b): On problems involving inequalities. Math. Mech. Res. Preprints 7: 1–13.

Wu, W.-t. (1992c): On a finiteness theorem about optimization problems. Math. Mech. Res. Preprints 8: 1–18.

Wu, W.-t. (1992d): A report on mechanical geometry theorem proving. Progr. Natural Sci. **2**: 1–17.

Wu, W.-t. (1992e): On the char-set method and the linear equations method of non-linear polynomial equations-solving. In: Proc. Int. Workshop Math. Mech. (Wu, W.-t., Cheng, M., eds.), Int. Academic Publ., Beijing, pp. 101–109.

Wu, W.-t. (1993a): On a linear equations method of non-linear polynomial equations-solving. Syst. Sci. Math. Sci. **6**/1: 1–12 [also in Math. Mech. Res. Preprints 6: 23–36].

Wu, W.-t. (1993b): On the development of polynomial equations solving in China. In: Computer mathematics (Tianjin, China, 1991), Nankai Ser. Pure Appl. Math. Theor. Phys. **5**, World Scientific, River Edge, pp. 1–7.

Wu, W.-t. (1993c): On a hybrid method of polynomial equations solving. Math. Mech. Res. Preprints 9: 1–10.

Wu, W.-t. (1993d): On surface-fitting problem in CAGD. Math. Mech. Res. Preprints 10: 1–10.

Wu, W.-t. (1994a): Mechanical theorem proving in geometries: Basic principles [translated from Wu (1984b) by X. Jin and D. Wang]. Springer, Wien, New York.

Wu, W.-t. (1994b): On a finiteness theorem about problems involving inequalities. Syst. Sci. Math. Sci. **7**/3: 193–200.

Wu, W.-t. (1994c): Some remarks on factorization and GCD of multivariate polynomials. Math. Mech. Res. Preprints 11: 1–14.

Wu, W.-t. (1994d): Polynomial equations-solving and its applications. In: Proc. ISAAC '94 (Du, D.-Z., Zhang, X.-S., eds.), LNCS **834**, Springer, Berlin, Heidelberg, pp. 1–9.

Wu, W.-t. (1995a): Central configurations in planet motions and vortex motions. Math. Mech. Res. Preprints 13: 1–14.

Wu, W.-t. (1995b): Char-set method of polynomial equation-solving and its applications. In: Proc. ASCM (Shi, H., Kobayashi, H., eds.), Scientific Inc., Tokyo, pp. 11–22.

Wu, W.-t. (1995c): Geometry problem-solving and its contemporary significance. In: Proc. ATCM '95 (Singapore, December 18–21, 1995), Assoc. of Math. Educators, Singapore, pp. 67–72.

Wu, W.-t. (1999): Mathematics mechanization. Science Press, Beijing.

Wu, W.-t., Wang, D.-K. (1994): On surface-fitting problem in CAGD. Math. Practice Theory 3: 26–31 [in Chinese].

Xiao, L., Liang, C.-G. (1992): Forward displacement of 5-4 and 6-4 Stewart platform. In: Proc. 5th Mech. Transm. Conf. China (Shanghai, China, 1992), pp. 320–323.

Xu, C., Shi, Q., Cheng, M. (1996a): Applications of Wu's method to stero vision. Acta Math. Appl. Chinese Universities. **11(2)**127–138.

Xu, C., Shi, Q., Cheng, M. (1996b): New development of stero vision. Chinese Sci. Bull. **41**/11: 949–953.

Yamamoto, Y., Takizawa, E.-I. (1981): On a solution on non-linear time-evolution of fifth order. J. Phys. Soc. Japan **50**: 1421–1422.

Yang, C. N. (1967): Some exact results for the many-body problem in one dimension with repulsive delta-function interaction. Phys. Rev. Lett. **19**: 1312–1314.

Yang, C. N. (1968): S-matrix for one-dimensional N-body problem with repulsive or attractive δ-function interaction. Phys. Rev. Lett. **168**: 1920–1923.

Yang, H. (1998): Clifford algebra and mechanical geometry theorem proving. Ph.D thesis, Jilin Univ., China.

Yang, H., Zhang, S., Feng, G. (1998): Clifford algebra and mechanical geometry theorem proving. In: Proc. ASCM '98 (Li, Z., ed.), Lanzhou Univ. Press, Lanzhou, pp. 49–64.

Yang, H., Zhang, S., Feng, G. (1999): A Clifford algebraic method for geometric reasoning. In: Automated deduction in geometry (Gao, X.-S., Wang, D., Yang, L., eds.), LNAI **1669**, Springer, Berlin, Heidelberg, pp. 111–129.

Yang, L. (1989): Computer-aided proving for new theorems of non-Euclidean geometry. Res. Rep. no. 4, Math. Res. Section, Australian National Univ., Australia.

Yang, L. (1991): A new method of automated theorem proving. In: The mathematical revolution inspired by computing (Johnson, J. H., Loomes, M. J., eds.), Oxford Univ. Press, Oxford, pp. 115–126.

Yang, L. (1998): Practical automated reasoning on inequalities. In: Proc. ATCM '98 (Tsukuba, Japan, August 24–28, 1998), Springer, Singapore, pp. 24–35.

Yang, L., Fu, H., Zheng, Z. (1997): A practical symbolic algorithm for inverse kinematics of 6R manipulators with simple geometry. In: Proc. CADE-14 (Townsville, Australia, July 13–17, 1997), LNAI **1249**, Springer, Berlin, Heidelberg, pp. 73–88.

Yang, L., Gao, X.-S., Chou, S.-C., Zhang, J.-Z. (1997): Automated production of readable proofs for theorems in non-Euclidean geometries. In: Automated deduction in geometry (Wang, D., ed.), LNAI **1360**, Springer, Berlin, Heidelberg, pp. 171–188.

Yang, L., Hou, X., Xia, B. (1999): Automated discovering and proving for geometric inequalities. In: Automated deduction in geometry (Gao, X.-S., Wang, D., Yang, L., eds.), LNAI **1669**, Springer, Berlin, Heidelberg, pp. 30–46.

Yang, L., Zhang, J.-Z. (1994): Searching dependency between algebraic equations: An algorithm applied to automated reasoning. In: Artificial intelligence in mathematics (Johnson, J., McKee, S., Vella, A., eds.), Oxford Univ. Press, Oxford, pp. 147–156 [also in Math. Mech. Res. Preprints 7: 105–114 (1992)].

Yang, L., Zhang, J.-Z., Hou, X.-R. (1992): A criterion of dependency between algebraic equations and its applications. In: Proc. Int. Workshop Math. Mech. (Wu, W.-t., Cheng, M., eds.), Int. Academic Publ., Beijing, pp. 110–134.

Yang, L., Zhang, J.-Z., Hou, X.-R. (1995): An efficient decomposition algorithm for geometry theorem proving without factorization. In: Proc. ASCM (Beijing, China, August 18–20, 1995), Scientists Inc., Tokyo, pp. 33–41.

Yang, L., Zhang, J.-Z., Hou, X.-R. (1996): Nonlinear algebraic equation system and automated theorem proving. Shanghai Sci. Tech. Education Publ., Shanghai [in Chinese].

Yang, L., Zhang, J.-Z., Li, C. (1992): A prover for parallel numerical verification of a class of constructive geometry theorems. In: Proc. Int. Workshop Math. Mech. (Wu, W.-t., Cheng, M., eds.), Int. Academic Publ., Beijing, pp. 244–250.

Yin, J.-P., Liang, C.-G. (1994): The forward displacement analysis of a kind of special platform manipulator mechanisms. Mech. Mach. Theory **29**/1: 1–9.

Yin, J.-P., Matthew, G. K., Duffy, J. (1996): The inverse force analysis of a four-bar mechanism with compliant input and output links. In: Proc. 1996 ASME Des. Eng. Tech. Conf. Comput. Eng. Conf. (Irvine, USA, August 18–22, 1996), 96-DETC/MECH-1141.

Zamolodchikov, A. B. (1979): Factorized S-matrices in two dimensions as the exact solutions of certain relativistic quantum field theory models. Ann. Phys. **120**: 253–291.

Zassenhaus, H. (1969): On Hensel factorization I. J. Number Theory **1**: 291–311.

Zhang, H. (1978): A united theory on general solutions of system of elasticity. J. Dalian Univ. Tech. **18**: 25–47.

Zhang, H. (1981a): Algebraic constructions for general solutions of linear operator systems. Acta Mech. Sinica (Special Issue): 152–161.

Zhang, H. (1981b): Superfluous order and the proper solution of the Maxwell equation. Appl. Math. Mech. **2**: 349–360.

Zhang, H. (1988): Reduction, superposition and induction: Automated reasoning in an equational logic. Ph.D thesis, Rensselaer Polytechnic Inst., USA.

Zhang, H. (1991): The method for constructing general solution of system of partial differential equations. In: Proc. Comput. Math. Tianjin Congr. (Tianjin, China, 1991), pp. 110–112.

Zhang, H. (1994a): A new method for the Boolean ring based theorem proving. J. Symb. Comput. **17**: 189–211.

Zhang, H. (1994b): Hamiltonian representation for linear selfadjoint partial operators. In: Thirty years for nonholonomic mechanics in China (Chen B., Mei, F., eds.), Henan Univ. Press, Kaifeng, pp. 182–186.

Zhang, H. (1997): The algebraization, mechanization, symplectication and geometrization for mechanics. In: Modern Math. and Mech. VIII (Dai, S., ed.), Shanghai Univ. Press, Shanghai, pp. 20–25.

Zhang, H., Chao, L. (1996): Operational form Hilbert Nullstellensatz and symbolic algorithm for construction of general solution to system in elasticity. J. Dalian Univ. Tech. **36**: 373–379.

Zhang, H., Chao, L. (1997): Mathematica program package to compute symmetries of PDEs and its applications. Comput. Phys. **14**: 375–379.

Zhang, H., Chao, L. (1998): Exact algorithm of Taylor polynomial for symmetries of nonlinear partial differential equations. Appl. Math. Mech. **19**: 195–202.

Zhang, H., Fan, E. (1997): Bäcklund transformation and exact solution for (2+1) dimensional KP equation. J. Dalian Univ. Tech. **37**: 624–626.

Zhang, H., Fan, E. (1998): Linearization, similarity reduction and soliton solutions of Kupershmidt equation in shallow water. J. Nonlinear Dynamics **5**: 236–239.

Zhang, H., Feng, H. (1994): Algebraic structure of general solutions to system of non-homogeneous linear operator equations. J. Dalian Univ. Tech. **34**: 249–255.

Zhang, H., Wang, Z. (1986): The completeness and approximation of Hu Haichang's solution. Kexue Tongbao **10**: 667–670

Zhang, H., Wu, F. (1992a): Mechanical method to construct the general solution of a system of partial differential equations. In: Proc. Int. Workshop Math. Mech. (Wu, W.-t., Cheng, M., eds.), Int. Academic Publ., Beijing, pp. 280–285.

Zhang, H., Wu, F. (1992b): The computational differential algebraic geometrical method for constructing the fundamental solution of partial differential equations. In: Proc. 3rd Congr. Finite Element Method China (Hunan, China, 1992), pp. 183–191.

Zhang, H., Wu, F. (1992c): General solution for a class of system of partial differential equations and its application in the theory of shells. Acta Mech. Sinica **24**: 700–707.

Zhang, H., Wu, F. (1993): General method for constructing general solution of theory of plane and shell theory. Kexue Tongbao **13**: 671–672.

Zhang, J., Yu, K. (1998): An algorithm of blending two quadratic algebraic surfaces. Acta Sci. Naturalium Universitatis Jilinensis **3**: 27–37.

Zhang, J.-Z., Chou, S.-C., Gao, X.-S. (1992): Automated production of traditional proofs for theorems in Euclidean geometry I: The Hilbert intersection point theorems. Tech. Rep. TR-92-3, Dept. of Comput. Sci., Wichita State Univ., USA [also in Ann. Math. Artif. Intell. **13**: 109–137 (1995)].

Zhang, J.-Z., Yang, L. (1989): A method to overcome the reducibility difficulty in mechanical theorem proving. ICTP preprint IC/89/263, ICTP, Italy.

Zhang, J.-Z., Yang, L., Deng, M. (1988): Parallel numerical method in mechanical theorem proving. Preprint, Chengdu Branch, Academia Sinica, China.

Zhang, J.-Z., Yang, L., Deng, M. (1990): The parallel numerical method of mechanical theorem proving. Theor. Comput. Sci. **74**: 253–271.

Zhang, Y., Liang, C.-G., Duffy, J., Primrose, E. J. (1997): A reverse force analysis of a spatial three-spring system. Mech. Mach. Theory **32**/6: 667–678.

Zhang, Z., Ding, T., Huang, W., Dong, Z. (1985): Qualitative theory of differential equations. Science Press, Beijing [in Chinese].

Zhi, L. (1996): Polynomial factorization over algebraic fields and its applications. Ph.D thesis, Academia Sinica, China.

Zhi, L. (1997): An optimal method for algebraic factoring. J. Comput. Sci. Tech. **12**: 1–9.

Zhou, J. (1990): On the degree of extensions generated by finitely many algebraic numbers. J. Number Theory **34**/2: 133–141.

Zhou, Y., Wu, T. (1995): Blending algebraic surfaces of lowest degree. In: Proc. ASCM (Shi, H., Kobayashi, H., eds.), Scientists Inc., Tokyo, pp. 125–131.

Zhu, S.-M., Shi, Q.-Y. (1997): Applications of symbolic computation and Wu-elimination in the analysis of singular points of Painleve test. In: Modern Math. and Mech. VIII (Dai, S., ed.), Shanghai Univ. Press, Shanghai, pp. 482–485.

Zoładek, H. (1993): The solution of the center-focus problem. Manuscript, Inst. Math., Univ. of Warsaw, Poland.

Zoładek, H. (1995): Eleven small limit cycles in a cubic vector field. Nonlinearity **8**: 843–860.

Acknowledgments. The editors thank Changcai Zhu for his help during the preparation of this long list of references.

INDEX